The Prindle, Weber & Schmidt Series in Mathematics

Althoen and Bumcrot, *Introduction to Discrete Mathematics*
Boye, Kavanaugh, and Williams, *Elementary Algebra*
Boye, Kavanaugh, and Williams, *Intermediate Algebra*
Burden and Faires, *Numerical Analysis*, Fourth Edition
Cass and O'Connor, *Fundamentals with Elements of Algebra*
Cullen, *Linear Algebra and Differential Equations*, Second Edition
Dick and Patton, *Calculus, Volume I*
Dick and Patton, *Calculus, Volume II*
Dick and Patton, *Technology in Calculus: A Sourcebook of Activities*
Eves, *In Mathematical Circles*
Eves, *Mathematical Circles Adieu*
Eves, *Mathematical Circles Squared*
Eves, *Return to Mathematical Circles*
Fletcher, Hoyle, and Patty, *Foundations of Discrete Mathematics*
Fletcher and Patty, *Foundations of Higher Mathematics*, Second Edition
Gantner and Gantner, *Trigonometry*
Geltner and Peterson, *Geometry for College Students*, Second Edition
Gilbert and Gilbert, *Elements of Modern Algebra*, Third Edition
Gobran, *Beginning Algebra*, Fifth Edition
Gobran, *Intermediate Algebra*, Fourth Edition
Gordon, *Calculus and the Computer*
Hall, *Algebra for College Students*
Hall, *Beginning Algebra*
Hall, *College Algebra with Applications*, Third Edition
Hall, *Intermediate Algebra*
Hartfiel and Hobbs, *Elementary Linear Algebra*
Humi and Miller, *Boundary-Value Problems and Partial Differential Equations*
Kaufmann, *Algebra for College Students*, Fourth Edition
Kaufmann, *Algebra with Trigonometry for College Students*, Third Edition
Kaufmann, *College Algebra*, Second Edition
Kaufmann, *College Algebra and Trigonometry*, Second Edition
Kaufmann, *Elementary Algebra for College Students*, Fourth Edition
Kaufmann, *Intermediate Algebra for College Students*, Fourth Edition
Kaufmann, *Precalculus*, Second Edition
Kaufmann, *Trigonometry*
Kennedy and Green, *Prealgebra for College Students*
Laufer, *Discrete Mathematics and Applied Modern Algebra*
Nicholson, *Elementary Linear Algebra with Applications*, Second Edition
Pence, *Calculus Activities for Graphic Calculators*
Pence, *Calculus Activities for the TI-81 Graphic Calculator*
Plybon, *An Introduction to Applied Numerical Analysis*
Powers, *Elementary Differential Equations*
Powers, *Elementary Differential Equations with Boundary-Value Problems*
Proga, *Arithmetic and Algebra*, Third Edition
Proga, *Basic Mathematics*, Third Edition
Rice and Strange, *Plane Trigonometry*, Sixth Edition
Schelin and Bange, *Mathematical Analysis for Business and Economics*, Second Edition
Strnad, *Introductory Algebra*
Swokowski, *Algebra and Trigonometry with Analytic Geometry*, Seventh Edition
Swokowski, *Calculus*, Fifth Edition
Swokowski, *Calculus*, Fifth Edition (Late Trigonometry Version)
Swokowski, *Calculus of a Single Variable*
Swokowski, *Fundamentals of College Algebra*, Seventh Edition
Swokowski, *Fundamentals of Algebra and Trigonometry*, Seventh Edition
Swokowski, *Fundamentals of Trigonometry*, Seventh Edition
Swokowski, *Precalculus: Functions and Graphs*, Sixth Edition
Tan, *Applied Calculus*, Second Edition
Tan, *Applied Finite Mathematics*, Third Edition
Tan, *Calculus for the Managerial, Life, and Social Sciences*, Second Edition
Tan, *College Mathematics*, Second Edition
Trim, *Applied Partial Differential Equations*

Venit and Bishop, *Elementary Linear Algebra*, Third Edition
Venit and Bishop, *Elementary Linear Algebra*, Alternate Second Edition
Wiggins, *Problem Solver for Finite Mathematics and Calculus*
Willard, *Calculus and Its Applications*, Second Edition
Wood and Capell, *Arithmetic*
Wood and Capell, *Intermediate Algebra*
Wood, Capell, and Hall, *Developmental Mathematics*, Fourth Edition
Zill, *A First Course in Differential Equations with Applications*, Fourth Edition
Zill and Cullen, *Advanced Engineering Mathematics*
Zill, *Calculus*, Third Edition
Zill, *Differential Equations with Boundary-Value Problems*, Second Edition

The Prindle, Weber & Schmidt Series in Advanced Mathematics

Brabenec, *Introduction to Real Analysis*
Ehrlich, *Fundamental Concepts of Abstract Algebra*
Eves, *Foundations and Fundamental Concepts of Mathematics*, Third Edition
Keisler, *Elementary Calculus: An Infinitesimal Approach*, Second Edition
Kirkwood, *An Introduction to Real Analysis*
Ruckle, *Modern Analysis: Measure Theory and Functional Analysis with Applications*
Sieradski, *An Introduction to Topology and Homotopy*

AN INTRODUCTION TO APPLIED NUMERICAL ANALYSIS

Benjamin F. Plybon

Miami University

PWS-KENT PUBLISHING COMPANY

Boston

PWS-KENT
Publishing Company

20 Park Plaza
Boston, Massachusetts 02116

PWS-KENT Publishing Company is a division of Wadsworth, Inc.

Library of Congress Cataloging-in-Publication Data

Plybon, Benjamin F.
    An introduction to applied numerical analysis / Benjamin F. Plybon.
        p.   cm.
    Includes bibliographical references and index.
    ISBN 0-534-92284-8
    1. Numerical analysis.   I. Title.
QA297.P58 1992   91-25118
519.4—dc20   CIP

International Student Edition ISBN: 0-534-97205-5

Printed in the United States of America.

92 93 94 95 96 — 10 9 8 7 6 5 4 3 2 1

 This book is printed on recycled, acid-free paper.

Sponsoring Editor: Steve Quigley
Production Coordinator: Susan M. C. Caffey
Production: Carol Dondrea/Bookman Productions
Manufacturing Coordinator: Marcia A. Locke
Interior Designer: Joseph R. di Chiarro
Interior Illustrator: Carl Brown
Cover Designer: Henry A. Besanceney III
Typesetter: Santype International Ltd.
Cover Printer: John P. Pow Company, Inc.
Printer and Binder: R. R. Donnelley & Sons Company

CONTENTS

PREFACE

This text evolved from a collection of lecture notes amassed over nearly three decades of teaching numerical analysis to undergraduates. The author has taught such courses numerous times since 1960. Over that time, the available computing machinery has changed dramatically. Mainframes are much more accessible for today's students than they were in 1960. Minicomputers and micros abound. We even have pocket calculators that can perform computations that we once reserved for large computers. In the 1950s, many of us were required to use IBM 650s to do our projects in numerical analysis. Programs were written in machine language. This situation was reflected in the textbooks, such as the 1956 edition of the well-known text by F. B. Hildebrand or the text by C. Lanczos. These texts contained very little numerical data and were written in the style of an advanced text in analysis. The exercises were designed for hand computation or possibly a desk calculator. All of that has changed.

Texts written in the past 20 years have reflected the increased availability of computers. Most contain some machine output in examples. Many exercises are designed to be done using a pocket calculator or some type of computer. Most of these texts are still mainframe oriented, however. In the 1980s, we experienced a revolution in the computing world with the arrival of the microcomputer. Many of us now have in our homes and offices machines that will outperform the machines we had in our university computing center in 1965. Most numerical analysis instructors have micros or terminals available in the classroom for teaching purposes. Our students have access to micros and minis as well as mainframes to carry out computing projects.

The present text has been written with that situation in mind. Hundreds of examples are included with illustrations of numerical output from micros or minicomputers. The operation of every algorithm is illustrated with some numerical output from a machine. Most of the exercises require some machine computation. This is a machine-oriented text. The theory is there, but not to stand alone. Theory is developed as needed to understand the operation of algorithms or to analyze

ix

errors in the machine output. This has been the decisive factor with regard to choice of theoretical ideas. There is very little theory for theory's sake in this text.

A variety of very basic problems are discussed: solution of a nonlinear equation, solution of a linear system, solution of initial-value problems for ordinary differential equations, and so on. In each case, the problem is introduced as a purely mathematical problem. We then develop algorithms for numerical approximation of the solution to the problem.

Each algorithm is written in a step-by-step form that can easily be converted to the programming language with which the reader is acquainted. Since virtually all microcomputers come with some version of BASIC, the micro-oriented user may wish to use that language. No program listings are included in the text, but BASIC program listings are available upon request from the author. A diskette for IBM PC–compatible computers containing such programs is available also. It is important to execute the algorithms in this text on some machine to receive the full benefit of the presentation. Most of the student exercises require the use of a computer. All of the machine output in the examples was produced by converting the algorithms as written in this text to either BASIC or FORTRAN programs. These were then executed on a micro or minicomputer. Students should be able to verify the output presented in the examples by running their own programs on their choice of machine.

The text does not include any material on computer programming methods or languages except in very general terms. Several of the texts on the market do include condensed treatments of FORTRAN or other programming languages as well as program listings. Such material has not been of any value to this instructor. Any student intending to take a course in numerical methods should enter the course with some programming experience. Each college or university will have its own computing facilities, and the student (possibly with some assistance from the instructor) should become familiar with that system. The variety of systems and programming languages is too great to include specific information about any particular computing machine or language in a text such as this.

There is a difference between a course in numerical analysis and a course in computer science. This text is a mathematics text to be used in a mathematics department with the objective of introducing the students to numerical analysis as a branch of applied mathematics. At the end of each chapter, there is a section on applied problems using the methods studied up to that point in the course. Numerical methods are used extensively in modern applied mathematics. Some exposure to these applications is valuable to the students. The applied problems consist of a combination of case studies of some length and shorter student exercises.

In this text, there is a heavy emphasis on readability and usability for the student. The text is not written for specialists in numerical analysis but rather for beginners under the guidance of a specialist. Any topic or algorithm that is too complex or theoretical to be grasped by a competent junior- or senior-level mathematics student has been omitted. Such material has been left for those who wish to pursue the topic in later course work. A beginner needs to become thoroughly familiar with concepts and gain experience from actual computing before tackling the more difficult material.

All of the standard material normally included in a first course in numerical analysis is included in this text. There is adequate material for a two-semester course.

The following diagram indicates the interdependency of material in the various chapters. A variety of course sequences is possible, as can be seen from the diagram.

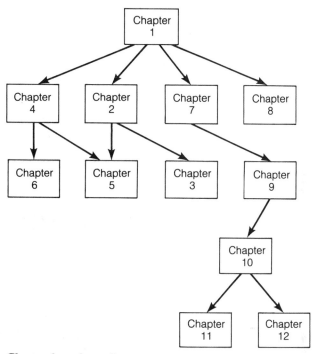

Chapter dependency diagram

The author has taught the material in the sequence given in the text as a two-semester course (18-week semesters). When only two quarters (ten weeks per quarter) were available, Chapters 1, 2, 3, and 4 or Chapters 1, 2, 4, and 6 were taught in the first quarter followed by Chapters 7, 8, 9, and 10 in the second quarter. These are workable choices. Many other course sequences are possible, of course, depending on time available and instructor preferences.

I would like to thank the following individuals for their many helpful comments and suggestions: Edmund I. Deaton, *San Diego State University*; Bruce Edwards, *University of Florida*; Leslie Foster, *San Jose State University*; Scott R. Fulton, *Clarkson University*; Nicolas D. Goodman, *State University of New York at Buffalo*; David R. Hill, *Temple University*; Gary J. Kurowski, *University of California—Davis*; John E. Lange, *St. John's University*; Greg Maybury, *Hope College*; Jim McKinney, *California State Polytechnic University*; Gary K. Rockswold, *Mankato State University*; Paul Saylor, *University of Illinois at Champaign—Urbana*; Paul S. Schnare, *Eastern Kentucky University*; David A. Sibley, *Pennsylvania State University*; Robert T. Smith, *Millersville University of Pennsylvania*; Thiab R. Taha, *University of Georgia*.

B. F. Plybon

■

INTRODUCTION

Since the mid-fifties, a great many books have been written with the title *An Introduction to Numerical Analysis, An Introduction to Numerical Computing,* or some such variation. Numerical analysis, in one form or another, has been studied since before the time of Newton. Prior to World War II, these studies were concerned with numerical procedures for approximating solutions to problems that could not be conveniently solved by theoretical methods leading to analytical solutions expressed by formulas. So we have, for example, Newton's method for approximating a solution of an equation $f(x) = 0$. For centuries, this has been a widely used method for solving transcendental equations. The method is well known to every calculus student. We also have the Gaussian elimination method for solving linear systems of equations. And of course, all students of differential equations are familiar with the Euler and Runge-Kutta methods for numerical approximation of the solution of an initial-value problem. These methods are very old and are part of the roots of modern numerical analysis.

Until the arrival of high-speed computing machines, such methods were difficult to use, and their full potential could not be realized. But during their development, attempts to use such numerical methods in an efficient manner produced much of the conceptual material we call numerical analysis. The concept of an iterative algorithm leading to a sequence of estimates of the solution of a problem was introduced; the basic ideas of error analysis were conceived; and the concepts of stability of algorithms, convergence of algorithms, and rate of convergence were defined and studied.

With the advent of twentieth-century analysis and linear algebra, leading eventually to the concept of a normed linear space, mathematicians were capable of producing very elegant theories for understanding and analyzing, in a theoretical sense, all the old numerical methods. The computational difficulties were still there, however, prior to the late forties.

With the arrival of modern computing machines, the whole character of numerical analysis has changed, and the value of this area of mathematics has increased tremendously. Iterative methods can now be used with much greater ease and effectiveness. Very large linear systems can be studied numerically. Numerical solutions of differential equations can be obtained using very small step sizes, or even variable step sizes, with thousands of steps. Thus, we can now approximate the orbit of a satellite or the trajectory of a missile with great accuracy in a short period of time. Numerical methods are now an integral part of modeling techniques used to simulate the behavior of complex interacting phenomena in scientific or engineering problems. We are no longer restricted to mathematical studies of physical phenomena in applied mathematics. Computer simulations are useful in the social sciences, business, military science, and many other areas. Most of these simulations make some use of the algorithms studied in numerical analysis.

Numerical analysis can be studied at various levels ranging from a simple introduction to numerical methods, with little or no discussion of theory, to a sophisticated study using the mathematics of normed linear spaces or other advanced mathematical theory to develop theoretical structures covering convergence theory, stability theory, and error analysis. Neither of these extremes is desirable for a beginning undergraduate course in numerical analysis.

In this text, an intermediate path has been taken. We present numerical methods for solving various standard mathematical problems in a sufficiently clear fashion that any upper-level undergraduate student of mathematics should be able to understand how and why the methods work. In fact, it is expected that most of the methods presented in this text will be converted to machine programs by the student or instructor and then executed on some computing machine, demonstrating how the methods work on sample problems. That kind of experience is an essential part of learning numerical analysis. One of the criteria for evaluation of algorithms is their suitability for present-day machines. Can the algorithm described be conveniently programmed? You should gain some firsthand experience with this criterion.

It is important to realize that even the best algorithms can be difficult to use with certain problems of the type they were designed to solve. It is helpful to experience this difficulty through numerical studies of sample problems that lead to such difficulties.

However, merely knowing some methods and running them on machines is not sufficient. Numerical analysis is a branch of mathematics and should be studied as such. If we wish to learn how to solve nonlinear equations on a machine, for example, we should begin by learning some of the theory concerning roots of an equation of the form $f(x) = 0$. This is particularly useful when the function f is a polynomial. A great deal of useful theory has been developed concerning roots of polynomial equations. The basics of this theory should be known by anyone interested in doing numerical analysis. Most of the known root-finding algorithms for polynomial equations have grown out of this theoretical base.

Numerical methods should never be presented as a substitute for a sound theoretical understanding of the problem at hand, whether we are solving a single polynomial equation or estimating the eigenvalues of a large matrix. Numerical methods are tools that can provide numerical approximations of solutions when that is the goal. They can also help test theoretical ideas by providing large amounts of numerical data for problems under theoretical investigation. The mathematics student needs to be concerned with both of these aspects of numerical mathematics.

There is, in fact, a large body of theory concerning numerical methods. Some theorems deal with error estimation or error propagation. Others are concerned with convergence rates for iterative methods. Advanced courses in numerical analysis concentrate on that body of knowledge. In this text, more emphasis is placed on concepts and methods, with enough examples and problems to provide a sound basis for further study of numerical analysis if you wish to continue in that direction. Theory studied for theory's sake without the aid of concrete problems would be undesirable in a first course. However, enough theory is included to enable you to understand machine output and to provide a good understanding of the problems of error analysis and convergence of algorithms.

It is conventional in a beginning course such as this one to include a discussion of certain basic problems. The following problems have been selected for this text:

- Solution of a nonlinear equation $f(x) = 0$
- Solution of a polynomial equation
- Solution of a linear system
- Solution of a nonlinear system
- Estimation of eigenvalues of a matrix
- Polynomial approximation of a function
- Polynomial interpolation in a table
- Estimation of $f'(a)$ for a function f
- Estimation of a definite integral
- Numerical solution of an initial-value problem for a first-order differential equation
- Numerical solution of a partial differential equation
- Numerical solution of a two-point boundary value problem for an ordinary differential equation

The algorithms developed to approximate solutions for these problems form the core of numerical analysis. Study each one carefully. Pay particular attention to the logic of the computational procedure and the possible sources of error. To develop some appreciation for the usefulness of these algorithms, you should also study some of the applied problems at the end of each chapter.

1 PREPARATION FOR COMPUTING

In this chapter, you will be introduced to a viewpoint concerning the numerical solution of mathematical problems that is shared by most people who use computing machines to analyze such problems. This discussion will include some important facts about machine number systems, error types and sources, and certain theoretical problems associated with the algorithms used in numerical work.

1.1 Introduction to Computing

This text has been written for students who have completed two years of university-level mathematics, including calculus and linear algebra or the equivalent. Thus, you have been exposed to enough mathematics to acquaint you with traditional approaches to solving many mathematical problems. You have studied the problem of finding roots of an equation $f(x) = 0$ or the problem of solving a linear system of equations. You have been exposed to many key mathematical concepts and theorems in calculus, linear algebra, and possibly differential equations. All of this study is necessary preparation for doing numerical mathematics. The methods discussed in this book for solving problems numerically have been developed and analyzed by using the mathematical theories developed in these prerequisite courses of study. Hence, the first essential point to observe is that numerical methods are not a substitute for theoretical mathematics. They are tools to be used along with all the other mathematics you know.

As each type of problem is introduced, we will review the theoretical ideas connected with such problems. Numerical methods will be developed from this theoretical base and contrasted with purely theoretical or so-called analytical solutions when such solutions exist. In some cases, numerical algorithms have been developed for certain problems because no analytical solution is known. In other cases, numerical methods are preferred to solve a problem because the analytical solution is not suitable for the machine computation of numerical answers or is less convenient to use than some simple numerical algorithm. In all cases, theoretical results are useful in the interpretation or analysis of machine output. Without any theoretical framework to fall back on, some machine output would be incomprehensible.

In fact, in some of the recent work in applied mathematics, researchers are forced to rely on numerical results from machines because an adequate theory has not been developed yet. While this is an undesirable situation, it is unavoidable owing to the complexity of the problems. In these cases, it is usually hoped that machine results will aid in the development of some theoretical results.

In practice, the following general procedure is used repeatedly:

1. A mathematical problem is presented.
2. Some mathematical analysis is performed.
3. A numerical algorithm is developed.
4. The algorithm is converted to a machine program.
5. Specific cases of the problem are solved by machine.
6. Numerical results are analyzed as to accuracy and suitability for the purpose at hand.

As an example of this procedure, consider the problem of solving the quadratic equation

$$ax^2 + bx + c = 0$$

A theoretical solution exists and is given by the quadratic formula

$$x = \frac{-b \pm \sqrt{b^2 - 4ac}}{2a}$$

We know from theory that we have three cases to consider:

1. There may be two distinct real roots.
2. There may be one repeated real root.
3. There may be two complex roots.

And we also know from theory that by use of the discriminant $d = b^2 - 4ac$, we can determine in advance which of the three cases we have for a specific quadratic equation.

We can develop a numerical algorithm based on this theoretical information. The steps of the algorithm might be the following if only real roots are desired:

1. Input coefficients a, b, and c.
2. Compute discriminant $d_1 = b^2 - 4ac$.
3. If $d_1 < 0$, then print "ROOTS COMPLEX" and stop.
4. If $d_1 = 0$, then print "DOUBLE ROOT VALUE $= -b/2a$" and stop.
5. Compute $d = \sqrt{d_1}$.
6. Compute $r_1 = (-b + d)/2a$.
7. Compute $r_2 = (-b - d)/2a$.
8. Print "ROOTS ARE:".
9. Print r_1 and r_2.
10. Stop.

Such an algorithm could be used on any programmable machine capable of displaying or printing results. A personal computer was used to produce the following numerical results using the preceding algorithm.

For the equation $x^2 - x - 6 = 0$, the machine produced root estimates 3.000000 and -2.000000, which are exactly correct.

For the equation $x^2 - 110x + 1 = 0$, the machine produced estimates

$$R1 = 109.991 \quad \text{and} \quad R2 = 9.09424E - 3$$

How accurate are these root estimates? We could draw some conclusions by substituting R1 and R2 into the quadratic equation to see how close these numbers come to satisfying the equation. However, it is possible to have a small value for $ax^2 + bx + c$ even when the number x is some distance away from any actual root. In the present case, substituting R1 yields 0.0100810 and substituting R2 yields $-2.83695E - 4$. Does this mean that estimate R2 is more accurate than R1?

We might also recall from theory that we should have $R1 + R2 = 110$ and that the product of R1 and R2 should equal 1 for the given quadratic equation. We find for our machine values that $R1 + R2 = 110$ and the product of R1 and R2 is 1.00028, so the agreement is good.

Consider a similar analysis for the equation

$$x^2 - (1.111111 \times 10^9)x + 1 = 0$$

The same machine yielded estimates $R1 = 1.11111E + 9$ and $R2 = -448$ for this equation. In this case, $R1 + R2 = 1.11111E + 9$, and the product is $-4.97778E + 11$. The sum is satisfactory, but the product clearly is not. This arithmetic indicates great inaccuracy in the root estimates. Hence, an apparently correct algorithm yielded very inaccurate results.

Anyone who uses machines to solve mathematical problems must always be on guard for such inaccurate results. In this case, we have an algorithm that is correct in theory and yields satisfactory numerical values for most equations, but apparently fails for some equations. In such cases, we must learn how to find the source of the inaccuracy and modify the algorithm, if possible, to reduce or eliminate the errors being produced. You will see later that the difficulty arises in this case because b and $\sqrt{b^2 - 4ac}$ have nearly the same value. Subtraction of numbers that are close together is a frequent source of inaccuracies in machine results. (More about this later, in Section 1.4.)

For the present, it is enough to note that since

$$r_1 + r_2 = 1.111111 \times 10^9 \quad \text{and} \quad r_1 r_2 = 1$$

for the theoretical roots r_1 and r_2, then one root must be extremely small while the other root is extremely large. Hence, we can assume that

$$r_1 = 1.111111 \times 10^9 \quad \text{and} \quad r_2 = 1/r_1 = 9.0000 \times 10^{-10}$$

So machine value R1 is satisfactory while value R2 is greatly in error. Clearly, checking machine results is always necessary.

Now consider the problem of computing $\sqrt{N}$ for a positive number N. There is a numerical procedure for estimating $\sqrt{N}$ often taught to public school students (but seldom understood) and which is unsuitable for machine use. In this case, conventional theory does not seem to assist us very much. However, note that if x_0 is a guess at the value of $\sqrt{N}$, then $\sqrt{N}$ lies between x_0 and N/x_0. Therefore,

$$x_1 = \frac{1}{2}\left(x_0 + \frac{N}{x_0}\right)$$

should be a better estimate of $\sqrt{N}$ than x_0. This argument suggests a numerical procedure as follows:

1. Input N and x_0.
2. Compute $x_1 = \frac{1}{2}\left(x_0 + \frac{N}{x_0}\right)$ and display.
3. If $|x_0 - x_1|$ is sufficiently small, then stop.
4. Set $x_0 = x_1$.
5. Go back to step 2.

This algorithm can even be used on any programmable pocket calculator. Using a programmable calculator to perform step 2 automatically and entering $N = 3$ with $x_0 = 1.5$, we found the following results:

1.5
1.75
1.73214287
1.732050810
1.732050809

In this case, the iterations were stopped when successive iterates agreed through seven decimal places.

The algorithm rapidly produced an accurate result. Accuracy can be verified in this case by squaring the final estimate. No precise error estimate will be attempted at this point, but after error estimation techniques have been discussed in Chapter 2, you will see that an error estimate can easily be made that confirms the apparent accuracy of our estimate of $\sqrt{3}$.

This algorithm uses an iterative procedure. Such iterative procedures are ideally suited for computer work and are frequently employed. A large number of numerical algorithms are based on the idea of making an initial rough estimate of the solution to a problem and then employing an iterative procedure that produces a sequence of better estimates. In such cases, we need to be concerned with questions of convergence or erratic behavior of the estimates due to inherent features of computer arithmetic or mathematical features of the algorithm itself that might tend to magnify errors when they occur.

■ *Exercise Set 1.1*

1. Estimate $\sqrt{2}$ by using the algorithm given in this section for estimation of $\sqrt{N}$. Use a pocket calculator to carry out the arithmetic. Experiment with various initial estimates of the square root.

2. Given the formula $x_{n+1} = (2x_n^3 + N)/(3x_n^2)$, do each of the following.
   a. Compute x_n for $n = 1$ through $n = 5$ with $N = 3$ and $x_0 = 1.5$.
   b. By setting x_n and $x_{n+1} = x$ and solving for x, show that the iterations must converge to $\sqrt[3]{N}$ if the sequence of estimates is convergent.
   c. Find the error in estimating $\sqrt[3]{3}$ by the value of x_5 in part (a).
   d. Repeat part (a) with $x_0 = 1.3$. Is the value of x_5 obtained in this case a better estimate of $\sqrt[3]{3}$?

3. Estimate the real roots of each of the following quadratic equations by using the quadratic formula. Try to decide whether your estimates are satisfactory by considering the sum and product of the estimates. You might also compute the value of $y = ax^2 + bx + c$ for each of your estimates to see if the value of y is close to zero.
   a. $x^2 - 500000x + 3 = 0$ **b.** $2x^2 - 99999x + 1 = 0$

4. Outline a step-by-step procedure for estimating the solution of the linear system

$$ax + by = e$$
$$cx + dy = f$$

   by using a calculator or computer. Be sure to include every step in enough detail so that another person could follow your outline and arrive at the desired estimates.

5. Use the procedure you outlined in exercise 3 to estimate the solution of the following system:

$$\frac{x}{3} + \frac{y}{7} = \frac{1}{11}$$

$$\frac{x}{9} + \frac{y}{13} = \frac{1}{17}$$

   Now check your solution by substitution into the system. Are you satisfied? Find the exact solution of this system using rational arithmetic and hand computations. Convert to decimal equivalents and compare with your calculated estimates. Do they agree? Would you say your estimated solution was an accurate one?

6. When we want to compute the value of a polynomial on a machine, it is better to use a process based on nested multiplications. For example, to compute the value of

$$P(x) = 2x^3 + 5x^2 - 2x + 7$$

   we compute

$$x(x(2x + 5) - 2) + 7$$

   beginning with the computation of $2x + 5$ and then working outward through the parentheses.
   a. Generalize this procedure for a polynomial of degree n.
   b. Use this nested multiplication procedure to compute the value of $P(1.05)$ for $P(x) = 3x^3 + 4x^2 - 5x + 10$ using a pocket calculator.

1.2 Computer Arithmetic

Now let's investigate computer arithmetic. How are numbers stored in a computer? How are they combined? What effect do these machine features have on numerical results? The answers to these questions are fundamental to our understanding of the behavior of machines (sometimes strange to us) when we instruct them to execute our numerical algorithms.

Recall that every real number can be represented by an infinite decimal of the form

$$d_n d_{n-1} \ldots d_1 d_0.d_{-1} d_{-2} d_{-3} \ldots$$

where each d_k is an integer from 0 to 9, with $d_n \neq 0$. The digit d_k represents a multiple of 10^k, and the number represented is the sum of the infinite series

$$d_n 10^n + \cdots + d_1 10 + d_0 + d_{-1} 10^{-1} + d_{-2} 10^{-2} + \cdots$$

In our decimal system, we express numbers as a sum of multiples of powers of 10, so we have a base 10 numeration system. Most computers, however, represent numbers as sums of multiples of powers of 2 or 16. If base 2 is used, we have a binary number system.

If the base is b, then the machine treats each number as having the base b form

$$d_n b^n + d_{n-1} b^{n-1} + \cdots + d_1 b + d_0 b^0 + d_{-1} b^{-1} + \cdots$$

where each d_k is an integer with $0 \leq d_k \leq b - 1$. To remember a number, the machine needs to store the digits d_k corresponding to the base b. Only a finite number of digits can be stored, of course, so a great many real numbers cannot be represented exactly in machine memory. This fact of machine arithmetic has its effect on the numerical results produced when we use numerical algorithms on machines.

Since this is not a text on computer science, we will not be concerned with the details of machine construction. It is enough to know that each machine uses a number system in which numbers are restricted to the form

$$\pm .d_1 d_2 \ldots d_p \times b^E$$

The number b is the **base** for the machine number system. Digits $d_1, d_2, \ldots, d_p$ define the **mantissa** of the number. Each d_k is an integer between 0 and $b - 1$, with $d_1 \neq 0$. The mantissa is of some finite length p. Exponent E is restricted to some interval $[M_1, M_2]$. We say that the numbers are stored in **floating-point form**.

For a typical hexadecimal mainframe machine, we might have $b = 16$, $p = 6$, $M_1 = -64$, and $M_2 = 63$. In this case, the digits d_1 to d_6 are integers from 0 to 15 and are normally stored as binary numbers 0000 to 1111.

Binary machines store each number as a succession of 1s and 0s, since those are the only possible digits.

EXAMPLE 1.1   In binary notation, the number 8 becomes 1000, the number 13 becomes 1101, the number $\frac{1}{2}$ becomes 0.1, and the number $\frac{3}{8}$ becomes 0.011.   ∎

A positive integer N written in base 10 form can be converted to base 2 form by the following procedure.

Determine quotients Q_i and remainders R_i such that

$$N = 2Q_1 + R_1$$
$$Q_1 = 2Q_2 + R_2$$
$$Q_2 = 2Q_3 + R_3$$
$$\vdots \qquad \vdots \qquad \vdots$$
$$Q_k = 2 \cdot 0 + R_{k+1}$$

The remainders are 0s and 1s, and the binary form for number N is

$$R_{k+1} R_k R_{k-1} \cdots R_1$$

We simply write down the remainders from last to first. For example, if $N = 30$, then we find quotients

$$Q_1 = 15, \quad Q_2 = 7, \quad Q_3 = 3, \quad Q_4 = 1, \quad Q_5 = 0$$

and remainders

$$R_1 = 0, \quad R_2 = 1, \quad R_3 = 1, \quad R_4 = 1, \quad R_5 = 1$$

So the binary form of 30 is 11110.

In the case of a number n between 0 and 1, we note that if the binary form is $.d_1 d_2 d_3 \ldots d_k$, then

$$n = \frac{d_1}{2} + \frac{d_2}{2^2} + \frac{d_3}{2^3} + \cdots + \frac{d_k}{2^k}$$

So

$$2n = d_1 + \frac{d_2}{2} + \frac{d_3}{2^2} + \cdots + \frac{d_k}{2^{k-1}}$$

It follows that binary digit d_1 is the integer part of $2n$. Also,

$$2(2n - d_1) = d_2 + \frac{d_3}{2} + \cdots + \frac{d_k}{2^{k-2}}$$

So binary digit d_2 is the integer part of $2(2n - d_1)$. Clearly, by continuing this doubling process for the number n while at each step removing the integer part and retaining it as the next digit in the sequence $d_1, d_2, d_3, \ldots$, we can compute the binary digits for number n. In some cases, the binary expansion is an infinite sum, so the process does not always terminate in a finite number of steps, as in the outlined procedure.

For example, consider the case of $n = 0.8125$.

$$
\begin{aligned}
2 \times 0.8125 &= 1.6250 \quad \text{so} \quad d_1 = 1 \\
2 \times 0.6250 &= 1.2500 \quad \text{so} \quad d_2 = 1 \\
2 \times 0.2500 &= 0.5000 \quad \text{so} \quad d_3 = 0 \\
2 \times 0.5000 &= 1.0000 \quad \text{so} \quad d_4 = 1
\end{aligned}
$$

All further products are 0s, so all the remaining binary digits are 0s. It follows that the binary form of 0.8125 is 0.1101.

Note that these procedures for converting base 10 numbers to binary form can easily be programmed.

Now let us assume that we are using a binary machine ($b = 2$) and consider some effects of machine arithmetic. If we enter 0.25 as the value of a variable, then the machine records this number as 0.1×2^{-1}, that is, with mantissa 0.1 and exponent -1. If we enter 0.26 as the value of a variable, then difficulties arise because as a binary number,

$$0.26 = 0.0100001010001 \ldots$$

(Can you verify this?) The number 0.26 can only be represented exactly by an infinite, nonterminating binary number. If the machine uses mantissa length $p = 6$, then number 0.26 is stored with mantissa 0.100001 and exponent -1. So 0.26 is replaced by the machine representation of

$$\frac{0.5 + 0.015625}{2} = 0.2578125 \ldots$$

Hence, an error is introduced. More on this later in Section 1.3, when we turn to error analysis.

The fundamental fact of machine arithmetic illustrated here is that machines do not use the real number system or even our rational number system. Each machine uses its own finite subset of the rational numbers determined by the values of p, b, M_1, and M_2 for that machine. Hence, each number entered into the machine is replaced by the closest machine number from the finite set of numbers available in that particular machine. If the machine uses base b with mantissa length p and exponent range from M_1 to M_2, we denote the set of machine numbers by the sumbol $F(b, p, M_1, M_2)$. This set contains

$$1 + 2(b - 1)b^{p-1}(M_2 - M_1 + 1)$$

distinct numbers. It is a worthwhile exercise to try to verify this fact.

Table 1.1   BASE 2 NUMBER SYSTEM
MANTISSA LENGTH 4
EXPONENT RANGE IS −3 TO 3

Mantissa						Exponent					
				−3	−2	−1	0	1	2	3	
1	0	0	0	0.0625	0.125	0.25	0.5	1	2	4	
1	0	0	1	0.0703125E − 01	0.140625	0.28125	0.5625	1.125	2.25	4.5	
1	0	1	0	0.078125	0.15625	0.3125	0.625	1.25	2.5	5	
1	0	1	1	0.0859375E − 01	0.171875	0.34375	0.6875	1.375	2.75	5.5	
1	1	0	0	0.09375	0.1875	0.375	0.75	1.5	3	6	
1	1	0	1	0.101563	0.203125	0.40625	0.8125	1.625	3.25	6.5	
1	1	1	0	0.109375	0.21875	0.4375	0.875	1.75	3.5	7	
1	1	1	1	0.117188	0.234375	0.46875	0.9375	1.875	3.75	7.5	

Let us now consider the floating-point number system $F(2, 4, -3, 3)$. A computer-generated listing of the positive elements is shown in Table 1.1. If a computer were using this number system and the input received was the number 0.19, the computer would use approximation 0.1875, which is the closest number available in this number system. Suppose such a computer was instructed to compute the sum $1.72 + 0.35$. The machine would use approximations 1.75 and 0.34375 from its number supply and form the sum, which is 2.09375. But the computer must also replace this actual sum by the closest number available, which is 2.000000. So we obtain the approximate result 2.0000 for a sum that in fact is 2.07. Similar inaccuracies would arise in all arithmetic calculations because of the finiteness of the available numbers in the machine number system.

The numbers in set $F(b, p, M_1, M_2)$ are not equally spaced along the line. There is a largest number B_2 and a smallest positive number B_1 in the set. Numbers larger than B_2 in absolute value fall into the "overflow" region, while numbers with absolute value smaller than B_1 fall into the "underflow" region. Each machine has some built-in design procedure for handling overflow or underflow numbers. The machine will normally issue an overflow or underflow warning message. It then might proceed to replace the overflow number with B_2, while underflow numbers might be replaced with 0. Consult the reference manual for the machine you are using to see what these default procedures are for your machine.

As a consequence of the limitations of machine arithmetic, we lose some familiar algebraic features of arithmetic as we know it. Addition and multiplication are not associative in machine arithmetic. However, both operations are commutative. The effect is that machine arithmetic performed in different orders can produce different results. When we add a large number of numbers, for example, the sum we obtain depends on the order in which they are added because of the nonassociative feature of machine addition.

For a more detailed discussion of machine arithmetic and some very interesting illustrations of its effects, refer to Forsythe (1970).

■ *Exercise Set 1.2*

1. Convert the number given in decimal form to binary form.
  a. 128   **b.** 25
  c. 0.5   **d.** 8.125
  e. 0.640625   **f.** 0.05

2. In each of the following, carry out the indicated arithmetic using number system $F(2, 4, -3, 3)$. Use the values given in Table 1.1.
  a. $0.55 + 0.69$ **b.** 0.3×0.8
  c. $(0.37 + 0.45) \times 0.43$ **d.** $(0.2 + 0.33) \div (0.15 + 0.72)$

3. Find the first seven digits in the binary equivalent of each of the following decimals.
  a. 0.17   **b.** 0.23   **c.** 1.023
  d. 0.33   **e.** 0.51

4. Given a hypothetical machine with $b = 2$, $p = 4$, $M_1 = -3$, and $M_2 = 3$, find each of the following.
  a. The machine approximation to 1.126.
  b. The machine result when asked to compute the sum of 1.38 and 2.72.
  c. The machine result when asked to subtract 1.38 from 2.72.

5. In hexadecimal arithmetic, the letters A, B, C, D, E, and F are used to denote the numbers 10, 11, 12, 13, 14, and 15. Use this information to find the decimal equivalent of each of the following hexadecimal numbers.
  a. A02   **b.** FF   **c.** 2000   **d.** FFFF

6. Convert each of the following decimal numbers to base 16 form.
  a. 256   **b.** 1024   **c.** 3000   **d.** 40

7. Addresses in memory for a personal computer such as the Apple IIe or the IBM PC are usually given in hexadecimal form. Find the decimal equivalent of each PC address.
  a. FFFF   **b.** CF0B   **c.** 800

8. How many numbers are there in the machine number system $F(2, 6, -38, 38)$? Find the largest and smallest positive numbers in this system.

9. Assume that a certain machine uses the number system $F(2, 3, -6, 6)$.
  a. Find the machine equivalent of $\frac{1}{3}$.
  b. Find the machine equivalent of $\frac{1}{5}$.
  c. What is the value assigned by this machine to the sum of these two numbers? Compare with the value obtained by hand computation of $\frac{1}{3} + \frac{1}{5}$ using rational arithmetic.

10. Using any form of computing machine available, calculate the sequence of numbers

$$x = 1 + kh \qquad k = 1, 2, 3, \ldots$$

using a small value of h such as $h = 0.01$ or $h = 0.0001$. The computed results should all be multiples of h, but eventually they will not be. Explain why this happens.

11. You will be using a computer to execute some of the algorithms in this text. For the computer you will be using, find the values of the parameters b, p, M_1, and M_2 for the associated number system. Consult your system manual or consult with your computer center manager if necessary. Compute the number of numbers available in your system.

Find the overflow and underflow values and the procedure followed by your computer when underflow or overflow values are encountered.

12. Write a program for converting base 10 numbers to base 2 (binary) form. Use the procedures outlined in this section to determine the binary digits.

1.3   Some Error Analysis

It is safe to say that all machine output contains some error. There are several reasons for this. Some error is a continuation of, or even magnification of, errors in the input. Some errors are produced by the machine because the machine number system is a finite set. Other errors result from the fact that the solutions of some problems would theoretically require an infinite number of machine operations to produce exact results. This is unattainable, of course.

EXAMPLE 1.2   Compute $\dfrac{1}{23689} - \dfrac{1}{23690}$.

Solution   Using a certain microcomputer, the result obtained was $1.78191328E - 09$. But the answer is the reciprocal of the product of 23689 and 23690. The same computer calculated this reciprocal to be $1.78192004E - 09$. The results differ noticeably. In fact, the second result is more accurate. Clearly, there is some error involved in the first calculation.   ∎

EXAMPLE 1.3   Let $S_n = \sum_{k=1}^{n} a_k$, where $a_k = 0.1$ for all k. Using a microcomputer to compute sums S_n for several values of n, it was found that

$$S_{98} = \quad 9.80000001$$

$$S_{500} = 49.9999993$$

$$S_{1000} = 99.9999963$$

All of these results clearly contain some error.   ∎

EXAMPLE 1.4   Find $f(0.001)$ and $g(0.001)$ for

$$f(x) = \frac{\sin^2 x}{1 + \cos x} \quad \text{and} \quad g(x) = 1 - \cos x$$

Solution   Our microcomputer gives us

$$f(0.001) = 4.99999958E - 07 \quad \text{and} \quad g(0.001) = 5.00192982E - 07$$

Theoretically, the results should be exactly the same, but they are not.   ∎

EXAMPLE 1.5   Using a microcomputer, we ask for the value of $104348 \div 33215$ and obtain the result 3.141593. The true value is easily verified to be $3.1415926539\ldots$. The computer division is close but not exactly correct, so some error has been introduced.

Even though the error is small, a computer carries out hundreds, perhaps thousands, of such arithmetic operations in the execution of a program, and such small errors accumulate to disastrous values unless guarded against.   ∎

Most of the error encountered in numerical analysis is some combination of three basic types: round-off error, truncation error, and propagated error. Let us now consider these types of error in relation to three criteria: the source of the error, how serious it is, and what can be done to minimize its effects.

Round-off error is produced when exact values are replaced by machine approximations. Round-off error originates in the fact that machine number sets are finite. As the mantissa length p is increased and the exponent range is expanded, more numbers become available and the magnitude of round-off error decreases. Some round-off error will always be present, however, and all machine output should be suspected of containing some round-off error. Occasionally, because of cancellation effects when numbers are added or because of the absence of round-off error in the representation of some input numbers (such as 1.375 on binary machines), we can obtain error-free results from machine arithmetic. This, however, is not the usual situation. In fact, the error encountered in Example 1.5 is accumulated round-off error. It is also important to note that the error in the final output of an algorithm due to accumulated round-off error can be amplified by some arithmetic procedures.

Truncation error results when an infinite process, such as summing a series, must be terminated after a finite number of steps. The infinite process might be an iterative algorithm that would theoretically yield an exact solution to a problem if we could use an infinite number of iterations. Iterative algorithms (algorithms that repeat a finite set of mathematical operations repeatedly) are frequently used in computing, so this kind of error is quite common.

EXAMPLE 1.6   Estimate the limit of the sequence defined by

$$x_{n+1} = \frac{2x_n^3 + 1}{3x_n^2 - 1} \quad \text{with} \quad x_0 = 1$$

Solution   After three iterations, we obtain $x_3 = 1.3252004$. In fact, the limit of the sequence is the real root of the equation $x^3 - x - 1 = 0$. The root is 1.3247 to four decimal places. Our estimate of $\lim x_n$ is inaccurate. The value will improve as more iterations are carried out, but only an infinite number of iterations would produce an exact result. Also, these calculations would have to be done using errorless arithmetic. No machine can do that.   ∎

Truncation error can sometimes be described by error formulas derived from some analysis, so we have some control over the magnitude and can find numerical bounds. This is the case for the algorithms used to estimate definite integrals or numerical values of derivatives, as you will see in Chapter 9. In these problems, the error term is always a function of a step size h, and the truncation error goes to zero as h approaches zero. The following definition is often used in such cases.

Definition 1.1

If $g(h)$ is any function of h with the property that $|g(h)/h^p| \leq K$ for some constant K when h is sufficiently small, then we write $g(h) = O(h^p)$ and say that g is of the order of $O(h^p)$ near zero.

For example, it will be shown in Chapter 9 that

$$f'(a) = \frac{f(a+h) - f(a)}{h} - \tfrac{1}{2}f''(\xi)h$$

where ξ lies between a and $a + h$. If we assume that $|f''(x)| < M$ on the interval $[a, a + \delta]$ and $0 < h < \delta$, then we see that

$$\left|\tfrac{1}{2}f''(\xi)h\right| < \tfrac{1}{2}Mh$$

If we let $g(h) = -\tfrac{1}{2}f''(\xi)h$, then

$$|g(h)/h| \leq M/2 = K \quad \text{if} \quad 0 < h < \delta$$

So we say that $g(h) = O(h)$ in this case.

If $D(h)$ happens to be another approximation of $f'(a)$ and we discover that $f'(a) = D(h) + O(h^2)$, we can conclude that for sufficiently small values of h, the $D(h)$ approximation should provide a more accurate estimate of $f'(a)$ than the quotient above because an $O(h^2)$ term goes to zero faster than an $O(h)$ term.

This "big O" notation is also used in other places, such as in Taylor series approximations. You know from calculus that

$$\sin h = h - \frac{h^3}{3!} + \frac{h^5 \cos \xi}{5!}$$

so we write

$$\sin h = h - \frac{h^3}{3!} + O(h^5)$$

to indicate that the error depends on h^5 as in Definition 1.1.

So far, we have dealt with errors produced by a machine in performing arithmetic operations. These errors come about because the machine number system is finite and because the number of operations used must be finite. The next type of error we will discuss originates with machine input or as a result of other errors introduced early in the computational procedure. In general, if a machine is instructed to perform computations using estimates of the true values of the variables concerned, then the output will contain some error as a result of these input errors. The input error propagated through the process to the output is called **propagated error**.

For example, consider the multiplication of numbers x and y that is performed using estimates $x*$ and $y*$, where

$$x = x* + \varepsilon_1 \quad \text{and} \quad y = y* + \varepsilon_2$$

The machine computes $x*y*$. What is desired is xy. Since

$$xy = (x* + \varepsilon_1)(y* + \varepsilon_2)$$
$$= x*y* + \varepsilon_1 y* + \varepsilon_2 x* + \varepsilon_1 \varepsilon_2$$

we see that xy is estimated by product $x*y*$ with "error" $y*\varepsilon_1 + x*\varepsilon_2 + \varepsilon_1\varepsilon_2$. This last error is a type of propagated error. It is combined with other forms of error in machine output, of course.

In general, if $x*$ is entered into a machine to compute $f(x)$, where $x = x* + \delta$, then by using a Taylor series, we find that

$$f(x) = f(x* + \delta) = f(x*) + \delta f'(x*) + \frac{\delta^2}{2} f''(x*) + \cdots$$

Even if $f(x*)$ were computed exactly, the error in estimating $f(x)$ by $f(x*)$ would be the propagated error given by

$$\delta f'(x*) + \frac{\delta^2}{2} f''(x*) + \cdots = \delta f'(\bar{x}) \cong \delta f'(x*)$$

where $\bar{x}$ lies between x and $x*$.

EXAMPLE 1.7   To compute the area of a circle of radius 5, we need to compute the value of 25π. To do this, we use an estimate of π, such as 3.1416. In this case, the value of the area will be in error because we are introducing an error with the input of the calculation (π is being replaced by an approximation). This input error is propagated to the output and combined with other errors, such as round-off error.   ■

In practice, of course, machine output contains some combination of the three types of errors we have described. A detailed analysis of the error involved in machine computations is very difficult and will not be attempted here. It is sufficient for you to be aware that most machine output contains some error and to have some idea of why this is unavoidable.

We now turn to measures of error. The term *error* has been used to mean that some numbers are merely approximations of the numbers we desire in a given situation; they are not exactly correct. How do we measure the amount of "incorrectness"?

Definition 1.2

If x^* is an approximation to x, we say that the **absolute error** in x^* is $|x - x^*|$.

EXAMPLE 1.8   If $x = 1.34789$ and $x^* = 1.35$, then the absolute error in approximation $x \cong x^*$ is

$$|1.35 - 1.34789| = 2.11 \times 10^{-3}$$ ∎

In practice, we do not know x, of course, so we settle for estimates or bounds on $|x - x^*|$ obtained from analysis.

Another frequently used measure of error is given in Definition 1.3.

Definition 1.3

If x^* is an approximation to x, we say that the **relative error** is

$$\frac{|x - x^*|}{|x|}$$

In Definition 1.3, we try to take into account the difference in situations, such as where the absolute error is the same in two cases but we feel one approximation is still poorer than the other. Consider the approximation of $x = 2347.59$ by $x^* = 2347.6$ and the approximation of $x = 0.02$ by $x^* = 0.01$. In both approximations, we have an absolute error of $|x - x^*| = 0.01$. Yet the second approximation seems intuitively to be less satisfactory. In the first case, the relative error is $0.01/2347.59 = 4.26 \times 10^{-6}$. In the second case, the relative error is $0.01/0.02 = 0.5$, which indicates less accuracy in the second estimate. A variation of Definition 1.3 is sometimes preferable.

Definition 1.4

If x^* is an approximation to x, then

$$\frac{100|x - x^*|}{|x|}$$

is called the **percent error** in approximation x^*.

Hence, the approximation of $x = 2347.59$ by $x^* = 2347.6$ involves a percent error of 4.26×10^{-4}, while an approximation of $x = 0.02$ by $x^* = 0.01$ involves a percent error of 50—a very serious amount of error.

We sometimes measure the accuracy of an estimate by the number of **significant digits** relative to the correct value of the number being estimated. When a number is written in decimal form, some of the digits are significant in relation to accuracy, and some are not.

In this text, we will agree to define significant digits as follows. If a number is written in the standard decimal floating-point form

$$.d_1 d_2 d_3 d_4 \ldots d_k \times 10^E$$

with $d_1 \neq 0$ and $d_k \neq 0$, we say that number has k significant digits. For example, $0.005673 = 0.5673 \times 10^{-2}$ has four significant digits, $0.2004520 = 0.200452 \times 10^0$ has six significant digits, and $23.450078200 = 0.234500782 \times 10^2$ has nine significant digits. Now if x^* is an estimate of x and the first m significant digits of x^* are the same as the first m significant digits of x, while the $(m + 1)$th digits differ by less than 5, then we say the approximation x^* has m significant digits relative to x. If the $(m + 1)$th digits differ by 5 or more, then the approximation has $m - 1$ significant digits relative to the true value. So if

$$p^* = 3.14159 = 0.314159 \times 10$$

is given as an approximation to

$$\pi = 3.14159265389 \ldots = 0.314159265389 \ldots \times 10$$

then we say that p^* has six significant digits relative to π, that is, as an approximation of π.

Now suppose that we input numbers $x = 0.998765$ and $y = 0.998763$ and instruct the machine to compute $x - y$. The machine correctly outputs $0.2E - 5$. Input numbers x and y each had six significant digits, indicating a fair degree of precision. But the machine value of $x - y$ has only *one* significant digit. Hence, we have lost significant digits in this computation. So it would seem that the value of $x - y$ is less precise than the input numbers. This kind of error, which is a form of round-off error, will usually occur if a machine subtracts two positive numbers that are very close together. For that reason, we should avoid subtracting numbers that are close in value during a numerical procedure.

If this were the end of the calculation, then the loss of significant digits might not be quite so serious a matter to numerical analysts. However, if loss of significant digits occurs in an intermediate computation and the result must be combined with other numbers for additional computations, the effect on the final output can be serious. This is particularly true when a number smaller than one with very few significant digits is divided into other numbers.

Consider the problem of computing the value of $1/x - 1/y$ using large values for x and y. Loss of significant digits occurs here because the reciprocals are very close together. This was the source of error in Example 1.2.

All numerical calculations except the simplest ones lead to results that contain some amount of error. In our application of the algorithms in this text, we must never forget that fact. In this section, we have seen how these errors are generated and how we can classify them for purposes of analysis. We will deal with these issues repeatedly throughout the text.

Error analysis is an essential feature of numerical analysis. Unfortunately, we can rarely do more than bound the error in practical applications. An approximation with no error analysis has limited value, however, so we must continue to do the best we can to get some idea of the size and sources of the error in each case. We must at least find an upper bound on the error, and whenever possible, we should try to obtain a more accurate estimate of the size. Although it is not always possible to do this with our present knowledge of error analysis, it will be shown in the remainder of this text that it is sometimes possible to obtain error estimates that are satisfactory for the purpose at hand.

■ *Exercise Set 1.3*

1. Compute each of the following using any available computer (micro, mini, or mainframe). Compare results. Note that all of the answers should be the same.

a. $\dfrac{1}{17189} - \dfrac{1}{17190}$ **b.** $\dfrac{1}{(17189)(17190)}$ **c.** $\dfrac{\dfrac{3}{17189} - \dfrac{3}{17190}}{3}$

2. Using $x = 0.785398$, compute each of the following. (Theoretically, all values should be the same. Are they?)
a. $\cos^2 x - \sin^2 x$ **b.** $(\cos x - \sin x)(\cos x + \sin x)$
c. $\cos(2x)$

3. Discuss the error sources involved in each computation in exercise 2.

4. The rational number $104348/33215$ is a very good estimate of the number π. Compute the decimal equivalent to at least 12 decimal places. Compare with an accurate value of π from a handbook (ask your librarian for help if you need it). Determine the absolute error and the relative error in this approximation.

5. To estimate the value of

$$\lim_{x \to 0} \frac{\sqrt{x + 1} - 1}{x}$$

compute the value of the fraction for $x = 0.00001$, 0.000001, 0.0000001, and 0.00000001. The arithmetic should be carried out with a calculator or computer capable of producing a result with at least six accurate decimal places. Compare results with the theoretical value of the limit. Which estimate is most accurate?

6. Theoretically, we know that for all x,

$$e^x = 1 + x + \frac{x^2}{2!} + \cdots + \frac{x^n}{n!} + \cdots$$

so we can compute e^{-3} using this series with $x = -3$. Try this using the first 20 terms. Compare your result with an accurate value of e^{-3}. How much error do you have in your estimate? Now compute e^3 with the first 20 terms of the series and $x = 3$. Take the reciprocal of this result as your estimate of e^{-3}. How much error is in this estimate? Why does the first method produce such an inaccurate estimate?

7. We can approximate ln $(1 + x)$ by the rational expression

$$L(x) = \frac{8x(x^2 + 3x + 3)}{3(x + 2)^3} \quad 0 < x \le 1$$

Determine the absolute error and the relative error for $x = 0.3, 0.7$, and 1.

8. The quadratic formula applied to the equation

$$x^2 - 4x + 3.9999999 = 0$$

gave roots $r_1 = r_2 = 2.0000000$ when a certain microcomputer was used. Find the roots accurately to at least eight decimal places and determine the absolute error in the given root estimates. What is the source of the error in the estimates obtained from the quadratic formula?

9. In each of the following estimation problems, decide whether we should be more concerned about absolute error or relative error.
   a. Estimation of one's personal income for the coming year
   b. Estimation of the number of friends who will come to the party you are giving tomorrow night
   c. Estimation of the impact time for a lunar lander
   d. Estimation of the number of faculty parking spaces needed next year at your college
   e. Estimation of the tax revenue needed by the federal government next budget year to avoid a deficit
   f. Estimation of the age of the earth

1.4   Algorithms

In doing numerical computations on machines, we employ algorithms. A numerical **algorithm** is a well-defined sequence of computations that leads to an end result that approximates the solution of a given mathematical problem. A text such as this one is primarily concerned with the design, construction, and use of such algorithms. As the course develops, we will discuss many of the well-known algorithms for solving basic mathematical problems.

EXAMPLE 1.9   The following algorithm is useful when computing $P(x)$ for a polynomial function P. Assume that coefficients $a_n, \ldots, a_1, a_0$ are known for

$$P(x) = a_n x^n + a_{n-1}x^{n-1} + \cdots + a_1 x + a_0$$

and a value of x has been selected. Do the following:

Step 1 Set $Y_0 = a_n$.
Step 2 Compute sequence $Y_i = xY_{i-1} + a_{n-i}$ for $i = 1$ to $i = n$.
Step 3 Set $P(x) = Y_n$.

This algorithm is the familiar synthetic division algorithm from beginning algebra. It is easily programmed for machine use. ∎

EXAMPLE 1.10   The value of the number e can be estimated by using the first n terms of the series

$$e = 1 + 1 + \frac{1}{2} + \frac{1}{6} + \frac{1}{24} + \cdots + \frac{1}{n!} + \cdots$$

where the terms are reciprocals of factorials. Here is an efficient algorithm for performing the computations:

Step 1 Set $S_0 = 1$ and $D_0 = 1$.
Step 2 For $k = 1$ to $k = n$, let $D_k = kD_{k-1}$ and $S_k = S_{k-1} + 1/D_k$.

The last value of S_k is the approximation to e resulting from adding the first n terms of the series. The following table is a computer-generated output from this algorithm.

Estimation of number e

N	Nth ESTIMATE
2	2.5
3	2.66667
4	2.70833
5	2.71667
6	2.71806
7	2.71825
8	2.71828
9	2.71828

The algorithm generated an accurate estimate of the number e with only a few terms. The computer used in this example was a VAX with single-precision arithmetic. Using higher-precision arithmetic, even nine terms will yield an estimate accurate to eight decimal places. Accuracy of the final result depends on the amount of accuracy in the intermediate arithmetic computations as well as on the number of terms used.   ∎

It will be interesting to see how algorithms are suggested by theoretical results and how they differ from analytical solutions of problems when analytical solutions exist.

We must also develop criteria for deciding which algorithm is best when more than one is available for solving the same problem. For example, several algorithms are available for solving a nonlinear equation $f(x) = 0$. Which one should we choose in a given case?

Sometimes an algorithm developed for solving a certain type of mathematical problem will fail to produce a usable result in a specific problem. It may fail to converge to a solution if the algorithm is iterative. It may produce an estimate that is clearly inaccurate. We will need to find out why.

Some very complex theory has been developed concerning the behavior of machine algorithms. We will confine our attention to only three very basic concepts, which are to some extent interrelated. The first of these is the concept of **con-**

vergence. Many algorithms produce a sequence

$$x_1, x_2, x_3, \ldots, x_n, \ldots$$

of estimates of the solution of the problem being studied. If these "iterates" approach the solution x as a limit when the number of iterations increases, then we say that the algorithm converges for that problem.

EXAMPLE 1.11   In Section 1.1, we discussed an algorithm for estimating the principal square root of a number N. The estimates are calculated by the following algorithm:

Step 1 Choose any estimate x_0 for the root $\sqrt{N}$. Select a value for TOL.

Step 2 Compute the new estimate $x_1 = \dfrac{1}{2}\left(x_0 + \dfrac{N}{x_0}\right)$.

Step 3 If $|x_1 - x_0| >$ TOL, then set $x_0 = x_1$ and return to step 2.

Step 4 Output x_1 and stop.

This algorithm usually converges rapidly.   ■

If an algorithm converges, we need to consider the *rate of convergence*. Suppose $\{x_n\}$ is a convergent sequence of numbers generated by an algorithm, and the limit of the sequence is x. If

$$\lim_{n\to\infty} \frac{|x_{n+1} - x|}{|x_n - x|^p} = K$$

for some positive constants K and p, then we say that the sequence $\{x_n\}$ converges to x with p being the *order of convergence*. If $p = 1$, we say that the sequence (or algorithm) converges *linearly* to the solution. If $p = 2$, we say that the sequence converges *quadratically* to the solution. Larger values of p correspond to faster rates of convergence.

For example, consider the square root algorithm described in Example 1.11, where

$$x_{n+1} = \frac{1}{2}\left(x_n + \frac{N}{x_n}\right)$$

and the limit of the sequence generated is $\sqrt{N}$. A simple computation shows that

$$x_{n+1} - \sqrt{N} = \frac{(x_n - \sqrt{N})^2}{2x_n}$$

So

$$\lim_{n\to\infty} \frac{|x_{n+1} - \sqrt{N}|}{(x_n - \sqrt{N})^2} = \lim_{n\to\infty} \frac{1}{2|x_n|} = \frac{1}{2\sqrt{N}}$$

The sequence in this case converges quadratically to N.

In another type of algorithm, we must compute a sequence of numbers x_1, x_2, $x_3,\ldots, x_n,\ldots$, where the sequence itself is the solution to some mathematical problem. The terms of the sequence might represent estimates of the values of a function f at points $t_1, t_2,\ldots, t_n$ on the real line with $x_i = f(t_i)$. In these cases, x_i is computed from the value of x_{i-1} or possibly from some combination of the numbers x_1 to x_{i-1}. Each estimate x_i contains some error, and the amount of error normally increases as we move from estimation of x_{i-1} to estimation of x_i. In cases such as this, we must be concerned with the rate of growth of error as we move further out in the sequence of estimates.

Let e_n denote the error in the estimation of x_n resulting from error ε in the estimation of x_0.

Definition 1.5

If $|e_n| \cong Cn\varepsilon$ for some constant C independent of n, then we say that the growth of error is *linear*. If $|e_n| \cong K^n\varepsilon$ for some constant $K > 1$, then we say that the growth of error is *exponential*.

Note that if errors exist in each of the estimates $x_1, x_2,\ldots, x_{n-1}$, then the error in x_n resulting from these errors is a sum of terms of the type $Cn\varepsilon$ or $K^n\varepsilon$, depending on whether the growth of error is linear or exponential.

Linear growth of error is acceptable and usually unavoidable. Exponential growth of error can lead to disastrous results, however, when the algorithm is implemented. We will encounter both types of error growth in the examples studied in this text.

EXAMPLE 1.12   A recent college graduate, Amy Peterson, has decided to begin paying off one of her credit card balances. The presence balance is $3000, and she wants to pay $100 per month over the next 36 months, without making any further charges on the card. The annual interest rate is 18%. She wants to keep track of her progress. To compute the balance at the end of each month, she can use an algorithm based on the financial formula

$$\text{New balance} = \text{Old balance} + \text{Interest} - \text{Payments}$$

Specifically, the algorithm is as follows:

Step 1 Input old balance (OLDBAL), payment (PYMT), and annual interest rate (AIR). Set MONTH $= 0$.
Step 2 Compute MR $=$ AIR/1200.
Step 3 Compute NEWBAL $=$ OLDBAL $(1 + \text{MR}) -$ PYMT. Then set MONTH $=$ MONTH $+ 1$.
Step 4 Output MONTH, NEWBAL.

Step 5 If MONTH = 36, then go to step 7.
Step 6 Set OLDBAL = NEWBAL and return to step 3.
Step 7 Stop.

Amy writes a computer program to execute this algorithm and obtains the numbers shown in the NEWBAL column in the following table.

CREDIT CARD BALANCE

ENTER TERM IN MONTHS 36
ENTER INITIAL BALANCE 3000
ENTER INITIAL ERROR EPSILON 0.05
ENTER ANNUAL INT. RATE AS A PERCENT 18
ENTER MONTHLY PAYMENT 100

MONTH	NEWBAL	BANKBAL	ERROR	RATIO
1	2945.0000	2945.0510	0.0508	1.0156
2	2889.1750	2889.2270	0.0515	0.5151
3	2832.5130	2832.5650	0.0522	0.3483
4	2775.0000	2775.0530	0.0530	0.2649
5	2716.6250	2716.6790	0.0537	0.2148
6	2657.3750	2657.4290	0.0547	0.1823
7	2597.2350	2597.2910	0.0554	0.1583
8	2536.1940	2536.2500	0.0562	0.1404
9	2474.2360	2474.2940	0.0571	0.1270
10	2411.3500	2411.4080	0.0579	0.1157
11	2347.5200	2347.5790	0.0588	0.1070
12	2282.7330	2282.7930	0.0598	0.0997
13	2216.9740	2217.0350	0.0608	0.0935
14	2150.2280	2150.2900	0.0615	0.0879
15	2082.4820	2082.5440	0.0625	0.0833
16	2013.7190	2013.7820	0.0635	0.0793
17	1943.9250	1943.9890	0.0643	0.0757
18	1873.0830	1873.1490	0.0653	0.0726
19	1801.1800	1801.2460	0.0663	0.0698
20	1728.1970	1728.2650	0.0673	0.0673
21	1654.1200	1654.1890	0.0682	0.0650
22	1578.9320	1579.0010	0.0692	0.0629
23	1502.6160	1502.6860	0.0702	0.0610
24	1425.1550	1425.2270	0.0713	0.0594
25	1346.5330	1346.6050	0.0724	0.0579
26	1266.7310	1266.8040	0.0735	0.0565
27	1185.7320	1185.8060	0.0746	0.0552
28	1103.5170	1103.5930	0.0758	0.0541
29	1020.0700	1020.1470	0.0769	0.0530
30	935.3711	935.4492	0.0781	0.0521
31	849.4016	849.4809	0.0793	0.0512
32	762.1426	762.2231	0.0805	0.0503
33	673.5748	673.6565	0.0817	0.0495
34	583.6784	583.7614	0.0830	0.0488
35	492.4336	492.5178	0.0842	0.0481
36	399.8201	399.9055	0.0855	0.0475

The bank's computer uses the same algorithm to compute Amy's monthly balances, but the data in the bank computer contain a 5-cent error, so the bank's computer computes balances beginning with a $3000.05 balance. These erroneous balances are given in the BANKBAL column of our table. The ERROR column gives the difference between these balances. We are interested in the effect of this 5-cent error on the sequence of computed balances. Fortunately for Amy, the error grows only linearly, so the accumulated error at the end of 36 months is only about 9 cents. As a check on the error growth, the RATIO column in the table shows values of the ratio ERROR/(MONTH times EPSILON), where EPSILON = 0.05. This ratio is approximately constant in the last 12 months. According to Definition 1.5, the error is growing linearly.   ■

If a given algorithm, when applied to a specific problem, produces a sequence for which the growth of error is exponential, we say that the algorithm is *unstable* for that problem. If an algorithm is found to be unstable for certain problems, an effort is made to find out for which class of problems the algorithm is stable and for which it is unstable. Examples of this type of behavior by an algorithm will be illustrated later in the text.

Before we can use an algorithm on a machine, we need to convert the steps of the algorithm to machine instructions using some programming language. Select an available language with which you are familiar. All of the algorithms described in this text can be executed using a BASIC program on a microcomputer. Many of the tables of numerical results displayed in this text were generated in this manner. For most micro- or minicomputers, some version of BASIC is available, so that is a possible choice of programming language. BASIC program listings for the algorithms in this text are available from the publisher of this text. Software is available commercially that allows the more sophisticated programmers to employ some version of FORTRAN or Pascal on microcomputers. Most college computer centers have all these languages available.

Commercial software libraries, such as the IMSL MATH/LIBRARY are available now. The IMSL MATH/LIBRARY consists of over 400 FORTRAN subroutines for solving problems of the type studied in this text. A main program to call the desired routine must be written, so you need to have some knowledge of FORTRAN. Sample main programs for each subroutine are included in the IMSL MATH/LIBRARY documentation to ease this task. Some familiarity with the IMSL software is desirable. In the past, this software was available only on mainframes, but it has now been made available for use on personal computers. You might want to investigate the availability of this library at your college or university and make some use of some of the subroutines.

IMSL also distributes other more specialized software packages. The EISPACK package solves matrix eigenvalue problems of various types. The LINPACK package can be used to solve linear systems. The MINPACK software package can be used to solve nonlinear systems.

Every program is itself an algorithm, and programming is nothing more than constructing algorithms that cause a machine to perform certain operations in a desired order. It is essential that you experience using some of the algorithms in this text on a computer. The programs should be executed on a machine, and the numerical results should be studied. That is what numerical analysis is about. We do not provide program listings in this text because listings in any given language would have very limited use, owing to the wide variety of languages available. In all probability, you already have some knowledge of programming, and no attempt will be made in this text to teach a particular programming language.

We are now ready to begin a detailed study of various algorithms. Try to keep the basic ideas of this chapter in mind while proceeding through the remainder of the text.

■ *Exercise Set 1.4*

1. Use the algorithm in Example 1.9 to compute $P(x)$ in each case.
   a. $P(x) = x^4 + 3x^3 - 4x^2 - 6x + 3$ with $x = 2$
   b. $P(x) = x^3 - x^2 - 1$ with $x = 1.2$
   c. $P(x) = x^4 - 9$ with $x = 2.5$

2. Use the square root algorithm in Example 1.11 with the indicated x_0 to estimate $\sqrt{N}$ for each N. Subtract each estimate in the sequence of estimates from the final estimate (which we take to be a good estimate of the true value) to approximate the error in each of the estimates in the sequence generated. From examination of the errors, determine whether the sequence of estimates converges linearly or with a higher order of convergence.
   a. $N = 2, x_0 = 1.5$
   b. $N = 5, x_0 = 3$
   c. $N = 2.785, x_0 = 2$

3. Write a program to construct a square root table for numbers from 1 to 5, with a spacing of 0.1. Use the algorithm in Example 1.11.

4. If an object moves with variable velocity $v(t)$, then the distance moved in time interval $[0, T]$ can be estimated as follows. Partition the interval into short subintervals $[t_{i-1}, t_i]$, where $i = 1, \ldots, n$, with $t_0 = 0$ and $t_n = T$. Estimate the distance traveled during time interval $[t_{i-1}, t_i]$ by the approximate value $v(t_i)(t_i - t_{i-1})$. Now adding up all the estimates for the n subintervals gives a sum that approximates the distance moved in time interval $[0, T]$. Convert this idea to a usable algorithm for estimation of the desired distance. Write a program to carry out the computations, and display the result. Try your algorithm on the special case where $v = 32t$ (for a falling object), and compare with the theoretical value $16t^2$ for the distance. You may want to explore other possible choices for $v(t)$ and compare with known values for the distance.

5. Theoretically, the sequence

$$x_n = \left(1 + \frac{1}{n}\right)^n$$

converges to e, although the convergence is slow. Compute values of x_n as estimates of e. Use the approximation 2.718281828 as the true value of e to estimate the error in each x_n. Compute values of the error in x_n divided by the error in x_{n+1}. Is the convergence linear or of higher order?

6. The familiar long division process learned in arithmetic classes is an algorithm for computing the digits in the quotient of two numbers.
   a. Convert this algorithm to a machine program in some programming language.
   b. Use your program to compute the first 50 digits of the quotient when 2347 is divided by 113.
   c. Use your program to find the first 50 digits in the reciprocal of 131.

2 APPROXIMATION OF ROOTS OF $f(x) = 0$

Now that you have become acquainted with the basics of numerical mathematics, we will begin a study of selected types of problems that occur frequently in scientific work. In each case, we will review some basic mathematical ideas that should be familiar to you. We will then use this material as a starting point for developing numerical algorithms suitable for machine use. These algorithms can be used to approximate the solutions to such problems.

We begin with a very basic mathematical problem often found in both pure and applied science. It is a very old problem, much studied by people interested in the numerical estimation of solutions to mathematical problems. The problem is to estimate, as accurately as possible, the roots of an equation $f(x) = 0$ for a given real function f.

The function f may be given explicitly by a formula or it may be given implicitly. Bessel functions, for example, are defined as solutions of certain differential equations, but we need to know how to solve $f(x) = 0$ approximately when f is a Bessel function. The methods developed in this chapter are easier to use when f is given explicitly, but we should not assume that they are unusable otherwise.

2.1 Theory

If r is a real or complex number and $f(r) = 0$, then we say that r is a **root** of the equation $f(x) = 0$. We also say that r is a **zero** of function f. We will assume in the following discussion that f is continuous on some interval I (which may be the real line), and we are searching for roots in I. In this chapter, we will assume that f is a real function, and we are concerned only with real roots of $f(x) = 0$. In Chapter 3, where we study polynomial equations $P(x) = 0$, we will be concerned with both real and complex zeros of a polynomial. The methods developed in Chapter 2, however, can be used to find real zeros of a polynomial.

The *primary* theoretical fact about equation $f(x) = 0$ is that real roots may or may not exist. For example, the equation

$$x^2 + 1 = 0$$

clearly has no real roots. Not so obvious is the fact that equation

$$e^x - \ln x = 0$$

also has no real roots, but inspecting the graphs of $E(x) = e^x$ and $L(x) = \ln x$ will confirm this fact. If the following numerical methods are applied to such equations, nonsensical results will be obtained.

It should be noted that the solution of $f(x) = 0$ *usually is not unique* when a solution exists. Equations $x^3 - x - 1 = 0$ and $xe^x = 2$ have unique real roots. However, equation $x^2 = \sin x$ has two real distinct roots, while equation $x = \tan x$ has infinitely many distinct real roots.

For some special types of equations, such as polynomial equations, we have theoretical methods for determining how many distinct roots the equation has. There is, however, no known general procedure applicable to all equations. Graphical methods are frequently useful for establishing the existence of roots, the number of roots, and their approximate locations.

EXAMPLE 2.1   Determine the number of real roots for the equation $x^3 - x - 1 = 0$.

Solution   Using graphing techniques from calculus, we can establish that the graph has the form shown in Figure 2.1, so one real root exists in the interval $[1, 2]$. The approximate value is 1.3 or 1.4.   ■

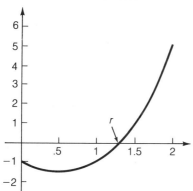

Figure 2.1

EXAMPLE 2.2   Determine the number of roots for equation $e^{-x} = \sin x$.

Solution   In this case, we sketch the graphs of the familiar functions $h(x) = e^{-x}$ and $g(x) = \sin x$ as in Figure 2.2. The roots are the values of x where these two graphs

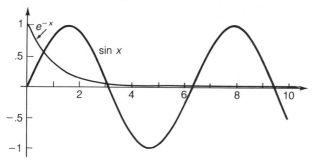

Figure 2.2

intersect. Clearly, an infinite number of roots exist. All are positive. The smallest positive root lies in the interval [0, 1].   ∎

The basic idea in such examples is that roots of the equation $f(x) = 0$ occur at points where the graph of f crosses the x axis or touches it, and roots of $h(x) = g(x)$ occur at points where the graphs intersect.

In the case of equation $f(x) = 0$, if r is a root, then the graph of f near r will look like one of the graphs shown in Figure 2.3, except for concavity differences, if f is sufficiently smooth. In cases I and II, we have $f'(r) > 0$ and $f'(r) < 0$, respectively. In cases III–VI, we have $f'(r) = 0$. When applying numerical methods to the problem of approximating r, this is an important difference. Cases III–VI cause difficulties for some root estimation algorithms. Estimates of roots in cases I and II may be noticeably inaccurate if $|f'(r)|$ is very small, however.

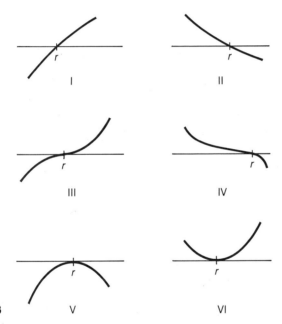

Figure 2.3

In cases I and II, we say that r is a *simple* root. However, if $f(r) = 0$ and $f'(r) = 0$, then r is a *multiple* root, and we can expect difficulties when trying to approximate r by the methods described in this chapter.

Definition 2.1

If $f^{(k)}(r) = 0$ for $k = 0, 1, \ldots, m - 1$, and $f^{(m)}(r) \neq 0$, then we say that r is a root of order m.

In cases I and II, root r has order 1. In cases III and IV, root r has order $m > 1$, where m is odd. In cases V and VI, root r has order $m > 1$, with m even. We can see that these claims are true by noting that if r has order $m > 1$, then

$$f(x) = \frac{(x - r)^m}{m!}\, f^{(m)}(r) + \frac{f^{(m+1)}(\xi)}{(m + 1)!}\,(x - r)^{m+1}$$

by Taylor's theorem (see Appendix 3). We assume that $f^{(m+1)}$ exists on an interval about r containing x. If $|x - r|$ is sufficiently small, the first term dominates the sum, so $f(x)$ behaves like the polynomial

$$\frac{f^{(m)}(r)}{m!}\,(x - r)^m$$

Algebraically speaking, r is a root of multiplicity m for this polynomial. We see that by inspecting the derivatives of f near r, we may be able to determine the multiplicity of root r if f is a polynomial.

To apply the preceding theory, we need to know that a root exists and we need an approximate location. We can use various methods to get this information. One frequently used tool is the intermediate value theorem from calculus. This theorem tells us that if f is continuous on $[a, b]$ with $f(a)$ and $f(b)$ of opposite sign, then equation $f(x) = 0$ has at least one root in (a, b). Clearly, this theorem is useful only in the case of roots of odd order, but many problems do involve simple roots or roots of higher odd order, so the theorem is useful in practice.

EXAMPLE 2.3   Verify that equation $xe^x - 2 = 0$ has a simple root in $[0, 1]$.

Solution   We have

$$f(x) = xe^x - 2$$
$$f(0) = -2 \quad \text{and} \quad f(1) = e - 2 > 0$$

Hence, the intermediate value theorem tells us that a root of odd order exists in $[0, 1]$. Now

$$f'(x) = (x + 1)e^x$$

So $f'(x) > 0$ in $[0, 1]$. Hence, $f'(r) \neq 0$ and r is a simple root.   ∎

EXAMPLE 2.4   Determine whether equation $\ln x + (1/ex) = 0$ has any roots in $(0, \infty)$, and if so, of what order.

Solution   Note that $f(x) = \ln x + 1/ex$ is clearly positive if $x \geq 1$. Substituting sample values of x from $(0, 1)$ into $f(x)$ also yields only positive values. That is, $f(\tfrac{1}{2}) > 0$, $f(\tfrac{1}{4}) > 0$, and so on. This suggests that $f(x) = 0$ has no roots of odd order. If r is a root of even order, then $f'(r) = 0$. But

$$f'(x) = \frac{1}{x} - \frac{1}{ex^2} = \frac{1}{x}\left(1 - \frac{1}{ex}\right)$$

So $f'(r) = 0$ if and only if $r = 1/e$. If $f(x) = 0$ has an even-order root, it must be $r = 1/e$. Substituting in f reveals that $f(1/e) = -\ln e + 1 = 0$. Hence, $r = 1/e$ is a root and of even order. Since

$$f''(x) = \frac{-1}{x^2} + \frac{2}{ex^3} = \frac{1}{x^2}\left(\frac{2}{ex} - 1\right)$$

and

$$f''(1/e) = e^2 \neq 0$$

then $r = 1/e$ is a root of order 2. Now $f'(x) = 0$ at $x = 1/e$ and $f''(x) > 0$ if $2/ex > 1$— that is, if $x < 2/e$. So f has a minimum value at $x = 1/e$. Since $f(1/e) = 0$, then $f(x) \geq 0$ on $(0, \infty)$. Hence, f has no odd-order roots.   ■

The preceding theoretical ideas are enough to allow us to determine the character of the roots of $f(x) = 0$ and approximate locations for a large number of equations. We turn now to the construction of machine algorithms for refining these approximate solutions.

■ *Exercise Set 2.1*

1. Use the intermediate value theorem for continuous functions to show that the given equation has a real root in the indicated interval.
  a. $x^4 + x^3 - x^2 - 2x - 2 = 0$ on $[1, 2]$
  b. $x^2 = e^{-x}$ on $[0, 1]$
  c. $x^3 - 5x^2 + x + 9 = 0$ on $[-2, -1]$
  d. $x^2 - 100x + 1 = 0$ on $[0, 0.1]$
  e. $\cos x = e^{-x}$ on $[1.29, 1.30]$
  f. $\sqrt{x} = 1 + x$ on $[0, 1]$
  g. $e^x = 1 + 1/x$ on $[0, 1]$
  h. $\cos^2 x = x^{2/3}$ on $[0, 1]$

2. Use graphical methods in each case to determine how many real roots the equation has, and find approximate locations for the roots.
  a. $x^2 = \sin x$ **b.** $x^3 - x + 1 = 0$
  c. $x^2 = e^{-x^2}$ **d.** $x = \tan x$ on $[0, 2\pi]$
  e. $xe^{-x} = \ln x$ **f.** $\sin x = \ln x$
  g. $2\cos x = e^x$ **h.** $\arctan x = 1 - x^2$
  i. $e^{2x} = 2e^x(x + 1) - 1$ **j.** $\arcsin x = \sqrt{x}$

3. For each equation, use derivatives to determine the order of the root r as indicated.
   a. $x^4 - 2x^3 + 2x^2 - 2x + 1 = 0$; $r = 1$
   b. $\sec x + x^2 = 1$; $r = 0$
   c. $x + \ln x^2 = 1$; $r = 1$
   d. $x = \arctan x$; $r = 0$
   e. $2e^x = x + 2$; $r = 0$
   f. $e^{-x^2} = (e^x + e^{-x})/2$; $r = 0$

4. To solve max-min problems in calculus, we need to find the roots of the equation $f'(x) = 0$ for a function f. Estimate the location of the max-min points for each of the following functions.
   a. $f(x) = x^4 - 2x^2 - 4x + 4$
   b. $f(x) = x^2 + 2e^{-x}$
   c. $f(x) = e^{-x} \ln x$

   d. $f(x) = \dfrac{\sin x}{x}$ on $(0, 2\pi]$

   e. $f(x) = x^2 - 3 \cos x$ on $[0, 2\pi]$
   f. $f(x) = e^{-x} + \arctan x$
   g. $f(x) = e^{-x^2} \ln x$ on $(0, \infty)$

2.2 Bisection Method

Assuming we know that a root r exists in some interval (a, b) for equation $f(x) = 0$, we now move to the problem of producing a usable approximation to r by using computing machinery. There are a number of well-known iterative algorithms for accomplishing this. The simplest of these is the following method.

We assume that r is an odd-order root and that $f(a)f(b) < 0$. So we know that r lies in interval (a, b) as shown in Figure 2.4. Let m be the midpoint of the interval. Then the root must lie in one of the intervals $(a, m]$ or $[m, b)$. If $f(a)f(m) < 0$, the root lies in interval $(a, m]$; otherwise, the root must lie in interval $[m, b)$. In either case, we have an interval half the size of the original interval containing the desired root. By continuing this argument, we can produce a sequence of intervals containing the root, with each interval half the length of the preceding one. This reasoning leads us to the following machine algorithm, known as the **bisection method**.

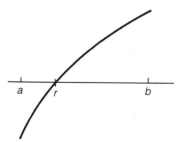

Figure 2.4

Algorithm 2.1

Step 1 Input a, b, and error tolerance E.

Step 2 Compute $m = (a + b)/2$.

Step 3 If $f(m) = 0$, then go to step 8.

Step 4 If $|b - a|/2 < E$, then go to step 8.

Step 5 Compute $f(m)f(a)$.

Step 6 If $f(m)f(a) < 0$, then set $b = m$ and return to step 2.

Step 7 If $f(m)f(a) > 0$, then set $a = m$ and return to step 2.

Step 8 Set $r = m$, and output root estimate r.

Number E is the permissible error in our root estimate. In selecting E, we must take into account the limitations of machine arithmetic and not select E unreasonably small. Each computed value of m is an estimate of the root r. It is clear that $|m - r| < (b - a)/2$ for each interval $[a, b]$ and associated m. So the error in estimate $r = m$ cannot exceed $(b - a)/2$.

If we denote the original interval by $[a_0, b_0]$ and succeeding ones as $[a_n, b_n]$ for $n = 1, 2, 3, \ldots$, then

$$(b_1 - a_1) = \tfrac{1}{2}(b_0 - a_0)$$

$$(b_2 - a_2) = \tfrac{1}{2}(b_1 - a_1) = \tfrac{1}{4}(b_0 - a_0)$$

and in general

$$|b_n - a_n| = \frac{|b_0 - a_0|}{2^n}$$

If $m = (a_n + b_n)/2$, then $r = m$ with error less than

$$\tfrac{1}{2}|b_n - a_n| = \frac{|b_0 - a_0|}{2^{n+1}}$$

So the theoretical error has a limit of zero as n approaches infinity. But again note that machine arithmetic limits the accuracy attainable in practice.

EXAMPLE 2.5   Equation $xe^x - 2 = 0$ has a simple root r in $[0, 1]$. Use the bisection method to estimate r.

Solution   We have $a = 0$, $b = 1$, $f(a) = -2$, and $f(b) = e - 2$. So $f(b) > 0$. We compute $m = 0.5$. Then $f(0.5) = -1.1756$. So $f(0.5) < 0$. Since $f(m)f(b) < 0$ in this case, r lies in $[0.5, 1]$. We set $a = 0.5$ and $b = 1$ and compute $m = 0.75$. Then

$$f(0.75) = 0.75e^{0.75} - 2 = -0.4122 < 0$$

So $f(m)f(b) < 0$, and r lies in $[0.75, 1]$. We set $a = 0.75$ and $b = 1$. Then $m = 0.875$. Now $f(m) = 0.0990 > 0$. Hence, $f(a)f(m) < 0$, and root r lies in interval $[0.75, 0.875]$.

If we stopped at this point, we could say with certainty that $r = (0.75 + 0.875) = 0.8125$ with error less than $\frac{1}{2}(0.875 - 0.75) = 0.0625$. In fact, continuing the arithmetic with a pocket calculator shows that r lies in the closed interval $[0.85260550, 0.85260551]$, so we know that $r = 0.8526055$ to an accuracy of seven decimal places.

Here is an example of machine output from this algorithm.

Solution of equation X**3 − X − 1 = 0

```
SOLUTION OF F(X) = 0 BY BISECTION METHOD
ENTER INTERVAL LIMITS A, B 1, 1.5
ROOT IS 1.32471795724476E + 00 + OR − 5.00E − 14
NUMBER OF ITERATIONS WAS 44
```

Solution of equation X*X − SIN (X) = 0

```
SOLUTION OF F(X) = 0 BY BISECTION METHOD
ENTER INTERVAL LIMITS A, B 0.5, 1
ROOT IS 8.76726215395053E − 01 + OR − 5.00E − 14
NUMBER OF ITERATIONS WAS 44
```

Solution of equation X**3 + 3*X − 3 = 0

```
SOLUTION OF F(X) = 0 BY BISECTION METHOD
ENTER INTERVAL LIMITS A, B 0, 1
ROOT IS 8.17731673886811E − 01 + OR − 5.00E − 14
NUMBER OF ITERATIONS WAS 45
```

Listing all iterations is not advisable, since so many iterations are required. Owing to the speed of the machine, the root estimate listed in each case was almost instantly obtained. An error estimate is included in each case. ■

The bisection method is slow but sure. Also, note that an interval estimate is obtained for r. If we select the midpoint of this interval as our estimate of r, then we know that the error cannot be more than half the length of our interval.

However, the bisection method can only be used to locate roots of odd order. A root of order 2 can be estimated by applying the bisection method to the equation $f'(x) = 0$, of course, but this requires computation of $f'(x)$.

Note that if we have just one equation to solve and the roots are all simple roots with approximate locations known, then the bisection method is a very usable method in practice. The slow convergence of the iterations is not a serious problem because of the speed of modern computers. Since this text is concerned with practical features of computation, we should not discard the bisection method simply because faster, more sophisticated methods are available. The algorithm is easy to program and is used by serious workers in the sciences to solve equations.

In practical root-finding problems, however, life is not always so simple. Root-finding algorithms are frequently only subroutines in large complex programs analyzing a complex scientific problem. The root-finding subroutine may need to be called a large number of times during execution of the program. For example, in orbit computation, the Kepler equation

$$m = x - E \sin x$$

must be solved repeatedly for various values of m and E, which change as the program execution proceeds. Much research has been put into developing rapid methods for solving this equation. This is not an unusual situation in modern science and technology. So you, the student, must avoid any tendency to take too narrow a view of each computing problem as we study it. Where, and exactly how, a given computing problem will occur in practice is not always predictable. The proper approach must always be to develop the fastest and most accurate methods we can for each problem presented.

■ *Exercise Set 2.2*

1. In each case, verify that the given equation has an odd-order root in the indicated interval. Then estimate the root accurately to four decimal places.
   a. $x^2 e^x = 1$ on $[0, 1]$ **b.** $x^3 - x - 1 = 0$ on $[1, 2]$
   c. $(5 - x)e^x = 5$ on $[4.9, 5]$ **d.** $x^2 = \sin x$ on $[0.5, 1]$
   e. $x^4 + 5x^3 - 9x^2 - 85x - 136 = 0$ on $[4, 4.5]$

2. Given that the equation

$$x^4 - 26x^2 + 24x + 21 = 0$$

   has a root near $x_0 = 1.6$, use the bisection method to estimate the root accurately to five decimal places.

3. For each equation, find an interval containing an odd-order root, and estimate the root accurately to five decimal places using the bisection method.
   a. $x^2 - 2x - 7 = 0$ **b.** $x^3 - 2x - 5 = 0$
   c. $x = 2e^{-x}$ **d.** $x + \sin x = 1$

4. Use the bisection method to find, as accurately as you can, all real roots for each equation.
   a. $x^3 - x^2 - x - 1 = 0$
   b. $x^2 = e^{-x^2}$
   c. $\ln |x| = \sin x$

5. A certain technical problem requires solution of the equation

$$21.13 - \frac{3480}{T} - 5.08 \log_{10} T = 0$$

   for a temperature T. Technical information indicates that the temperature should lie between $400°$ and $500°$. Use the bisection method to estimate the desired temperature to the nearest degree.

6. Use the bisection method with some calculus to find the minimum value of

$$f(x) = \sin x/x   \text{ on interval } [\pi, 2\pi]$$

2.3   Regula Falsi, Secant, and Newton Methods

We will now consider some well-known methods for speeding up the root approximation procedure. Let us first consider a very simple geometric procedure for obtaining an approximation of root r, known to be in interval $[a, b]$, which should be a better approximation than the midpoint used in the bisection method.

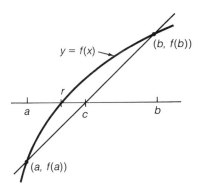

Figure 2.5

Consider the graph of f near root r of $f(x) = 0$. If r is a simple root in $[a, b]$ with $f(a) < 0$ and $f(b) > 0$, then the graph might look like the one in Figure 2.5. The straight line passing through points $(a, f(a))$ and $(b, f(b))$ intersects the x axis near r at point c. In fact, if $b - a$ is small, the graph of f frequently lies very close to this straight-line approximation. Hence, $c = r$ should be a usable approximation of r that has very little error if interval $[a, b]$ is short. This suggests that we solve the equation

$$y - f(a) = \frac{f(b) - f(a)}{b - a}(x - a)$$

with $y = 0$. This yields

$$x = a - \frac{(b - a)f(a)}{f(b) - f(a)} = \frac{af(b) - bf(a)}{f(b) - f(a)}$$

EXAMPLE 2.6   The equation $x^3 - x - 1 = 0$ has a simple root in $[1, 2]$. Make an approximation of this root.

Solution   Using $a = 1$ and $b = 2$, we find that

$$r \cong \frac{f(2) - 2f(1)}{f(2) - f(1)} = \frac{5 - 2(-1)}{5 - (-1)} = \frac{7}{6} = 1.16$$

The root in this case is $r = 1.3247$ to four decimal places.

It is easily established that $f(1.5) = 0.875 > 0$, so r is in $[1, 1.5]$. Then our method tells us that

$$r \cong \frac{f(1.5) - 1.5f(1)}{f(1.5) - f(1)} = \frac{0.875 - 1.5(-1)}{0.875 - (-1)} = \frac{2.3750}{1.875}$$

So $r \cong 1.2667 \cong 1.3$. This is a better approximation because the initial interval is smaller.   ∎

The preceding analysis suggests a possible algorithm.

Algorithm 2.2

Given that the equation $f(x) = 0$ has a simple root r in $[a, b]$, we do the following:

Step 1 Compute $c = \dfrac{af(b) - bf(a)}{f(b) - f(a)}$

Step 2 Compute $f(c)f(a)$. If $f(c)f(a) < 0$, then set $b = c$ and compute new c from step 1. Otherwise proceed to step 3.

Step 3 Set $a = c$ and return to step 1.

We repeat this procedure "until satisfied." We may decide to stop when successive values of c are virtually unchanged owing to the limitations of machine arithmetic. Or we may decide to stop when $|f(c)|$ is sufficiently small. We will discuss various "stopping rules" later for iterative procedures such as this one.

The method described is known as the **method of regula falsi**. In practice, we think of the successive values of c as a sequence $x_1, x_2, x_3, \ldots$ of approximations of r. A question we must consider later is whether this sequence necessarily converges to root r, or at least under what conditions we can be sure that the sequence will converge to r. When convergence is assured, we need to be concerned with the speed of convergence and estimation of $|x_n - r|$. Note that the last approximation x_n lies in a known interval, but the intervals do not shrink to zero in length, so the size of the final interval gives no useful information concerning the accuracy of the root approximation.

The regula falsi method is easily programmed for machine use. A sample output for estimation of the root in $[-2, -1]$ for equation $x^3 - x + 5 = 0$ is shown in Table 2.1. No error estimate is available. Some idea of the accuracy is given by the small value of $f(r)$. Convergence is rapid.

A method closely related to the regula falsi method is the **secant method**. Suppose that the equation $f(x) = 0$ has a root r, and estimates a and b are known. The root need not be in the interval $[a, b]$, but that is frequently the case. The step-by-step procedure is given in Algorithm 2.3.

Algorithm 2.3

Step 1 Set $x_1 = a$ and $x_2 = b$. Select a value for E.

Step 2 Compute $x_3 = x_2 - f(x_2) \dfrac{x_2 - x_1}{f(x_2) - f(x_1)}$

Step 3 If $|(x_3 - x_2)/x_2| < E$, then go to step 5.

Step 4 Set $x_1 = x_2$ and $x_2 = x_3$. Go to step 2.

Step 5 Output the value of x_3.

Table 2.1   SOLUTION OF F(X) = 0 BY REGULA FALSI METHOD

ENTER INTERVAL LIMITS A, B −2, −1.8

ESTIMATE OF X	VALUE OF F(X)
−1.898373983739837L + 00	5.696867374186265L − 02
−1.903851439702193L + 00	3.055739146430092L − 03
−1.904144349565632L + 00	1.630714780742615L − 04
−1.904159978339623L + 00	8.700038556908041L − 06
−1.904160812144252L + 00	4.641496433865910L − 07
−1.904160856627956L + 00	2.476250227090304L − 08
−1.904160859001173L + 00	1.321085685290768L − 09
−1.904160859127785L + 00	7.048028827227881L − 11
−1.904160859134540L + 00	3.760103339800480L − 12
−1.904160859134900L + 00	2.005062782473033L − 13
−1.904160859134920L + 00	1.088018564132653L − 14

FINAL ESTIMATE IS −1.904160859134920L + 00

F(R) =   1.088018564132653L − 14

The computation in Algorithm 2.3 is similar to the regula falsi procedure. The iteration formula used in step 2 is equivalent to

$$x_3 = \frac{x_1 f(x_2) - x_2 f(x_1)}{f(x_2) - f(x_1)}$$

This is simply the formula used in step 1 of Algorithm 2.2 with a change of labels for the variables. The form used in Algorithm 2.3 is less likely to lead to numerical difficulties as the difference $f(x_2) - f(x_1)$ becomes small.

In Algorithm 2.3, we omit the check for a change in sign found in step 2 of Algorithm 2.2 and make no attempt to produce a bounding interval. If the initial interval used in the regula falsi method is small enough, then the concavity of the graph of f does not change over that interval. In that case, the regula falsi root approximations converge on root r from one side only, and the check for a change in sign becomes pointless. The secant method takes advantage of that fact. Hence, the secant method is slightly faster than the regula falsi method.

EXAMPLE 2.7   Find the first two secant estimates for the equation $x^3 - x - 1 = 0$ using initial root estimates 1 and 2.

Solution   Set $x_1 = 1$ and $x_2 = 2$. Then

$$x_3 = 2 - f(2) \frac{2 - 1}{f(2) - f(1)} = 1.166667$$

Now set $x_1 = 2$ and $x_2 = 1.166667$. Then

$$x_3 = 1.166667 - f(1.166667) \frac{1.166667 - 2}{f(1.166667) - f(2)}$$

So $x_3 = 1.253112$ is the second estimate of the root.   ∎

Here is some computer-generated output for the root estimation problem in Example 2.7.

SECANT ALGORITHM FOR F(X) = 0

ENTER TWO REAL ROOT ESTIMATES ? 1,2

  ESTIMATES

  1.166666666666667
  1.253112033195021
  1.337206445841656
  1.323850096387641
  1.324707936532088
  1.324717965353818
  1.32471795724467
  1.324717957244746
  1.324717957244746

ROOT IS 1.324717957244746

F(X) = 2.775557561562891D — 17

One of the most popular and widely used root estimation algorithms is **Newton's method**. You very likely have already encountered this method in calculus. But let us review the derivation here and take a closer look at the method.

Suppose that x_0 is an estimate of root r of the equation $f(x) = 0$. We assume that f is differentiable in some interval I containing r and x_0. The mean value theorem for derivatives tells us that for points x and x_0 in I,

$$f(x) - f(x_0) = f'(\xi)(x - x_0)$$

where ξ lies between x and x_0. Then setting $x = r$, we see that

$$-f(x_0) = f'(\xi)(r - x_0)$$

where ξ lies between r and x. Hence,

$$r \cong x_0 - \frac{f(x_0)}{f'(\xi)}$$

If x_0 is sufficiently close to r, we reason that

$$r \cong x_0 - \frac{f(x_0)}{f'(x_0)}$$

if $f'(x_0)$ is not zero.

Algorithm 2.4   Newton's Method

Step 1 Select an estimate x_0 of a root of $f(x) = 0$ and set $n = 1$. Select a value for bound E.

Step 2 Compute $x_n = x_{n-1} - \dfrac{f(x_{n-1})}{f'(x_{n-1})}$.

Step 3 If $|x_n - x_{n-1}| < E$, then go to step 5.
Step 4 Set $n = n + 1$. Return to step 2.
Step 5 Output root estimate x_n.

This algorithm generates a sequence of approximations $\{x_n\}$ that converges rapidly to a root r in a large number of problems. If r is a simple root and $|f'(r)|$ is not too small, then convergence is very rapid and stable. If $f'(r) = 0$, r is a multiple root, and the iterations may move toward r slowly at first, but eventually they will show erratic behavior with no improvement in accuracy. This is not surprising, since the denominator $f'(x_{n-1})$ is approaching zero in this case.

A number of convergence theorems have been produced for the Newton algorithm, and we will examine some of this theory later. For now, we understand enough of the method to use it intelligently.

EXAMPLE 2.8   Beginning with estimate $x_1 = 1.3$, use Newton's method to estimate the root r of $x^3 - x - 1 = 0$ in $[1, 2]$.

Solution
$$f(x) = x^3 - x - 1$$

so

$$f'(x) = 3x^2 - 1$$

$$x_2 = 1.3 - \frac{f(1.3)}{f'(1.3)} = 1.325$$

$$x_3 = 1.325 - \frac{f(1.325)}{f'(1.325)} = 1.324718$$

x_3 agrees with x_4 through six decimal places.   ∎

Newton's method can be used to produce a general iteration formula for machine estimations of $\sqrt{N}$. We use $f(x) = x^2 - N$. Then $f'(x) = 2x$ and we get

$$x_{n+1} = x_n - \frac{x_n^2 - N}{2x_n}$$

This can be rearranged to the form

$$x_{n+1} = \frac{1}{2}\left(x_n + \frac{N}{x_n}\right)$$

This iteration formula produces approximations $x_1, x_2, x_3, \ldots$, which converge rapidly to $\sqrt{N}$. We examined the computations described here in the case of $N = 3$ in Section 1.1.

More generally, note that Algorithm 2.4 can be used to approximate the principal pth root of N for any positive number N. The principal pth root of N is a root of the equation $f(x) = 0$ for $f(x) = x^p - N$. Applying Algorithm 2.4, we find that

$$x_{n+1} = \frac{(p-1)x_n^p + N}{px_n^{p-1}} = \frac{1}{p}\left[(p-1)x_n + \frac{N}{x_n^{p-1}}\right]$$

To approximate the cube root of 10, for example, we use $p = 3$, $N = 10$, and $x_1 = 2$. The iteration formula gives $x_2 = 2.166667$, $x_3 = 2.154504$, $x_4 = 2.154435$, and $x_5 = 2.154435$. We conclude that $\sqrt[3]{10}$ has the approximate value 2.154435.

The principal difficulty, or weakness, of Newton's method lies in the need to compute $f'(x)$. When the equation $f(x) = 0$ involves a sufficiently complicated function, this can be a real problem.

There are difficulties when r is a multiple root, but all of the methods discussed have difficulties in that case.

In Sections 2.1 through 2.3, you have been introduced to the best-known algorithms for the approximation of roots of equations. Each of these algorithms is simple to convert to a computer program. Results are quite satisfactory for most equations.

■ *Exercise Set 2.3*

1. Estimate to five decimal places the root of the equation $x^3 + 3x - 5 = 0$ in $[1, 1.5]$.
   a. Use the method of regula falsi.
   b. Use the secant method.
   c. Use Newton's method.

2. Repeat exercise 1 using equation $xe^x - 2 = 0$ on $[0.5, 1]$.

3. Explain why Newton's method is not a satisfactory method for estimating the root of equation $x^3 - 3x^2 + 4 = 0$ in interval $[1.5, 2.5]$. Try to estimate the root using Newton's method with $x_0 = 1.1$ to see what actually happens numerically. List iterative estimates x_1 to x_{10}.

4. Use the method of regula falsi to estimate the root of $xe^x = 2$ in the interval $[0.5, 1]$. Continue the iterations until successive estimates agree through the first five decimal places. Do you think your root estimate is accurate to five decimal places? Support your conclusion.

5. Each of the following transcendental equations has exactly one positive real root. In each case, choose one of the methods described in this section to estimate the root to five decimal places. Give a reason for your choice of methods.

   a. $\sin x = \ln x$ b. $\dfrac{\sin x}{x} = \dfrac{1}{2}$

   c. $e^{-x} = \arctan x$ d. $e^x + x = 2$
   e. $\cos x = \sqrt{x}$ f. $x^2 \ln x = \frac{1}{2}$
   g. $xe^x = 2$ h. $e^{-x} = \sqrt{x}$

2.4 Error Estimation and Stopping Procedures

Each of the algorithms discussed so far is an iterative procedure. Hence, we must instruct the machine as to when the iterations should stop if we use these algorithms on a computing machine. There are several simple procedures for doing this.

1. We might instruct the machine to perform an assigned number N of iterations and then print out the final result.

2. We might select a small number $E > 0$ and instruct the machine to stop when $|x_n - x_{n-1}| < E$. At this point, root estimate x_n would be printed.

3. We might select a small number $E > 0$ and instruct the machine to stop when $|f(x_n)| < E$. At this point, root estimate x_n would be printed.

4. We might select a small number $E > 0$ and instruct the machine to stop when

$$\left| \frac{x_n - x_{n-1}}{x_n} \right| < E$$

In practice, procedure 1 combined with 2, 3, or 4 is often used. The number N places an upper limit on the number of iterations to be performed.

It should be noted that in procedures 2, 3, and 4, the number E is not a measure of the error in the root estimate x_n. In particular, note that procedure 2 is based on the practical idea that if successive estimates are very close, then further iterations will provide little or no improvement. However, successive terms in a sequence may become very close while x_n is still far away from the limit of the sequence if the limit exists. In fact $|x_n - x_{n-1}|$ may become arbitrarily small, even for a divergent sequence. This is a well-known fact from real analysis. Hence, in procedure 2, we see that $|x_n - x_{n-1}| < E$ *does not* imply that $|x_n - r| < E$.

In procedure 3, we also employ a "practical fact." Sometimes we want to find r so that we can in fact reduce $f(x)$ to zero. Hence, if $|f(x_n)|$ is extremely small, we have practically accomplished this.

Also note that it is unlikely that we will ever attain $f(x_n) = 0$ for any root estimate; so we settle for $|f(x_n)| < E$ for some selected small positive number E. But we can attain $|f(x_n)| < E$ even when $|x_n - r|$ is relatively large. Consider the equation

$$f(x) = \frac{x^3}{1000}$$

The equation $f(x) = 0$ has root $r = 0$. Suppose $x_n = 0.01$ is an estimate of this root. Then $|f(x_n)| = 10^{-9}$, which would be considered small enough by the machine if we used $E = 10^{-8}$. However, root estimate $x_n = 0.01$ is not very accurate. This estimate is certainly not as accurate as $|f(x_n)| = 10^{-9}$ might suggest.

In procedure 4, we are using an estimate of the relative error in our root estimate to provide a stopping criterion. This is an improvement over procedure 2, but the machine can still be satisfied with a root estimate x_n further away from root r than one might wish.

We can make good error estimates in some root-finding problems, however. For example, if we are using the bisection method, we normally stop the iterations when $|b_n - a_n| < E$ for some small number E. In this case, the root estimate is

$$x_n = \frac{b_n + a_n}{2}$$

and the error is less than $(b_n - a_n)/2$.

EXAMPLE 2.9   An application of the bisection method to equation $(5 - x)e^x = 5$ yielded a final interval estimate (4.96511, 4.96515) with root estimate

$$x_n = \frac{4.96511 + 4.96515}{2} = 4.96513$$

We can say with confidence that the error is less than

$$\frac{4.96515 - 4.96511}{2} = 2.00 \times 10^{-5}$$

In fact, the root is 4.965114 correct to six decimal places. Hence, the true absolute error is

$$|4.96513 - 4.965114| = 1.6 \times 10^{-5} \qquad \blacksquare$$

In the general case, let

$$|x_n - r| = \varepsilon$$

and suppose that ε is small but unknown. Then

$$|f(x_n) - f(r)| = |f'(c_n)(x_n - r)|$$

for some c_n between r and x (by the mean value theorem). We are assuming that function f is continuous on the closed interval with end points x_n and r, and we are assuming that f' exists on the open interval between x and r. Now

$$f'(x_n) \cong f'(c_n)$$

in this case. Also, $f(r) = 0$. So we conclude that

$$\varepsilon \cong \left| \frac{f(x_n)}{f'(x_n)} \right|$$

if $f'(x_n) \neq 0$.

This approximation formula supplies a usable, and surprisingly accurate, error estimate in practice.

EXAMPLE 2.10   Given root estimate $x_n = 1.32471817$ for the root of $x^3 - x - 1 = 0$ in (1, 2), make an error estimate.

Solution   We find that

$$f(x_n) = 9.1 \times 10^{-7} \quad \text{and} \quad f'(x_n) = 4.264635$$

Hence,

$$\varepsilon \cong \frac{9.1 \times 10^{-7}}{4.264635} = 2.1 \times 10^{-7}$$

We conclude that $r = 1.324718$ correct to six decimal places. The true value is $1.3247179524\ldots$. ∎

EXAMPLE 2.11   Given root estimate $x_n = 0.8526055$ for the root of equation $xe^x = 2$ lying in (0, 1), make an error estimate.

Solution   We have

$$f(x) = xe^x - 2 \quad \text{and} \quad f'(x) = (x + 1)e^x$$

So

$$f(x_n) = -3.8 \times 10^{-10}$$
$$f'(x_n) = 4.34575$$

Therefore,

$$\varepsilon \cong \frac{3.8 \times 10^{-10}}{4.34575} = 0.87 \times 10^{-10} = 8.7 \times 10^{-11}$$

In this case, we should use caution, since x_n was apparently computed using seven-decimal arithmetic. Hence, the precision indicated by our value of ε seems unlikely. It is true, however, that $r = 0.8526055$ correct to seven decimal places. ∎

This error estimation procedure gives us another stopping procedure provided that we are willing to compute $f'(x)$. We then instruct the machine to terminate our iterative procedure when

$$\left| \frac{f(x_n)}{f'(x_n)} \right| < E$$

for some error tolerance E. We can then conclude that the error in x_n is less than E in absolute value if E is not smaller than the precision used in computing x_n.

Note that the computation described for estimating the absolute error $|x_n - r|$ is built into Newton's method. Hence, when the Newton algorithm is used, an estimate of the error in x_n is $|x_{n+1} - x_n|$. We can regard Newton's algorithm as the generation of improved estimates by correcting each x_n for the approximate error, as estimated by our formula. In this sense, the Newton algorithm is a preview of a basic numerical technique known as extrapolation. Extrapolation techniques are often employed in iterative numerical methods designed to improve on existing estimates of a number we wish to determine. If x_n is the nth estimate in a sequence of estimates of a number c, and ε is an estimate of $c - x_n$, then a better estimate of c should be $x_{n+1} = x_n + \varepsilon$. This idea will be explored in later chapters to develop iterative improvement techniques for numerical solutions to various problems.

■ Exercise Set 2.4

1. Given $f(x) = e^{-x} - \cos x$, we find that $f(1.292) = -4.78\text{E} - 4$ and $f(1.293) = 2.1\text{E} - 4$. We conclude that the equation $f(x) = 0$ has a root between 1.292 and 1.293.

   a. Explain why we cannot conclude from the values of f alone that the root is in fact 1.293 correct to three decimal places, even though the fact that $|f(1.293)| < |f(1.292)|$ does suggest this.

   b. Show by use of the error estimate $|f(1.293)|/|f'(1.293)|$ that the root is in fact 1.293 correct to three decimal places.

2. A student finds that the equation $xe^x - 2 = 0$ has a root in $[0.5, 1]$. Applying the method of regula falsi, she finally arrives at root estimates

$$x_7 = 0.8526055 \quad \text{and} \quad x_8 = 0.8526055$$

so her best estimate is $r \cong 0.8526055$. But regula falsi does not provide a useful error estimate. Make an error estimate, and determine the accuracy of the final root estimate obtained.

3. Using the iterative formula $x_{n+1} = \frac{1}{2}(x_n + 3/x_n)$ with $x_0 = 1.5$, a student arrives at the following estimates

$$x_3 = x_4 = 1.7320508 \quad \text{for } \sqrt{3}$$

   a. Using the fact that $\sqrt{3}$ is a root of the equation $x^2 = 3$, use the methods described in this section to make an error estimate and determine the accuracy of the root estimate

$$\sqrt{3} \cong 1.7320508$$

   b. Using a handbook or table, find a very precise value (accurate to at least eight decimal places) for $\sqrt{3}$. Use this value to estimate the absolute error and the relative error in the estimate of $\sqrt{3}$ given in this exercise.

4. Let $S_n = 1 + \frac{1}{2} + \ldots 1/n$ for $n = 1, 2, 3, \ldots$. To estimate the limit of this sequence, we might compute successive values of terms S_n and stop when $|S_n - S_{n-1}| < 10^{-6}$, for example.

   a. Explain why $|S_n - S_{n-1}| < 10^{-6}$ for $n = 1000001$.

   b. Explain why $S_{1000001}$ is not an accurate estimate of the limit of the sequence $\{S_n\}$.

5. Find all real roots of each equation with an accuracy of at least six decimal places. Make error estimates to confirm the accuracy of your root estimates.

a. $x^2 = e^x$

b. $e^{-x} = \dfrac{\cos x}{x}$

c. $\ln x = \sin x$

d. $x^3 - x - 1 = 0$

e. $x^3 + 4x^2 - 7 = 0$

f. $5x - \sin x = 4$

g. $x^4 - 26x^2 + 24x + 21 = 0$

2.5   Fixed-Point Iteration Method

Given an equation $f(x) = 0$, suppose we can find an equivalent equation $x = g(x)$. So $f(r) = 0$ if and only if $r = g(r)$. We can use the function g to estimate roots of $f(x)$ in this case.

EXAMPLE 2.12   The equation $xe^x - 2 = 0$ is equivalent to the equation $x = 2e^{-x} = g(x)$. ∎

EXAMPLE 2.13   The equation $x^3 - x - 1 = 0$ is equivalent to the equation $x = \sqrt[3]{x + 1} = g(x)$. Other equivalent equations are $x = x^3 - 1$ and $x = (x + 1)/x^2$. ∎

A given equation may be equivalent to several other equations of the form $x = g(x)$.

Since every root of $x = g(x)$ is a root of $f(x) = 0$, we can concentrate on finding solutions of the equation $x = g(x)$.

Definition 2.2

If $c = g(c)$, then we say c is a **fixed point** for function g.

Here is the key theorem about fixed points.

Theorem 2.1

Suppose that g is differentiable on a finite open interval I and $g(x)$ is in I for every x in I. Assume that $|g'(x)| \le K < 1$ on I. Assume that c in I is a fixed point for g. Then if x_0 is any point in I, the sequence

$$x_{n+1} = g(x_n) \qquad n = 0, 1, 2, \ldots$$

converges to the point c.

Proof   Let

$$e_n = c - x_n \qquad n = 0, 1, 2, \ldots$$

Now

$$e_n = g(c) - g(x_{n-1}) \qquad n = 1, 2, 3, \ldots$$

So

$$e_n = g'(\bar{x}_n)(c - x_{n-1})$$

where $\bar{x}_n$ lies between c and x_{n-1}. Hence,

$$e_n = g'(\bar{x}_n)e_{n-1} \qquad n = 1, 2, 3, \ldots$$

Then

$$|e_n| \le K|e_{n-1}| \le K^2|e_{n-2}| \le K^3|e_{n-3}|$$

and so on. So clearly,

$$|e_n| \le K^n|e_0|$$

But $|e_0| = |c - x_0|$ is less than the length of I, which is a finite number. Also, $0 \le K < 1$. So $|e_n| \to 0$ as $n \to \infty$. Since $\lim |c - x_n| = 0$, $\lim x_n = c$, as asserted.   □

The proof of Theorem 2.1 also shows that under the assumed conditions, g can only have one fixed point in I.

Now suppose that we know that the equation $f(x) = 0$ has a root in a finite open interval I. Then r is a fixed point for function g if $x = g(x)$ is an equivalent equation. If g is differentiable on I with $|g'(x)| \le K < 1$ on I, then Theorem 2.1 shows that we can estimate r by selecting a point x_0 in I that seems to be a fair estimate of r and then generating the sequence

$$x_{n+1} + g(x_n) \qquad n = 0, 1, 2, \ldots$$

The numbers x_n form a sequence that converges to root r as n increases.

EXAMPLE 2.14   Solve $f(x) = x^3 - x - 1 = 0$.

Solution   We note that

$$f(1) = -1 \quad \text{and} \quad f(2) = 5$$

so a root exists in interval $I = (1, 2)$. We choose $g(x) = \sqrt[3]{1 + x}$, since $x = \sqrt[3]{1 + x}$ is an equivalent equation. Now $g'(x) = \frac{1}{3}(1 + x)^{-2/3}$. So on I we have

$$\frac{1}{3(1 + 2)^{2/3}} < g'(x) < \frac{1}{3(1 + 1)^{2/3}}$$

So

$$0 < g'(x) < \frac{1}{3(2^{2/3})} = K$$

and $|g'(x)| \leq K < 1$ on $I = (1, 2)$. If we choose $x_0 = 1.3$ and $x_{n+1} = \sqrt[3]{1 + x_n}$, then this sequence should converge to the desired root. Machine computation gives the result

$$x_0 = 1.3$$

$$x_1 = 1.320006122$$

$$x_2 = 1.323822354$$

$$\vdots \qquad \vdots$$

$$x_{11} = 1.324717957 = x_{12} = x_{13} = \cdots$$

Hence the sequence converges to the root estimate $r = 1.324717957$ as the final estimate of r. ■

Such a **fixed-point iteration** scheme is useful provided that:

1. Estimates $x_1, x_2, x_3, \ldots, x_n$ can actually be computed from initial choice x_0.
2. The sequence converges.
3. The sequence does not converge too slowly.

It is important to note in particular that even if $x = g(x)$ is equivalent to $f(x) = 0$, the iterations may not converge.

EXAMPLE 2.15 Given $f(x) = x^3 - 2 = 0$, we could choose $g(x) = 2/x^2$, since $x = 2/x^2$ is equivalent to $x^3 - 2 = 0$. If we choose $x_0 = 1.2$ (which is a fair estimate of the root $\sqrt[3]{2}$) and generate sequence

$$x_{n+1} = \frac{2}{(x_n)^2} \quad \text{for} \quad n = 0, 1, 2, \ldots$$

then we obtain a divergent sequence (as you can easily verify). The reason, of course, lies in the fact that no $K < 1$ exists for which the condition $|g'(x)| \leq K$ is satisfied in any open interval about fixed point $\sqrt[3]{2}$. ■

It is also important to note that even when the sequence of iterates does converge, the convergence may be very slow.

EXAMPLE 2.16 Consider the equation $f(x) = x^3 - 2 = 0$ again. This time, let us choose $g(x) = \sqrt{2} x^{-1/2}$. Then $|g'(\sqrt[3]{2})| = \frac{1}{2}$. So $|g'(x)| \leq K < 1$ on some open interval about the root because of the continuity of g' at the root. We can expect the sequence generated by iterations

$$x_{n+1} = \sqrt{\frac{2}{x_n}} \quad n = 0, 1, 2, \ldots$$

to converge to the root if we choose x_0 sufficiently close to $\sqrt[3]{2}$. Choosing $x_0 = 1.2$, we obtain

$$x_1 = 1.290994$$

$$x_2 = 1.244666$$

$$x_3 = 1.267619$$

$$x_4 = 1.256090$$

$$\vdots \qquad \vdots$$

$$x_{14} = 1.259917$$

$$x_{15} = 1.259923$$

$$x_{16} = 1.259920$$

The sequence is converging to the fixed point very slowly.   ∎

For a given equation $f(x) = 0$, there may be (and usually are) several choices for g yielding equations equivalent to $f(x) = 0$. Some choices for g will be better than others, of course. Some may produce divergent sequences, which are useless for root estimation. Some may produce very slowly converging sequences, and these are to be avoided if possible.

EXAMPLE 2.17 Given $e^{-x} - \cos(x) = 0$ with a root known to be near $x_0 = 1.29$, we need an accurate estimate of that root. Equivalent equations are

$$x = \ln(\sec x) \quad \text{and} \quad x = \arccos(e^{-x})$$

Choosing $g(x) = \ln(\sec x)$, we find that $g'(x) = \tan x$. So $g'(1.29) = 3.47$ and $|g'(x)| > 1$ near 1.29. Such a choice of g will produce a divergent sequence of estimates. We turn to the choice $g(x) = \arccos(e^{-x})$. We now find that

$$g'(x) = \frac{e^{-x}}{\sqrt{1 - e^{-2x}}}$$

So $|g'(x)| \leq 0.4$ if we stay sufficiently near 1.29 (using the fact that $g'(x)$ is a decreasing function in $[1, 2]$ so that the value does not exceed $g'(1)$). The iterations generated by

$$x_{n+1} = \arccos(e^{-x_n}) \quad \text{with} \quad x_0 = 1.29$$

should converge. Computation shows that the iterations do in fact converge to root estimate 1.292695719 in 15 iterations.   ∎

The iteration process discussed in this section is a model for many root estimation algorithms, including Newton's method. So the theory introduced here has wide application. Such iteration schemes are particularly convenient when estimating a root for a transcendental equation using a pocket calculator.

■ *Exercise Set 2.5*

1. Find an equivalent equation of the form $x = g(x)$ for each of the following equations.
   a. $x^3 - 3x + 1 = 0$ **b.** $xe^x = \sin x$
   c. $\ln x = \sin x$ **d.** $e^{-x} = \cos x$

2. Use the iteration formula given in Example 2.17 with $x_0 = 1.3$ to compute iterations x_1, $x_2, x_3, \ldots$ and verify convergence to the root between 1.2 and 1.3.

3. The iteration formula

$$x_{n+1} = \tfrac{1}{2}\left(x_n + \frac{1}{x_n}\right)$$

   with $x_0 = 2$ is used to generate a sequence of numbers $x_1, x_2, x_3, \ldots$.
   a. Show that the sequence of iterates converges.
   b. Determine the number to which the iterates should converge. (*Hint:* If x is that number, then $x = (x + 1/x)/2$.)
   c. Confirm the conclusion in part (b) by actually computing the terms of the sequence to determine the number to which they converge.

4. The equation $5x - \sin x = 4$ has a root between 0.9 and 1. Use the fixed-point iteration method with an appropriately chosen $g(x)$ to estimate the root as accurately as you can. Make an error estimate.

5. The equation $x = \tan x$ has a root just below 1.5π, as we can verify graphically.
   a. Show that the iteration scheme $x_{n+1} = \tan x_n$ does not converge for any choice of x_0.
   b. Show that the iteration scheme $x_{n+1} = \arctan x_n$ does not converge to the desired root for any choice of x_0.
   c. Find a fixed-point iteration scheme that does converge to the desired root.
   d. Use the iteration scheme found in part (c) to estimate the desired root to at least six decimal places.

6. For each function g, find a fixed point in the domain of g if any fixed points exist.
   a. $g(x) = \sin x$ **b.** $g(x) = e^{-x}$
   c. $g(x) = \ln x$ **d.** $g(x) = 1 - x^2$

7. Show that if $g(x)$ is a polynomial and g has infinitely many fixed points on the real line, then $g(x) = x$ for all real numbers x.

8. Show that the iteration scheme

$$x_{n+1} = \tfrac{1}{2}\left(x_n + \frac{N}{x_n}\right)$$

   converges for every x_0 and the iterations converge to $\sqrt{N}$.

9. Use the iteration scheme in exercise 8 to estimate the following.
   a. $\sqrt{5}$ **b.** $\sqrt{7}$ **c.** $\sqrt{37}$

2.6   Rate of Convergence

In Section 1.4, we defined the rate of convergence of an algorithm. We now apply that idea to algorithms that approximate roots of an equation.

Suppose that an algorithm based on fixed-point iteration is employed. We generate a sequence of estimates of root r defined by the equation $r = g(r)$. We use the iteration scheme

$$x_{n+1} = g(x_n) \qquad n = 0, 1, 2, \ldots$$

Suppose that the sequence of estimates converges to r as desired. We need to investigate the rate of convergence. We want to find positive constants K and p such that

$$\lim_{n \to \infty} \frac{|x_{n+1} - r|}{|x_n - r|^p} = K$$

if such numbers exist. In Section 1.4, we called p the *order of convergence*.

Let $e_n = |x_n - r|$. Then

$$e_{n+1} = |x_{n+1} - r| = |g(x_n) - g(r)|$$

Assume that g is continuous on an interval $[a, b]$ and differentiable on (a, b), where (a, b) is an interval containing r and all root estimates $x_0, x_1, x_2, \ldots$. The mean value theorem from calculus tells us that

$$|g(x_n) - g(r)| = |g'(c_n)(x_n - r)| = |g'(c_n)| |e_n|$$

where c_n lies between r and x_n. Then

$$\frac{|x_{n+1} - r|}{|x_n - r|} = |g'(c_n)|$$

If g' is continuous at r, then

$$\lim_{n \to \infty} |g'(c_n)| = |g'(r)|$$

So

$$\lim_{n \to \infty} \frac{|x_{n+1} - r|}{|x_n - r|} = |g'(r)| = K$$

We have $p = 1$ and $K = |g'(r)|$, so the convergence is linear in this case if $K > 0$.

For a linearly convergent iteration scheme, we must have $K < 1$. If K is very small, the convergence is more rapid than if K is near 1. We will call the number K the *convergence factor*; it is also sometimes called the asymptotic error constant.

EXAMPLE 2.18   In the case of the iteration scheme used in Example 2.14, we have

$$g'(x) = \tfrac{1}{3}(1 + x)^{-2/3} \quad \text{and} \quad r \cong 1.32471796$$

Hence, g' is continuous at r as required. The value of K is approximately

$$0.333333(2.32471796)^{-2/3} \cong 0.18995 \cong 0.19$$

This value of K is small enough to indicate a usable iteration scheme.   ∎

EXAMPLE 2.19   Solve the equation $xe^x = 2$.

Solution   We can use the equivalent equation $x = 2e^{-x} = g(x)$. Note that $f(x) = xe^x - 2$ in this case. Computation gives $f(0.8) = -0.219567$ and $f(0.9) = 0.21364$, so root r lies between 0.8 and 0.9. In fact, $r \cong x_0 = 0.85$. Now $g'(x) = -2e^{-x}$, and on interval $(0.5, 1)$, we see that $|g'(x)| = 2e^{-x}$ varies from $|g'(0.5)| = 1.2131$ to $|g'(1)| = 0.73576$. Near r we see that $|g'(x)| \cong |g'(0.85)| = 0.855 < 1$. So the convergence factor is $K \cong 0.855$. Then iterations $x_{n+1} = g(x_n)$, with $x_0 = 0.85$, should converge, but the convergence can be expected to be slow. Computations confirm this expectation.   ∎

Clearly, convergence is most rapid when $|g'(r)|$ is very small. This suggests that we consider the case of $g'(r) = 0$. What then? In this case, the preceding analysis does not apply. However, if g has a continuous second derivative on some open interval centered on r, and x_0 is selected from this interval, then we may conclude that

$$e_{n+1} = |g(x_n) - g(r)|$$

where

$$g(x_n) - g(r) = g'(r)(x_n - r) + \tfrac{1}{2}g''(c_n)(x_n - r)^2$$

by Taylor's theorem from calculus. If $g'(r) = 0$, we see that

$$e_{n+1} = |g(x_n) - g(r)| = |\tfrac{1}{2}g''(c_n)| e_n^2$$

where c_n lies between x_n and r. Then

$$\frac{e_{n+1}}{e_n^2} = \tfrac{1}{2}|g''(c_n)|$$

It follows that

$$\lim_{n \to \infty} \frac{e_{n+1}}{e_n^2} = \tfrac{1}{2}|g''(r)|$$

Assuming that $g''(r) \neq 0$, we have convergence of order $p = 2$ and convergence factor $K = |g''(r)|/2$. The fixed-point iteration scheme converges *quadratically* in this case.

EXAMPLE 2.20   Solve $x^2 = N$, where N is a positive real number. We choose

$$g(x) = \tfrac{1}{2}\left(x + \frac{N}{x}\right)$$

The equation $x = g(x)$ is equivalent to $x^2 = N$, as you can verify. This fixed-point iteration scheme was illustrated in Section 1.1 and used to estimate $\sqrt{3}$. We now investigate the rate of convergence. We find that

$$g'(x) = \tfrac{1}{2}\left(1 - \frac{N}{x^2}\right)$$

So

$$g'(r) = g'(\sqrt{N}) = 0$$

Hence, the iteration scheme converges quadratically.   ∎

The preceding theory can also be applied to the Newton algorithm discussed in Section 2.3. In this case, we have

$$g(x) = x - \frac{f(x)}{f'(x)}$$

so

$$g'(x) = \frac{f(x)f''(x)}{f'(x)^2}$$

If $f'(r) \neq 0$, then $g'(r) = 0$ because $f(r) = 0$. Hence, if for some $\delta > 0$,

$$\frac{|f(x)f''(x)|}{f'(x)^2} \leq K < 1$$

over interval $(r - \delta, r + \delta)$, and $x_0 \in (r - \delta, r + \delta)$, then the Newton algorithm converges quadratically to root r.

If we assume that f'' is continuous on an open interval centered on r, and $f'(r) \neq 0$, then g' is continuous at r. Since $g'(r) = 0$ and g' is continuous at r, then we can be sure that constants K and δ exist for which

$$x \in (r - \delta, r + \delta) \rightarrow |g'(x)| \leq K < 1$$

This leads to the following theorem.

Theorem 2.2   Suppose that f'' is continuous on the interval $(r - c, r + c)$ for some $c > 0$ and $f(r) = 0$. Assume that $f'(x) \neq 0$ on $(r - c, r + c)$. Then there exists some $\delta > 0$ such that $x_0 \in (r - \delta, r + \delta)$ implies that the Newton algorithm converges to root r quadratically.

Many such "convergence theorems" for the Newton algorithm can be found in the current literature. There are also convergence theorems for regula falsi and the other root estimation algorithms, but we do not discuss them here because they have very little practical use. They are primarily of theoretical interest.

Users of root estimation algorithms know that results from the Newton algorithm are usually unsatisfactory when estimating a root r for which $f'(r) = 0$, that is, for estimation of a multiple root. As an application of the preceding theory, we will investigate this phenomenon. Of course, since the algorithm requires a division by $f'(x_n)$ at each step, the problem is at least partly due to the fact that $f'(x_n)$ is approaching zero as x_n approaches the root r. So the ratio of $f(x_n)$ to $f'(x_n)$ can, and usually does, begin to behave erratically as the root estimate approaches r.

Since it is the value of $|g'(r)|$ that determines the rate of convergence for fixed-point iteration, and for the Newton algorithm

$$|g'(x)| = \frac{|f(x)f''(x)|}{|f'(x)|^2}$$

the convergence is slowed down because this ratio is not small near the root r. We may in particular expect trouble if $f'(r) = 0$ and $f''(r) \neq 0$ because the ratio of $f''(x)$ to $f'(x)$ will grow as we approach the root, and the decrease in absolute value of $f(x)$ may not be able to compensate for this growth. We can also see that if

$$\lim_{x \to r} \frac{f''(x)}{f'(x)^2} = C$$

does exist, then the iterations should still converge rapidly.

Some alternatives to the Newton algorithm have been proposed in the past to avoid the difficulties that arise from $f'(r) = 0$.

Let us consider the algorithm

$$x_{n+1} = x_n - \frac{f(x_n)}{m} = g(x_n)$$

where m is a constant. Clearly, if $f(r) = 0$, then $g(r) = r$, as required. For this algorithm to work properly, the value of m should be an approximation to the value of $f'(r)$. In fact, in this case, we have

$$g'(x) = 1 - \frac{f'(x)}{m}$$

so the closer m is to $f'(r)$, the smaller $|g'(r)|$ will be. This algorithm is best suited to the case where we want to avoid repeated computation of $f'(x_n)$, but $f'(r) \neq 0$. If $f'(r) = 0$, then clearly, $g'(r) = 1$, so we have linear convergence with a convergence factor close to 1.

In the very common case where r is a double root of $f(x) = 0$, so that $f'(r) = 0$ but $f''(r) \neq 0$, there is a very simple alternative to the basic Newton algorithm. We use iterations

$$x_{n+1} = x_n - \frac{f'(x_n)}{f''(x_n)}$$

That is, we apply the Newton algorithm to $f'(x)$ rather than $f(x)$ because r is a simple root of $f'(x) = 0$. The need to compute $f''(x)$ is undesirable, but many modifications of the Newton algorithm intended for estimation of multiple roots require computation of $f''(x)$. There is a way to avoid this if we know the order of the multiple root.

Suppose that r is a root of order p for $f(x) = 0$. Then

$$f(x) = (x - r)^p h(x) \qquad h(r) \neq 0$$

It is also true that $f^p(r) \neq 0$ in this case, while $f^k(r) = 0$ if $k < p$. Now

$$f'(x) = p(x - r)^{p-1} h(x) + (x - r)^p h'(x)$$

so

$$p \frac{f(x)}{f'(x)} = \frac{p(x - r)h(x)}{ph(x) + (x - r)h'(x)}$$

Now $h(r) \neq 0$ and $(x - r)$ approaches zero as x approaches r, so the denominator approaches $ph(r)$ as x approaches r. We conclude that

$$p \frac{f(x)}{f'(x)} \cong \frac{p(x - r)h(x)}{ph(x)} = x - r$$

and

$$r \cong x - p \frac{f(x)}{f'(x)}$$

for x near r. This suggests the iteration scheme

$$x_{n+1} = x_n - p \frac{f(x_n)}{f'(x_n)} = g(x_n)$$

in this case. We can verify that $g'(r) = 0$ in this case and that the iterations converge quadratically. To use this scheme, we need to know the order p of the root. If we have good reason to believe that the desired root is a double root, then the method is satisfactory. In some cases, determining p could be difficult.

We will now derive a modification of the Newton algorithm by combining a sequence of approximations. The resulting algorithm gives satisfactory results for double roots, as we shall see. First note that

$$f'(r) \cong \frac{f(x_n) - f(r)}{x_n - r} = \frac{f(x_n)}{x_n - r}$$

so

$$r \cong x_n - \frac{f(x_n)}{f'(r)} \tag{1}$$

Also,

$$f'(x_n) \cong \frac{f(r) - f(x_n)}{r - x_n} = \frac{f(x_n)}{x_n - r}$$

so

$$x_n - r \cong \frac{f(x_n)}{f'(x_n)} \tag{2}$$

Now

$$f'(r) \cong f'(x_n) + (r - x_n)f''(x_n)$$

So

$$f'(r) \cong f'(x_n) - \frac{f(x_n)}{f'(x_n)} f''(x_n)$$

follows from (2), and

$$f'(r) \cong \frac{f'(x_n)^2 - f(x_n)f''(x_n)}{f'(x_n)}$$

Substituting in Equation 1 yields

$$r \cong x_n - \frac{f(x_n)f'(x_n)}{f'(x_n)^2 - f(x_n)f''(x_n)}$$

This suggests the iteration scheme

$$x_{n+1} = x_n - \frac{f(x_n)f'(x_n)}{f'(x_n)^2 - f(x_n)f''(x_n)}$$

This "derivation" is, of course, very nonrigorous, but the resulting algorithm gives good numerical results for estimating double roots, and that is the real test of a numerical algorithm. An alternate derivation can be found in Burden and Faires (1985). Consider the following example of computer output.

EXAMPLE 2.21 The equation $x^4 - 4x^2 + 4 = 0$ has a double root $r = \sqrt{2}$. Beginning with $x_0 = 2$, the results were as shown in the following computer output table.

EST. NO.	MOD. NEWTON	NEWTON
1	1.33333333	1.75
2	1.41176471	1.59821429
3	1.41421147	1.51150988
4	1.41421147	1.46442747
5	1.41421147	1.43975096
6	1.41421147	1.42709551
7	1.41421147	1.4206836
8	1.41421147	1.41745594
9	1.41421147	1.41583661
10	1.41421147	1.41502556
11	1.41421147	1.41461961
12	1.41421147	1.41441656
13	1.41421147	1.41431537
14	1.41421147	1.41426506
15	1.41421147	1.41423907
16	1.41421147	1.41422538
17	1.41421147	1.41422538

■

■ *Exercise Set 2.6*

1. In Example 2.19 it was shown that the iteration scheme $x_{n+1} = 2e^{-x_n}$ with $x_0 = 0.85$ will converge to the root of the equation $xe^x = 2$ between 0.8 and 0.9, but the convergence will be slow. Confirm this prediction by actually computing iterates $x_1, x_2, x_3, \ldots$ to observe the convergence rate numerically.

2. The real solution of equation $x^3 = 2$ can be estimated using either of the following iteration schemes:
   I. $x_{n+1} = \sqrt{2} x_n^{-1/2}$; $x_0 = 1.25$
   II. $x_{n+1} = \frac{1}{2}(x_n + 2/x_n^2)$; $x_0 = 1.25$
   a. Show that I and II both have the same theoretical convergence factor.
   b. Perform the calculations, and determine for each iteration scheme how many iterations are required to produce an estimate of $\sqrt[3]{2}$ correct to six decimal places. The value of $\sqrt[3]{2}$ is 1.25992105 correct to eight decimal places.

3. The equation $8x^3 - 12x^2 + 6x - 1 = 0$ has a triple root in $[0, 1]$. Using $x_0 = 1$, estimate the root by the following methods.
   a. Use Newton iterations $x_{n+1} = x_n - 3f(x_n)/f'(x_n)$
   b. Use the modified Newton algorithm described at the end of this section.

4. We want to estimate the root of the equation $x^3 - x - 1 = 0$ that lies in $[1, 1.5]$. Fixed-point iterations given by

$$x_{n+1} = \sqrt[3]{1 + x_n} \quad \text{with } x_0 = 1$$

converge linearly to the root. Newton's method converges quadratically, and the secant method has order of convergence p between 1 and 2. Apply each of these methods to this problem. Observe the actual numerical rates of convergence. Use $x_0 = 1$ for Newton's method. Use $x_0 = 1$ and $x_1 = 1.3$ for the secant method.

5. Iterations $x_{n+1} = (2x_n^2 + 1)/(4x_n + 1)$ with $x_0 = 1$ should converge quadratically to the root $r = 0.5$ for the equation $2x^2 + x - 1 = 0$, as you should verify. Numerically, quadratic convergence means that if x_n has k correct digits, then x_{n+1} should have $2k$ correct

digits. Compute x_1, x_2, x_3,... using a calculating machine capable of displaying nine or more digits. Do your numerical results confirm this interpretation of quadratic convergence?

6. Determine in each case whether the iterations converge linearly or quadratically.

   a. $x_{n+1} = \ln(2 + \sin x_n)$; $x_0 = 2$

   b. $x_{n+1} = \dfrac{x_n^2 + 2e^{-x_n}}{x_n + 1}$; $x_0 = 1$

   c. $x_{n+1} = 1 - \sqrt{x_n}$; $x_0 = 0.5$

2.7 Acceleration of Convergence

It is not unusual when estimating roots by an iteration algorithm to find that the algorithm employed does converge to the desired root, but very slowly. In some cases, it is possible to speed up the convergence. Even with high-speed computers, slow convergence of algorithms is undesirable, so there will always be a need for acceleration techniques. In this section, we will explore a technique for accelerating the convergence of root-finding iteration schemes that converge linearly, but this section is only an introduction to the problem of "speeding up" a numerical procedure. Other such techniques will be developed for other numerical problems as we proceed through the text.

Suppose that r is a root of the equation $f(x) = 0$ and the iteration scheme $x_{n+1} = g(x_n)$ converges linearly to root r. This will happen if $0 < |g'(x)| \le K < 1$ on some open interval I containing r and if the initial estimate x_0 is selected in I.

Let us set $E_n = r - x_n$. We found in previous arguments that $E_{n+1} = g'(c_n)E_n$, where c_n is between x_n and r. When x_n is close to r, we may assume that $E_{n+1} \cong KE_n$, where $K = g'(r)$. Then

$$r - x_{n+1} \cong K(r - x_n)$$

and

$$r - x_n \cong K(r - x_{n-1})$$

So

$$\frac{r - x_{n+1}}{r - x_n} \cong \frac{r - x_n}{r - x_{n-1}}$$

Solving this equation for r, we obtain

$$r \cong \frac{x_{n+1}x_{n-1} - x_n^2}{x_{n+1} - 2x_n + x_{n-1}} = \hat{x}_n$$

The estimate $\hat{x}_n$ should be a better estimate of r than either x_n or x_{n+1}. Numerical computations confirm that $\hat{x}_n$ is usually better than the last iterative estimate x_{n+1} used in the computation of $\hat{x}_n$.

The equation for $\hat{x}_n$ can be rearranged into a form better suited for repeated calculations on machines (or even pocket calculators). Note that

$$\hat{x}_n = x_{n+1} - \frac{x_{n+1}^2 - 2x_n x_{n+1} + x_n^2}{x_{n+1} - 2x_n + x_{n-1}}$$

Now let

$$\Delta x_n = x_{n+1} - x_n$$

We define

$$\Delta^2 x_n = \Delta x_{n+1} - \Delta x_n = x_{n+2} - 2x_{n+1} + x_n$$

Then we find that

$$\hat{x}_n = x_{n+1} - \frac{(\Delta x_n)^2}{\Delta^2 x_{n-1}}$$

This formula is the basis for an acceleration algorithm commonly known as the **Aitken Δ^2 process**. The steps of the algorithm are as follows:

Step 1 Compute at least three consecutive estimates: x_{n-1}, x_n, and x_{n+1}.
Step 2 Compute differences: Δx_n and $\Delta^2 x_{n-1}$.

Step 3 Compute: $\hat{x}_n = x_{n+1} - \dfrac{(\Delta x_n)^2}{\Delta^2 x_{n-1}}$.

EXAMPLE 2.22   The following computer output shows the results of applying the Aitken algorithm to the fixed-point iteration results (column 1) obtained using the iteration $x_{n+1} = 2e^{-x_n}$ to solve the equation $xe^x = 2$. The final Aitken estimate is close to the root value. The corresponding iteration in column 1 is still clearly very far from the root.

ENTER VALUE OF X0 0.85

X	DELTA X	DELTASQ X	AITKEN EST	RATIO R
0.85	4.82986379E − 03	−8.94862111E − 03	0.852606836	−0.8528
0.854829864	−4.11875732E − 03	7.62985577E − 03	0.852606472	−0.8525
0.850711107	3.51109845E − 03	−6.5050975E − 03	0.852606207	−0.8528
0.854222205	−2.99399905E − 03	5.5463945E − 03	0.852606015	−0.8526
0.851228206	2.55239545E − 03	−4.72880248E − 03	0.852605874	−0.8527
0.853780601	−2.17640703E − 03	4.03186283E − 03	0.852605773	−0.8526
0.851604194	1.8554558E − 03	−3.43754445E − 03	0.852605699	−0.8527
0.85345965	−1.58208865E − 03	2.93090125E − 03	0.852605645	−0.8526
0.851877562	1.3488126E − 03	−2.49887933E − 03	0.852605606	−0.8527
0.853226374	−1.15006673E − 03	2.13057501E − 03	0.852605578	−0.8526
0.852076307	9.80508281E − 04	−1.81652769E − 03	0.852605557	−0.8527
0.853056816	−8.36019404E − 04	1.54879061E − 03	0.852605542	−0.8526
0.852220796	7.12771201E − 04	−1.32050109E − 03	0.852605531	−0.8527
0.852933567	−6.0772989E − 04	1.12587097E − 03	0.852605523	−0.8526
0.852325838	5.18141082E − 04	−9.59919998E − 04	0.852605517	−0.8527
0.852843979	−4.41778917E − 04	8.18435336E − 04	0.852605513	−0.8526 ∎

Note that in the derivation of the formula for $\hat{x}_n$, we have assumed that the ratio of E_{n+1} to E_n is approximately K. Now

$$\Delta x_n = x_{n+1} - x_n = (r - E_{n+1}) - (r - E_n) = E_n - E_{n+1}$$

So

$$\Delta x_n \cong (1 - K)E_n \quad \text{and} \quad \Delta x_{n-1} = (1 - K)E_{n-1}$$

Then

$$R_n = \frac{\Delta x_n}{\Delta x_{n-1}} \cong \frac{E_n}{E_{n-1}} = K$$

So as a test of the reliability of the algorithm when used on a machine, it is advisable to print values of R_n at each step to observe the behavior. If values of R_n approach a nearly constant value as the computations proceed, then the acceleration algorithm should provide reliable results. If values of R_n fluctuate, however, then results from the Aitken algorithm become suspect because the hypotheses underlying the derivation are apparently not satisfied in that case. When the calculations are proceeding well, the values of R_n will normally approach a steady value for a few iterations and then begin to vary. At that point, machine errors have become large enough to interfere with the acceleration effort. The best estimate of the root in this case is the estimate x obtained at the point where values of R_n were steady.

EXAMPLE 2.23   The following machine results were obtained for the equation $x^3 - x - 1 = 0$ using iterations $x_{n+1} = \sqrt[3]{x_n + 1}$. Note the steady value of R_n and the rapid convergence of the Aitken algorithm to the root at 1.32471796.

ENTER VALUE OF X0 1.5

X	DELTA X	DELTASQ X	AITKEN EST	RATIO R
1.5	−0.142791192	0.116443342	1.32489918	0.1845
1.35720881	−0.0263478495	0.0213706652	1.3247246	0.1889
1.33086096	−4.97718435E − 03	4.0327739E − 03	1.3247182	0.1897
1.32588378	−9.44410451E − 04	7.65057746E − 04	1.32471797	0.1899
1.32493936	−1.79352704E − 04	1.45286787E − 04	1.32471796	0.1899
1.32476001	−3.40659171E − 05	2.75950879E − 05	1.32471796	0.1899
1.32472595	−6.47082925E − 06	5.24194911E − 06	1.32471796	0.1899
1.32471948	−1.22888014E − 06	9.95583832E − 07	1.32471796	0.1898
1.32471825	−2.33296305E − 07	1.89058483E − 07	1.32471796	0.1896
1.32471801	−4.42378223E − 08	3.53902579E − 08	1.32471796	0.2

EXAMPLE 2.24   The following results were obtained from the iteration $x_{n+1} = \pi/2 + \arctan x_n$ used to locate the root of $x = \tan x$ in the interval (2, 3). The Aitken algorithm arrives at an accurate root estimate very rapidly in this case. Values of R_n are quite constant for the first three iterations, but then begin to drift.

ENTER VALUE OF X0 2.8

X	DELTA X	DELTASQ X	AITKEN EST	RATIO R
2.8	$-1.43128727E - 03$	$1.26930419E - 03$	2.79838606	0.1131
2.79856871	$-1.61983073E - 04$	$1.43641606E - 04$	2.79838605	0.1132
2.79840673	$-1.83414668E - 05$	$1.62646174E - 05$	2.79838605	0.1132
2.79838839	$-2.07684934E - 06$	$1.84122473E - 06$	2.79838605	0.1134
2.79838631	$-2.35624611E - 07$	$2.09547579E - 07$	2.79838605	0.1106 ■

In this section, the Aitken algorithm has been applied only to the sequence of estimates derived from a fixed-point iteration scheme. However, this algorithm has broader application. In theory, the Aitken algorithm is applicable to any linearly convergent sequence of estimates, regardless of the source.

■ *Exercise Set 2.7*

1. The iteration scheme $x_{n+1} = \arccos(e^{-x_n})$ with $x_0 = 1.29$ converges to the root of the equation $e^{-x} = \cos x$ near 1.29.
   a. Compute iterates x_1 through x_6.
   b. Compute accelerated estimates $\hat{x}_1$ through $\hat{x}_5$ with the associated values of R_n.
   c. State the value of the root with as much accuracy as your calculations justify.

2. After x_1 and x_2 have been computed using fixed-point iteration, we can apply the Aitken process to generate $\hat{x}_1$. Using this value for x_0, we can restart the iterations to generate new x_1 and x_2. Then we can apply the Aitken process again to generate a new $\hat{x}_1$, which we use as x_0 again, and so forth. Write a program based on this algorithm, and use it to estimate the indicated roots for each of the following equations.
   a. $xe^x = 2$;   root near 0.85   b. $e^{-x} = \cos x$;   root near 1.29
   c. $x^3 - x - 1 = 0$;   root near 1.32
   Compare your results with the results from exercise 1 and Examples 2.22 and 2.23.

3. We know from calculus that $\lim x_n = e$ for $x_n = (1 + 1/n)^n$.
   a. Compute x_1 through x_{20} for this sequence.
   b. Apply the Aitken Δ^2 process to accelerate the convergence.
   c. Note that the "accelerated" sequence $\hat{x}_1, \hat{x}_2, \ldots$ does not seem to be converging to $e = 2.718281828\ldots$, as we might have expected. Explain why the Aitken process should not be expected to accelerate convergence for this sequence.

4. Recall that the values of $R_n = \Delta x_n/\Delta x_{n-1}$ are estimates of $g'(r)$ at each step in the Aitken process. Calculate the values of $g'(r)$ for Examples 2.22 and 2.23, and compare with the tabulated values of R_n (ratio R).

2.8   Putting It All Together (Optional)

In this chapter, you have been introduced to a variety of algorithms that can be used to find roots of an equation $f(x) = 0$ from root estimates. The problem now is how to apply this information when confronted by a specific equation.

Each algorithm has advantages and disadvantages. When approximating a simple root, bisection converges linearly while the Newton algorithm converges quadratically, but the bisection algorithm is simpler and does not require knowledge of the derivative of f. The order of convergence of the secant method can be shown to be about 1.62, so the convergence is almost quadratic. The secant method does not require knowledge of the derivative, but it can produce erratic results when the approximations become close together. Regula falsi is usually faster than the bisection method, but the bisection method yields estimates with usable known error bounds, while regula falsi does not.

In theory, we could simply use the Newton algorithm in all cases, but in practice, this is not advisable. The derivative can be difficult or virtually impossible to compute. The value of the derivative may be zero at the root or very small.

One might advocate that we use bisection at all times. But besides the fact that the bisection method is usually slower than Newton's method, or even regula falsi, we must also take into account that bisection will not work if we need to estimate a root of even order.

The fixed-point iteration algorithm is a surprisingly useful method for estimating roots of transcendental equations. It is especially useful when used in conjunction with a programmable pocket calculator. But there can be difficulties in determining a suitable iteration function $g(x)$, and convergence can be slow or nonexistent.

Assuming that an algorithm has been selected for producing a sequence of root estimates, we must then choose some "stopping method" from those discussed in Section 2.4. All are usable. Some are better than others with regard to simplicity of use, but which one we should use in a given instance depends on the circumstances.

In every case after a root estimate x_n has been found, some attempt to estimate the accuracy of the root estimate should be made. If $f'(x)$ is not difficult to compute and $f'(r) = 0$, then the estimate given by

$$\frac{|f(x_n)|}{|f'(x_n)|}$$

is usable. If the root is of odd order, we should be able, by some experimentation, to find a $\delta > 0$ such that $f(x_n - \delta)$ and $f(x_n + \delta)$ have opposite signs. Then x_n and root r both lie in the interval $(x_n - \delta, x_n + \delta)$, so we know that $|x_n - r| < \delta$, and we have an error estimate. There are a number of ways to arrive at error estimates in each situation.

So we see that using numerical methods is both an art and a science. Some experience and imagination is required to select an appropriate algorithm for a given problem. The science of numerical analysis provides us with a collection of usable algorithms with some analysis of their operation. Selecting the best one for a given problem is very much an art developed through experience with their use.

■ *Exercise Set 2.8*

Write a general computer program that can be used to find all real roots in the specified interval for each of the following equations if appropriate input is supplied.

1. $e^{-x} = \cos x$; $[0, 10]$
2. $x^2 = \sin x$; $(-\infty, \infty)$

  3. $x^4 - 56x^3 + 490x^2 + 11112x - 117495 = 0$; $(-\infty, \infty)$
  4. $2x = (x^2 - 1) \tan x$; $[0, 2]$
  5. $0.6 = x - 0.2 \sin x$; $[0, \pi]$
  6. $x^3 + 4x^2 - 7 = 0$; $(-\infty, \infty)$
  7. $x = \tan x$; $[0, 2\pi]$
  8. $x^4 - 26x^2 + 24x + 21 = 0$; $(-\infty, \infty)$
  9. $922.85x^8 - 942.85x^7 - 1000x + 1020 = 0$; $[0, 2]$
  10. $x \ln x = 2 - x$; $(0, 2)$

2.9   Applied Problems

In this section, we will look at some practical root-finding problems.

In celestial mechanics, the problem of computing the position of a planet in its orbit leads to a transcendental equation known as the Kepler equation. We must determine x from the equation

$$m = x - E \sin x$$

where m and E are known positive numbers. Both m and E have values between 0 and 1, and x is a positive real number. To illustrate how this equation can be solved by the methods described in this chapter, let us consider the case where $m = 0.8$ and $E = 0.2$. We have

$$x = m + E \sin x = g(x)$$

and $g'(x) = E \cos x$. So $|g'(x)| \leq E < 1$ for all x. Then the iterations

$$x_{n+1} = 0.8 + 0.2 \sin x_n$$

with $x_0 = 1$ should converge. Applying this algorithm, we obtain

$$x_1 = 0.968294197$$
$$x_2 = 0.964780448$$
$$x_3 = 0.9643851857$$
$$x_4 = 0.9643397351$$
$$x_5 = 0.9643345543$$
$$x_6 = 0.9643339637$$
$$x_7 = 0.9643338964$$

Clearly, $x \cong 0.964334$ to six decimal places. As a check on this estimate, let

$$f(x) = m - x + E \sin x = 0.8 - x + 0.2 \sin x$$

Then

$$f(0.9643345) = -5.4\text{E} - 7$$

while

$$f(0.9643335) = +3.4\text{E} - 7$$

so the root certainly lies between 0.9643335 and 0.9643345. Then the root is approximately 0.964334, with absolute error less than $5\text{E} - 7$.

Fixed-point iteration gives very satisfactory convergence in this example. We could also use Newton's method. Note that fixed-point iteration works well in the numerical example selected because $|g'(x)|$ is small for the selected value of E and the resulting root x. If, however, the root were near zero and E were near 1, the results might not be quite so satisfactory. You might want to consider the results when $m = 0.01$ and $E = 0.95$, for example. Various specialized methods have been developed for this equation for serious use, since in orbit computation, the equation must be solved repeatedly with various combinations of m and E. It is essential to have a method that converges rapidly and reliably in all these cases.

Next let us consider a problem from radiation theory. Physicist Max Planck discovered that the intensity of radiation at wavelength λ for a blackbody radiator with temperature T is given by the equation

$$\psi_\lambda = \frac{8\pi ch}{\lambda^5} \cdot \frac{1}{e^p - 1}$$

where $p = ch/\lambda kT$. The numbers c, h, and k are physical constants. This intensity function has a maximum value at some wavelength λ_m that is the solution of equation

$$\frac{d\psi_\lambda}{d\lambda} = 0$$

After simplification, this equation reduces to

$$\left(1 - \frac{x}{5}\right)e^x = 1$$

where $x = ch/\lambda kT$. This is equivalent to

$$(5 - x)e^x = 5$$

We need to solve this final equation for x.

Before doing this, note that if r is the solution, then

$$r = \frac{ch}{\lambda_m kT}$$

implies that

$$\lambda_m T = \frac{ch}{kr}$$

is a constant. The equation

$$\lambda_m T = \text{Constant}$$

is known as the *Wien radiation law* and is a verified experimental fact about radiation. The temperature of the blackbody radiator is T (in degrees Kelvin), and λ_m (in centimeters) is the wavelength of the radiation produced with greatest intensity by the radiator. If you are not familiar with the concept of a blackbody radiator, it might be well to note that in astrophysics, stars are always treated as blackbody radiators for valid physical reasons. The Wien radiation law explains the relationship between the color of a star and the surface temperature. From the measurement of λ_m, the wavelength at which the radiation is most intense, the surface temperature of a star can be estimated via the Wien radiation law.

From experimental work, it is known that r lies between 4.9 and 5.0. So we must solve the equation

$$(5 - x)e^x = 5$$

with approximate root $x_0 = 4.95$ known. We can use the iterations

$$x_{n+1} = 5(1 - e^{-x_n}) = g(x_n) \quad \text{with} \quad x_0 = 4.95$$

We find that $g'(x) = 5e^{-x}$. So

$$g'(r) \cong 5e^{-4.95} = 0.035$$

Then our iteration algorithm should converge nicely. We do, in fact, obtain

$$x_1 = 4.964582955$$
$$x_2 = 4.965095693$$
$$x_3 = 4.965113585$$
$$x_4 = 4.965114209$$
$$x_5 = 4.965114231$$

An approximate value of 4.9651142 is indicated. Evaluation of

$$f(x) = (5 - x)e^x - 5$$

at the points $4.965114231 \pm 5E - 8$ confirms that the root is 4.9651142, with absolute error less than $5E - 8$.

Also note in this case that

$$f''(x) = (4 - x)e^x$$

and the error in root estimate $r = 4.9651142$ should be about $|f(r)/f'(r)|$ by previous theory. We find that

$$\frac{|f(r)|}{|f'(r)|} = \frac{(5-r)e^r - 5}{(4-r)e^r} = \frac{(5-r) - 5e^{-r}}{4 - r}$$

$$= \frac{3.06E - 8}{0.9651142} = 3.17E - 8$$

This computation also confirms that $r = 4.9651142$ correct to seven decimal places.

Finally, let us consider the polynomial equation

$$x^4 - 56x^3 + 490x^2 + 11112x - 117495 = 0$$

which was derived by D. M. Dennison (1931) as a part of an investigation of quantum mechanical energy levels. Dennison observed that the roots were all real with values near 37, 22, −13, and 10. It is now an easy matter to establish by machine computation using bisection or regula falsi that precise values of the roots are 36.89601854, 22.34112775, −13.66677360, and 10.42962731, with each root estimate correct to eight decimal places. You should confirm this fact.

The methods developed in this chapter have widespread use in the sciences and technology. The examples given are intended only to give some idea of the possibilities. These methods can be applied to any type of equation, including polynomial equations, as the last application shows. However, there are algorithms better suited to the estimation of polynomial roots than those described so far. These algorithms take into account the known theory about polynomial equations and the special form of such equations. In the next chapter, we will investigate some of these root estimation algorithms for polynomial equations.

■ *Exercise Set 2.9*

1. Solve the Kepler equation $m = x - E \sin x$ for each of the following cases.
   a. $m = 0.9, E = 0.15$ **b.** $m = 0.01, E = 0.9$

2. The differential equation $N'(t) = BN(t) + A$ describes the growth of a population with growth rate B and migration rate A. The population size at time t is $N(t)$.
   a. Verify that

$$N(t) = N_0 e^{Bt} + \frac{A}{B}(e^{Bt} - 1)$$

is the solution of the differential equation if $N(0) = N_0$.
   b. Given $N(0) = 100000$, $N(5) = 500000$, and $A = 25000$, we find that

$$500000 = 100000e^{5B} + \frac{25000}{B}(e^{5B} - 1)$$

Solve this equation for the growth rate B.

3. A sphere of radius r floats in a pool of water. The volume of the submerged portion is $\pi x^2(r - x/3)$, where x is the depth to which the sphere sinks. From physics we know that the floating sphere displaces its own weight of water, so we conclude that

$$\frac{4}{3} \pi r^3 \rho = \pi x^2 \left(r - \frac{x}{3} \right)$$

if the sphere material has specific gravity ρ. Given $r = 1$ ft and $\rho = 0.6$, find the depth to which the sphere sinks (value of x).

4. The sound level (in decibels, db) at distance r meters from a source is

$$L = L_0 - 20 \log_{10} r - \beta r$$

where L_0 is the decibel level 1 m from the source and β is an attenuation coefficient whose value is determined by the physical conditions (viscosity, temperature, humidity, etc.) of the intervening air. Given $L_0 = 80$ db and $\beta = 1.15E - 3$ db/m, find the value of r for which the level L is 20 db. Such a computation needs to be made when investigating the range for birdcalls in ecological studies of bird communication.

5. A projectile fired with muzzle velocity V and inclination angle A follows a parabolic path with equation

$$y = x \tan A - \left(\frac{g}{2} V^2 \cos A \right) x^2$$

Given $V = 1500$ ft/sec and $g = 32.2$ ft/sec^2, find the value of A needed so that the projectile will strike a target 500 ft downrange at an altitude of 50 ft. (*Hint:* Find A such that the parabola passes through point (500, 50).)

3 POLYNOMIAL EQUATIONS

We will now apply the discussion of the preceding chapter to the special case where function f is a polynomial function of degree $n > 0$. We want to find real and complex roots of the equation

$$a_n x^n + a_{n-1} x^{n-1} + \cdots + a_1 x + a_0 = 0 \qquad a_n \neq 0, \quad n > 0$$

In this case, we cannot restrict our attention to real roots only. In most instances where we encounter the problem of finding the roots of a polynomial equation, we need to find all the roots, not just the real roots. In applied problems, the coefficients $a_n, a_{n-1}, \ldots, a_1, a_0$ are usually real numbers. However, most of the algorithms designed for the numerical estimation of roots of a polynomial equation are still applicable in theory when the coefficients are not real. Also, most theorems concerning roots assume that the coefficients are complex.

It is assumed that the reader is familiar with complex numbers; if not, the basics are presented in Appendix 1. It is not assumed that the reader is familiar with the basic algebraic theorems concerning roots of polynomial equations, however, and a discussion of these theorems is useful here because we are again concerned with the mathematical process of moving from theoretical facts to useful numerical algorithms.

We can tell a great deal about the roots of a polynomial equation by use of theorems. That is an advantage we have in this special case that is not available in the general case of a nonlinear equation. If we know the degree and coefficients of a polynomial, the theorems can tell us how many roots exist, their approximate locations, upper bounds for the roots, and so on. We can even determine the roots exactly without using machine algorithms when the roots are rational and the coefficients are integers.

As stated previously, machine computation is not a substitute for theory. Mathematical theory and analysis should be used in conjunction with machine computation. The combination is much more powerful than theory or machine computation alone. This is particularly true in the case of polynomial equations.

3.1 Algebraic Theory

In this section, we will summarize and review the basic algebraic theorems concerning roots of polynomial equations.

Theorem 3.1

Fundamental Theorem of Algebra
If $P(z)$ is a polynomial of degree $n > 0$, then equation $P(z) = 0$ has at least one complex root.

This is a very important theorem and of some historical interest. It was first proved by Karl Gauss when he was quite young, after other mathematicians had failed to find a proof. Elementary proofs are quite lengthy, and other, shorter proofs require the use of theorems from complex function theory, so they are beyond the level of this text.

In this theorem, the coefficients of the polynomial are assumed to be any complex numbers, and the polynomial is assumed to be a nonconstant polynomial. That is all we need to guarantee that the equation has at least one root in the complex numbers. The fundamental theorem is an existence theorem. Existence theorems are important to numerical analysts. In this case, for instance, we now know that all polynomial equations of degree 1 or greater have at least one root. If it were possible for some such equations to fail to have any roots, we would need to be on guard for that situation when computing numerical estimates of roots. Sometimes a computer algorithm will generate apparent solutions to problems even when none exist. Sometimes nonexistence of a solution for the problem at hand will simply cause peculiar output or failure of an algorithm to converge. In the case of polynomial equations, however, if results are not satisfactory, we know that the fault lies in the algorithm or its implementation.

Theorem 3.2

Let $P(z)$ and $G(z)$ be polynomials with degrees n and m, respectively. Assume that $0 < m < n$. Then unique polynomials $Q(z)$ and $R(z)$ exist such that $P(z) = G(z)Q(z) + R(z)$, where $\deg R(z) < m$.

This theorem asserts existence and uniqueness of the quotient $Q(z)$ and the remainder $R(z)$ when $P(z)$ is divided by $G(z)$. This innocent-looking and quite believable theorem is a key tool when proving theorems about polynomials, as we will soon see. The algebraic algorithm for actually finding $Q(z)$ and $R(z)$ is familiar to all algebra students. We will soon look at a numerical procedure for determining the coefficients of these two polynomials.

Definition 3.1

If $P(z) = G(z)Q(z)$, then we say that $G(z)$ and $Q(z)$ are *factors* of $P(z)$.

Theorem 3.3   **Remainder Theorem**
If a is any complex number and deg $P(z) > 0$, then the remainder $R(z)$ when $P(z)$ is divided by $G(z) = (z - a)$ is $P(a)$.

Proof   We know by Theorem 3.2 that

$$P(z) = (z - a)Q(z) + R(z)$$

where deg $R(z) = 0$, so that $R(z) = R$ is a constant. Then

$$P(a) = (a - a)Q(a) + R = R \qquad \square$$

Theorem 3.4   **Factor Theorem**
If a is a root of the polynomial equation $P(z) = 0$, then $z - a$ is a factor of $P(z)$.

Proof   Since $R = P(a) = 0$, then the result follows at once. $\square$

Combining the preceding results, we can now prove the following very important theorem concerning roots of polynomial equations.

Theorem 3.5   If $P(z)$ is a polynomial of degree $n > 0$, then equation $P(z) = 0$ has precisely n complex roots (not necessarily distinct).

Proof   The fundamental theorem implies that at least one root r_1 exists. Then, by Theorem 3.4, we see that $P(z) = (z - r_1)Q_1(z)$, where $Q_1(z)$ has a degree $n - 1$. If $n = 1$, we are finished. If $n > 1$, then $Q_1(z) = 0$ has some root r_2 and $Q_1(z) = (z - r_2)Q_2(z)$, where the degree of $Q_2(z)$ is $n - 2$. Clearly, we can continue this argument, eventually arriving at

$$P(z) = (z - r_1)(z - r_2)(z - r_3) \ldots (z - r_n)Q_n(z)$$

where deg $Q_n(z) = 0$, so $Q_n(z)$ is a constant. By the factor theorem, the equation $P(z) = 0$ has n complex roots $r_1, r_2, \ldots, r_n$. $\square$

Note that the n roots need not be distinct. So in general, we have the following useful corollary.

Corollary 3.1   If $P(z)$ is a polynomial with degree $n > 0$, then the equation $P(z) = 0$ has at most n distinct roots.

Theorem 3.5 states that the equation has n roots, some of which may be real and some of which may not be real. The following theorem gives more information concerning the character of the roots.

Theorem 3.6   If polynomial $P(z)$ has real coefficients and c is a root of $P(z) = 0$, then $\bar{c}$ (the conjugate of c) is also a root.

Proof   We have

$$P(c) = a_n c^n + a_{n-1}c^{n-1} + \cdots + a_1 c + a_0 = 0$$

where coefficients $a_n, a_{n-1}, \ldots, a_1, a_0$ are real numbers. The conjugate of $P(c)$ is

$$\overline{P(c)} = \overline{a_n c^n} + \overline{a_{n-1}c^{n-1}} + \cdots + \overline{a_1 c} + \bar{a}_0 = \bar{0} = 0$$

But

$$\overline{a_k c^k} = \bar{a}_k (\bar{c})^k$$

for $k = n, n-1, \ldots, 0$ and $\bar{a}_k = a_k$ in each case. Hence,

$$a_n(\bar{c})^n + a_{n-1}(\bar{c})^{n-1} + \cdots + a_1(\bar{c}) + a_0 = 0$$

This is $P(\bar{c}) = 0$. Hence, $\bar{c}$ is a root as claimed.   □

This theorem asserts that in the case of polynomials with real coefficients, the complex roots must occur in conjugate pairs. So, for example, a polynomial of degree 3 with real coefficients must have either one or three real roots. In general, a polynomial with real coefficients and of odd degree must necessarily have at least one real root. This is a useful fact to remember. Note, however, that a polynomial of even degree may have all of its roots complex (nonreal). This frequently happens when root-finding algorithms are being employed, so we need to develop algorithms that can estimate both real and nonreal complex roots. It is useful to have a machine capable of doing complex arithmetic, but some routines do not require that facility.

Theorem 3.7   If $P(z)$ has integer coefficients and p/q is a rational root of $P(z) = 0$, then p must be a divisor of a_0 and q must be a divisor of a_n.

Since the proof of this theorem requires some knowledge of divisibility theorems for integers, it will be omitted here. Because most polynomial equations have integer coefficients, this is a widely applicable theorem.

EXAMPLE 3.1 Show that the equation $x^3 - x - 1 = 0$ has no rational roots.

Solution By Theorem 3.7, the only possible rational roots would be 1 or -1. It is easily verified that neither of these is a root. By Theorem 3.6, this equation has one real root and two complex roots. The real root is irrational. ∎

EXAMPLE 3.2 Find all roots of the equation

$$2x^3 - x^2 - 4x + 2 = 0$$

Solution By Theorem 3.7, the possible rational roots are ± 1, ± 2, and $\pm\frac{1}{2}$. Substituting in the equation shows that $\frac{1}{2}$ is a root. Hence, $2x - 1$ is a factor. Then division by $2x - 1$ reveals that $P(x) = (2x - 1)(x^2 - 2)$. Hence, the other two roots are $\pm\sqrt{2}$. ∎

The theorems covered in this section form a theoretical base for most of the numerical work done with polynomial equations. We are now ready to proceed to numerical methods.

■ *Exercise Set 3.1*

1. Use Theorem 3.7 to find all roots for each equation.
   a. $x^3 - 2x^2 - 3x + 6 = 0$ b. $x^3 - 3x^2 - 10x + 24 = 0$
   c. $x^3 + 5x^2 + 2x - 8 = 0$ d. $x^3 - 2x^2 + 3x - 6 = 0$

2. Use Theorem 3.4 to determine whether $Q(x)$ is a factor of $P(x)$ in each case.
   a. $P(x) = x^3 + 2x^2 + 3x - 6$ and $Q(x) = x - 1$
   b. $P(x) = 2x^4 - 3x^2 + x - 18$ and $Q(x) = x + 2$
   c. $P(x) = 2x^4 - 3x^2 + x - 18$ and $Q(x) = x - 3$
   d. $P(x) = 2x^3 + 5x^2 + 5x + 3$ and $Q(x) = 2x + 3$

3. Explain why the equation

   $$a_n x^n + a_{n-1} x^{n-1} + \cdots + a_1 x + a_0 = 0$$

   cannot have any positive roots if the coefficients $a_n, \ldots, a_0$ all have the same sign.

4. Suppose that

   $$P(x) = (x - c)(b_0 + b_1 x + \cdots + b_k x^k) + b_{k+1}$$

   where numbers $b_0, b_1, \ldots, b_{k+1}$ all have the same sign. Show that $P(x) = 0$ cannot have any root greater than c. (In this case, we say that c is an upper bound for the roots.)

5. Use the result in exercise 4 to show the following.
   a. $x^3 - x - 1 = 0$ has no root greater than 2.
   b. $x^3 - 2x^2 - 3x + 6 = 0$ has no root greater than 3.

6. Show that the cubic equation

   $$x^3 + a_2 x^2 + a_1 x + a_0 = 0$$

can always be transformed into a "reduced cubic" equation

$$z^3 + pz + q = 0$$

by choosing $z = x + a_2/3$.

7. Transform each cubic equation into a reduced cubic.
   a. $x^3 + 3x^2 - x + 7 = 0$ **b.** $x^3 + x^2 + 2x - 9 = 0$

8. Given that c is a real root of the equation $x^3 + px + q = 0$, show that the other two roots are real if and only if $4p + 3c^2 < 0$.

9. The equation $x^3 - 5x - 1 = 0$ has a real root near 2.330. Are the other two roots real? Why or why not?

10. Show that if the equation

$$x^3 + a_2 x^2 + a_1 x + a_0 = 0$$

has roots r_1, r_2, and r_3, then

$$a_0 = -r_1 r_2 r_3 \quad \text{and} \quad a_2 = -(r_1 + r_2 + r_3)$$

Also, find a formula expressing a_1 in terms of the roots.

11. Generalize the results in exercise 10 to the case of a polynomial of degree n.

3.2 Newton Algorithm for Polynomial Equations

In this section, we will begin to study numerical methods for estimating roots of polynomial equations. We are concerned with finding algorithms that can be used efficiently on computing machines. We know, for example, how to find the quotient and remainder when $P(z)$ is to be divided by $G(z)$ using algebraic methods. Some computers can do algebraic computations, but we are concerned in this text with numerical procedures to be used on machines. For the division of polynomial $P(z)$ by linear polynomial $z - c$, for example, we need an algorithm that can compute the coefficients of the quotient and determine the remainder R from the given values of c and the coefficients of $P(z)$. We will begin by developing an algorithm for precisely that problem. The resulting algorithm is surprisingly useful for numerically handling several other problems concerning polynomials.

Our first problem is as follows. Given the polynomial

$$P(z) = a_n z^n + a_{n-1} z^{n-1} + \cdots + a_1 z + a_0$$

and linear polynomial $z - c$, we want to find polynomial $Q(z)$ and number R such that

$$P(z) = (z - c)Q(z) + R$$

We know that

$$Q(z) = b_{n-1} z^{n-1} + b_{n-2} z^{n-2} + \cdots + b_1 z + b_0$$

for some numbers b_k, $k = n - 1, \ldots, 0$. Then

$$(z - c)Q(z) = b_{n-1}z^{n-1} + (b_{n-2} - cb_{n-1})z^{n-1} + \cdots + (b_0 - cb_1)z + (R - rb_0)$$

If this is to be $P(z)$, then we must have

$$
\begin{aligned}
b_{n-1} &= a_n \\
b_{n-2} &= a_{n-1} + cb_{n-1} \\
&\vdots \qquad \vdots \qquad \vdots \\
b_0 &= a_1 + cb_1 \\
R &= a_0 + cb_0
\end{aligned}
$$

Briefly,

$$b_k = a_{k+1} + cb_{k+1} \qquad k = n - 2, \ldots, 2, 1, 0$$

where $b_{n-1} = a_n$ and $R = a_0 + cb_0$.

This algorithm is nothing more than the well-known synthetic division method familiar to all high school algebra students, although the form presented here is different. For computer algorithms, iterative formulas as shown here are preferable, while for pencil-and-paper computations, a tabular layout such as the one normally used when doing synthetic division may be best.

EXAMPLE 3.3   Find the quotient and remainder when

$$P(z) = z^4 + 4z^3 - 3z^2 + 5z - 3$$

is divided by $z - 2$.

Solution   We have $n = 4$ and $c = 2$ with $a_4 = 1$, $a_3 = 4$, $a_2 = -3$, $a_1 = 5$, and $a_0 = -3$. Hence,

$$
\begin{aligned}
b_3 &= 1 \\
b_2 &= 4 + 2(1) = 6 \\
b_1 &= -3 + 2(6) = 9 \\
b_0 &= 5 + 2(9) = 23 \\
R &= -3 + 2(23) = 43
\end{aligned}
$$

So the quotient is $z^3 + 6z^2 + 9z + 23$ with remainder 43.   ∎

The synthetic division algorithm is easily programmed and makes a very useful subroutine when working with polynomials on computers.

Now suppose that we simply want to find the value of a polynomial for the number c. In machine calculations, we should never directly compute the powers and linear combinations given in the algebraic formula for the polynomial. By the remainder theorem, the value of the polynomial $P(c)$ is the remainder when $P(z)$ is divided by $z - c$. Hence, to find $P(c)$, we perform synthetic division by $z - c$ and use the fact that $R = P(c)$.

EXAMPLE 3.4   Compute $P(2)$ and $P(0.12)$ for

$$P(z) = 3z^3 - 4z^2 + z - 3$$

Solution   With $c = 2$, we have

$$b_2 = 3$$
$$b_1 = -4 + 2(3) = 2$$
$$b_0 = 1 + 2(2) = 5$$
$$R = -3 + 2(5) = 7$$

Hence, $P(2) = 7$.

Similarly, with $c = 0.12$, we have

$$b_2 = 3$$
$$b_1 = -4 + 0.12(3) = -3.64$$
$$b_0 = 1 + 0.12(-3.64) = 0.5632$$
$$R = -3 + 0.12(0.5632) = -2.932416$$

Hence, $P(0.12) = -2.932416$. ∎

Note the ease with which $P(0.12)$ can be calculated by this algorithm using a pocket calculator as compared with the arithmetic involved in direct calculation of the cube, square, and so on, when substituting into the formula for $P(z)$. The real value of the algorithm, however, is the ease with which it can be used in computer programs.

Note that the polynomial

$$P(z) = a_3 z^3 + a_2 z^2 + a_1 z + a_0$$

can be written in the nested multiplication form

$$z(z(a_3 z + a_2) + a_1) + a_0$$

The steps we have used to compute $P(z)$ can be easily identified as corresponding to the evaluation of the polynomial by computing the values in the nested multiplication form beginning with the innermost parentheses and working our way out. This result is easily generalized to polynomials of degree n. Hence, the computation of

$P(c)$ by using synthetic division and then identifying R as $P(c)$ is frequently called the **nested multiplication algorithm**. Some authors also call this algorithm **Horner's scheme**.

The nested multiplication algorithm is also useful in computing the derivative of a polynomial. This is important when estimating roots of polynomial equations by Newton's method.

If

$$P(z) = (z - c)Q(z) + R$$

then

$$P'(z) = Q(z) + (z - c)Q'(z)$$

Hence,

$$P'(c) = Q(c)$$

That is, to compute $P'(c)$, we need only compute $Q(c)$ where $Q(z)$ is the quotient when polynomial $P(z)$ is divided by $z - c$. Clearly, this can be done efficiently by the synthetic division algorithm.

EXAMPLE 3.5 Given

$$P(z) = 2z^4 - z^3 + 3z^2 + 2z + 5$$

find $P'(0.1)$ and $P'(0.25)$.

Solution Using $c = 0.1$, we have

$b_3 = 2$	$c_3 = 2$
$b_2 = -1 + 0.1(2) = -0.8$	$c_2 = -0.8 + 0.1(2) = -0.6$
$b_1 = 3 + 0.1(-0.8) = 2.92$	$c_1 = 2.92 + 0.1(-0.6) = 2.86$
$b_0 = 2 + 0.1(2.92) = 2.292$	$c_0 = 2.292 + 0.1(2.86) = 2.578$
$R = 5 + 0.1(2.292) = 5.2292$	

Then $P(0.1) = 5.2292$ and $P'(0.1) = c_0 = 2.578$. If only $P'(c)$ is desired, then we do not need to compute R, of course.

Using $c = 0.25$, we have

$b_3 = 2$	$c_3 = 2$
$b_2 = -1 + 0.25(2) = -0.5$	$c_2 = -0.5 + 0.25(2) = 0$
$b_1 = 3 + 0.25(-0.5) = 2.875$	$c_1 = 2.875 + 0.25(0) = 2.875$
$b_0 = 2 + 0.25(2.875) = 2.71875$	$c_0 = 2.71875 + 0.25(2.875) = 3.4375$

Hence, $P'(0.25) = 3.4375$. ∎

Since we are able to compute $P(c)$ and $P'(c)$ for a polynomial by the preceding algorithm, we can apply the Newton algorithm to polynomial equations without actually computing the derivative analytically. All work can be performed numerically, working from the coefficients of the polynomial and the initial root estimate. We have the following:

Algorithm 3.1   Newton Method for Polynomials

Given coefficients $a_n, a_{n-1}, \ldots, a_0$ and root estimate x_0, do the following:

Step 1 Compute $P(x_0)$ and $P'(x_0)$ by nested multiplication.
Step 2 Compute $x_1 = x_0 - P(x_0)/P'(x_0)$.
Step 3 If $|x_1 - x_0|$ is sufficiently small, go to step 5.
Step 4 Set $x_0 = x_1$ and go to step 1.
Step 5 Output x_1.

There is, of course, no difference between this Newton algorithm and the one discussed in Chapter 2 other than the method for evaluating the polynomial and its derivative.

EXAMPLE 3.6   The polynomial equation $x^3 - 3x^2 + 3 = 0$ has a root near $x_0 = 2.5$. Use Algorithm 3.1 to estimate this root to at least three decimal places.

Solution   We have $a_3 = 1$, $a_2 = -3$, $a_1 = 0$, and $a_0 = 3$. We compute $P(2.5)$ and $P'(2.5)$ as in Algorithm 3.1.

$$b_2 = 1 \qquad\qquad\qquad c_2 = 1$$
$$b_1 = -3 + 2.5 = -0.5 \qquad\qquad c_1 = -0.5 + 2.5(1) = 2.0$$
$$b_0 = 0 + 2.5(-0.5) = -1.25 \qquad c_0 = -1.25 + 2.5(2.0) = 3.75$$

and

$$R = 3 + 2.5(-1.25) = -0.125$$

Hence, $P(2.5) = -0.125$ and $P'(2.5) = 3.75$. Then

$$x_1 = 2.5 - \left(\frac{-0.125}{3.75}\right) = 2.533 \ldots$$

We use $x_1 = 2.53$. We use Algorithm 3.1 again to compute $P(2.53)$, $P'(2.53)$, and x_2.

$$b_2 = 1 \qquad\qquad\qquad\qquad c_2 = 1$$

$$b_1 = -3 + 2.53(1) = -0.47 \qquad\qquad c_1 = -0.47 + 2.53(1) = 2.06$$

$$b_0 = 0 + 2.53(-0.47) = -1.1891 \qquad c_0 = -1.1891 + 2.53(2.06) = 4.0227$$

and

$$R = 3 + 2.53(-1.1891) = -0.008423$$

Hence, $P(2.53) = -0.008423$ and $P'(2.53) = 4.0227$. Then

$$x_2 = 2.53 - \left(\frac{-0.008423}{4.0227} \right) = 2.5320938$$

We use $x_2 = 2.532$ as our final estimate. ∎

Suppose that $P(z)$ is a polynomial of degree n, where n is greater than or equal to 2. Suppose that r is a root of the polynomial equation $P(z) = 0$. Then

$$P(z) = (z - r)Q(z)$$

The equation $Q(z) = 0$ is a polynomial equation of degree $n - 1$. When r is known, the synthetic division algorithm can be used to determine the coefficients of polynomial $Q(z)$. This process by which we replace the equation $P(z) = 0$ by the lower-degree equation $Q(z) = 0$ is called **deflation**. All roots of the deflated polynomial equation are also roots of the original equation $P(z) = 0$.

In theory, the Newton algorithm could be used to estimate both real and nonreal roots of polynomial equations. If complex roots are to be estimated, we need a method to provide initial estimates. In Section 3.6, some methods are discussed that are capable of providing such estimates. We would then need a computing machine that could do complex arithmetic. Most of today's college and university computing centers have machines with software capable of doing complex arithmetic. However, the Newton algorithm is rarely used to estimate nonreal roots because we have better methods; these include the Muller algorithm described in Section 3.4.

■ *Exercise Set 3.2*

1. For each polynomial equation, use the estimate of the given root, together with Algorithm 3.1, to find an improved root estimate x_1.
   a. $x^3 - x^2 - x - 1 = 0$; $x_0 = 1.8$
   b. $x^3 - x - 1 = 0$; $x_0 = 1.3$
   c. $x^4 + 5x^3 - 9x^2 - 85x - 136 = 0$; $x_0 = 4$
   d. $x^3 - 2x - 5 = 0$; $x_0 = 2.09$
   e. $x^3 - 36x - 100 = 0$; $x_0 = 7.1$

2. Find an estimate with error less than 0.01 for the real root in the specified interval for each polynomial equation.

   a. $x^3 + 2x - 4 = 0$;   root in $(1, 1.5)$
   b. $x^3 + x - 3 = 0$;   root in $(1, 1.5)$
   c. $x^4 - 26x^2 + 24x + 21 = 0$;   root in $(1.5, 2)$
   d. $x^3 + 18x - 30 = 0$;   root in $(1, 1.5)$

3. Consider the equation

   $$x^5 + x^4 - 3x^2 + 2x - 0.001 = 0$$

   a. Explain why a root with very small absolute value must exist.
   b. Explain why 0.0005 is a good estimate of this root.
   c. Use the Newton algorithm to find the root with error less than 10^{-5}.

4. For each cubic equation, a real root is given. Use that root to deflate the equation. Then solve the deflated equation to determine the remaining roots.

   a. $x^3 - 5x - 1 = 0$; $r_1 = 2.330058$
   b. $x^3 + x^2 - 2x - 1 = 0$; $r_1 = 1.24698$
   c. $y^3 + 3y^2 - 1 = 0$; $r_1 = 0.53209$
   d. $z^3 + 4z^2 - 7 = 0$; $r_1 = 1.16425$

5. Consider the equation

   $$f(x) = x^4 - 2x^3 + x^2 - x + 5$$

   Estimate with error less than 0.01 the critical points for f. (Critical points of f are real solutions of $f'(x) = 0$.)

6. Write a computer program to implement the Newton algorithm for polynomial equations. Use your program to find *all* roots of the following equations with as much accuracy as possible.

   a. $x^3 + x^2 - 2x - 1 = 0$ **b.** $x^4 - 6x^2 - 64x - 39 = 0$
   c. $x^5 + x^3 + x - 1 = 0$ **d.** $x^4 - 5x^2 + 60x - 26 = 0$
   e. $x^3 - 36x - 96 = 0$ **f.** $x^4 - 11x^2 - 44x - 24 = 0$

3.3   Deflation Procedures

We will now consider some algorithms for determining all roots of a polynomial equation.

Let us assume that polynomial $P(z)$ is of degree $n > 0$ and that the equation $P(z) = 0$ has a root r_1. Then $P(z) = (z - r_1)P_1(z)$, where $P_1(z)$ is a polynomial of degree $n - 1$. If the equation $P_1(z) = 0$ has a root r_2, then r_2 is also a root of $P(z) = 0$. So after determining one root of $P(z) = 0$, we can replace that equation by the *deflated* equation $P_1(z) = 0$. We can, of course, continue this process until reaching a deflated equation of degree 1. Solving this linear equation completes the problem of determining all the roots of the equation $P(z) = 0$. The deflation process can also be terminated after a polynomial of degree 2 is reached. The quadratic formula can be used to find the remaining two roots.

Given the coefficients of $P(z)$ and real root r, the coefficients of the deflated polynomial $P_1(z)$ can be determined by using the nested multiplication algorithm.

EXAMPLE 3.7   Given $P(z) = z^3 - z - 1 = 0$ and root estimate 1.324718, find the deflated polynomial equation and complete the solution.

Solution   Using nested multiplication, we find

$$b_2 = 1$$
$$b_1 = 0 + 1.324718 = 1.324718$$
$$b_0 = -1 + 1.324718(1.324718) = 0.7548777$$
$$R = -1 + 1.324718(0.7548777) = 0 \quad \text{(7 decimal places)}$$

Hence, the deflated equation is

$$P_1(z) = z^2 + 1.324718z + 0.7548777 = 0$$

By the quadratic formula, we find that the roots of this equation are $-0.6623590 \pm 0.5622795i$. So the complete set of roots of the equation $P(z) = 0$ is 1.324718 and $-0.6623590 \pm 0.5622795i$.   ■

The initial root can be determined by the Newton algorithm for polynomials (Algorithm 3.1) or by any of the root estimation algorithms given in Chapter 2. In all cases, the nested multiplication algorithm should be used to evaluate the polynomials and to determine the deflated polynomial equations.

It is now fairly easy to write a computer program that will accept as input the coefficients and degree of the polynomial equation $P(z) = 0$ and estimates of the real roots and then proceed to provide as output accurate approximations of all the roots. The procedure is a simple one. Beginning with the first root estimate, we must compute an accurate root approximation using Newton, regula falsi, bisection, or whatever method we wish to employ. Of course, bisection and regula falsi are not usable for nonreal roots. After a root has been located, we find the deflated polynomial by nested multiplication. Now using the next root estimate, we proceed to determine an accurate approximation of this root as a root of the deflated polynomial equation. We repeat this procedure until all real roots are found. In principle, this could be done with all complex roots. Since many computing machines are restricted to real arithmetic, there may be practical difficulties with nonreal root estimation, however.

To illustrate the ease with which this procedure can be applied, a BASIC program was written for a microcomputer, with the results as shown in Example 3.8.

EXAMPLE 3.8   Find all roots of the equation

$$z^4 - 26z^2 + 24z + 21 = 0$$

using root estimates -5.5, -0.5, 1.5, and 4.5, which are easily determined from sign changes at integer values.

Solution   Using a BASIC program that employs the Newton algorithm for polynomials (Algorithm 3.1) with deflation, we get the following computer output:

COEFFICIENTS OF POLYNOMIAL ARE:
1   0   −26   24   21
ROOT 1 IS −5.44948974

COEFFICIENTS OF DEFLATED POLYNOMIAL ARE:
1   −5.44948974   3.69693845   3.85357182
ROOT 2 IS −0.550510259

COEFFICIENTS OF DEFLATED POLYNOMIAL ARE:
1   −6   7.00000001
ROOT 3 IS 1.58578644

COEFFICIENTS OF DEFLATED POLYNOMIAL ARE:
1   −4.41421356
LAST ROOT IS 4.41421356   ■

EXAMPLE 3.9   Find all roots for the polynomial equation

$$z^4 - 3z^2 + 18z - 20 = 0$$

using root estimates −3.2 and 1.3.

Solution   The computer output is as follows:

COEFFICIENTS OF POLYNOMIAL ARE:
1   0   −3   18   −20
ROOT 1 IS −3.23606798

COEFFICIENTS OF DEFLATED POLYNOMIAL ARE:
1   −3.23606798   7.47213596   −6.1803399
ROOT 2 IS 1.23606798

COEFFICIENTS OF DEFLATED POLYNOMIAL ARE:
1   −2   5

In this case, two root estimates are available from graphical estimates. Examination of the graph in Figure 3.1 indicates that the other two roots are complex. These roots are roots of the quadratic equation $x^2 - 2x + 5 = 0$ and are easily found to be $1 \pm 2i$.   ■

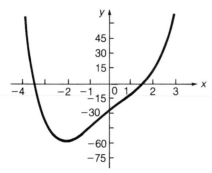

Figure 3.1

EXAMPLE 3.10 Find all roots of the cubic equation

$$z^3 + 84z - 84 = 0$$

Solution Substituting $z = 1$ into the polynomial yields 1, while substituting $z = 0$ yields -84, so there is a real root in the interval $(0, 1)$ very near 1. We choose this as our root estimate. Computer output is as follows:

```
COEFFICIENTS OF POLYNOMIAL ARE:
1  0  84  -84
ROOT  1 IS 0.988501205

COEFFICIENTS OF DEFLATED POLYNOMIAL ARE:
1   0.988501205   84.9771347
```

To find the remaining roots, we solve the quadratic equation

$$z^2 + 0.988501205z + 84.9771347 = 0$$

Using the quadratic formula, we obtain $-0.49425060 \pm 9.2050447i$ for the remaining roots. ∎

It would be desirable to have an algorithm that did not require initial root estimates and that could determine complex roots as well as real roots. Such algorithms do exist and will be discussed in Sections 3.4 and 5.5. But when we are interested in real roots only, the methods discussed in Section 2.1 are usually sufficient for providing good initial estimates.

■ *Exercise Set 3.3*

1. One root r_1 is given for each of the following equations. In each case, use the given root to deflate the equation and determine all remaining roots.
 a. $x^3 + 2x - 3 = 0$; $r_1 = 1$
 b. $x^4 - 2x^3 - x^2 - 2x - 2 = 0$; $r_1 = i$
 c. $x^5 - 2x^4 + x - 2 = 0$; $r_1 = 2$
 d. $x^3 + 2.1x^2 + 2.1x + 1.1 = 0$; $r_1 = 1.1$

2. The cubic equation $x^3 - 6x^2 - 2x + 5 = 0$ has a real root in $(0, 1)$. Make an estimate x_0. Then use the Newton algorithm for polynomials to determine the root with an error less than 0.0001. Solve the resulting deflated equation to determine the remaining roots.

3. Write a computer program that employs the Newton algorithm to estimate a real root after an estimate is provided. The output should be the final root estimate and the coefficients of the deflated polynomial. Apply your program to each of the following equations to determine all the roots. You must provide initial estimates of the real roots.
 a. $x^4 - 15x^2 - 12x - 2 = 0$
 b. $x^4 - 10x^2 - 8x + 5 = 0$
 c. $x^4 - 27x^2 + 30x + 14 = 0$
 d. $x^4 + 32x - 60 = 0$
 e. $x^4 - 5x^2 + 18x - 20 = 0$
 f. $x^5 - 3x^3 + x^2 - 3 = 0$

4. Verify that the equation $x^4 - 10^{-3}x^3 - 10^{-4} = 0$ has a real root in the interval $(0.1, 0.2)$. Use your computer program from exercise 3 to estimate this root and deflate the polynomial. Find all the roots of this equation.

3.4   The Muller Algorithm

Most of the algorithms discussed in this text are very old. The original ideas date back to Newton, Lagrange, Gauss, and others. The algorithms were merely numerical procedures for estimating solutions of mathematical problems, and their use was somewhat limited owing to the practical difficulties of computing without machine assistance. Since the advent of computers in the 1940s, these algorithms have been given new life and can now be employed to their fullest without the practical limitations of doing the arithmetic by hand.

It has also become feasible to develop and explore some new algorithms that may involve computations that would be extremely difficult to perform without computer assistance. Many such algorithms have been developed in recent years. This section is concerned with such an algorithm developed by D. E. Muller (1956) to find the roots of a polynomial equation.

Given a polynomial equation $P(z) = 0$, suppose that numbers x_{n-2}, x_{n-1}, and x_n are three distinct approximations to a root r. We first find a quadratic function $Q(x)$ whose graph passes through the points $(x_{n-2}, f(x_{n-2}))$, $(x_{n-1}, f(x_{n-1}))$, and $(x_n, f(x_n))$. Now we let r_1 and r_2 be the roots of the quadratic equation $Q(x) = 0$. Selecting x_{n+1} as the root closest to x_n, we repeat the process using x_{n+1}, x_n, and x_{n-1} as the root estimates. We continue the process until $|x_{n+1} - x_n|$ is sufficiently small.

There are some computational difficulties in carrying out the details of this general procedure. We proceed as follows.

We let $y_k = f(x_k)$ for $k = n - 2$, $n - 1$, and n. Now we define quadratic $Q(x)$ by

$$Q(x) = L_n y_n + L_{n-1} y_{n-1} + L_{n-2} y_{n-2}$$

where

$$L_n = \frac{(x - x_{n-1})(x - x_{n-2})}{(x_n - x_{n-1})(x_n - x_{n-2})}$$

$$L_{n-1} = \frac{(x - x_n)(x - x_{n-2})}{(x_{n-1} - x_n)(x_{n-1} - x_{n-2})}$$

and

$$L_{n-2} = \frac{(x - x_n)(x - x_{n-1})}{(x_{n-2} - x_{n-1})(x_{n-2} - x_n)}$$

It is now easy to verify that

$$Q(x_k) = y_k \quad \text{for } k = n - 2, n - 1, \text{ and } n$$

Now we define $h = x - x_n$, $h_n = x_n - x_{n-1}$, and $q_n = h_n/h_{n-1}$, $q = h/h_n$. After changing the variables as described and simplifying the results, we find that

$$Q(x) = \frac{A_n q^2 + B_n q + C_n}{1 + q_n}$$

where

$$A_n = q_n y_n - q_n(1 + q_n)y_{n-1} + q_n^2 y_{n-2}$$
$$B_n = (1 + 2q_n)y_n - (1 + q_n)^2 y_{n-1} + q_n^2 y_{n-2}$$
$$C_n = (1 + q_n)y_n$$

To solve $Q(x) = 0$, it is sufficient to solve the equation

$$A_n q^2 + B_n q + C_n = 0$$

We are looking for a very small root. So we use the quadratic formula in the form

$$q = \frac{-2C_n}{B_n \pm \sqrt{B_n^2 - 4A_n C_n}}$$

We choose the sign in the denominator that yields the maximum absolute value for the denominator when we compute q. Here is the complete **Muller algorithm.**

Algorithm 3.2

Step 1 Choose approximations E_1, E_2, E_3 to a root.
Step 2 Compute $y_k = f(E_k)$ for $k = 1, 2, 3$.
Step 3 Compute $h = E_3 - E_2$ and $q = h/(E_2 - E_1)$.
Step 4 Compute coefficients

$$A = qy_3 - q(1 + q)y_2 + q^2 y_1$$
$$B = (1 + 2q)y_3 - (1 + q)^2 y_2 + q^2 y_1$$
$$C = (1 + q)y_3$$

Step 5 Compute $D = \sqrt{B^2 - 4AC}$, $F_1 = B + D$, and $F_2 = B - D$.
Step 6 Choose F as the larger in absolute value of F_1 and F_2.
Step 7 Let $W = -2C/F$ and $E_4 = E_3 + Wh$.
Step 8 If $|Wh|$ is sufficiently small, then choose E_4 as the root estimate and stop.
Step 9 Otherwise set $E_1 = E_2$, $E_2 = E_3$, and $E_3 = E_4$.
Step 10 Go to step 2.

To find all the roots of a polynomial equation, we need to implement this algorithm on a computer that can handle complex arithmetic, since all the variables must be treated as complex numbers. Note that the function f in the algorithm does not have to be a polynomial. Algorithm 3.2 can also be used to find roots of non-polynomial equations.

To start the algorithm, we do not have to locate accurate root estimates. For most root-finding problems, the standard set of values $E_1 = -1$, $E_2 = 0$, and $E_3 = 1$ will do the job. If a real root is known to be in the interval $[a, b]$, however, estimates a, b, and $(a + b)/2$ are better. If $f(E_1) = f(E_2) = f(E_3)$, the procedure fails, since $Q(x)$ is a linear function in this case. This does, in fact, happen for $P(x) = x^3 - x - 1$ if $E_1 = -1$, $E_2 = 0$, and $E_3 = 1$. In such a case, new choices for the initial estimates must be made.

The algorithm converges for a wide variety of choices of the initial estimates, and you can verify this fact by doing some experimenting. In theory, the method will always converge if the initial estimates are sufficiently close to a root. Some choices lead to more rapid convergence than others. In theory, the rate of convergence is always nearly quadratic for simple roots and better than linear for double roots.

Combining the Muller algorithm with the deflation procedures described in Section 3.3 leads to a computer program that is quite capable of handling a wide variety of polynomial root-finding problems. All complex roots, real and nonreal, can be located rapidly with properly designed software based on this idea.

We are now ready to consider some examples of Algorithm 3.2 using a VAX minicomputer. Double-precision complex arithmetic was used in a FORTRAN program to produce the computer output. Although it is not essential to use double-precision arithmetic when using the Muller algorithm, it is advisable. Most current versions of FORTRAN permit this. Note that complex numbers are treated as ordered pairs of real numbers, so complex roots on all computer printouts simply appear as ordered pairs of real numbers.

EXAMPLE 3.11   Apply the Muller algorithm to the problem of approximating a root of the polynomial equation $x^3 - x - 1 = 0$.

Solution   Initial estimates -1, 0, and 0.5 are used. As the following computer printout shows, the algorithm produces an approximation of the complex root

$$-0.6623589786 + 0.5622795121i$$

ENTER E0, E1, E2

−1.   0.   0.5

ROOT ESTIMATE		POLYNEST	
REAL PART	IMAGINARY	REAL PART	IMAGINARY
−0.5000000000	1.3228756555	2.0000000000	−2.6457513111
−0.4133197109	0.3931729205	−0.4656099962	−0.2524505900
−0.5906285289	0.4964472585	−0.1787088333	−0.0992565388
−0.6591559010	0.5538995220	−0.0205420938	−0.0018528909
−0.6624472509	0.5623526689	0.0002193082	0.0001510235
−0.6623589484	0.5622794813	−0.0000000879	−0.0000000481
−0.6623589786	0.5622795121	0.0000000000	0.0000000000
−0.6623589786	0.5622795121	0.0000000000	0.0000000000

ROOT FOUND IS   −0.6623589786   0.5622795121

Note that for this polynomial, the standard initial estimates -1, 0, and 1 cannot be used because the "quadratic" approximation used in the first step of the algorithm would reduce to a linear approximation, and the algorithm would fail.   ∎

EXAMPLE 3.12   Consider the same polynomial equation used in Example 3.11, but with different initial estimates.

Solution   It is easy to see by the methods of Section 2.1 that this equation has a root between 1 and 1.5, so it seems reasonable to use three estimates near that interval. Doing this speeds up the iteration process considerably, as the following printout shows.

```
ENTER E0, E1, E2

1.   1.5   2.
```

ROOT ESTIMATE		POLYNEST	
REAL PART	IMAGINARY	REAL PART	IMAGINARY
1.3333333333	0.0000000000	0.0370370370	0.0000000000
1.3244715050	0.0000000000	−0.0010507869	0.0000000000
1.3247182939	0.0000000000	0.0000014357	0.0000000000
1.3247179572	0.0000000000	0.0000000000	0.0000000000

```
ROOT FOUND IS     1.3247179572      0.0000000000
```

It is also interesting to note that all intermediate estimates are real, just as the final estimate is. That is not necessarily the case for all real roots of polynomial equations.   ∎

EXAMPLE 3.13   Polynomials of high degree present no difficulty. Consider the polynomial equation

$$x^{14} - 6x^{10} + x^8 - 16x^6 - 15x^4 - 16 = 0$$

Solution   In cases such as this, it is difficult to make initial estimates of the roots, but for the Muller algorithm, that presents no problem. The standard initial estimates are usable here. Convergence is very stable and rapid, as the computer results show, even though the initial estimates are not really very close to the root that is located.

```
ENTER DEGREE OF POLYNOMIAL

14

ENTER E0, E1, E2

−1.   0.   1.
```

ROOT ESTIMATE		POLYNEST	
REAL PART	IMAGINARY	REAL PART	IMAGINARY
0.000000000	0.676123404	−17.446869804	0.000000000
0.245977424	1.155263606	−28.450540566	−48.711286363
−0.156800231	0.770161062	−18.826715519	0.025811867
−0.156491455	1.046554773	−14.887137862	16.994879981
−0.021601295	1.138249764	11.895670469	4.999506575
−0.014642765	1.073515625	−0.283704067	2.255279683
−0.000520768	1.075257436	0.097996723	0.081246798
0.000001903	1.074630808	0.000582630	−0.000295635
0.000000000	1.074627058	0.000000012	−0.000000039
0.000000000	1.074627058	0.000000000	0.000000000

ROOT FOUND IS   0.0000000000   1.0746270576   ∎

The preceding examples illustrate the operation of the algorithm alone as various sets of preliminary estimates are used. The result is a final approximation of *one root* of the polynomial equation. In practice, it is desirable to have a computer program that will produce accurate approximations of all the roots from some kind of initial estimates. That can easily be accomplished by following the approximation of each root with deflation of the polynomial equation. All polynomial evaluation in a program such as the one used here should be combined with a subroutine employing the nested multiplication algorithm discussed in Section 3.2. Such a subroutine automatically produces the coefficients of the deflated polynomial each time it is called to evaluate the polynomial being employed at that moment. After a root has been located with sufficient precision, it is a trivial matter to replace the polynomial coefficients with the coefficients of the deflated polynomial and restart the algorithm using the same initial estimates repeatedly. The degree of the polynomial is decreased by 1 with each deflation, of course. When the degree reaches 1, the last root is determined directly from the linear equation, and we do not at that point return to the Muller algorithm. The following examples illustrate the results of such a computer program.

EXAMPLE 3.14   Consider the polynomial equation $x^3 - x - 1 = 0$ again, this time using initial estimates -1, 0, and 2.

Solution   The final root approximations are printed without intermediate iterations in this case. All three roots are produced quickly. All listed digits can be verified to be accurate in this case.

ENTER DEGREE OF POLYNOMIAL 3

ENTER E0, E1, E2

−1.   0.   2.

ROOT ESTIMATES FOLLOW:

	REAL PART	IMAGINARY PART
ROOT 1	1.3247179572	0.0000000000
ROOT 2	−0.6623589786	−0.5622795121
LAST ROOT:	−0.6623589786	0.5622795121

∎

EXAMPLE 3.15 We want to find all roots of the equation

$$x^6 + x^5 + x^4 + x^3 + x^2 + x + 1 = 0$$

Solution All the roots are complex, so this would be a very difficult problem to handle without the Muller algorithm. For this example, the standard estimates -1, 0, and 1 are employed.

ENTER DEGREE OF POLYNOMIAL 6

ENTER E0, E1, E2

-1. 0. 1.

	REAL PART	IMAGINARY PART
ROOT 1	-0.2225209340	-0.9749279122
ROOT 2	-0.2225209340	0.9749279122
ROOT 3	-0.9009688679	-0.4338837391
ROOT 4	0.6234898019	0.7818314825
ROOT 5	0.6234898019	-0.7818314825
LAST ROOT:	-0.9009688679	0.4338837391

Complex roots in this case should occur in conjugate pairs, and this is easily seen to be the case here. It is also worth noting that the conjugate pairs have real and imaginary parts that agree through all ten digits displayed. In this case, the two roots in the pairs are each estimated independently. Such agreement should certainly be the case theoretically, but does not always happen when making estimates. ■

The accuracy of these results demonstrates the precision available in modern computing machines when used properly. As pointed out in Chapter 1, the organization of the steps in the computation can be important if errors are to be minimized. Some student programs illustrate that fact clearly, as you may discover from your own efforts.

EXAMPLE 3.16 We want to compute the sixth roots of unity by solving the equation $x^6 - 1 = 0$.

Solution The values of the roots are easily obtained theoretically as the values of

$$\cos\left(\frac{k\pi}{3}\right) + i\sin\left(\frac{k\pi}{3}\right)$$

for $k = 0, 1, 2, 3, 4,$ and 5. We get ± 1 and $\pm\frac{1}{2} \pm \sqrt{3}/2i$. The computer printout gives accurate decimal approximations of these numbers.

ENTER DEGREE OF POLYNOMIAL 6

ENTER E0, E1, E2

-1. 0. 1.

ROOT ESTIMATES FOLLOW:

	REAL PART	IMAGINARY PART
ROOT 1	1.0000000000	0.0000000000
ROOT 2	−0.5000000000	−0.8660254038
ROOT 3	−0.5000000000	0.8660254038
ROOT 4	−1.0000000000	0.0000000000
ROOT 5	0.5000000000	−0.8660254038
LAST ROOT:	0.5000000000	0.8660254038

If we compare these numerical approximations with numerical values obtained from the theoretical values of the roots, we find that all the digits displayed in this computer output are correct. This is an example of the accuracy that can be obtained when we are using the Muller method. ∎

Since the deflation process introduces polynomial equations with approximated coefficients, we have an added potential source of error. When extreme accuracy is desired, it is advisable to reiterate with the original polynomial coefficients as a precaution against this source of error. In the preceding examples, very little error was introduced by the deflation process, but life is not always so simple. In the next section, we will explore the dangers lurking in this situation.

■ *Exercise Set 3.4*

1. Verify the details of the derivation leading to the formulas for A_n, B_n, and C_n in this section.

2. Compute by hand the first iteration in the Muller algorithm for the equation in Example 3.11 using initial estimates −1, 0, and 0.5 as stated. Try to verify the machine value given in that example.

3. During implementation of the Muller algorithm, quadratic equations with complex coefficients may arise. Use the quadratic formula to solve each of the following quadratic equations of that type.
  a. $z^2 + 2iz + (1 + i) = 0$
  b. $z^2 + iz + (3 + 4i) = 0$
  c. $z^2 + iz + 1 = 0$
  (*Hint:* To find the square root of complex number z, write z in polar form $re^{i\theta}$. Then the principal square root of z is the number $\sqrt{r}e^{i\theta/2}, 0 \le \theta < 2\pi$.)

4. Write a program that implements the Muller algorithm to find all roots of a polynomial equation with real coefficients. After each root is found, the computer should automatically deflate the polynomial and proceed to apply the Muller algorithm to the deflated polynomial. Continue until all roots are found. Apply this program to each of the polynomial equations in exercise 3 of Exercise Set 3.3.

5. Use the program written for exercise 4 to find all roots of each of the following equations.
  a. $x^{10} - 1 = 0$
  b. $x^7 + x^6 + x^5 + x^4 + x^3 + x^2 + x + 1 = 0$
  c. $x^4 + 8x^2 + x^3 + 5x + 15 = 0$
  d. $3x^{12} - 37x^5 + 123x^3 - 41x + 5 = 0$
  e. $x^7 - 9.86x^5 + 11.05x^4 + 0.051x^3 - 2.86 = 0$

6. Solve each of the equations.
   a. $x^8 + 6x^7 - 5x^4 + x^3 + 3x^2 - x + 1 = 0$
   b. $x^8 + 5.99x^7 - 5.01x^4 + x^3 + 2.999x^2 - x + 1 = 0$
   Compare the solutions. Which roots are shifted most?

7. Leonardo of Pisa claimed in one of his writings, in the year 1225, that 1.3688081075 is a root of the third-degree polynomial equation $x^3 + 2x^2 + 10x - 20 = 0$. It is uncertain today how he obtained this result, although some historians think that he used geometric arguments. Verify that his solution is correct, and determine the remaining roots.

8. Find all roots of the equation $\sum_{k=0}^{19} a_k x^k = 0$, where $a_{2k} = 0$ and $a_{2k+1} = (-1)^k/(2k + 1)!$ for $k = 0, 1, \ldots, 9$. Can you explain why the four smallest (in absolute value) nonzero roots are approximately $\pm\pi$ and $\pm 2\pi$?

3.5 Ill-Conditioned Polynomial Equations

In applied mathematics, it is frequently assumed that small variations in the data, or parameters, of a problem will produce small changes in the solution. Unfortunately, this is not always true. Solutions of some mathematical problems change drastically in value, or even in character, when the problem is slightly modified. Many such problems are encountered by numerical analysts and are said to be *ill-conditioned*. In this section, we are concerned with the case of ill-conditioned polynomial equations.

Specifically in the case of the equation

$$a_n x^n + a_{n-1} x^{n-1} + \cdots + a_1 x + a_0 = 0$$

we are concerned with those equations for which a small change in one or more of the coefficients $a_n, a_{n-1}, \ldots, a_1, a_0$ can produce large changes in one or more of the roots. A detailed analysis of this problem, with some numerical examples, was given in well-known papers by J. H. Wilkinson (1959) and F. W. J. Olver (1952).

In the Wilkinson example, we consider the polynomial equation

$$(x + 1)(x + 2) \ldots (x + 20) = 0$$

The roots are $-1, -2, \ldots, -20$, of course. But suppose that we add 2^{-23} to the coefficient of x^{19}. Wilkinson found that the roots were now as follows: $-1, -2, -3,$ -4 (exactly as before), $-4.999999928, -6.000006944, -6.999697234, -8.007267603,$ $-8.917250249,$ $-20.846908101,$ $-10.09526614 \pm 0.64350090i,$ -11.79363388 $\pm 1.65232973i,$ $-13.99235814 \pm 2.51883007i,$ $-16.73073747 \pm 2.81262489i,$ and $-19.502439400 \pm 1.940330347i$. The dramatic shift in the root values is clear, even though the shift in the coefficients is almost negligible. The equation even has five pairs of complex roots now.

Let us explore some normal results of coefficient shifting (or approximation) for comparison.

EXAMPLE 3.17   Consider the equation

$$x^6 + x^5 + x^4 + x^3 + x^2 + x + 1 = 0$$

Using the Muller algorithm, the root approximations are as follows:

	REAL PART	IMAGINARY PART
ROOT 1	−0.2225209340	−0.9749279122
ROOT 2	−0.2225209340	0.9749279122
ROOT 3	−0.9009688679	−0.4338837391
ROOT 4	0.6234898019	0.7818314825
ROOT 5	0.6234898019	−0.7818314825
ROOT 6	−0.9009688679	0.4338837391

Adding 0.0001 to the coefficient of x produces new roots:

	REAL PART	IMAGINARY PART
ROOT 1	−0.2225427116	−0.9749328836
ROOT 2	−0.2225427116	0.9749328836
ROOT 3	−0.9009567822	−0.4338586414
ROOT 4	0.6234994938	0.7818392114
ROOT 5	0.6234994938	−0.7818392114
ROOT 6	−0.9009567822	0.4338586414

There is some shift in the roots, of course, but nothing so drastic as in the Wilkinson example.   ∎

Ill conditioning is not unusual for polynomials of high degree. It is important to realize also that even in the case of polynomials of low degree, the shift in roots produced by shifting or approximating a coefficient may be more than expected.

EXAMPLE 3.18   Consider the equation $x^3 - 3x + 2 = 0$. The roots are 1, 1, and −2. For the slightly different equation $x^3 - 3.001x + 2 = 0$, the roots are found to be

	REAL PART	IMAGINARY PART
ROOT 1	0.9818540265	0.0000000000
ROOT 2	1.0183681875	0.0000000000
ROOT 3	−2.0002222140	0.0000000000

∎

It is important to understand that the problem of ill-conditioned polynomial equations is a real one; that is, the difficulty does not occur only in mathematically pathological cases. For example, the characteristic equation for matrix eigenvalue problems (see Chapter 6) is frequently an ill-conditioned polynomial equation. Olver (1952) discusses in detail an example of an ill-conditioned polynomial that occurred in an engineering problem.

Some insight into these problems can be gained from the following analysis. Let the equation $P(x) = 0$ have roots $r_1, r_2, \ldots, r_n$. Let

$$P(x, \varepsilon) = P(x) + \varepsilon q(x) = 0$$

have roots $z_1(\varepsilon), \ldots, z_n(\varepsilon)$, where $r_j = z_j(0)$. We assume that $q(x)$ is a polynomial of degree n. It can be shown that each $z_j(\varepsilon)$ is an *analytic function* in some neighborhood of $\varepsilon = 0$. This means that $z_j(\varepsilon)$ can be approximated by $z_j(0) + \varepsilon z'(0)$ near $\varepsilon = 0$. We can approximate the derivative $z_j'(0)$ as follows. We know that $P(z_j(\varepsilon)) + \varepsilon q(z_j(\varepsilon)) = 0$ (approximately). Differentiation yields

$$P'(z_j(\varepsilon))z_j'(\varepsilon) + q(z_j(\varepsilon)) + \varepsilon q'(z_j(\varepsilon))z_j'(\varepsilon) = 0$$

Hence,

$$z_j'(\varepsilon) = \frac{-q(z_j(\varepsilon))}{P'(z_j(\varepsilon)) + \varepsilon q'(z_j(\varepsilon))}$$

We now assume that r is a simple root, so that derivative $P'(r_j)$ is not zero. Then

$$z_j'(0) = \frac{-q(r_j)}{P'(r_j)}$$

So

$$z_j(\varepsilon) \cong r_j - \varepsilon\, \frac{q(r_j)}{P'(r_j)}$$

The shift in the root is

$$|z_j(\varepsilon) - r_j| = \frac{|\varepsilon|\,|q(r_j)|}{|P'(r_j)|}$$

EXAMPLE 3.19   Suppose that

$$P(x) = (z - 1)(z - 2) \ldots (z - 7)$$

Let us choose $q(x) = x^6$ and $\varepsilon = -0.002$ and consider the effect on root $r = 2$. The new root is

$$z(-0.002) \cong 2 - \frac{(-0.002)(2^6)}{P'(2)}$$

Now

$$P'(2) = (x - 1)(x - 3) \ldots (x - 7)\Big|_{x=2} = -120$$

So

$$z(-0.002) \cong 2 + 0.002 \left(\frac{64}{-120} \right) = 1.998933$$

In comparison, the machine estimate provided by the Muller algorithm for the root $z(-0.002)$ is 1.9989382.   ∎

The analysis shows that one cause of ill conditioning is a small value of the derivative of the polynomial at the root in question. Also note that if $q(x)$ is a power of x and r_j is a large root, then $|q(r_j)|$ may be large, so the shift in r_j can be several times larger than ε.

If for any reason one root estimate is inaccurate, then the coefficients of all deflated polynomials from that point on in the root-finding algorithm will also be inaccurate, and the error is propagated as the computations proceed. We thus have a case of propagated error, which amplifies the effect of the ill conditioning of the roots. Consider the following example of a polynomial equation suggested by G. E. Forsythe (1970).

EXAMPLE 3.20   We want to solve the equation

$$x^4 - 6.7980x^3 + 2.9948x^2 - 0.043686x + 0.000089248 = 0$$

The Newton algorithm with deflation is used. Root estimates 6, 0.5, 0.01, and 0.002 are supplied, so the roots are computed in descending order of magnitude. Computer output is as follows:

COEFFICIENTS OF POLYNOMIAL ARE:
1   −6.798   2.9948   −0.043686   8.9248E − 05
ROOT 1 IS 6.32565422

COEFFICIENTS OF DEFLATED POLYNOMIAL ARE:
1   −0.472345779   6.90393115E − 03   −1.41187966E − 05
ROOT 2 IS 0.457316679

COEFFICIENTS OF DEFLATED POLYNOMIAL ARE:
1   −0.0150290999   3.08731034E − 05
ROOT 3 IS 0.0125737354

COEFFICIENTS OF DEFLATED POLYNOMIAL ARE:
1   −2.45536415E − 03
LAST ROOT IS 2.45536415E − 03

Now we repeat using estimates 0.002, 0.01, 0.5, and 6, so the roots are computed from smallest to largest (in absolute magnitude). Computer results are:

COEFFICIENTS OF POLYNOMIAL ARE:
1   −6.798   2.9948   −0.043686   8.9248E − 05
ROOT 1 IS 2.45321555E − 03

COEFFICIENTS OF DEFLATED POLYNOMIAL ARE:
1   −6.79554678   2.97812906   −0.0363800075
ROOT 2 IS 0.0125759368

COEFFICIENTS OF DEFLATED POLYNOMIAL ARE:
1   −6.78297085   2.89282685
ROOT 3 IS 0.457316627

COEFFICIENTS OF DEFLATED POLYNOMIAL ARE:
1 −6.32565422
LAST ROOT IS 6.32565422

In summary, the results are:

Descending Order	*Ascending Order*
ROOT 4 IS 2.45536415E − 03	ROOT 1 IS 2.45321555E − 03
ROOT 3 IS 0.0125737354	ROOT 2 IS 0.0125759368
ROOT 2 IS 0.457316679	ROOT 3 IS 0.457316627
ROOT 1 IS 6.32565422	ROOT 4 IS 6.32565422

The order in which the roots were calculated in Example 3.20 affected all the estimates except the estimate of the largest root. Coefficients of the deflated polynomials were all approximate, and this fact introduced some error, as we found previously. This is the source of most of the error in these root estimates, since the root estimates for the polynomials were obtained by the Newton algorithm. In this case, the results would be accurate to eight decimal places if the coefficients were exact. Different deflation polynomials arise when the order of calculation of the roots is changed. So the two estimates of the root near 0.01257 were obtained from two different polynomials. Which values are more accurate?

One way to minimize propagation of error is to return to the original polynomial to refine each root estimate and determine a corresponding deflated polynomial with maximum possible accuracy. In the preceding case, when the root estimates were used to estimate the roots directly from the original polynomial equation, again using the Newton algorithm but without deflation, the new root estimates were 2.45321E − 03, 0.01257593, 0.45731662, and 6.32565422, with an error less than 1E − 08 in each estimate. The agreement with the root estimates obtained in the second run is very good. In the first run, where the largest root was estimated first, there is a steady deterioration in accuracy as we proceed through the smaller roots. When using root approximation algorithms that employ deflation as a part of the algorithm, it is generally true that more accurate results are obtained if the roots are computed in ascending order of magnitude.

■ *Exercise Set 3.5*

1. Apply the method of Example 3.19 to the equation in Example 3.18, and estimate the theoretical shift in each root. Compare with the machine output given in Example 3.18.

2. Verify that the equation

$$x^3 - 20x + 34.4265 = 0$$

has real roots with approximate values -5.16, 2.58, and 2.58. Now find the roots of equation

$$x^3 - 19.99x + 34.4265 = 0$$

What has happened to the roots?

3. Given the equation $x^3 + px + q = 0$, let $D = -4p^3 - 27q^2$. It can be shown that if $D > 0$, three distinct real roots exist; if $D < 0$, one real and two conjugate imaginary roots exist;

and if $D = 0$, at least two equal real roots exist. Apply this theory to the first equation in exercise 2 to explain the observed root shifts.

4. From what you know about the discrimant for quadratic equations, show that we can find two quadratic equations of the form $x^2 + px + q = 0$ with only slightly different values for p such that one has distinct real roots while the other has a conjugate pair of complex roots.

5. Verify that the equation

$$x^4 - 4x^3 + 6x^2 - 4x + 1 = 0$$

has one real root of multiplicity 4. Show that the equation

$$x^4 - 4x^3 + 5.9999x^2 - 4x + 1 = 0$$

has two real roots and two complex roots. Find precise values of these roots by factoring the second equation and solving the resulting quadratic equations by hand. Compare with machine values obtained by use of the Muller algorithm.

3.6   Survey of Some Other Algorithms (Optional)

The problem of approximating the roots of a polynomial equation occurs frequently in both pure and applied mathematics. So it is not surprising that numerous algorithms have been developed, over a long period of time, for doing this. We have methods associated with Newton, Laguerre, and Bernoulli, as well as more recent algorithms developed by Muller (1956), Bairstow (1914), and Lin (1943). In the preceding sections, we encountered two easy-to-understand methods: one very old and one relatively new. These two methods are quite adequate to solve a large number of polynomial equations completely. Both are easy to program and use by a beginning student.

Later, in Section 5.5, the method of Bairstow will be introduced in connection with the study of nonlinear systems of equations. The Bairstow method is capable of finding all real and complex roots of most polynomial equations encountered in applications. Only real arithmetic is required, in contrast to the Muller algorithm. The technique is based on approximating quadratic factors of the polynomial, so the theory behind the method is quite different from the theory that produced the Muller algorithm.

In this section, we will discuss some methods, old and new, that a student of numerical analysis should be acquainted with—at least in principle. The emphasis is on just the key ideas in each case, since most of these algorithms require tedious and lengthy derivations if treated in detail. All of the following methods were designed to produce rough estimates of the roots of a polynomial equation so that the analyst could then proceed to a method that converges rapidly and surely to the final root approximation. Hence, none of the following could be used as a replacement for the Muller algorithm or the method of Bairstow and should not be regarded as such in their present form. But each involves an interesting idea worth preserving. Perhaps some future student of numerical methods will find a way to derive a usable programmable root-finding algorithm from these beginnings that does not suffer from the defects the present algorithms have.

The **Graeffe root-squaring method** is a very old algorithm designed to estimate all the roots of a polynomial equation simultaneously. The origin lies in the algebraic relationship between the roots of a polynomial equation and its coefficients. It is well known that if r_1 and r_2 are roots of the quadratic equation

$$x^2 + a_1 x + a_0 = 0$$

then $a_1 = -(r_1 + r_2)$ and $a_0 = r_1 r_2$. In the case of the cubic equation

$$x^3 + a_2 x^2 + a_1 x + a_0 = 0$$

with roots r_1, r_2, and r_3, we have

$$a_2 = -(r_1 + r_2 + r_3) = -\sum_j r_j$$

$$a_1 = r_1 r_2 + r_1 r_3 + r_2 r_3 = \sum_{i<j} r_i r_j$$

$$a_0 = -r_1 r_2 r_3 = -\sum_{i<j<k} r_i r_j r_k$$

as you can easily verify.

These formulas are special cases of the general result that for the general polynomial equation of degree n with roots $r_1, r_2, \ldots, r_n$, we have

$$a_{n-1} = -\sum_j r_j$$

$$a_{n-2} = \sum_{i<j} r_i r_j$$

$$a_{n-3} = -\sum_{i<j<k} r_i r_j r_k$$

$$\vdots \qquad \vdots$$

$$a_0 = (-1)^n r_1 r_2 \ldots r_n$$

Now suppose that we are interested in estimating the roots of the fourth-degree equation

$$x^4 + a_3 x^3 + a_2 x^2 + a_1 x + a_0 = 0$$

Assume that the roots are well separated in absolute value, with

$$|r_1| > |r_2| > |r_3| > |r_4|$$

Then keeping only the largest term in each sum, we have approximately

$$a_0 = r_1 r_2 r_3 r_4$$

$$a_1 = -r_1 r_2 r_3$$

$$a_2 = r_1 r_2$$

$$a_3 = -r_1$$

Then we have approximately

$$r_4 = -\frac{a_0}{a_1}$$

$$r_3 = -\frac{a_1}{a_2}$$

$$r_2 = -\frac{a_2}{a_3}$$

$$r_1 = -a_3$$

EXAMPLE 3.21   Consider the polynomial equation

$$x^4 - 11101x^3 + 11111100x^2 - 1011100000x + 1000000000 = 0$$

The approximations become

$$r_4 \cong \frac{1000000000}{1011100000} = 0.989022$$

$$r_3 \cong \frac{1011100000}{11111100} = 90.99909$$

$$r_2 \cong \frac{11111100}{11101} = 1000.91$$

$$r_1 \cong 11101$$

The correct values of the roots are 1, 100, 1000, and 10,000, so the approximations are fair. If we try the same approach on the equation

$$x^4 - 10x^3 + 35x^2 - 50x + 24 = 0$$

we get approximations

$$r_4 \cong \frac{24}{50} = 0.48$$

$$r_3 \cong \frac{50}{35} = 1.43$$

$$r_2 \cong \frac{35}{10} = 3.5$$

$$r_1 \cong 10$$

The correct values are 1, 2, 3, and 4. So the approximations are not quite so good this time. The reason, of course, is that the roots are not sufficiently separated in this case. ∎

These Graeffe approximations are of very little use in the present form when some roots are complex or when some roots have the same absolute value. Modifications can be made for making approximations in such cases, however. Of course, the preceding approximation is only part of the Graeffe method. The second part is a procedure for generating new polynomials from the given polynomial in such a way that the new polynomials may have roots that can be estimated by the earlier procedure even when the method is very inaccurate when applied directly to the given polynomial equation.

Suppose that $P(x) = 0$ has roots $r_1, r_2, \ldots, r_n$. Then

$$P(x) = (x - r_1)(x - r_2) \ldots (x - r_n)$$

and

$$P(-x) = (-1)^n(x + r_1)(x + r_2) \ldots (x + r_n)$$

So

$$(-1)^n P(x)P(-x) = (x^2 - r_1^2)(x^2 - r_2^2) \ldots (x^2 - r_n^2)$$

Let us call this polynomial $P_2(x^2)$. The polynomial equation $P_2(x) = 0$ has roots r_1^2, $r_2^2, \ldots, r_n^2$, which are the squares of the roots of the given polynomial equation, and $P_2(x)$ is a polynomial of degree n also. Similarly, if we let

$$P_4(x^2) = (-1)^n P_2(x)P_2(-x)$$

then the equation $P_4(x) = 0$ has for its roots the fourth powers of the roots of $P(x) = 0$. By continuing this root-squaring process at the kth step, we arrive at a polynomial equation whose roots are $r_1^{2k}, r_2^{2k}, \ldots, r_n^{2k}$.

Now consider the case of a fourth-degree equation with roots 1, 2, 3, and 4 (not widely separated). By the procedure already outlined, we could, in theory, generate a polynomial equation of degree 4 whose roots are 1, 2^{2k}, 3^{2k}, and 4^{2k}, where k can be as large as we wish. If $k = 5$, for example, the roots are 1, 1024, 59,049, and 1,048,576. These roots are sufficiently separated so that the coefficient ratios approximation should yield fair approximations.

To make the preceding ideas useful, we now need an algorithm for generating coefficients of polynomials $P_{2k}(x)$ from the coefficients of given polynomial $P(x)$. It would be impractical to carry out the indicated multiplication of polynomials $P(x)$ and $P(-x)$ directly in each case. It is not difficult to produce a simple algorithm for generating coefficients of polynomial $P_2(x)$ from those of $P(x)$. We simply need to repeat that algorithm iteratively k times to arrive at the coefficients of polynomial $P_{2k}(x)$.

Let

$$P(x) = \sum_{k=0}^{\infty} a_k x^k$$

where $a_k = 0$ if $k > n$. Define $a_k = 0$ if $k < 0$. Then

$$(-1)^n P(-x) = \sum_{k=0}^{\infty} (-1)^{k+n} a_k x^k$$

So

$$P_2(x^2) = \sum_{j,\,k=0}^{\infty} (-1)^{k+n} a_k a_j x^{k+j}$$

Let $p = k + j$. Then

$$P_2(x^2) = \sum_{p=0}^{\infty} \sum_{j=0}^{p} (-1)^{j+n} a_{p-j} a_j x^p$$

Denoting the coefficient of x^p by c_p, we have

$$P_2(x^2) = \sum_{p=0}^{\infty} c_p x^p$$

where

$$c_p = \sum_{j=0}^{p} (-1)^{j+n} a_{p-j} a_j$$

Now $c_p = 0$ if p is odd, so in fact, we have

$$P_2(x) = \sum_{m=0}^{\infty} c_{2m} x^{2m}$$

Let $b_i = c_{2i}$. Then

$$P_2(x) = \sum_{i=0}^{\infty} b_i x^i$$

where

$$b_i = \sum_{j=0}^{2i} (-1)^{j+n} a_{2i-j} a_j$$

It should be noted that $b_i = 0$ if $i > n$, however, and in fact, in the summation defining b_i, the terms are zero if $j > n$, so we actually have

$$b_i = \sum_{j=0}^{n} (-1)^{j+n} a_{2i-j} a_j$$

If $P(x)$ has degree 4 and $a_4 = 1$, the preceding formulas for the coefficients of polynomial $P_2(x)$ become

$$b_0 = a_0^2$$
$$b_1 = 2a_0 a_2 - a_1^2$$
$$b_2 = a_2^2 + 2a_0 - 2a_1 a_3$$
$$b_3 = 2a_2 - a_3^2$$
$$b_4 = 1$$

Verify that for the polynomial

$$P(x) = x^4 - 10x^3 + 35x^2 - 50x + 24$$

we obtain

$$P_2(x) = x^4 - 30x^3 + 273x^2 - 820x + 576$$

Now let us consider some examples of computer implementation of the root-squaring algorithm.

EXAMPLE 3.22   Estimate the roots of the equation

$$x^4 - 10x^3 + 35x^2 - 50x + 24 = 0$$

Solution   Computer output is:

COEFFICIENTS: 1 -10 35 -50 24

K	ROOT 1	ROOT 2	ROOT 3	ROOT 4
1	5.47722558	3.01662063	1.7331079	0.838116355
2	4.33761314	2.94091708	1.91737934	0.981227257
3	4.04979294	2.97878258	1.99049151	0.999491923
4	4.00249773	2.99841581	1.99980985	0.999999045

The exact roots are 1, 2, 3, and 4. So the approximations are quite good after only four applications of root squaring. ■

EXAMPLE 3.23   Estimate the roots of the equation

$$x^4 - 26x^2 + 24x + 21 = 0$$

Solution   Computer output is:

COEFFICIENTS: 1 0 -26 24 21

K	ROOT 1	ROOT 2	ROOT 3	ROOT 4
1	7.21110256	3.71587033	1.52417839	0.514187212
2	5.96732668	4.05529846	1.58217253	0.548482554
3	5.5665841	4.32153723	1.58576308	0.550495738
4	5.46100569	4.40490508	1.58578643	0.550510256

In this case, the roots are -5.44948974, 4.41421356, 1.58578644, and -0.550510257, so the estimates are satisfactory except for the algebraic sign. ∎

In an algorithm designed to use estimates as input, it is a simple matter to build in an evaluation of the polynomial for each root estimate. If $P(r)$ is large for some root estimate r, then we try $-r$ as an estimate. We must verify that $P(-r)$ is sufficiently small, of course.

   We will now discuss a method that has been attributed to Daniel Bernoulli and which has been much discussed by writers on numerical methods for centuries. The **Bernoulli method** uses a relation between roots of polynomial equations and solutions of difference equations. For example, suppose that we want to determine a sequence of numbers $\{x_n\}$ defined by the recurrence relation

$$a_2 x_n + a_1 x_{n-1} + a_0 x_{n-2} = 0 \qquad a_2 \neq 0$$

where x_0 and x_1 are given. Technically, this recurrence relation is a second-order difference equation. So we are trying to solve a difference equation with initial values of x_0 and x_1 given. For more information about difference equations, see Henrici (1956). Many texts on differential equations also include some discussion of difference equations because there are strong similarities in the theory of linear nth-order difference equations and linear nth-order differential equations. However, the following discussion does not require any previous knowledge of such equations.

   Solving for x_n, we obtain

$$x_n = -\frac{a_1 x_{n-1} + a_0 x_{n-2}}{a_2}$$

By substituting $n = 2, 3, 4, \ldots$, we could routinely compute x_2, x_3, and so on, in succession and generate the terms of the sequence numerically. In some cases, these numerical calculations might prove to be unstable, of course, but in theory we could determine the solution sequence in this way. In the theory of difference equations, it is shown, however, that if z_1 and z_2 are distinct roots of the quadratic equation

$$a_2 z^2 + a_1 z + a_0 = 0$$

then the general solution of the difference equation is

$$x_n = c_1 z_1^n + c_2 z_2^n$$

Then

$$\frac{x_{n+1}}{x_n} = \frac{c_1 z_1^{n+1} + c_2 z_2^{n+1}}{c_1 z_1^n + c_2 z_2^n} = z_1 \frac{c_1 + c_2 (z_2/z_1)^{n+1}}{c_1 + c(z_2/z_1)^n}$$

Now if it happens that $|z_1| > |z_2|$, then ratio x_{n+1}/x_n approaches z_1 as n grows. This suggests that if we compute ratios x_{n+1}/x_n for $n = 1, 2, 3, \ldots$, we should obtain a sequence of numbers approaching the largest root of the quadratic equation as a limiting value.

EXAMPLE 3.24 Estimate the largest root of the equation $x^2 - 8x + 12 = 0$.

Solution We choose $x_0 = 1$ and $x_1 = 1$ for convenience. Then we compute the sequence

$$x_n = 8x_{n-1} - 12x_{n-2} \qquad n = 2, 3, 4, \ldots$$

As the terms of the sequence are obtained, we compute ratios x_{n+1}/x_n. The work can be done with a pocket calculator in this case. The following table summarizes the results of doing this.

x_n	x_{n+1}/x_n
1	*
1	1
-4	-4
-44	11
-304	6.91
-1904	6.26
-11584	6.08
-69824	6.03

Clearly, the column of ratios is converging on the largest root, which is 6. ■

In the case of the general polynomial of degree n, we compute terms of the sequence using the recurrence relation

$$a_n x_k + a_{n-1} x_{k-1} + \cdots + a_0 x_{k-n} = 0$$

for the polynomial equation

$$a_n x^n + a_{n-1} x^{n-1} + \cdots + a_1 x + a_0 = 0$$

EXAMPLE 3.25 Estimate the largest root of $x^3 + 4x^2 - 7 = 0$.

Solution We compute numbers

$$x_n = -4x_{n-1} + 7x_{n-3} \qquad n = 3, 4, 5, \ldots$$

with $x_0 = x_1 = x_2 = 1$. Here is a summary of the calculation:

x_n	x_{n+1}/x_n
3	3
−5	−1.6667
27	−5.4444
−87	−3.2222
313	−3.5977
−1063	−3.3962
3643	−3.4271
−12381	−3.3986
42083	−3.3990
−142831	−3.3940
484657	−3.3932
−1644047	−3.3922

It seems that the estimates are converging on a root near −3.39. There is, in fact, a root at −3.39138, and this is the root of maximum absolute value, since the other roots are −1.772866 and 1.1642479. So the Bernoulli idea works nicely in this example.   ∎

EXAMPLE 3.26   Estimate the largest root of $x^3 + x^2 - 2x - 1 = 0$.

Solution   Using $x_0 = 1$, $x_1 = 2$, and $x_2 = 3$, we obtain the results tabulated.

Iteration	x_n	Ratio
1	2	0.66667
2	6	3.
3	1	0.16667
4	13	13
5	−5	−0.3846
⋮	⋮	⋮
12	809	−2.0175
13	−1350	−1.6687
⋮	⋮	⋮
14	2567	−1.9015
⋮	⋮	⋮
21	−154697	−1.7946
22	279540	−1.8070

Convergence is slow, since we need 20 or more iterations to reach a useful root estimate. The ratios are converging on a root near −1.8070, however. The given equation has roots −0.4450, 1.2470, and −1.8019, in fact, so the algorithm does provide an estimate of the largest root. The convergence is slow because the roots are not widely separated.   ∎

EXAMPLE 3.27   Estimate a root for $x^3 - x - 1 = 0$.

Solution Selected iterations are shown starting from initial values of 1, 1, and 1.

Iteration	x_n	Ratio
4	4	1.3333
7	9	1.2857
10	21	1.3125
15	86	1.3231
17	151	1.3246

The results are interesting, since the equation has a real root at 1.32472 and two complex roots at $-0.662359 \pm 0.5622795i$. It is interesting to note that existence of the complex roots does not cause the algorithm to fail, because the real root has maximum absolute value. The convergence is slow, but this is usually the case with the Bernoulli aglorithm. ■

EXAMPLE 3.28 Estimate a root of $x^3 - x^2 + x - 1 = 0$.

Solution Using initial values of 4, 3, and 2, we obtain the results shown.

x_n	x_{n+1}/x_n
3	1.5
4	1.3333
3	0.75
2	0.6667
3	1.5
4	1.3333
3	0.75

The values of x_n and the ratios are cyclic and do not seem to be converging on anything. The roots of the equation are 1 and $\pm i$. All have equal absolute value. So we should expect trouble in this case. A cyclic pattern always signals the existence of dominant complex roots for the equation in question.

Perhaps you are wondering if we can cause the ratios to converge on 1 by the right choice of starting values. We can do that by choosing values 1, 1, and 1, as you can verify. All iterates are 1 in this case. There is no root of largest absolute value, so convergence to the real root, as in Example 3.25, does not happen. ■

There are many difficulties with the Bernoulli algorithm as a general root-finding algorithm. But with appropriate modifications, it can be used to estimate roots in a large number of problems and can even estimate complex conjugate pairs in some cases. For further details, consult Henrici (1956), which includes a detailed analysis of this algorithm. Rutishauser (1954) published an improved version of the Bernoulli algorithm known as the *quotient-difference (QD) algorithm*, which overcomes many of the difficulties with Bernoulli's form and also provides estimates of all roots simultaneously. The QD algorithm can also be used to estimate complex roots. The Henrici text also provides an excellent discussion of the QD algorithm and is highly recommended.

If you are interested in pursuing the subject of root approximation for polynomial equations, you might want to investigate the *Laguerre method* and the *Lehmer-Schur method*. An introduction to the Lehmer-Schur method can be found in Ralston (1965) or Todd (1962). The more advanced reader should consult Lehmer (1961). For details concerning the very powerful Laguerre method, refer to Householder (1970). A recent addition to the collection of methods for solving polynomial equations is the *Jenkins-Traub method* (1970). Other methods are surveyed in Greenspan (1957/1958).

Several IMSL subroutines are available for machine solution of polynomial equations. The subroutine ZPLRC uses the Laguerre method to find the zeros of a polynomial with real coefficients. The subroutine ZPORC uses the Jenkins-Traub algorithm to perform the same task. If you want to use the Muller method, the routine ZANLY can be used. This subroutine is not restricted to polynomial equations. You might try solving some of the polynomial equations in this chapter using one or more of these subroutines.

■ *Exercise Set 3.6*

1. For each of the following polynomials $P(x)$, find the corresponding polynomial $P_2(x)$ as described in the Graeffe root-squaring method.
   a. $P(x) = x^4 - 10x^3 + 35x^2 - 50x + 24$
   b. $P(x) = x^4 - 26x^2 + 24x + 21 = 0$
   c. $P(x) = x^3 + x + 1 = 0$

2. Let $P(x) = (x - 1)(x - 2)(x - 3)$.
   a. Compute $P_2(x)$ by actual multiplication of the factors of $P_2(x)$.
   b. Find the coefficients of $P_2(x)$ by the formulas given in this section for the Graeffe root-squaring algorithm. Compare with your results in part (a).

3. Consider the equation
$$x^4 - 526x^3 + 13125x^2 - 62600x + 50000 = 0$$
   a. Estimate the roots by use of the ratios $-a_0/a_1$, $-a_1/a_2$, $-a_2/a_3$, and $-a_3$.
   b. Using the estimates obtained in part (a) as initial estimates in the Newton algorithm (computer program), find accurate values for the roots. How good are the Graeffe estimates?

4. Verify the first five values of x_n in Example 3.24. Also, compute the ratios x_{n+1}/x_n and compare with the machine results listed in the example.

5. Write down the difference equation used in the Bernoulli algorithm for the equation $x^3 - 3x^2 + x - 3 = 0$. Now beginning with $x_0 = x_1 = x_2 = 1$, compute the terms x_3, x_4, and x_5 and corresponding ratios. What root is being estimated in this case?

6. Write a computer program that employs the Graeffe root-squaring method to produce estimates of real roots for cubic equations and then proceeds to approximate the roots by use of the Newton algorithm. Use your program to approximate all real roots for each of the following equations.
   a. $x^3 + 4x^2 - 7 = 0$
   b. $x^3 + x^2 - 2x - 1 = 0$
   c. $x^3 + 3x^2 - 1 = 0$
   d. $x^3 - 1 = 0$
   e. $x^3 - 7x^2 + 7x + 15 = 0$

3.7 Applications

Now we will consider some applied problems that require the solution of a polynomial equation. In each case, we will see how the polynomial arises in the given situation. Then we will determine the roots by one of our root-finding algorithms.

Let us begin with a financial problem. Mr. Sharp has a boat for sale. He is asking $50,000. A buyer offers to pay $70,000 in seven annual installments of $10,000 each, beginning one year from the date of sale. Mr. Sharp knows that he can safely invest the $50,000 and earn 10% interest compounded annually. Should he insist on a cash deal or take $70,000 in seven installments as offered?

We need to find the annual interest rate equivalent to receiving $10,000 each year for the next seven years; that is, we need to find interest rate r such that

$$50,000x^7 = 10,000x^6 + 10,000x^5 + \cdots + 10,000x + 10,000$$

where $x = 1 + r$ and r is the decimal equivalent of the interest rate ($r = 0.1$ if the interest rate is 10%). The polynomial equation simplifies to

$$5x^7 - x^6 - x^5 - x^4 - x^3 - x^2 - x - 1 = 0$$

We know that there is a root around 1.1, so we can use the Newton algorithm with $x = 1.1$. We can easily determine all the roots by using the Muller algorithm. It is more practical to simply use the Newton algorithm, but it is interesting to see what all the roots look like as a matter of curiosity. Using the Muller algorithm, we obtain the following computer output:

ENTER DEGREE OF POLYNOMIAL 7

ENTER REAL COEF. FOR YOUR POLYNOMIAL

$-1.$ $-1.$ $-1.$ $-1.$ $-1.$ $-1.$ $-1.$ $5.$

ENTER E0, E1, E2

$-1.$ $0.$ $1.$

ROOT ESTIMATES FOLLOW:

	REAL PART	IMAGINARY PART
ROOT 1	-0.2135825985	0.7128121176
ROOT 2	1.0919613667	0.0000000000
ROOT 3	-0.2135825985	-0.7128121176
ROOT 4	-0.6609428131	-0.2999064268
ROOT 5	-0.6609428131	0.2999064268
ROOT 6	0.4285447283	-0.6665270782
ROOT 7	0.4285447283	0.666527078

The desired root is apparently 1.09191. This corresponds to an interest rate of about 9.19%. Since Mr. Sharp can earn 10% on the $50,000 if he has the cash, he should not accept the installment plan described.

A surprising number of financial problems lead to polynomial equations when they are analyzed algebraically. Such problems are usually handled in normal business practice by use of tables or pocket calculator estimates. Precise solutions are easily found, however, as in the preceding example. A not-so-well-known problem of this type concerns the yield to maturity (YTM) of a coupon bond. If you are not familiar with the concept of YTM or present-value calculations, consult Homer and Leibowitz (1972).

Homer and Leibowitz (1972) define the YTM to be the discount rate that makes the present value of a bond's cash flow equal to its market price. Suppose that a $1000 face-value coupon bond has a market price M and coupon rate c with T semiannual interest periods remaining until maturity. Let $r = $ YTM/2. Then

$$M(1 + r)^T = 1000 + 500c(1 + r)^{T-1} + 500c(1 + r)^{T-2}$$
$$+ 500c(1 + r)^{T-3} + \cdots + 500c$$

Hence,

$$M(1 + r)^T = 1000 + 500c\left(\frac{(1 + r)^T - 1}{r}\right)$$

It follows that

$$M = \frac{500c}{r} + \frac{1}{(1 + r)^T}\left(1000 - \frac{500c}{r}\right)$$

Suppose that M, T, and c are known. We must determine $r = $ YTM/2. There are YTM tables (or we can work from present-value tables) that provide the desired YTM for stockbrokers or other interested financial managers. However, an algebraic solution is possible.

If we let $x = 1 + r$, then it is easily seen that

$$Mr(1 + r)^T = 500c(1 + r)^T + 1000r - 500c$$

or

$$M(1 + r)^{T+1} - M(1 + r)^T - 500c(1 + r)^T - 1000(1 + r) + 1000 + 500c = 0$$

So

$$Mx^{T+1} - Mx^T - 500cx^T - 1000x + 1000 + 500c = 0$$

Finally,

$$Mx^{T+1} - (M + 500c)x^T - 1000x + 1000 + 500c = 0$$

Now suppose that the market price of a certain bond is $922.85, and we have 3.5 years left before redemption. Also, the coupon rate is 4%. We want to find the

YTM. We have $M = 922.85$, $T = 7$, and $c = 0.04$. We must solve the equation

$$922.85x^8 - 942.85x^7 - 1000x + 1020 = 0$$

We know that there is a real root near 1.04. Solving by any of the methods available to us, we find that the equation has a real root at 1.0325000. So the YTM is apparently

$$2r = 2(0.0325) = 0.065, \quad \text{or } 6.5\%$$

Now we turn to a physical problem. A wooden sphere 1.8 ft in diameter floats on calm water. The specific gravity of the wood is 0.7. We want to find the depth to which the sphere will sink. We reason as follows. It is a well-known physical fact that the floating sphere displaces its own weight of water. In the following, let ρ denote the density of water. If a sphere with radius r sinks to depth x ft, then the volume of the submerged part is

$$\pi x^2 \left(r - \frac{x}{3} \right)$$

The weight of the water displaced is ρ times this volume. The weight of the sphere is

$$0.7\rho(\tfrac{4}{3})\pi r^3$$

Equating these two weights, we find that

$$0.7\rho(\tfrac{4}{3})\pi r^3 = \pi x^2 \rho \left(r - \frac{x}{3} \right)$$

Simplifying, and substituting $r = 0.9$, we obtain the equation

$$x^3 - 2.7x^2 + 2.0412 = 0$$

We look for a real solution between 0 and 1.8, with a probable value near 1. We find real roots 1.14614, -0.76727, and 2.32113. The desired value of x is apparently 1.146 ft.

Polynomial equations frequently arise in physical problems in connection with eigenvalue problems. Such an example was discussed in Section 2.9. Now that we have become acquainted with methods for finding roots of polynomial equations, it is worthwhile to review the problem. In a certain problem in quantum mechanics concerning energy levels of a system, there was a need to find the roots of the equation

$$x^4 - 56x^3 + 490x^2 + 11112x - 117495 = 0$$

In 1931, when the problem arose, finding precise values of the roots was extremely difficult. The author of the paper quoted did manage to find approximate locations of the roots, which were all real, and proceeded from there to more accurate estimates.

This problem is discussed in Margenau and Murphy (1943). Using the Graeffe root-squaring method, these authors obtain estimates 36.89, 22.36, −13.65, and 10.45. Then, using the Newton method, the estimates are refined to 36.8960, 22.3410, −13.6669, and 10.4302. These approximations are claimed to be accurate through the fourth decimal place. The calculations are presented as hand calculations aided by log tables and a desk calculator if available.

With present-day computers and an algorithm such as the Muller algorithm, the problem is much simpler and easier to handle. Without any preliminary estimates, the Muller algorithm produces the following results.

ENTER DEGREE OF POLYNOMIAL 4

ENTER REAL COEF. FOR YOUR POLYNOMIAL

−117495.   11112.   490.   −56.   1.

ENTER E0, E1, E2

−1.   0.   1.

ROOT ESTIMATES:

	REAL PART	IMAGINARY PART
ROOT 1	10.4296273071	0.0000000000
ROOT 2	22.3411277525	0.0000000000
ROOT 3	−13.6667736021	0.0000000000
ROOT 4	36.8960185425	0.0000000000

These values do not agree exactly through four decimal places with those quoted by Margenau and Murphy, but their values are close. We now have the machines and software needed to produce more accurate values than researchers were able to obtain in 1940.

It is easily seen that polynomial equations arise in a wide variety of applied problems. These polynomials rarely have a degree exceeding 10, and most have at least one real root. It is also true that from the origin of the equation, we can often set limits on the size of the root or even produce an estimate sufficiently accurate to use with the Newton method. If no estimate is available, we can use the Muller algorithm or possibly root squaring. There is no serious problem with finding accurate values for all roots, real and complex, for most polynomials of degree 20 or less using the methods given in this chapter, and that covers virtually all polynomial equations arising in practice.

■ *Exercise Set 3.7*

1. Suppose that you have won a national contest and you must choose between receiving your $100,000 prize now, in one payment, or receiving $150,000 in five equal annual installments. Your banker has told you that you can earn 10% on $100,000 invested *at this time*. Would you prefer to have $100,000 cash now, or $30,000 in five annual installments over the next five years, with first payment today? You have no assurance that you can earn 10% on the annual installments. Determine the minimum interest rate required so that the income for the next four years is the same for the two investment possibilities. Ignore possible tax complications. Only compare the totals of interest earned and cash received.

2. A sphere 3 ft in diameter floats on calm water. The sphere is made of material with average specific gravity 0.95. Find the depth to which the sphere sinks.

3. Find all intersection points for the graphs of equations

$$y = x^2 \quad \text{and} \quad xy - 2x - y + 1 = 0$$

Sketch the graphs to find approximate locations of intersection points. Use the Newton or Muller algorithm to find precise values.

4. Find the dimensions of a right circular cylinder if the surface area is 5.57 ft^2 and the volume is 0.98 ft^3.

5. The system of orthogonal polynomials known as Legendre polynomials is defined by the recursion formula

$$P_{n+1}(x) = \frac{(2n + 1)xP_n(x) - nP_{n-1}(x)}{n + 1} \qquad n = 1, 2, \dots$$

where $P_0(x) = 1$ and $P_1(x) = x$. It is important to know the zeros of these polynomials. Precise values are listed in various mathematics handbooks. Using the methods of this chapter, find the zeros and make a table of the results for the Legendre polynomials with degrees $n = 2$ through $n = 6$.

6. The Bessel function of order 0 is defined by the infinite series

$$J_0(x) = 1 - \sum_{n=1}^{\infty} \frac{(-1)^n x^{2n}}{2^{2n}(n!)^2}$$

Estimate the two smallest positive zeros of $J_0(x)$ by truncating the series to a polynomial of degree 15 and finding the roots of this polynomial. Compare your values with tabulated values from a mathematics handbook in your library. How good are your estimates?

4 SOLVING LINEAR SYSTEMS OF EQUATIONS

In Chapters 4 and 5, we will consider the problem of solving systems of equations. Because some understanding of the theory of linear systems is needed to study non-linear systems, we will discuss linear systems first. A well-developed theory of linear systems exists. Also, in the linear case, we have easily understood numerical methods for estimating solutions. There are difficulties, of course, and we must explore these both analytically and numerically. It is the numerical methods, the errors that arise, and associated problems with which we shall be primarily concerned.

It is assumed that you are familiar with the elementary definitions and theorems concerning linear spaces, including the concepts of basis and spanning set. In this chapter, we only consider vectors in R^n.

Although you are expected to have some understanding of the general theory of linear systems, we will begin with a brief review of some of this theory, since it forms the foundation for the development and analysis of numerical algorithms.

4.1 General Theory

We wish to solve the linear system

$$\sum_{j=1}^{n} a_{ij} x_j = b_i \qquad i = 1, 2, \ldots, n$$

Explicitly,

$$a_{11}x_1 + a_{12}x_2 + \cdots + a_{1n}x_n = b_1$$
$$a_{21}x_1 + a_{22}x_2 + \cdots + a_{2n}x_n = b_2$$
$$\vdots \qquad\qquad\qquad \vdots \quad \vdots$$
$$a_{n1}x_1 + a_{n2}x_2 + \cdots + a_{nn}x_n = b_n$$

Coefficients a_{ij} and constants b_i are given. The object is to find the numbers x_1, $x_2,\ldots,x_n$. In matrix form, our system is $\mathbf{Ax} = \mathbf{b}$, where $\mathbf{A} = (a_{ij})$, $\mathbf{b}$ is the column matrix with elements b_i, and $\mathbf{x}$ is the column matrix with elements x_i. We will restrict our study to systems where $n = m$, so matrix $\mathbf{A}$ is always $n \times n$. Matrix $\mathbf{A}$ will be called the **coefficient matrix** for the system, and the matrix $\mathbf{W} = [\mathbf{A}\,|\,\mathbf{b}]$, formed by attaching $\mathbf{b}$ as an $(n + 1)$th column, will be called the **augmented matrix**. Numerical algorithms normally require the augmented matrix $\mathbf{W}$ as input, and all operations are performed on the rows or columns (usually the rows) of matrix $\mathbf{W}$.

If $\mathbf{b} = 0$, the system becomes $\mathbf{Ax} = 0$. Such a system is called a **homogeneous system**. Obviously, one solution of any homogeneous system is $\mathbf{x} = \mathbf{0}$. We call this the **trivial solution**.

The first question to consider is existence and uniqueness of solutions for $\mathbf{Ax} = \mathbf{b}$. Clearly, if $\mathbf{A}^{-1}$ exists, then the unique solution is $\mathbf{x} = \mathbf{A}^{-1}\mathbf{b}$. From previous study of linear algebra, we know the following.

Theorem 4.1   If $\mathbf{A}$ is an $n \times n$ matrix, then $\mathbf{A}^{-1}$ exists iff one of the following is true:

1. The determinant of $\mathbf{A}$ is not zero.
2. The rows of $\mathbf{A}$ are linearly independent.
3. The columns of $\mathbf{A}$ are linearly independent.
4. Homogeneous system $\mathbf{Ax} = 0$ has only the trivial solution.

If $\mathbf{A}^{-1}$ exists, we say that matrix $\mathbf{A}$ is *nonsingular*; otherwise $\mathbf{A}$ is a *singular* matrix. If $\mathbf{x} = \mathbf{y}$ is any solution of $\mathbf{Ax} = \mathbf{b}$, and $\mathbf{z}$ is any solution of $\mathbf{Az} = 0$, then it follows at once that $\mathbf{x} = \mathbf{y} + \mathbf{z}$ is also a solution of $\mathbf{Ax} = \mathbf{b}$. So we can construct more than one solution of our system if the associated homogeneous system has any nontrivial solutions. Now, system $\mathbf{Az} = 0$ has only the trivial solution iff $\mathbf{A}$ is nonsingular. So the following theorem is seen to be true.

Theorem 4.2   If $\mathbf{A}$ is an $n \times n$ matrix, then the linear system $\mathbf{Ax} = \mathbf{b}$ has a unique solution iff $\mathbf{A}$ is nonsingular.

We also know that every linear system can be transformed into an *equivalent* system (same solution) by use of the following *elementary row operations*, which are to be performed on the rows of the augmented matrix $\mathbf{W}$.

1. Interchange of any two rows
2. Multiplication of any row by a nonzero number
3. Replacement of any row by the sum of that row and any multiple of another row.

It is a well-known fact that any $n \times n$ matrix **A** can be transformed into an upper triangular matrix

$$\hat{\mathbf{A}} = \begin{bmatrix} \hat{a}_{11} & \hat{a}_{12} & \hat{a}_{13} & \cdots & \hat{a}_{1n} \\ 0 & \hat{a}_{22} & \hat{a}_{23} & \cdots & \hat{a}_{2n} \\ 0 & 0 & \hat{a}_{33} & \cdots & \hat{a}_{3n} \\ & & & \vdots & \\ 0 & 0 & & \cdots & \hat{a}_{nn} \end{bmatrix}$$

by use of elementary row operations. In this case, the determinant of **A** is

$$\det \mathbf{A} = (\text{const})\, \hat{a}_{11}\hat{a}_{22}\,\hat{a}_{33} \cdots \hat{a}_{nn}$$

Hence, $\det \mathbf{A} = 0$ and $\mathbf{A}^{-1}$ exists iff no $\hat{a}_{jj} = 0$. A matrix can be easily tested for singularity by this procedure. If the matrix can be converted to triangular form with no zero elements on the diagonal, then **A** is nonsingular.

For example, if **A** is the matrix

$$\begin{bmatrix} 1 & 0 & 1 & 2 \\ 0 & -1 & 2 & 1 \\ 2 & 1 & 1 & 0 \\ 3 & 0 & 1 & 2 \end{bmatrix}$$

we find that **A** is row-equivalent to each of the following matrices:

$$\begin{bmatrix} 1 & 0 & 1 & 2 \\ 0 & -1 & 2 & 1 \\ 0 & 1 & -1 & -4 \\ 0 & 0 & -2 & -4 \end{bmatrix}, \quad \begin{bmatrix} 1 & 0 & 1 & 2 \\ 0 & -1 & 2 & 1 \\ 0 & 0 & 1 & -3 \\ 0 & 0 & -2 & -4 \end{bmatrix}, \quad \text{and} \quad \begin{bmatrix} 1 & 0 & 1 & 2 \\ 0 & -1 & 2 & 1 \\ 0 & 0 & 1 & -3 \\ 0 & 0 & 0 & -10 \end{bmatrix}$$

Since the last matrix is triangular with no zero elements on the diagonal, **A** is nonsingular. In fact, $\det \mathbf{A} = 10$ in this case.

It follows that if $\mathbf{W} = [\mathbf{A}\,|\,\mathbf{b}]$ is the augmented matrix of a linear system, then by use of elementary row operations, **W** can be converted to the form $\hat{\mathbf{W}} = [\mathbf{U}\,|\,\mathbf{c}]$, where **U** is upper triangular. But $\hat{\mathbf{W}}$ is the augmented matrix for a system equivalent to $\mathbf{Ax} = \mathbf{b}$. If **A** is nonsingular, then the matrix **U** has no zeros on the diagonal, and the system $\mathbf{Ux} = \mathbf{c}$ has a unique solution that can be easily obtained by solving the equations one at a time, beginning with the nth equation and proceeding backward to the first. This is the theoretical basis for the Gauss elimination algorithm discussed in the next section.

As an aid to understanding existence and uniqueness of solutions for system $\mathbf{Ax} = \mathbf{b}$, consider the following. Denote the columns of **A** by $C_1, C_2, \ldots, C_n$, so that

$$\mathbf{A} = [\mathbf{C}_1 \mathbf{C}_2 \cdots \mathbf{C}_n]$$

Then

$$\mathbf{Ax} = x_1 \mathbf{C}_1 + x_2 \mathbf{C}_2 + \cdots + x_n \mathbf{C}_n$$

Hence,

$$\mathbf{Ax} = \mathbf{b} \quad \text{iff} \quad x_1\mathbf{C}_1 + x_2\mathbf{C}_2 + \cdots + x_n\mathbf{C}_n = \mathbf{b}$$

Then a solution of the system exists iff $\mathbf{b}$ can be expressed as a linear combination of the columns of $\mathbf{A}$. For example, suppose that we want to solve $\mathbf{Ax} = \mathbf{b}$, where

$$\mathbf{A} = \begin{bmatrix} 3 & -1 & 1 \\ 2 & 4 & 3 \\ 1 & 5 & -2 \end{bmatrix} \quad \text{and} \quad \mathbf{b} = \begin{bmatrix} 2 \\ -1 \\ 3 \end{bmatrix}$$

Letting $\mathbf{x}$ denote the transpose of matrix $[x_1, x_2, x_3]$, we find that

$$\mathbf{Ax} = \begin{bmatrix} 3x_1 - x_2 + x_3 \\ 2x_1 + 4x_2 + 3x_3 \\ x_1 + 5x_2 - 2x_3 \end{bmatrix} = x_1\begin{bmatrix} 3 \\ 2 \\ 1 \end{bmatrix} + x_2\begin{bmatrix} -1 \\ 4 \\ 5 \end{bmatrix} + x_3\begin{bmatrix} 1 \\ 3 \\ -2 \end{bmatrix}$$

So $\mathbf{Ax} = \mathbf{b}$ is equivalent to

$$x_1\begin{bmatrix} 3 \\ 2 \\ 1 \end{bmatrix} + x_2\begin{bmatrix} -1 \\ 4 \\ 5 \end{bmatrix} + x_3\begin{bmatrix} 1 \\ 3 \\ -2 \end{bmatrix} = \begin{bmatrix} 2 \\ -1 \\ 3 \end{bmatrix}$$

A solution to the system exists iff vector $\mathbf{b}$ is a linear combination of the columns of $\mathbf{A}$.

If every vector $\mathbf{b}$ in R^n can be expressed as a linear combination of the columns of $\mathbf{A}$, we say that the columns of $\mathbf{A}$ *span* R^n. So a solution exists for every choice of $\mathbf{b}$ iff the columns of $\mathbf{A}$ span R^n.

If vectors $\{\mathbf{C}_1, \mathbf{C}_2, \ldots, \mathbf{C}_n\}$ span R^n and are linearly independent, we say that $\{\mathbf{C}_1, \mathbf{C}_2, \ldots, \mathbf{C}_n\}$ is a **basis** for R^n. In this case, each $\mathbf{b}$ in R^n can be expressed as a *unique* linear combination of the vectors $\{\mathbf{C}_1, \mathbf{C}_2, \ldots, \mathbf{C}_n\}$. We conclude that the solution of $\mathbf{Ax} = \mathbf{b}$ is unique iff the columns of $\mathbf{A}$ form a basis for R^n. The following theorem is useful in applying this result.

Theorem 4.3   If $\{\mathbf{C}_1, \mathbf{C}_2, \ldots, \mathbf{C}_n\}$ is a set of n vectors in R^n, then the following are equivalent:

1. Vectors $\{\mathbf{C}_1, \mathbf{C}_2, \ldots, \mathbf{C}_n\}$ span R^n.
2. Vectors $\{\mathbf{C}_1, \mathbf{C}_2, \ldots, \mathbf{C}_n\}$ are linearly independent.
3. Vectors $\{\mathbf{C}_1, \mathbf{C}_2, \ldots, \mathbf{C}_n\}$ form a basis for R^n.

This concludes our review of those theorems and definitions from linear algebra that are particularly relevant when studying numerical methods for approximating the solution of a linear system. The discussion is brief and assumes some previous study of linear algebra. If you have understood this section, however, you should be ready to study that part of numerical linear algebra that is concerned with computation of the numerical solution of a linear system.

■ *Exercise Set 4.1*

1. Determine whether the matrix is singular or nonsingular in each case.

a. $\begin{bmatrix} 3 & 1 \\ 4 & 2 \end{bmatrix}$
b. $\begin{bmatrix} 2 & 1 & 3 \\ 3 & 1 & 2 \\ 1 & 0 & -1 \end{bmatrix}$

c. $\begin{bmatrix} 1 & 0 & 1 \\ 0 & 1 & 0 \\ 1 & 0 & 1 \end{bmatrix}$
d. $\begin{bmatrix} 0 & 0 & 1 \\ 0 & 1 & 0 \\ 1 & 0 & 1 \end{bmatrix}$
e. $\begin{bmatrix} 1 & 1 & 1 \\ 1 & -1 & 1 \\ -1 & 1 & -1 \end{bmatrix}$

2. Given the 2×2 matrix

$$\mathbf{A} = \begin{bmatrix} a & b \\ c & d \end{bmatrix}$$

with $D = ad - bc \neq 0$, show that the inverse exists and is the matrix

$$\begin{bmatrix} d/D & -b/D \\ -c/D & a/D \end{bmatrix}$$

3. Use the result in exercise 2 to find the inverse of each of the following matrices.

a. $\begin{bmatrix} 3 & 2 \\ 2 & 3 \end{bmatrix}$
b. $\begin{bmatrix} 3 & -2 \\ 5 & 1 \end{bmatrix}$
c. $\begin{bmatrix} 1/2 & 1/3 \\ -2/3 & 2/5 \end{bmatrix}$

4. Given an $n \times n$ matrix $\mathbf{A}$, the adjoint of $\mathbf{A}$ is the transpose of the matrix of cofactors of $\mathbf{A}$. That is, adj $\mathbf{A} = (b_{ji})$, where $b_{ij} = A_{ij}$ and A_{ij} is the cofactor of a_{ij} in matrix $\mathbf{A}$. Using this definition of adj $\mathbf{A}$, do the following.
a. Verify that (adj $\mathbf{A}$)$\mathbf{A}$ = det $\mathbf{A}$ for 3×3 matrices.
b. Use the fact that the product of $\mathbf{A}$ and adj $\mathbf{A}$ is the determinant of $\mathbf{A}$ to write a general formula for the inverse of a nonsingular matrix.
c. Verify that the formula for the inverse of the 2×2 matrix given in exercise 2 is a special case of the result in part (b) above.

5. Use the result in exercise 4b to find the inverse of the following matrices.

a. $\begin{bmatrix} 1 & -2 & 3 \\ 1 & -1 & -1 \\ 2 & 3 & 5 \end{bmatrix}$
b. $\begin{bmatrix} 1 & 2 & 3 \\ 2 & 3 & 4 \\ 3 & 4 & 5 \end{bmatrix}$
c. $\begin{bmatrix} 1 & 0 & -1 \\ 0 & 2 & 1 \\ 1 & 1 & 0 \end{bmatrix}$

6. Solve each linear system using the result in exercise 2 and the formula $\mathbf{x} = \mathbf{A}^{-1}\mathbf{b}$.
a. $\begin{aligned} 2x + 3y &= 1 \\ 4x - 5y &= -1 \end{aligned}$
b. $\begin{aligned} x + y &= 3 \\ 2x - y &= 5 \end{aligned}$
c. $\begin{aligned} 0.1x + 0.25y &= 0.3 \\ 0.3x - 0.45y &= 1.2 \end{aligned}$

7. Use elementary row operations to convert each matrix to upper triangular form.

a. $\begin{bmatrix} 1 & 1 & -1 \\ 2 & 0 & 1 \\ 3 & 1 & 0 \end{bmatrix}$
b. $\begin{bmatrix} 2 & 1 & 1 & 2 \\ 1 & 1 & 0 & 1 \\ -1 & 1 & 1 & 1 \\ -2 & 3 & 2 & 1 \end{bmatrix}$
c. $\begin{bmatrix} 1 & -1 & 3 & 4 \\ 0 & 1 & 0 & 2 \\ 2 & 0 & -1 & 3 \\ 3 & -3 & 2 & 1 \end{bmatrix}$

8. If the $n \times n$ matrix $\mathbf{A}$ can be transformed to upper triangular form with no zero entries appearing on the diagonal, then the rows of $\mathbf{A}$ are independent. Use that fact to decide whether the rows are independent for each of the following matrices.

a. $\begin{bmatrix} 1 & 0 & 1 & 0 \\ 0 & 1 & 2 & 1 \\ 2 & 0 & 1 & 0 \\ -1 & 1 & 2 & 1 \end{bmatrix}$
b. $\begin{bmatrix} 1 & 2 & 3 \\ 4 & 1 & -1 \\ 2 & 0 & 1 \end{bmatrix}$
c. $\begin{bmatrix} 1 & 1 & -1 & 1 \\ 3 & 9 & -5 & 4 \\ 2 & 0 & 2 & 3 \\ 1 & 3 & 0 & 2 \end{bmatrix}$

9. Determine which sets of vectors form a basis for R^3.

a. $\left\{ \begin{bmatrix} 1 \\ 1 \\ -1 \end{bmatrix}, \begin{bmatrix} 2 \\ 1 \\ 0 \end{bmatrix}, \begin{bmatrix} 1 \\ 2 \\ 3 \end{bmatrix} \right\}$
b. $\left\{ \begin{bmatrix} 1 \\ 2 \\ 3 \end{bmatrix}, \begin{bmatrix} 4 \\ 5 \\ 6 \end{bmatrix}, \begin{bmatrix} 7 \\ 8 \\ 9 \end{bmatrix}, \begin{bmatrix} 10 \\ 11 \\ 12 \end{bmatrix} \right\}$
c. $\left\{ \begin{bmatrix} 1 \\ 1 \\ 0 \end{bmatrix}, \begin{bmatrix} 0 \\ 1 \\ -1 \end{bmatrix} \right\}$

10. Given vectors

$$\mathbf{C}_1 = \begin{bmatrix} 1 \\ 0 \\ 1 \end{bmatrix}, \quad \mathbf{C}_2 = \begin{bmatrix} 2 \\ 1 \\ 0 \end{bmatrix}, \quad \mathbf{C}_3 = \begin{bmatrix} -1 \\ 2 \\ 1 \end{bmatrix}, \quad \mathbf{C}_4 = \begin{bmatrix} 0 \\ 1 \\ 1 \end{bmatrix}$$

express $\mathbf{C}_4$ as a linear combination of $\mathbf{C}_1$, $\mathbf{C}_2$, and $\mathbf{C}_3$.

4.2 Gauss Elimination Method

We will now look at our first numerical algorithm for finding solutions of linear systems. The basic idea of the method is to convert the given system to triangular form by use of elementary row operations and then solve the resulting system by "back substitution." That is, we solve the last equation (the only unknown is x_n), substitute the result in the $(n - 1)$th equation, and solve for x_{n-1}; then we substitute x_n and x_{n-1} in the next equation and solve for x_{n-2}; and so on. In principle, the method is very straightforward, very old, and familiar to students of linear algebra for 3×3 and 4×4 systems at least. It is called the **Gauss elimination method**.

We need to examine the algorithm in detail and then study its implementation on a computer, where rounding errors and propagated error become a consideration. For a 3×3 system with augmented matrix

$$\mathbf{W} = \begin{bmatrix} a_{11} & a_{12} & a_{13} & \vdots & b_1 \\ a_{21} & a_{22} & a_{23} & \vdots & b_2 \\ a_{31} & a_{32} & a_{33} & \vdots & b_3 \end{bmatrix}$$

the algorithm takes the following form:

Part I: Forward Elimination
   We assume that $w_{11} \neq 0$.

Step 1 Set $w_{1j}^{(1)} = w_{1j}, j = 1, 2, 3, 4$.
Step 2 Set $m_{21} = w_{21}/w_{11}$ and $m_{31} = w_{31}/w_{11}$.
Step 3 Compute $w_{ij}^{(1)} = w_{ij} - m_{i1}w_{1j}$ for $i = 2, 3$ and $j = 1, 2, 3, 4$.

Step 4 If $w_{22}^{(1)} = 0$, then go to step 8.
Step 5 Set $m_{32} = w_{32}^{(1)}/w_{22}^{(1)}$.
Step 6 Compute $w_{3j}^{(2)} = w_{3j}^{(1)} - m_{32}\, w_{2j}^{(1)}$ for $j = 2, 3, 4$.
Step 7 Set $w_{ij}^{(2)} = w_{ij}^{(1)}$ for $i = 1, 2$ and $j = 1, 2, 3, 4$. Computations end.
Step 8 Matrix **A** is singular. Stop.

The result is a new augmented matrix **W** with the form

$$\mathbf{W}^{(2)} = \begin{bmatrix} w_{11}^{(2)} & w_{12}^{(2)} & w_{13}^{(2)} & w_{14}^{(2)} \\ 0 & w_{22}^{(2)} & w_{23}^{(2)} & w_{24}^{(2)} \\ 0 & 0 & w_{33}^{(2)} & w_{34}^{(2)} \end{bmatrix}$$

which is the augmented matrix for an equivalent system.

Part II: Back Substitution

Step 1 Compute $x_3 = w_{34}^{(2)}/w_{33}^{(2)}$.
Step 2 Compute $x_2 = (w_{24}^{(2)} - w_{23}^{(2)} x_3)/w_{22}^{(2)}$.
Step 3 Compute $x_1 = (w_{14}^{(2)} - w_{12}^{(2)} x_2 - w_{13}^{(2)} x_3)/w_{11}^{(2)}$.

This completes the solution.

EXAMPLE 4.1   Solve the system with augmented matrix

$$\mathbf{W} = \begin{bmatrix} 1 & -4 & 1 & 2 \\ 3 & 1 & 2 & 1 \\ 2 & 3 & 0 & 1 \end{bmatrix}$$

Solution   Forward elimination produces the **W** matrix

$$\begin{bmatrix} 1 & -4 & 1 & 2 \\ 0 & 1 & -\frac{1}{13} & -\frac{5}{13} \\ 0 & 0 & -\frac{15}{13} & \frac{16}{13} \end{bmatrix}$$

Back substitution produces

$$x_3 = -\frac{16}{15}$$
$$x_2 = -\frac{5}{13} + (\frac{1}{13})(-\frac{16}{15}) = -\frac{91}{195}$$
$$x_1 = 2 + 4(-\frac{91}{195}) - (-\frac{16}{15}) = \frac{234}{195}$$

and this completes the solution.   ∎

The algorithm is easily generalized to the case of an $n \times n$ system.

Algorithm 4.1   Gauss Elimination Method

Step 1 Initialize variables. Set $w_{ij} = a_{ij}$ for $i = 1, \ldots, n$ and $j = 1, \ldots, n$. Set $w_{i,\,n+1} = b_i$ for $i = 1, \ldots, n$. Set $k = 1$.

Step 2 If $w_{kk} = 0$, then locate the smallest p for which $k < p \leq n$ and $w_{pk} \neq 0$. Interchange rows k and p.

Step 3 For $i = k + 1, \ldots, n$, compute $m_{ik} = w_{ik}/w_{kk}$.

Step 4 For $i = k + 1, \ldots, n$ and $j = 1, \ldots, n + 1$, replace w_{ij} by $w_{ij} - m_{ik} w_{kj}$.

Step 5 Replace k by $k + 1$.

Step 6 If $k = n$, then go to step 8.

Step 7 Go to step 2.

Step 8 Compute $x_n = w_{n,\,n+1}/w_{nn}$.

Step 9 For $i = n - 1, \ldots, 1$, compute

$$x_i = \frac{w_{i,\,n+1} - \sum_{k=i+1}^{n} w_{ik} x_k}{w_{ii}}$$

EXAMPLE 4.2   Solve the linear system with augmented matrix

$$\left[\begin{array}{cccc|c}
1 & 1 & -1 & 1 & 7 \\
2 & 2 & 3 & 1 & 6 \\
1 & -2 & 4 & 2 & -1 \\
1 & 0 & 1 & -2 & -6
\end{array}\right]$$

Solution   The first stage of elimination gives

$$\left[\begin{array}{cccc|c}
1 & 1 & -1 & 1 & 7 \\
0 & 0 & 5 & -1 & -8 \\
0 & -3 & 5 & 1 & -8 \\
0 & -1 & 2 & -3 & -13
\end{array}\right]$$

Following step 2 of the algorithm, we must interchange rows 2 and 3 before proceeding. We obtain

$$\left[\begin{array}{cccc|c}
1 & 1 & -1 & 1 & 7 \\
0 & -3 & 5 & 1 & -8 \\
0 & 0 & 5 & -1 & -8 \\
0 & -1 & 2 & -3 & -13
\end{array}\right]$$

Now proceeding with the algorithm, we obtain

$$\begin{bmatrix} 1 & 1 & -1 & 1 & \vdots & 7 \\ 0 & -3 & 5 & 1 & \vdots & -8 \\ 0 & 0 & 5 & -1 & \vdots & -8 \\ 0 & 0 & \frac{1}{3} & -\frac{10}{3} & \vdots & -\frac{31}{3} \end{bmatrix}$$

and finally

$$\begin{bmatrix} 1 & 1 & -1 & 1 & \vdots & 7 \\ 0 & -3 & 5 & 1 & \vdots & -8 \\ 0 & 0 & 5 & -1 & \vdots & -8 \\ 0 & 0 & 0 & -\frac{49}{15} & \vdots & -\frac{147}{15} \end{bmatrix}$$

Back substitution gives $x_4 = 3$, $x_3 = -1$, $x_2 = 2$, and $x_1 = 1$. This completes the solution.   ∎

Theoretically, the Gauss algorithm will always produce the exact solution for the system in a finite number of steps if a unique solution exists. If rational arithmetic is used, then the exact solution will also be produced in practice. Later we will see that when we use the approximate arithmetic of a computer, the solution is no longer exact and can be quite inaccurate.

It is interesting to observe what happens when the system is inconsistent or has infinitely many solutions.

EXAMPLE 4.3   Solve the system with augmented matrix

$$\mathbf{W} = \begin{bmatrix} 1 & 3 & -5 & \vdots & 8 \\ 3 & 2 & -1 & \vdots & 4 \\ 1 & -4 & 9 & \vdots & -12 \end{bmatrix}$$

Solution   The new **W** matrices are

$$\begin{bmatrix} 1 & 3 & -5 & \vdots & 8 \\ 0 & -7 & 14 & \vdots & -20 \\ 0 & -7 & 14 & \vdots & -20 \end{bmatrix} \quad \text{and} \quad \begin{bmatrix} 1 & 3 & -5 & \vdots & 8 \\ 0 & -7 & 14 & \vdots & -20 \\ 0 & 0 & 0 & \vdots & 0 \end{bmatrix}$$

The last is the augmented matrix for a system with one arbitrary variable. Back substitution gives

$$x_2 = \tfrac{20}{7} + 2x_3 \quad \text{and} \quad x_1 = -\tfrac{4}{7} - x_3$$

with x_3 arbitrary. So this system has infinitely many solutions.   ∎

EXAMPLE 4.4 Solve the system with augmented matrix

$$W = \begin{bmatrix} 1 & -1 & 1 & -1 & \vdots & 1 \\ 2 & 0 & 3 & -1 & \vdots & -3 \\ 3 & 1 & -1 & 4 & \vdots & 2 \\ 4 & 2 & 1 & 4 & \vdots & 2 \end{bmatrix}$$

Solution The new augmented matrices are

$$\begin{bmatrix} 1 & -1 & 1 & -1 & \vdots & 1 \\ 0 & 2 & 1 & 1 & \vdots & -5 \\ 0 & 4 & -4 & 7 & \vdots & -1 \\ 0 & 6 & -3 & 8 & \vdots & -2 \end{bmatrix}, \quad \begin{bmatrix} 1 & -1 & 1 & -1 & \vdots & 1 \\ 0 & 2 & 1 & 1 & \vdots & -5 \\ 0 & 0 & -6 & 5 & \vdots & 9 \\ 0 & 0 & -6 & 5 & \vdots & 13 \end{bmatrix}$$

and

$$\begin{bmatrix} 1 & -1 & 1 & -1 & \vdots & 1 \\ 0 & 2 & 1 & 1 & \vdots & -5 \\ 0 & 0 & -6 & 5 & \vdots & 9 \\ 0 & 0 & 0 & 0 & \vdots & 4 \end{bmatrix}$$

The last equation in the last system is $0 = 4$, so the system is clearly inconsistent. ∎

When we work with a calculator and scratch pad, even inconsistent or dependent systems can be handled with the Gauss algorithm. Computer programs can be written to detect these cases and act accordingly. But in this text, we are primarily interested in the problem of accurately determining the solution by machine when a unique solution exists. In the next section, we turn to that problem. What happens when we implement the Gauss algorithm on a machine where the arithmetic is no longer exact because of rounding errors?

■ *Exercise Set 4.2*

1. Use Algorithm 4.1 to find the solution of the system with the given augmented matrix.

a. $\begin{bmatrix} 1 & 2 & 1 & 1 \\ 2 & -1 & 0 & 3 \\ 3 & 1 & -1 & 0 \end{bmatrix}$　　**b.** $\begin{bmatrix} 1 & 2 & 0 & 7 \\ 2 & 1 & 2 & 10 \\ 0 & 2 & 1 & 5 \end{bmatrix}$　　**c.** $\begin{bmatrix} 0.12 & 0.25 & -0.01 \\ 0.15 & 0.12 & 0.18 \end{bmatrix}$

d. $\begin{bmatrix} 1 & 1 & -1 & 1 & 3 \\ 1 & 0 & 1 & 1 & 2 \\ 0 & 1 & 1 & -1 & -1 \\ 2 & -1 & 1 & 0 & 2 \end{bmatrix}$　　**e.** $\begin{bmatrix} 2 & -3 & 2 & 5 & 3 \\ 1 & -1 & 1 & 2 & 1 \\ 3 & 2 & 2 & 1 & 0 \\ 1 & 1 & -3 & -1 & 0 \end{bmatrix}$

2. Determine whether the system with the given augmented matrix is consistent or inconsistent.

a. $\begin{bmatrix} 3 & 1 & 4 & 2 \\ 2 & -1 & 2 & 3 \\ 11 & 7 & 16 & 4 \end{bmatrix}$ b. $\begin{bmatrix} 3 & 1 & 4 & 2 \\ 2 & -1 & 2 & 3 \\ 11 & 7 & 16 & 1 \end{bmatrix}$ c. $\begin{bmatrix} 3 & 1 & 4 & 2 \\ 2 & -1 & 2 & 3 \\ 1 & 1 & 1 & 1 \end{bmatrix}$

3. Solve the homogeneous system $\mathbf{A}\mathbf{x} = \mathbf{0}$ with the given coefficient matrix $\mathbf{A}$.

a. $\begin{bmatrix} 2 & 1 & 3 \\ 1 & 0 & 1 \\ -2 & 1 & 1 \end{bmatrix}$ b. $\begin{bmatrix} 3 & 1 & 4 \\ 2 & -1 & 2 \\ 11 & 7 & 16 \end{bmatrix}$ c. $\begin{bmatrix} 1 & 0 & 1 & 1 \\ 0 & 1 & 1 & 0 \\ 5 & -4 & 4 & -1 \\ 1 & -1 & 1 & -1 \end{bmatrix}$

d. $\begin{bmatrix} 1 & 1 & 1 & 1 \\ 0 & -1 & 2 & 3 \\ 1 & 1 & 1 & 1 \\ 1 & 1 & 1 & 1 \end{bmatrix}$ e. $\begin{bmatrix} 1 & 2 & 3 & 4 \\ 1 & 2 & 3 & 5 \\ -1 & -2 & -3 & 6 \\ 2 & 4 & 6 & 8 \end{bmatrix}$

4. Find the quadratic function $f(x) = ax^2 + bx + c$ for which $f(1) = 4$, $f(-1) = 6$, and $f(2) = 12$.

5. Find the numbers A, B, and C so that

$$\frac{x^2 + 7x + 1}{(x-1)(x+2)^2} = \frac{A}{x-1} + \frac{B}{x+2} + \frac{C}{(x+2)^2}$$

is an identity.

6. Find the intersection point for the planes with equations $x + y + z = 1$, $2x - y - z = 5$, and $x + 2y + 3z = -4$.

4.3   Machine Implementation

Algorithm 4.1 is easily programmed. In this section, we will look at some machine results. Since computers generally use only approximate arithmetic, rounding errors are introduced. These errors are propagated as the computations proceed. So it is not unusual to obtain very inaccurate results, even when the algorithm is correctly programmed. We will return to that problem later; for now, we are primarily concerned with normal run-of-the-mill results.

To illustrate the ease with which the Gauss algorithm can be programmed and used to obtain usable numerical results, we created the following examples with a microcomputer. Similar results could be obtained with virtually any other microcomputer on the market today. Large mainframes are not necessary to do this kind of numerical analysis. If we wanted to solve a large system rapidly, however, then a larger machine would be necessary.

EXAMPLE 4.5   Solve the system with augmented matrix

$$\begin{bmatrix} 1 & 3 & 5 & 7 & 50 \\ 7 & 5 & 3 & 1 & 30 \\ 5 & 7 & 3 & 1 & 32 \\ 3 & 5 & 1 & 7 & 44 \end{bmatrix}$$

Solution   Machine results are as follows.

INPUT MATRIX IS

```
1  3  5  7  50
7  5  3  1  30
5  7  3  1  32
3  5  1  7  44
```

STAGE NUMBER 1

```
1     3     5     7      50
0   -16   -32   -48   -320
0    -8   -22   -34   -218
0    -4   -14   -14   -106
```

STAGE NUMBER 2

```
1     3     5     7      50
0   -16   -32   -48   -320
0     0    -6   -10    -58
0     0    -6    -2    -26
```

STAGE NUMBER 3

```
1     3     5     7      50
0   -16   -32   -48   -320
0     0    -6   -10    -58
0     0     0     8     32
```

GAUSS ELIMINATION COMPLETED.

```
X(4) = 4
X(3) = 3
X(2) = 2
X(1) = 1
```

SOLUTION COMPLETED.   ■

Example 4.5 is interesting because the arithmetic is exact throughout. But such machine output is unusual. It can happen, as we see, but the approximate arithmetic of the machine normally requires some rounding of numbers, and an approximate solution results.

EXAMPLE 4.6   Solve the system

$$x + 2y + 3z = 4$$
$$2x + 4y - 5z = 1$$
$$3x + y - z = 2$$

Solution   Machine results are as follows.

```
INPUT MATRIX IS

1    2    3    4
2    4   -5    1
3    1   -1    2

STAGE NUMBER 1

1    2    3     4
0    0  -11   -7
0   -5  -10  -10

STAGE NUMBER 2

1    2    3     4
0   -5  -10  -10
0    0  -11   -7

GAUSS ELIMINATION COMPLETED.

X(3) = 0.636363636,   X(2) = 0.727272727,   X(1) = 0.636363637

SOLUTION COMPLETED.
```
■

It is wise to program the machine to display the augmented matrix of the system to be sure the solution displayed is actually the solution for the correct system. It is very easy to make a mistake when supplying input to the machine, particularly when the input is typed on a keyboard. In Example 4.6, because of the particular numbers selected for the input matrix, all arithmetic performed in the forward elimination process was exact, and virtually no error was introduced there. In the back substitution part of the routine, however, rational numbers were converted to decimals, so some error was introduced there. In that example, the machine arithmetic was very similar to typical blackboard examples, where all arithmetic is rational and performed by hand.

EXAMPLE 4.7   Solve the system with augmented matrix

$$\begin{bmatrix} 2 & 3 & -1 & 0 & 5 & 17 \\ 1 & 6 & 2 & -3 & -1 & 10 \\ 2 & 0 & 1 & 4 & -2 & -24 \\ 0 & 5 & -2 & 1 & 3 & -2 \\ 3 & 1 & 4 & -2 & 7 & 48 \end{bmatrix}$$

Solution   INPUT MATRIX IS

```
2    3   -1    0    5    17
1    6    2   -3   -1    10
2    0    1    4   -2  -24
0    5   -2    1    3   -2
3    1    4   -2    7    48
```

STAGE NUMBER 1

2	3	−1	0	5	17
0	4.5	2.5	−3	−3.5	1.5
0	−3	2	4	−7	−41
0	5	−2	1	3	−2
0	−3.5	5.5	−2	−0.5	22.5

STAGE NUMBER 2

2	3	−1	0	5	17
0	4.5	2.5	−3	−3.5	1.5
0	0	3.666666	2	−9.333334	−40
0	0	−4.777778	4.333333	6.888888	−3.666667
0	0	7.444444	−4.333334	−3.222223	23.666666

STAGE NUMBER 3

2	3	−1	0	5	17
0	4.5	2.5	−3	−3.5	1.5
0	0	3.666666	2	−9.333334	−40
0	0	0	6.939393	−5.272728	−55.787879
0	0	0	−8.39394	15.727272	104.878787

STAGE NUMBER 4

2	3	−1	0	5	17
0	4.5	2.5	−3	−3.5	1.5
0	0	3.666666	2	−9.333334	−40
0	0	0	6.939393	−5.272728	−55.787879
0	0	0	0	9.349344	37.397379

GAUSS ELIMINATION COMPLETED.

$X(5) = 4$
$X(4) = −5$
$X(3) = 2$
$X(2) = −1$
$X(1) = 1$

SOLUTION COMPLETED. ∎

In Example 4.7, the arithmetic is not so simple. The machine cannot produce exact results for intermediate computations. Nevertheless, the final solution is exactly correct, as you can verify. It is clear that the decimals that do occur in the various stages are merely the decimal equivalents of rational numbers, and no other round-off error seems to be present.

EXAMPLE 4.8 Solve the system

$$0.2x_1 + 0.32x_2 + 0.12x_3 + 0.3x_4 = 0.94$$

$$0.1x_1 + 0.15x_2 + 0.24x_3 + 0.32x_4 = 0.81$$

$$0.2x_1 + 0.24x_2 + 0.46x_3 + 0.36x_4 = 1.26$$

$$0.6x_1 + 0.4x_2 + 0.32x_3 + 0.21x_4 = 1.53$$

Solution   INPUT MATRIX IS

0.2	0.32	0.12	0.3	0.94
0.1	0.15	0.24	0.32	0.81
0.2	0.24	0.46	0.36	1.26
0.6	0.4	0.32	0.21	1.53

STAGE NUMBER 1

0.2	0.32	0.12	0.3	0.94
0	−0.01	0.18	0.17	0.34
0	−0.08	0.34	0.06	0.32
0	−0.56	−0.04	−0.69	−1.29

STAGE NUMBER 2

0.2	0.32	0.12	0.3	0.94
0	−0.01	0.18	0.17	0.34
0	0	−1.100001	−1.300001	−2.400003
0	0	−10.12001	−10.21001	−20.33002

STAGE NUMBER 3

0.2	0.32	0.12	0.3	0.94
0	−0.01	0.18	0.17	0.34
0	0	−1.100001	−1.300001	−2.400003
0	0	0	1.75	1.75

GAUSS ELIMINATION COMPLETED.

$X(4) = 1.000000$
$X(3) = 1.000000$
$X(2) = 1.000003$
$X(1) = 0.999995$

SOLUTION COMPLETED.

   In this example, noticeable rounding errors first appear in stage 2 of the computations. The errors are small, but nevertheless present, even in such a small system. You can verify that the solution of the given system is $x_1 = x_2 = x_3 = x_4 = 1$. This system involves only four equations and four unknowns. Since the accumulated errors of the type illustrated here increase with the number of operations required, it is easy to see that for a large system, the "solution" might very well be useless. Yet systems with coefficients similar to those encountered here are quite common in applications of mathematics requiring the solution of linear systems.   ■

   Errors enter the implementation of the Gauss algorithm in a variety of ways. Rounding errors are almost always present to some extent. Sometimes the machine finds it necessary to round off the initial data. More often, rounding is required as the machine carries out the indicated arithmetic. Even if the forward elimination is carried out exactly, as in Example 4.5, the back substitutions involve arithmetic operations (in particular, subtraction and division) that lead to errors.
   There is another important source of error in this algorithm. The form of the algorithm requires the machine to eliminate terms by subtracting a fixed multiple of

elements of one row in the augmented matrix from a second row. Even in the case of elements in the second row that are not to be eliminated, the new element can be the result of calculating the difference of two numbers that are very close together. The result is a loss of significant digits, which eventually leads to large accumulated errors. This is particularly serious if a diagonal element in the final augmented matrix is the result of such a process. Small diagonal elements with only a few significant digits can produce large errors in the back substitution part of the algorithm.

EXAMPLE 4.9   Solve the system with augmented matrix

$$
\begin{bmatrix}
0.1 & 2 & 100 & 111 \\
1 & 19.99 & 50 & 159.95 \\
1.1 & 22 & 149 & 270
\end{bmatrix}
$$

Solution   INPUT MATRIX IS

0.1	2	100	111
1	19.989999	50	159.949999
1.1	22	149	270

STAGE NUMBER 1

0.1	2	100	111
0	−0.010001	−950	−950.05
0	0	−951	−951

STAGE NUMBER 2

0.1	2	100	111
0	−0.010001	−950	−950.05
0	0	−951	−951

GAUSS ELIMINATION COMPLETED.

X(3) = 1
X(2) = 4.99999255
X(1) = 10.000149

SOLUTION COMPLETED.

Note the small value of w_{22} in the final matrix. Small diagonal elements normally produce errors and must be watched for. In this case, the effect shows in that the value of X(3) is exactly correct, but the absolute error in X(2), the estimate of x_2, is 7.45×10^{-6}. This contributes to the error in X(1), which is 1.49×10^{-4}. The system in this case is only 3×3. In larger systems, errors such as the error in X(2) can produce very large errors in the estimates computed later using the erroneous value. We have, in this case, a loss of significant digits in the computation of x_2 owing to the computation of the difference of two numbers that are very close together. ∎

If you have not yet done so, write a program for implementation of the Gauss algorithm on some machine, and experiment with a variety of systems to gain firsthand experience with machine output. It is important to see for yourself the results produced by the algorithm in the form discussed in Section 4.2. As you will see, there are variations on Algorithm 4.1 that can sometimes reduce errors of the type encountered in the preceding examples. Such variations have been necessitated by the fact that machine implementation of the Gauss algorithm as stated in Section 4.2 sometimes produces undesirably large errors. Some error is present in virtually every machine calculation, of course, but wherever possible, we need to consider algorithm modifications that can reduce the amount of error. We turn to that problem in the next section.

■ Exercise Set 4.3

1. Write a program to execute Algorithm 4.1 on a computer. Use this program to find the solution of the system with the given augmented matrix.

a.
$$\begin{bmatrix} 2 & -3 & 2 & 5 & 3 \\ 1 & -1 & 1 & 2 & 1 \\ 3 & 2 & 2 & 1 & 0 \\ 1 & 1 & -3 & -1 & 0 \end{bmatrix}$$
b.
$$\begin{bmatrix} -14 & 37 & -3 & 1 & -5 \\ 17 & -45 & 4 & -1 & 6 \\ -5 & 13 & -1 & 0 & -2 \\ 18 & -47 & 4 & -1 & 7 \end{bmatrix}$$

c.
$$\begin{bmatrix} 3 & 3 & 4 & -1 \\ -2 & 2 & -4 & 6 \\ 3 & 2.99 & 6 & 1 \end{bmatrix}$$
d.
$$\begin{bmatrix} 0.267 & 5.67 & 5.136 \\ 0.023 & 3.94 & 3.894 \end{bmatrix}$$

e.
$$\begin{bmatrix} \frac{1}{3} & \frac{1}{5} & \frac{1}{7} & 1 \\ \frac{1}{2} & \frac{1}{4} & \frac{1}{3} & 1 \\ \frac{1}{7} & \frac{1}{9} & \frac{1}{11} & 2 \end{bmatrix}$$
f.
$$\begin{bmatrix} 2365023 & 13512 & 213 & 2378322 \\ 199 & 49652 & 3001005 & -2951154 \\ 32165 & 49 & 1990550 & -1958336 \end{bmatrix}$$

2. Use a pocket calculator and Algorithm 4.1 to solve each of the following systems. Round each arithmetic result to the nearest hundredth as you proceed. Compare your "machine" results with the exact solution, which is $x = y = 1$ in each case.
   a. $0.25x - 0.36y = -0.11$
      $0.49x + 0.13y = 0.62$
   b. $0.5x + 0.6y = 1.1$
      $x + 1.21y = 2.21$
   c. $\frac{1}{3}x + \frac{1}{6}y = \frac{1}{2}$ (Hint: use decimal equivalents.)
      $\frac{1}{2}x - \frac{1}{3}y = \frac{1}{6}$

3. Discuss error types and sources expected in machine implementation of the Gauss algorithm.

4. Consider the system with the augmented matrix

$$\begin{bmatrix} 1 & 1 & -1 & 1 \\ 2 & 1 & 1 & 4 \\ 3 & 2 & 0.0001 & 5.0001 \end{bmatrix}$$

Explain why you might expect a machine solution to the system to contain a sizable amount of error. Solve the system by machine using the program developed for exercise 1, and compare the machine solution with the exact solution, which is 1, 1, 1. With the aid of a pocket calculator, go through the application of the Gauss algorithm by hand to see exactly where the trouble arises in the calculation.

4.4 Pivoting Strategies

At each stage in the Gauss algorithm, a **pivot row** must be selected, multiples of which are added to other rows to perform the desired eliminations. Normally, we use the rows in order. If a zero diagonal element a_{kk} is encountered that would lead to division by zero during computation of the multipliers, then we simply interchange row k with another row so that the new a_{kk} is not zero. When this procedure is followed, we say that we are using the **trivial pivoting** strategy. A pivoting strategy is a planned procedure for selecting the pivot row to be used at each stage of the forward elimination process. There are a variety of standard pivoting strategies available that are expected to produce more accurate results from the Gauss algorithm than would be obtained by use of the trivial strategy only.

As a general rule, the arithmetic calculations involved in the Gauss algorithm can be expected to produce more accurate results when the diagonal elements in the resulting upper triangular matrix are larger than the other elements in that matrix, or at least comparable in size. There are other factors that also affect the accuracy of the final estimates, but if the upper triangular matrix produced by forward elimination has any very small elements along the diagonal, these elements probably arose during the elimination procedure from computing the difference of two numbers with approximately the same value, so loss of significant digits has occurred. When these elements are used in the back substitution algorithm to estimate the solution, the error is propagated and amplified.

We can develop a pivoting strategy that produces more accurate results in many such problems. At the kth stage of the elimination process, we compare absolute values of the elements a_{jk} for $k \leq j \leq n$. Then we choose for the pivot row the row with the maximum value for $|a_{jk}|$. This is called the **partial pivoting** strategy. Let us consider a simple system first to understand how this pivoting strategy works.

EXAMPLE 4.10   Solve the system

$$x + 5y + 4z = 23$$

$$2x + 0.5y + z = 6$$

$$5x + 2y - 3z = 0$$

Solution   The augmented matrix is

$$\begin{bmatrix} 1 & 5 & 4 & 23 \\ 2 & 0.5 & 1 & 6 \\ 5 & 2 & -3 & 0 \end{bmatrix}$$

We choose row 3 for the pivot row and interchange row 3 and row 1. After performing the elimination, we obtain

$$\begin{bmatrix} 5 & 2 & -3 & 0 \\ 0 & -0.3 & 2.2 & 6 \\ 0 & 4.6 & 4.6 & 23 \end{bmatrix}$$

In the next step, we use the new row 3 for the pivot because $|4.6| > |-0.3|$. After performing the elimination, we obtain

$$
\begin{bmatrix}
5 & 2 & -3 & 0 \\
0 & 4.6 & 4.6 & 23 \\
0 & 0 & 2.5 & 7.5
\end{bmatrix}
$$

Now, solving for z, y, and x, we find that $z = 3$, $y = 2$, and $x = 1$.   ■

In Example 4.10, all arithmetic was performed exactly, so no round-off errors were introduced, and any pivoting strategy would have produced the exact solution. To see the value of pivoting strategies, we need to use machine arithmetic on a more difficult system, as follows.

EXAMPLE 4.11   Using Gauss elimination with trivial pivoting, solve the system with augmented matrix

$$
\begin{bmatrix}
1 & 100 & 1000 & 10000 & 11101 \\
100 & 9999 & 500 & 1001 & 11600 \\
100 & 1000 & 99000 & 900 & 101000 \\
100 & 1000 & 999 & 1 & 2100
\end{bmatrix}
$$

Solution   The machine output is as follows.

INPUT MATRIX IS

```
  1     100     1000    10000     11101
100    9999      500     1001     11600
100    1000    99000      900    101000
100    1000      999        1      2100
```

STAGE NUMBER 1

```
1       100       1000       10000        11101
0        -1     -99500     -998999     -1098500
0     -9000      -1000     -999100     -1009100
0     -9000     -99001     -999999     -1108000
```

STAGE NUMBER 2

```
1    100        1000        10000                   11101
0     -1      -99500      -998999               -1098500
0      0   895499000    8.9899919E + 09    9.8854909E + 09
0      0   895400999    8.989991E + 09     9.885392E + 09
```

STAGE NUMBER 3

```
1    100        1000        10000                   11101
0     -1      -99500      -998999               -1098500
0      0   895499000    8.9899919E + 09    9.8854909E + 09
0      0           0      982941.234         982942.313
```

X(4) = 1.0000011
X(3) = 0.999988989
X(2) = 1
X(1) = 1.00004578

The exact solution is 1, 1, 1, 1, as you can verify, so the machine solution contains a sizable amount of error. ∎

As the following example shows, partial pivoting produces more accurate results.

EXAMPLE 4.12 Solve the system in Example 4.11 using partial pivoting.

Solution The machine output is as follows.

INPUT MATRIX IS

1	100	1000	10000	11101
100	9999	500	1001	11600
100	1000	99000	900	101000
100	1000	999	1	2100

STAGE NUMBER 1

100	9999	500	1001	11600
0	0.01	995	9989.99	10985
0	−8999	98500	−101	89400
0	−8999	499	−1000	−9500

STAGE NUMBER 2

100	9999	500	1001	11600
0	−8999	98500	−101	89400
0	0	995.109456	9989.98989	10985.0993
0	0	−98001	−899	−98900

STAGE NUMBER 3

100	9999	500	1001	11600
0	−8999	98500	−101	89400
0	0	−98001	−899	−98900
0	0	0	9980.86138	9980.86138

X(4) = 1
X(3) = 1
X(2) = 1
X(1) = 1

As you can see, we were able to improve the accuracy tremendously by changing to partial pivoting. We now have the exact solution. Note that the diagonal elements in the stage 3 matrix are all comparable in size to other elements in the same row. In Example 4.11, this was not the case for rows 2 and 3. ∎

Errors in the solution cannot always be eliminated simply by switching to partial pivoting, however.

EXAMPLE 4.13 Using the Gauss algorithm with trivial pivoting, solve the system with augmented matrix

$$
\begin{bmatrix}
0.2 & 0.32 & 0.12 & 0.3 & 0.94 \\
0.1 & 0.15 & 0.24 & 0.32 & 0.81 \\
0.2 & 0.24 & 0.46 & 0.36 & 1.26 \\
0.6 & 0.4 & 0.32 & 0.21 & 1.53
\end{bmatrix}
$$

Solution   INPUT MATRIX IS

0.2	0.32	0.12	0.3	0.94
0.1	0.15	0.24	0.32	0.81
0.2	0.24	0.46	0.36	1.26
0.6	0.4	0.32	0.21	1.53

STAGE NUMBER 1

0.2	0.32	0.12	0.3	0.94
0	−0.01	0.18	0.17	0.34
0	−0.08	0.34	0.06	0.32
0	−0.56	−0.04	−0.69	−1.29

STAGE NUMBER 2

0.2	0.32	0.12	0.3	0.94
0	−0.01	0.18	0.17	0.34
0	0	−1.100001	−1.300001	−2.400003
0	0	−10.120010	−10.210010	−20.330020

STAGE NUMBER 3

0.2	0.32	0.12	0.3	0.94
0	−0.01	0.18	0.17	0.34
0	0	−1.100001	−1.300001	−2.400003
0	0	0	1.75	1.75

X(4) = 1
X(3) = 1.000000
X(2) = 1.000003
X(1) = 0.999995

The correct solution is 1, 1, 1, 1, so our estimates contain a noticeable amount of error. ■

EXAMPLE 4.14 Solve the system in Example 4.13 using partial pivoting.

Solution   INPUT MATRIX IS

0.2	0.32	0.12	0.3	0.94
0.1	0.15	0.24	0.32	0.81
0.2	0.24	0.46	0.36	1.26
0.6	0.4	0.32	0.21	1.53

STAGE NUMBER 1

0.6	0.4	0.32	0.21	1.53
0	0.0833333	0.1866667	0.285	0.5550001
0	0.1066667	0.3533334	0.29	0.7500001
0	0.1866667	0.0133333	0.23	0.4300001

STAGE NUMBER 2

0.6	0.4	0.32	0.21	1.53
0	0.1866667	0.0133333	0.23	0.4300001
0	0	0.3457143	0.1585714	0.5042857
0	0	0.1807143	0.1823214	0.3630358

STAGE NUMBER 3

0.6	0.4	0.32	0.21	1.53
0	0.1866667	0.0133333	0.23	0.4300001
0	0	0.3457143	0.1585714	0.5042857
0	0	0	0.0994318	0.0994319

$X(4) = 1.000001$
$X(3) = 0.9999994$
$X(2) = 0.9999992$
$X(1) = 1.0000010$ ∎

The solutions obtained in Examples 4.13 and 4.14 are very similar with regard to amount of error. Using partial pivoting did not reduce the error.

We can obtain some increase in accuracy for some systems if we scale the pivot elements before selecting the pivot row. One popular procedure is to select s_i as the maximum of the absolute values of the elements in row i of the coefficient matrix. Then at stage j, we compute numbers

$$w_{ij} = |a_{ij}|/s_i \qquad i = j, \ldots, n$$

We select the pivot row to be the row for which w_{ij} is largest. In all stages of the algorithm, the same scale factors s_i are used. They are not recalculated at each new stage. This procedure is called **scaled partial pivoting**.

EXAMPLE 4.15   Using the Gauss algorithm with scaled partial pivoting, estimate the solution of the system

$$x + 2y + 5z = 10$$

$$2x + y + 4z = 8$$

$$5x + y + 2z = 5$$

Solution   INPUT MATRIX IS

1	2	5	10
2	1	4	8
5	1	2	5

SCALE FACTORS ARE 5, 4 AND 5

STAGE NUMBER 1

5	1	2	5
0	0.6	3.2	6
0	1.799999	4.599999	9

STAGE NUMBER 2

5	1	2	5
0	1.799999	4.599999	9
0	0	1.666666	3

X(3) = 1.8
X(2) = 0.400000001
X(1) = 0.2

This simple example illustrates the procedure. Verify each step by hand calculation. ∎

EXAMPLE 4.16 Solve the system in Example 4.11 using scaled partial pivoting.

Solution   INPUT MATRIX

1	100	1000	10000	11101
100	9999	500	1001	11600
100	1000	99000	900	101000
100	1000	999	1	2100

SCALE FACTORS ARE: 10000, 9999, 99000 AND 1000

STAGE NUMBER 1

100	1000	999	1	2100
0	8999	−499	1000	9500
0	0	98001	899	98900
0	90	990.009999	9999.99	11080

STAGE NUMBER 2

100	1000	999	1	2100
0	8999	−499	1000	9500
0	0	98001	899	98900
0	0	995.000554	9989.98889	10984.9894

STAGE NUMBER 3

100	1000	999	1	2100
0	8999	−499	1000	9500
0	0	995.000554	9989.98889	10984.9894
0	0	0	−983049.096	−983049.096

X(4) = 1
X(3) = 1
X(2) = 1
X(1) = 0.99999999 ∎

The system in Example 4.16 is a difficult one, and Gaussian elimination gives inaccurate results when trivial pivoting is used, as we saw in Example 4.11. Both partial pivoting and scaled partial pivoting produce satisfactory results, however.

EXAMPLE 4.17 Use scaled partial pivoting to estimate the solution of the system given in Example 4.13.

Solution INPUT MATRIX IS

0.2	0.32	0.12	0.3	0.94
0.1	0.15	0.24	0.32	0.81
0.2	0.24	0.46	0.36	1.26
0.6	0.4	0.32	0.21	1.53

STAGE NUMBER 1

0.6	0.4	0.32	0.21	1.53
0	0.0833333	0.1866667	0.285	0.5550001
0	0.1066667	0.3533334	0.29	0.7500001
0	0.1866667	0.0133333	0.23	0.4300001

STAGE NUMBER 2

0.6	0.4	0.32	0.21	1.53
0	0.1066667	0.3533334	0.29	0.7500001
0	0	−0.0893750	0.0584375	−0.0309376
0	0	−0.6050000	−0.2775000	−0.8825001

STAGE NUMBER 3

0.6	0.4	0.32	0.21	1.53
0	0.1066667	0.3533334	0.29	0.7500001
0	0	−0.6050000	−0.2775000	−0.8825001
0	0	0	0.0994318	0.0994318

$X(4) = 1$
$X(3) = 1$
$X(2) = 1$
$X(1) = 0.9999999$

These results are much more accurate than those obtained in Example 4.14 using partial pivoting without scaling. There is a noticeable improvement. These results are also better than those obtained by trivial pivoting. ■

There are other pivoting strategies you might want to try for the systems studied in this section. Look at exercise 5, for example. In many cases, a noticeable improvement can be obtained by using the proper pivoting strategy. So become familiar with different pivoting strategies—at least the ones in this section.

■ *Exercise Set 4.4*

1. Write a machine program that uses scaled partial pivoting with the Gauss algorithm to estimate the solution of a linear system. Use your program to find solutions for each of the systems in exercise 1 of Exercise Set 4.3. Note any differences in results.

2. Using only four-significant-digit arithmetic, solve the system

$$0.0004x + 1.760y = 1.764$$
$$0.3537x - 3.469y = 0.0680$$

First use trivial pivoting and then partial pivoting. Compare results with the exact solution $x = 10$ and $y = 1$.

3. Using a pocket calculator and retaining only six significant digits in each arithmetic operation, solve the system in Example 4.11 using trivial pivoting. The printed results in Example 4.11 were obtained with a microcomputer that used nine-significant-digit arithmetic, so your results should be different but comparable. At which points in the calculation do the largest errors enter?

4. Using the system in Example 4.13, find a solution using your own program with trivial pivoting. Now "scale" the system by multiplying all elements in the matrix by 100 to eliminate the decimals. Solve the system again with trivial pivoting and show that the results are the same as before scaling. Hence, the fact that the entries were decimals is irrelevant and not the reason for the inaccurate results.

5. There are other pivoting strategies used with the Gauss algorithm. One of the better known is called *total pivoting*; see Morris (1983). Find out what is meant by total pivoting, and use that strategy in solving the system with the augmented matrix

$$\begin{bmatrix} 33 & 16 & 72 & 23 \\ 24 & 10 & 57 & 23 \\ 8 & 4 & 17 & 5 \end{bmatrix}$$

Use six-significant-digit arithmetic in all calculations. Now solve the system by machine using trivial pivoting. Compare both of these approximate solutions with the exact solution, which is $-1, -1, 1$. Did total pivoting provide a noticeably more accurate solution?

6. Use total pivoting to solve the system in Example 4.13.

4.5   Forward and Back Substitution Algorithm

Suppose that we have the problem of solving the linear system $\mathbf{Ax} = \mathbf{b}$ for a variety of choices of vector $\mathbf{b}$ while $\mathbf{A}$ remains unchanged. We could, of course, simply solve each system by the Gauss algorithm, one at a time. In doing this, we would find ourselves repeating the arithmetic of the forward elimination procedure over and over unchanged. Vector $\mathbf{b}$ would be transformed to a new vector each time, but the operations on $\mathbf{A}$ would always be the same. It would be desirable to find some way to get the desired results while avoiding the repetition.

Assume for the following that the trivial pivoting strategy is used and no row interchanges are required. During the execution of the Gauss algorithm (Algorithm 4.1), we compute multipliers m_{ij}. Let us agree to store these multipliers in the $\mathbf{W}$ matrix as we proceed, storing m_{ij} in the ith row and jth column, of course, where we had a zero in the procedure described in Algorithm 4.1. The storage space in the machine is no longer wasted, and we have stored valuable information. At the completion of the forward elimination process, we arrive at a **W** matrix of the form

$$
\begin{bmatrix}
w_{11} & w_{12} & w_{13} & \cdots & w_{1n} & w_{1,\,n+1} \\
m_{21} & w_{22} & w_{23} & \cdots & w_{2n} & w_{2,\,n+1} \\
m_{31} & m_{32} & w_{33} & \cdots & w_{3n} & w_{3,\,n+1} \\
\vdots & \vdots & \vdots & & \vdots & \vdots \\
m_{n1} & m_{n2} & m_{n3} & \cdots & w_{nn} & w_{n,\,n+1}
\end{bmatrix}
$$

with the multipliers stored below the diagonal and the usual values above the diagonal. We define matrices as follows:

$$
\mathbf{U} = \begin{bmatrix}
w_{11} & w_{12} & w_{13} & \cdots & w_{1n} \\
0 & w_{22} & w_{23} & \cdots & w_{2n} \\
0 & 0 & w_{33} & \cdots & w_{3n} \\
\vdots & \vdots & \vdots & & \vdots \\
0 & 0 & 0 & \cdots & w_{nn}
\end{bmatrix},
\quad
\hat{\mathbf{b}} = \begin{bmatrix}
w_{1,\,n+1} \\
w_{2,\,n+1} \\
w_{3,\,n+1} \\
w_{4,\,n+1} \\
\vdots \\
w_{n,\,n+1}
\end{bmatrix},
$$

and

$$
\mathbf{L} = \begin{bmatrix}
1 & 0 & 0 & \cdots & 0 \\
m_{21} & 1 & 0 & \cdots & 0 \\
m_{31} & m_{32} & 1 & \cdots & 0 \\
\vdots & \vdots & \vdots & \vdots \\
m_{n1} & m_{n2} & m_{n3} & \cdots & m_{nn}
\end{bmatrix}
$$

So **U** is an upper triangular matrix, and **L** is a lower triangular matrix. The Gauss algorithm automatically computes the elements of these matrices when it is applied to the system $\mathbf{Ax} = \mathbf{b}$.

EXAMPLE 4.18   Find matrixes **L**, **U**, and $\hat{\mathbf{b}}$ for the system with augmented matrix

$$
\mathbf{W} = \begin{bmatrix}
2 & -4 & 1 & -3 \\
3 & 1 & 2 & 11 \\
5 & 3 & 0 & 11
\end{bmatrix}
$$

Solution   We see that $m_{21} = \frac{3}{2}$ and $m_{31} = \frac{5}{2}$. We store these values, and at stage 2, we have

$$
\begin{bmatrix}
2 & -4 & 1 & -3 \\
\frac{3}{2} & 7 & \frac{1}{2} & \frac{31}{2} \\
\frac{5}{2} & 13 & -\frac{5}{2} & \frac{37}{2}
\end{bmatrix}
$$

Now we see that $m_{32} = \frac{13}{7}$, and our next **W** matrix is

$$
\begin{bmatrix}
2 & -4 & 1 & -3 \\
\frac{3}{2} & 7 & \frac{1}{2} & \frac{31}{2} \\
\frac{5}{2} & \frac{13}{7} & -\frac{24}{7} & -\frac{72}{7}
\end{bmatrix}
$$

Hence, the desired matrices are

$$
U = \begin{bmatrix} 2 & -4 & 1 \\ 0 & 7 & \frac{1}{2} \\ 0 & 0 & -\frac{24}{7} \end{bmatrix}, \quad \hat{b} = \begin{bmatrix} -3 \\ \frac{31}{2} \\ -\frac{72}{7} \end{bmatrix}, \quad \text{and} \quad L = \begin{bmatrix} 1 & 0 & 0 \\ \frac{3}{2} & 1 & 0 \\ \frac{5}{2} & \frac{13}{7} & 1 \end{bmatrix} \quad \blacksquare
$$

There is a simple relationship between the matrices $\hat{b}$, b, and L. Following through on the computation of the elements of $\hat{b}$ during the forward elimination procedure, we discover that

$$
\begin{aligned}
\hat{b}_1 &= b_1 \\
\hat{b}_2 &= b_2 - m_{21}\hat{b}_1 \\
\hat{b}_3 &= b_3 - m_{31}\hat{b}_1 - m_{32}\hat{b}_2 \\
&\;\vdots \quad\;\; \vdots \quad\;\; \vdots \quad\;\; \vdots \\
\hat{b}_n &= b_n - m_{n1}\hat{b}_1 - m_{n2}\hat{b}_2 - \cdots - m_{n,\,n-1}\hat{b}_{n-1}
\end{aligned}
$$

Then

$$
\begin{aligned}
b_1 &= \hat{b}_1 \\
b_2 &= m_{21}\hat{b}_1 + \hat{b}_2 \\
b_3 &= m_{31}\hat{b}_1 + m_{32}\hat{b}_2 + \hat{b}_3 \\
&\;\vdots \quad\;\; \vdots \quad\;\; \vdots \\
b_n &= m_{n1}\hat{b}_1 + m_{n2}\hat{b}_2 + \cdots + m_{n,\,n-1}\hat{b}_{n-1} + \hat{b}_n
\end{aligned}
$$

Hence, $b = L\hat{b}$. It follows that $\hat{b} = L^{-1}b$. This means that $\hat{b}$ is the solution of a lower triangular linear system $L\hat{b} = b$.

EXAMPLE 4.19   Compute $\hat{b}$ as the solution of the system $L\hat{b} = b$, using L and b from Example 4.18.

Solution   We have

$$
\begin{bmatrix} 1 & 0 & 0 \\ \frac{3}{2} & 1 & 0 \\ \frac{5}{2} & \frac{13}{7} & 1 \end{bmatrix} \begin{bmatrix} \hat{b}_1 \\ \hat{b}_2 \\ \hat{b}_3 \end{bmatrix} = \begin{bmatrix} -3 \\ 11 \\ 11 \end{bmatrix}
$$

Hence,

$$
\hat{b}_1 = -3 \quad \text{and} \quad 1.5\hat{b}_1 + \hat{b}_2 = 11
$$

So

$$
\hat{b}_2 = 11 - 1.5(-3) = 15.5
$$

$$
2.5\hat{b}_1 + (\tfrac{13}{7})\hat{b}_2 + \hat{b}_3 = 11 \rightarrow \hat{b}_3 = 11 - 2.5(-3) - (\tfrac{13}{7})(\tfrac{31}{2}) = -\tfrac{72}{7} \quad \blacksquare
$$

Given matrices **L** and **U**, the system **Ax** = **b** can be solved by the following procedure.

1. Solve the triangular system **L$\hat{b}$** = **b** for **$\hat{b}$**.
2. Solve the triangular system **Ux** = **$\hat{b}$** for **x**.

In more detail, we have the following algorithm.

Algorithm 4.2   Forward and Back Substitution

To solve **Ax** = **b**, given matrices **L**, **U**, and **b**, do the following:

Step 1 Set $c_1 = b_1$.
Step 2 Compute $c_i = b_i - \sum_{k=1}^{i-1} m_{ik} c_k$ for $i = 2, \dots, n$.
Step 3 Set $x_n = c_n/u_{nn}$.
Step 4 For $i = n - 1, \dots, 1$, set

$$x_i = \frac{c_i - \sum_{k=i+1}^{n} u_{ik} x_k}{u_{ii}}$$

The algorithm in this form is easily programmed. For more details concerning the cases requiring row interchanges, see Conte and de Boor (1972). The extension to this general case is not difficult, but the computational details are tedious and somewhat lengthy. However, it is important to realize that such an extension is possible.

EXAMPLE 4.20   Assume that the system in Example 4.18 has been solved, and the matrices **U** and **L** are known. Solve the new system whose augmented matrix is

$$\begin{bmatrix} 2 & -4 & 1 & \frac{13}{2} \\ 3 & 1 & 2 & -1 \\ 5 & 3 & 0 & -2 \end{bmatrix}$$

Solution   This system has the same **A** matrix as the system in Example 4.18, so we may use the **U** and **L** matrices obtained there in applying Algorithm 4.2. We then have

$$\mathbf{U} = \begin{bmatrix} 2 & -4 & 1 \\ 0 & 7 & \frac{1}{2} \\ 0 & 0 & -\frac{24}{7} \end{bmatrix} \quad \text{and} \quad \mathbf{L} = \begin{bmatrix} 1 & 0 & 0 \\ \frac{3}{2} & 1 & 0 \\ \frac{5}{2} & \frac{13}{7} & 1 \end{bmatrix}$$

Applying Algorithm 4.2, we obtain

$$c_1 = b_1 = \tfrac{13}{2}$$

$$c_2 = b_2 - m_{21}c_1 = -1 - (\tfrac{3}{2})(\tfrac{13}{2}) = -\tfrac{43}{4}$$

$$c_3 = b_3 - m_{31}c_1 - m_{32}c_2$$

$$= -2 - (\tfrac{5}{2})(\tfrac{13}{2}) - (\tfrac{13}{7})(-\tfrac{43}{4}) = \tfrac{12}{7}$$

$$x_3 = \frac{\tfrac{12}{7}}{-\tfrac{24}{7}} = -\frac{1}{2}$$

$$x_2 = \frac{-\tfrac{43}{4} - (\tfrac{1}{2})(-\tfrac{1}{2})}{7} = -\frac{3}{2}$$

$$x_1 = \frac{\tfrac{13}{2} + 4(-\tfrac{3}{2}) - (1)(-\tfrac{1}{2})}{2} = \frac{1}{2}$$

This completes the solution.   ∎

There is an interesting spin-off from Algorithm 4.2. Suppose that linear system $\mathbf{Ax} = \mathbf{b}$ has a unique solution $\mathbf{x}$. Then $\mathbf{x}$ is also the solution of the upper triangular system $\mathbf{Ux} = \hat{\mathbf{b}}$. Since $\hat{\mathbf{b}} = \mathbf{L}^{-1}\mathbf{b}$, then $\mathbf{Ux} = \mathbf{L}^{-1}\mathbf{b}$ also. But this implies that $\mathbf{LUx} = \mathbf{b}$. This suggests (not a proof) that $\mathbf{LU} = \mathbf{A}$. You can verify that this is true for the $\mathbf{L}$, $\mathbf{U}$, and $\mathbf{A}$ matrices in Example 4.20.

It is not hard to see why $\mathbf{LU} = \mathbf{A}$ must be true in general. Each of the elementary row operations used in the Gauss elimination algorithm has a corresponding *elementary matrix* $\mathbf{E}$. Matrix $\mathbf{E}$ is the matrix produced when the desired row operation is performed on the $n \times n$ identity. The row operation desired can be carried out on matrix $\mathbf{A}$ by multiplying on the left by $\mathbf{E}$. Beginning with an augmented matrix $\mathbf{W} = [\mathbf{A} \vdots \mathbf{b}]$, we can view the forward elimination process as being carried out by a succession of multiplications of $\mathbf{W}$ on the left by properly chosen elementary matrices. So $\mathbf{A}$ is transformed successively into matrices $\mathbf{E}_1\mathbf{A}$, $\mathbf{E}_2\mathbf{E}_1\mathbf{A}, \ldots,$ $\mathbf{E}_k \mathbf{E}_{k-1}\mathbf{E}_{k-2} \ldots \mathbf{E}_1\mathbf{A} = \mathbf{U}$. Simultaneously, vector $\mathbf{b}$ is being transformed into $\mathbf{E}_1\mathbf{b}$, $\mathbf{E}_2\mathbf{E}_1\mathbf{b}$, $\mathbf{E}_3\mathbf{E}_2\mathbf{E}_1\mathbf{b}, \ldots,$ and finally $\mathbf{E}_k \mathbf{E}_{k-1}\mathbf{E}_{k-2} \ldots \mathbf{E}_1\mathbf{b} = \hat{\mathbf{b}}$. Suppose that $\mathbf{L}$ is the matrix constructed from the multipliers m_{ij} as described previously which correspond to this particular sequence of row operations. Then

$$\mathbf{L}\hat{\mathbf{b}} = \mathbf{b} \quad \text{and} \quad \mathbf{L}^{-1}\mathbf{b} = \hat{\mathbf{b}}$$

for every vector $\mathbf{b}$. It follows that

$$\mathbf{L}^{-1} = \mathbf{E}_k \mathbf{E}_{k-1}\mathbf{E}_{k-2} \ldots \mathbf{E}_1$$

But then $\mathbf{L}^{-1}\mathbf{A} = \mathbf{U}$ and $\mathbf{A} = \mathbf{LU}$ must follow.

So we see that the Gauss algorithm provides a factorization of matrix $\mathbf{A}$ into a product of a lower triangular matrix and an upper triangular matrix provided that $\mathbf{A}$ can be transformed into $\mathbf{U}$ with no row interchanges. This argument can be generalized to the case involving row interchanges, but **L** is no longer a triangular matrix as defined here. The matrix **U** must be multiplied by a matrix derived from **L** as defined here by permuting the rows to correspond to the row interchanges needed. An alternate way to view this is that if **L** and **U** are constructed by the procedure given in this section and if row interchanges are required, then **LU** is now a matrix derived from **A** by permuting the rows.

■ *Exercise Set 4.5*

1. Find matrices **L** and **U** for each matrix **A**, and verify that **LU** = **A**.

a. $\begin{bmatrix} 1 & 0 & 1 \\ 0 & 1 & 0 \\ 1 & 1 & 1 \end{bmatrix}$
b. $\begin{bmatrix} 1 & -4 & 1 \\ 3 & 1 & 2 \\ 2 & 3 & 5 \end{bmatrix}$

c. $\begin{bmatrix} 1 & 1 & -1 & 1 \\ 2 & 2 & 3 & 1 \\ 1 & -2 & 4 & 2 \\ 1 & 0 & 1 & -2 \end{bmatrix}$
d. $\begin{bmatrix} 1 & -1 & 1 & -1 \\ 2 & 0 & 3 & -1 \\ 3 & 1 & -1 & 4 \\ 4 & 2 & 1 & 4 \end{bmatrix}$

2. Given triangular matrices

$$\mathbf{L} = \begin{bmatrix} 1 & 0 & 0 \\ \frac{2}{3} & 1 & 0 \\ \frac{1}{3} & \frac{1}{5} & 1 \end{bmatrix} \quad \text{and} \quad \mathbf{U} = \begin{bmatrix} 1 & 2 & -3 \\ 0 & 2 & -1 \\ 0 & 0 & 3 \end{bmatrix}$$

a. Use Algorithm 4.2 to solve the system **LU** = **b** with

$$\mathbf{b} = \begin{bmatrix} 1 \\ -1 \\ 1 \end{bmatrix}$$

b. Repeat with

$$\mathbf{b} = \begin{bmatrix} 1 \\ 0 \\ 0 \end{bmatrix}$$

c. Find matrix **A** = **LU**.
d. Find **A**$^{-1}$ by use of **A**$^{-1}$ = **U**$^{-1}$**L**$^{-1}$.

3. A transformation $T : R^3 \rightarrow R^3$ is defined by $T\mathbf{x} = \mathbf{Ax}$, where

$$\mathbf{A} = \begin{bmatrix} 1 & 2 & -3 \\ 2 & 1 & -1 \\ 3 & 4 & 2 \end{bmatrix} \quad \text{and} \quad \mathbf{x} = \begin{bmatrix} x_1 \\ x_2 \\ x_3 \end{bmatrix}$$

Suppose that $\mathbf{y} = \mathbf{Ax}$ and $\mathbf{z} = \mathbf{Ay}$ with

$$\mathbf{z} = \begin{bmatrix} 1 \\ -2 \\ 5 \end{bmatrix}$$

Find points **y** and **x** in R^3.

4.6   Computation of Matrix Inverse

We could use the preceding algorithms to compute the inverse of a nonsingular matrix A. Let $\mathbf{A}^{-1}$ have columns $\mathbf{C}_1, \mathbf{C}_2, \ldots, \mathbf{C}_n$. Then $\mathbf{AA}^{-1}$ has columns $\mathbf{AC}_1, \mathbf{AC}_2, \ldots, \mathbf{AC}_n$. Since $\mathbf{AA}^{-1} = \mathbf{I}$, then $\mathbf{AC}_k = \mathbf{e}_k$ for $k = 1, 2, \ldots, n$, where $\mathbf{e}_k$ is the kth column of the $n \times n$ identity. So the columns of $\mathbf{A}^{-1}$ are solutions of a sequence of n linear systems.

One possible procedure would be to solve the linear system $\mathbf{AC}_1 = \mathbf{e}_1$ for $\mathbf{C}_1$, saving the multipliers, and then apply the forward and back substitution algorithm from the preceding section to determine the remaining columns.

It is more efficient to set up the $n \times 2n$ matrix $\mathbf{W} = [\mathbf{A} \vdots \mathbf{I}]$ and apply elementary row operations to this matrix to convert the left half to the identity $\mathbf{I}$. Clearly, in doing this, the right half will be converted to a matrix with columns $\mathbf{C}_1, \mathbf{C}_2, \ldots, \mathbf{C}_n$; that is, the matrix $\mathbf{A}^{-1}$ will automatically be constructed in the right half as the left half is converted to the identity. This procedure is normally used in linear algebra to find inverses for small matrices. This procedure can be programmed and implemented on a machine.

EXAMPLE 4.21   Find the inverse of the matrix

$$\begin{bmatrix} 2 & 3 & -1 \\ 4 & 4 & -3 \\ -2 & 3 & -1 \end{bmatrix}$$

Solution   First we set up the array

$$\begin{bmatrix} 2 & 3 & -1 & 1 & 0 & 0 \\ 4 & 4 & -3 & 0 & 1 & 0 \\ -2 & 3 & -1 & 0 & 0 & 1 \end{bmatrix}$$

Then, performing elementary row operations, we obtain the following succession of matrices:

$$\begin{bmatrix} 2 & 3 & -1 & 1 & 0 & 0 \\ 0 & -2 & -1 & -2 & 1 & 0 \\ 0 & 6 & -2 & 1 & 0 & 1 \end{bmatrix}$$

$$\begin{bmatrix} 2 & 3 & -1 & 1 & 0 & 0 \\ 0 & -2 & -1 & -2 & 1 & 0 \\ 0 & 0 & -5 & -5 & 3 & 1 \end{bmatrix}$$

$$\begin{bmatrix} 2 & 3 & 0 & 2 & -\frac{3}{5} & -\frac{1}{5} \\ 0 & -2 & 0 & -1 & \frac{2}{5} & -\frac{1}{5} \\ 0 & 0 & 1 & 1 & -\frac{3}{5} & -\frac{1}{5} \end{bmatrix}$$

$$\begin{bmatrix} 2 & 3 & 0 & 2 & -\frac{3}{5} & -\frac{1}{5} \\ 0 & 1 & 0 & \frac{1}{2} & -\frac{1}{5} & \frac{1}{10} \\ 0 & 0 & 1 & 1 & -\frac{3}{5} & -\frac{1}{5} \end{bmatrix}$$

$$\begin{bmatrix} 2 & 0 & 0 & \frac{1}{2} & 0 & -\frac{1}{2} \\ 0 & 1 & 0 & \frac{1}{2} & -\frac{1}{5} & \frac{1}{10} \\ 0 & 0 & 1 & 1 & -\frac{3}{5} & -\frac{1}{5} \end{bmatrix}$$

$$\begin{bmatrix} 1 & 0 & 0 & \frac{1}{4} & 0 & -\frac{1}{4} \\ 0 & 1 & 0 & \frac{1}{2} & -\frac{1}{5} & \frac{1}{10} \\ 0 & 0 & 1 & 1 & -\frac{3}{5} & -\frac{1}{5} \end{bmatrix}$$

So the inverse is

$$\begin{bmatrix} \frac{1}{4} & 0 & -\frac{1}{4} \\ \frac{1}{2} & -\frac{1}{5} & \frac{1}{10} \\ 1 & -\frac{3}{5} & -\frac{1}{5} \end{bmatrix}$$ ■

The machine algorithm is a modified form of the Gauss elimination algorithm (Algorithm 4.1). After the matrix size and matrix elements are entered, the elements of the identity in rows 1 through n and columns $n + 1$ through $2n$ must be defined. Then steps 1–7 of Algorithm 4.1 are performed (forward elimination). Instead of the back substitution steps, we must perform row operations that convert the left half of the array into the identity matrix. After this is accomplished, the elements of the inverse matrix are stored in columns $n + 1$ through $2n$ and can be outputted for inspection or other use.

EXAMPLE 4.22 Typical machine output is the following.

INPUT MATRIX

5.000000	−2.000000	4.000000	1.000000
−2.000000	1.000000	1.000000	−1.000000
4.000000	1.000000	0.000000	0.000000
1.000000	−1.000000	0.000000	1.000000

INVERSE MATRIX IS

0.083333	−0.333333	0.083333	−0.416666
−0.333334	1.333335	0.666667	1.666669
0.083333	0.666667	0.083333	0.583334
−0.416667	1.666668	0.583334	3.083335

The machine used was a personal computer. No intermediate calculations were displayed. A hand calculation shows that the element in the first row and fourth column should be $-\frac{5}{12}$, so the exact decimal equivalent is -0.416667 to six decimals. Most of the other elements in the matrix are recognizable decimal equivalents of the fractions $\frac{1}{3}$, $\frac{1}{12}$, $\frac{7}{12}$, and so on. The element in the last row and last column should be $\frac{37}{12} = 3.083333\ldots$. Some slight error is apparent. In general, the computation is quite accurate and the results are satisfactory. This is typical of results obtained for most small matrixes.   ∎

It can be shown that the number of arithmetic operations involved in computing A^{-1} by the preceding method is proportional to n^3 (see exercise 7). So doubling the size of the matrix increases the amount of arithmetic by a factor of 8. For large matrices, the number of arithmetic operations is so large that round-off error becomes a serious problem. Accurate results for A^{-1} are very difficult to obtain in such cases, and there may be storage problems as well. More sophisticated algorithms are available for such cases, but they will not be discussed here. For further information, consult Fox (1965) or Householder (1970).

It is of value to know that A^{-1} can be determined, and for small matrices, there is no great difficulty. We must consider the question, however, of why we might want to compute A^{-1}. It is seldom, if ever, wise to compute A^{-1} for the purpose of solving a linear system by use of the theoretical formula $x = A^{-1}b$. Much less arithmetic is involved in applying the Gauss algorithm. This is confirmed by counting the number of arithmetic operations required in each case. A detailed analysis of the arithmetic operations required by the Gauss algorithm is given in Chapter 7 of Fox (1965). The number of operations is found to be $n^3/3 + O(n^2)$. Fox also shows that computation of the inverse requires $n^3 + O(n^2)$ operations. To solve the linear system, we must then compute $A^{-1}x$. This requires another $n^2 + O(n)$ operations. Even if the system $Ax = b$ is to be solved repeatedly (an argument often presented in linear algebra textbooks), fewer operations are required if we first solve the system using the Gauss algorithm and then use the forward and back substitution algorithm. Each application of this algorithm requires $n^2 + O(n)$ operations, which is the same as the number required to multiply x by A^{-1}. So the total number of operations is less.

There are some mathematical problems where it is of value to know the elements of A^{-1}. In many such cases, approximate values suffice. In Section 4.9, for example, we need to know the elements of A^{-1} to determine the condition number of a matrix A. Approximate values will suffice, however.

∎ *Exercise Set 4.6*

1. Use the method of Example 4.21 to find the inverse of each matrix.

a. $\begin{bmatrix} 2 & 0 & 3 \\ 10 & 1 & 17 \\ 7 & 12 & -4 \end{bmatrix}$ **b.** $\begin{bmatrix} 1 & 1 & 3 & 0 & 2 \\ 3 & 1 & 0 & 1 & 2 \\ 0 & 1 & 3 & 0 & 2 \\ 5 & 1 & 0 & 0 & 6 \end{bmatrix}$ **c.** $\begin{bmatrix} 1 & -1 & 1 & -1 \\ 0 & 2 & 1 & 3 \\ 0 & 0 & 5 & 4 \\ 0 & 0 & 0 & 2 \end{bmatrix}$

2. In exercise 4 of Exercise Set 4.1, you found that the inverse of A is adj A/det A. Hence, the solution of linear system $Ax = b$ is

$$x = \frac{(\text{adj } A)b}{\det A}$$

This formula is known as *Cramer's rule*. Use Cramer's rule to find the solution of each of the following systems.

a. $2x + 3y - z = -3$ **b.** $2x + 5y - z = 5$
$\quad\ x - 2y + 3z =\ \ 9$ $\quad 4x + 4y - 3z = 1$
$\quad 2x +\ \ y + 4z =\ \ 9$ $\quad -2x + 3y - z = 2$

3. Explain why it would not be feasible to use Cramer's rule as a machine algorithm for solving linear systems. (Nevertheless, Cramer's rule is of theoretical interest and sometimes useful in hand computations.)

4. The matrix

$$\mathbf{H}_n = \begin{bmatrix} 1 & \frac{1}{2} & \frac{1}{3} & \cdots & 1/n \\ \frac{1}{2} & \frac{1}{3} & \frac{1}{4} & \cdots & 1/(n+1) \\ \frac{1}{3} & \frac{1}{4} & \frac{1}{5} & \cdots & 1/(n+2) \\ \vdots & \vdots & \vdots & & \vdots \\ 1/n & 1/(n+1) & 1/(n+2) & \cdots & 1/(2n-1) \end{bmatrix}$$

is called the *Hilbert matrix* of order n. Machine computation of the inverse of a Hilbert matrix usually produces very inaccurate results. Find the inverses for the cases $n = 2$ and $n = 3$ by hand computation.

5. Show that if $\mathbf{P} = (p_{ij})$ is a 3×3 lower triangular matrix and $\mathbf{Q} = (q_{ij})$ is the inverse of $\mathbf{P}$, then $\mathbf{Q}$ is lower triangular with elements. Furthermore, show that

$$q_{ii} = \frac{1}{p_{ii}} \quad \text{for } i = 1, 2, 3$$

$$q_{21} = -\frac{p_{21}}{p_{11}p_{22}}$$

$$q_{31} = \frac{\begin{vmatrix} p_{21} & p_{22} \\ p_{31} & p_{32} \end{vmatrix}}{p_{11}p_{22}p_{33}} \quad \text{and} \quad q_{32} = -\frac{p_{32}}{p_{22}p_{33}}$$

6. Use the result in exercise 5 to find the inverse of each matrix.

a. $\begin{bmatrix} 2 & 0 & 0 \\ 4 & 5 & 0 \\ 2 & -1 & 2 \end{bmatrix}$ **b.** $\begin{bmatrix} -2 & 0 & 0 \\ 4 & 1 & 0 \\ 8 & -2 & 4 \end{bmatrix}$

7. Count the number of arithmetic operations used to find the inverse of the matrix in Example 4.21. Now estimate the number of operations required to find the inverse of an $n \times n$ matrix.

4.7   Errors and Residuals

Given a system $\mathbf{Ax} = \mathbf{b}$ with exact solution $\mathbf{x}$ and approximate solution $\hat{\mathbf{x}}$, how accurate is this approximation? We need to agree on what we mean when we say that vector $\hat{\mathbf{x}}$ is "close to" vector $\mathbf{x}$.

Definition 4.1

The error in the approximation of $\mathbf{x}$ by $\hat{\mathbf{x}}$ is the vector $\mathbf{E} = \hat{\mathbf{x}} - \mathbf{x}$.

We could agree that $\hat{\mathbf{x}}$ is a good approximation of $\mathbf{x}$ if all the components of vector $\mathbf{E}$ are small in absolute value. The absolute value of the kth component is $|\hat{x}_k - x_k|$, and this is the absolute error in the approximation of x_k by $\hat{x}_k$, of course. So if $|E_k|$ is small for each k, or if $\max_k |E_k|$ is small, it is reasonable to claim that $\hat{x}_k$ is a good approximation to x_k for each k.

If $\mathbf{A}$ is a 2×2 matrix then $\mathbf{x}$ and $\hat{\mathbf{x}}$ correspond to points X and $\hat{X}$ in the plane. From a geometric viewpoint, it might seem preferable to consider $\hat{\mathbf{x}}$ as a good approximation to $\mathbf{x}$ if points $\hat{X}$ and X are close together in the plane. Since the distance from point $\hat{X}$ to point X is

$$d(\hat{X}, X) = \sqrt{(\hat{x}_1 - x_1)^2 + (\hat{x}_2 - x_2)^2}$$

we might say that $\hat{\mathbf{x}}$ is a good approximation to $\mathbf{x}$ if $d(\hat{X}, X)$ is small. The extension to 3×3 systems is obvious. In the $n \times n$ case, vector $\mathbf{x}$ corresponds to a point X in R^n, and the distance between points X and Y in R^n is usually defined by

$$d(X, Y) = \sqrt{\sum_{k=1}^{n} (x_k - y_k)^2}$$

So in the $n \times n$ case, it also makes sense to claim that $\hat{\mathbf{x}}$ is a good approximation to $\mathbf{x}$ if the distance $d(X, \hat{X})$ is small.

Theoretically, the use of $\mathbf{E}$ to measure the accuracy of $\hat{\mathbf{x}}$ is satisfactory; but in practice, it is useless, because we do not know the exact solution $\mathbf{x}$, so we cannot actually compute $\mathbf{E}$. At best, we would have to settle for approximate values for $\mathbf{E}$, or bounds on the size of the components of $\mathbf{E}$ developed from some theory. That is a common approach in numerical problems, of course.

An alternate approach is to consider the question of how well vector $\hat{\mathbf{x}}$ satisfies the requirements of the equation. How close is $\mathbf{A}\hat{\mathbf{x}}$ to $\mathbf{b}$?

Definition 4.2

The **residual** for vector $\hat{\mathbf{x}}$ is $\mathbf{R} = \mathbf{A}\hat{\mathbf{x}} - \mathbf{b}$.

We can say that $\hat{\mathbf{x}}$ is a good approximation to the solution if the residual for $\hat{\mathbf{x}}$ is small. This approach does have the obvious advantage that $\mathbf{R}$ can be computed without knowing the exact solution $\mathbf{x}$. We are left with the question of what is meant by saying that $\mathbf{R}$ is small. Small $\mathbf{R}$ could mean either that all components of $\mathbf{R}$ are small in absolute value or that point R in R^n is close to zero, that is, $\sqrt{\sum_{k=1}^{n} R_k^2}$ is small. (R_i is the ith component of $\mathbf{R}$.) Note that

$$\sqrt{\sum_{i=1}^{n} R_i^2} \geq \sqrt{R_k^2} = |R_k|$$

for every k, so the second condition implies the first.

In general, $d(X, Y) \geq |x_k - y_k|$ for any two vectors **X** and **Y**. So in the case of the error vector **E**, we also have $d(X, \hat{X}) \geq \max_k |E_k|$. Hence, small $d(X, \hat{X})$ guarantees that $\max_k |E_k|$ is also small.

It is now clear that before we can decide in any precise manner whether an approximate solution is satisfactory or not, we need to make some agreement about how we measure distance between vectors and size of vectors. We use the concept of a norm to do this.

Definition 4.3

Suppose that a rule has been given for assigning to each vector **x** in R^n a real number $\|\mathbf{x}\|$. We say that such a rule defines a **norm** on R^n if each of the following is satisfied for all vectors **x** and **y** in R^n:

  1. $\|\mathbf{x}\| \geq 0$ and $\|\mathbf{x}\| = 0$ iff $\mathbf{x} = \mathbf{0}$.
  2. $\|r\mathbf{x}\| = |r| \|\mathbf{x}\|$ for each real r.
  3. $\|\mathbf{x} + \mathbf{y}\| \leq \|\mathbf{x}\| + \|\mathbf{y}\|$.

Definition 4.4

If a norm on R^n has been defined, then the distance between vectors **x** and **y** is $\|\mathbf{x} - \mathbf{y}\|$.

A wide variety of vector norms on R^n have been constructed by mathematicians. Two norms in particular are widely used in numerical analysis. The **Euclidean norm** of **x** is defined by

$$\|\mathbf{x}\| = \sqrt{x_1^2 + x_2^2 + \cdots + x_n^2}$$

and the **max norm** of **x** is defined by

$$\|\mathbf{x}\| = \max_{1 \leq k \leq n} |x_k|$$

It can be verified that each of these rules satisfies the three conditions in Definition 4.3. We will use the max norm in this text. It is the most frequently used vector norm in introductory discussions of numerical analysis. The reasons are not surprising. Computations using the Euclidean norm are more complex, and the max norm gives us an easy-to-use practical measure of the size of a vector. In practice, we usually think of $\|\mathbf{x}\|$ as the size of vector **x**.

Now we can say that error **E** is small and $\hat{\mathbf{x}}$ is a good approximation to **x** if $\|\mathbf{E}\|$ is small. If we prefer to use residuals, then we might claim that $\hat{\mathbf{x}}$ is a good approximation to **x** if $\|\mathbf{R}\|$ is small. Unfortunately, as the following example shows, these two approaches to deciding when an approximate solution is a good one do not always agree.

EXAMPLE 4.23   Given the system

$$0.780x_1 + 0.563x_2 = 0.217$$
$$0.913x_1 + 0.659x_2 = 0.254$$

with exact solution $x_1 = 1$ and $x_2 = -1$. Let

$$\hat{\mathbf{x}} = \begin{bmatrix} 0.999 \\ -1.001 \end{bmatrix}$$

Then

$$\mathbf{E} = \begin{bmatrix} -0.001 \\ -0.001 \end{bmatrix} \quad \text{and} \quad \mathbf{R} = \begin{bmatrix} -0.001243 \\ -0.001572 \end{bmatrix}$$

We have $\|\mathbf{E}\| = 0.001$ and $\|\mathbf{R}\| = 0.001572$ if we use the max norm. Both are small, so we might say that $\hat{\mathbf{x}}$ is a reasonably good approximate solution. Suppose, however, that

$$\hat{\mathbf{x}} = \begin{bmatrix} 0.341 \\ -0.087 \end{bmatrix}$$

Then

$$\mathbf{E} = \begin{bmatrix} -0.659 \\ +0.913 \end{bmatrix} \quad \text{and} \quad \mathbf{R} = \begin{bmatrix} -0.000001 \\ 0.000000 \end{bmatrix}$$

Then $\|\mathbf{E}\| = 0.913$ and $\|\mathbf{R}\| = 0.000001$. Using $\|\mathbf{E}\|$ as the criterion, the approximation is terrible. Using $\|\mathbf{R}\|$ as the criterion, the approximation is very good. The value of $\|\mathbf{E}\|$ is the better criterion, of course. In this case, we can compute $\|\mathbf{E}\|$ because we know the exact solution. In general, this is not possible. ■

Example 4.23 was intended to show the dangers in using $\|\mathbf{R}\|$, the readily available criterion, without further knowledge about the relation between $\|\mathbf{E}\|$ and $\|\mathbf{R}\|$. We will explore that relationship in the next section.

■ *Exercise Set 4.7*

1. Prove that $\|\mathbf{x}\| = \max_k |x_k|$ satisfies all the conditions in Definition 4.3.

2. Define $\|\mathbf{x}\| = \sum_{k=1}^{n} |x_k|$. Does this rule satisfy the conditions for a norm on R^n?

Use the max norm in exercises 3–6.

3. Find the distance between each pair of vectors in R^3.

   a. $\begin{bmatrix} 1 \\ 2 \\ 3 \end{bmatrix}$ and $\begin{bmatrix} 1.50 \\ 2.15 \\ 3.25 \end{bmatrix}$ **b.** $\begin{bmatrix} 1.10 \\ 0.98 \\ -2.30 \end{bmatrix}$ and $\begin{bmatrix} 1.05 \\ 0.93 \\ -2.50 \end{bmatrix}$

4. The system with

$$A = \begin{bmatrix} 33 & 16 & 72 \\ 24 & 10 & 57 \\ 8 & 4 & 17 \end{bmatrix} \quad \text{and} \quad b = \begin{bmatrix} 359 \\ 281 \\ 85 \end{bmatrix}$$

has the exact solution

$$x = \begin{bmatrix} -1 \\ 2 \\ 5 \end{bmatrix}$$

Find the error **E** and residual **R** for the approximate solution

$$\hat{x} = \begin{bmatrix} -0.99 \\ 1.98 \\ 5.01 \end{bmatrix}$$

5. Two different machine algorithms produced approximate solutions

$$x = \begin{bmatrix} 0.6097 \\ 0.7637 \\ 0.8951 \end{bmatrix} \quad \text{and} \quad y = \begin{bmatrix} 0.6100 \\ 0.7640 \\ 0.8950 \end{bmatrix}$$

for the system with

$$A = \begin{bmatrix} 9.38 & 3.04 & -2.44 \\ 3.04 & 6.20 & 1.22 \\ -2.44 & 1.22 & 8.45 \end{bmatrix} \quad \text{and} \quad b = \begin{bmatrix} 9.23 \\ 8.20 \\ 3.90 \end{bmatrix}$$

a. Compute the residual for each approximate solution.
b. Which solution do you think is more accurate?

6. A machine solution of the system with

$$A = \begin{bmatrix} \frac{1}{5} & \frac{1}{6} & \frac{1}{7} \\ \frac{1}{6} & \frac{1}{7} & \frac{1}{8} \\ \frac{1}{7} & \frac{1}{8} & \frac{1}{9} \end{bmatrix} \quad \text{and} \quad b = \begin{bmatrix} \frac{47}{210} \\ \frac{31}{168} \\ \frac{79}{504} \end{bmatrix}$$

produced

$$\hat{x} = \begin{bmatrix} 0.99946 \\ 1.00140 \\ -1.00089 \end{bmatrix}$$

The exact solution is 1, 1, -1. Verify this solution. Then find $\|E\|$ and $\|R\|$ for solution $\hat{x}$.

4.8 Errors and Matrix Norms

In this section, we will investigate the relation between the norm (size) of the error vector and the norm of the residual vector for an approximate solution of the system $Ax = b$. The key is the concept of a matrix norm. Suppose that a norm has been selected for vectors in R^n. Let **A** be an $n \times n$ matrix. We define a norm for **A** as follows.

Definition 4.5

The norm of matrix **A** is

$$\|\mathbf{A}\| = \max_{x \neq 0} \frac{\|\mathbf{A}\mathbf{x}\|}{\|\mathbf{x}\|}$$

Given a matrix **A**, the product **Ax** is a new vector in R^n for each choice of nonzero vector **x**. The idea is to compare the size of **Ax** with the size of vector **x**. The norm of the matrix **A** is merely the maximum of the ratios of lengths of vectors **Ax** and **x**. One immediate but useful consequence of this definition is that $\|\mathbf{A}\mathbf{x}\| \leq \|\mathbf{A}\| \cdot \|\mathbf{x}\|$ for every nonzero vector **x**. It can be shown that $\|\mathbf{A}\mathbf{x}\| = \|\mathbf{A}\| \cdot \|\mathbf{x}\|$ for at least one nonzero vector **x** for any choice of the vector norm.

It would be extremely difficult to determine $\|\mathbf{A}\|$ directly from the preceding definition. However, if we choose the max norm as our vector norm, there is an easy way to determine $\|\mathbf{A}\|$ for any matrix **A**.

Note that

$$\|\mathbf{A}\mathbf{x}\| = \max_i \left| \sum_{j=1}^{n} a_{ij} x_j \right| \leq \max_i \sum_{j=1}^{n} |a_{ij}| \cdot |x_j|$$

But $|x_j| \leq \|\mathbf{x}\|$ for every j. So

$$\|\mathbf{A}\mathbf{x}\| \leq \max_i \sum_{j=1}^{n} |a_{ij}| \cdot \|\mathbf{x}\|$$

Let

$$C = \max_i \sum_{j=1}^{n} |a_{ij}|$$

Then

$$\|\mathbf{A}\mathbf{x}\| \leq C\|\mathbf{x}\|$$

Since

$$\frac{\|\mathbf{A}\mathbf{x}\|}{\|\mathbf{x}\|} \leq C$$

then

$$\|\mathbf{A}\| \leq C$$

Now

$$C = \sum_{j=1}^{n} |a_{kj}|$$

for some k. We define nonzero vector $\mathbf{x}$ to be the vector with components $x_j = 1$ if $a_{kj} \geq 0$ and $x_j = -1$ if $a_{kj} < 0$. For this choice of $\mathbf{x}$, we find that

$$\|\mathbf{Ax}\| = \max_i \left| \sum_{j=1}^{n} a_{ij} x_j \right| = \max_i \left| \sum_{j=1}^{n} |a_{ij}| \right|$$

But this is $\max_i \sum_j |a_{ij}| = C$. Since $\|\mathbf{x}\| = 1$ in this case,

$$C = \|\mathbf{Ax}\| = \frac{\|\mathbf{Ax}\|}{\|\mathbf{x}\|}$$

We have shown that $\|\mathbf{A}\| \leq \mathbf{C}$ and that $\|\mathbf{Ax}\|/\|\mathbf{x}\| = C$ for one nonzero vector $\mathbf{x}$. It follows that

$$C = \max_{x \neq 0} \frac{\|\mathbf{Ax}\|}{\|\mathbf{x}\|}$$

Hence,

$$\|\mathbf{A}\| = \max_i \sum_{j=1}^{n} |a_{ij}|$$

EXAMPLE 4.24   Given matrices

$$A = \begin{bmatrix} 4 & 2 & 3 \\ 5 & -4 & 6 \\ 2 & -3 & 1 \end{bmatrix} \quad \text{and} \quad B = \begin{bmatrix} 4 & 0 & -2 & 1 \\ 8 & -5 & 3 & -2 \\ 7 & -3 & 1 & 1 \\ 10 & 0 & -3 & 0 \end{bmatrix}$$

A has norm 15 and $\|\mathbf{B}\| = 18$.   ∎

To see the purpose of this definition of a matrix norm, consider the following. Let $\hat{\mathbf{x}}$ be an approximate solution for the system $\mathbf{Ax} = \mathbf{b}$, and assume that $\mathbf{x}$ is the exact solution. By earlier definitions,

$$\mathbf{E} = \hat{\mathbf{x}} - \mathbf{x} \quad \text{and} \quad \mathbf{R} = \mathbf{A}\hat{\mathbf{x}} - \mathbf{b}$$

So

$$\mathbf{R} = \mathbf{A}(\hat{\mathbf{x}} - \mathbf{x}) = \mathbf{AE}$$

Then

$$\|\mathbf{R}\| \leq \|\mathbf{A}\| \cdot \|\mathbf{E}\|$$

Also, if $\mathbf{A}$ is nonsingular, we may write $\mathbf{E} = \mathbf{A}^{-1}\mathbf{R}$. So $\|\mathbf{E}\| \leq \|\mathbf{A}^{-1}\| \cdot \|\mathbf{R}\|$ also. Clearly, by use of matrix norms, we can relate the norms (sizes) of the error and residual vectors $\mathbf{E}$ and $\mathbf{R}$.

Combining the results

$$\|\mathbf{R}\| \leq \|\mathbf{A}\| \cdot \|\mathbf{E}\| \quad \text{and} \quad \|\mathbf{E}\| \leq \|\mathbf{A}^{-1}\| \cdot \|\mathbf{R}\|$$

we see that

$$\frac{\|\mathbf{R}\|}{\|\mathbf{A}\|} \leq \|\mathbf{E}\| \leq \|\mathbf{A}^{-1}\| \cdot \|\mathbf{R}\|$$

But $\mathbf{A}\mathbf{x} = \mathbf{b}$ and $\mathbf{x} = \mathbf{A}^{-1}\mathbf{b}$, so

$$\|\mathbf{b}\| \leq \|\mathbf{A}\| \cdot \|\mathbf{x}\| \quad \text{and} \quad \|\mathbf{x}\| \leq \|\mathbf{A}^{-1}\| \cdot \|\mathbf{b}\|$$

Then

$$\frac{\|\mathbf{b}\|}{\|\mathbf{A}\|} \leq \|\mathbf{x}\| \leq \|\mathbf{A}^{-1}\| \cdot \|\mathbf{b}\|$$

The inequalities for $\|\mathbf{E}\|$ and $\|\mathbf{x}\|$ jointly imply that

$$\frac{\|\mathbf{R}\|/\|\mathbf{A}\|}{\|\mathbf{A}^{-1}\| \cdot \|\mathbf{b}\|} \leq \frac{\|\mathbf{E}\|}{\|\mathbf{x}\|} \leq \frac{\|\mathbf{A}^{-1}\| \cdot \|\mathbf{R}\|}{\|\mathbf{b}\|/\|\mathbf{A}\|}$$

or

$$\frac{1}{\|\mathbf{A}\| \cdot \|\mathbf{A}^{-1}\|} \cdot \frac{\|\mathbf{R}\|}{\|\mathbf{b}\|} \leq \frac{\|\mathbf{E}\|}{\|\mathbf{x}\|} \leq \|\mathbf{A}\| \cdot \|\mathbf{A}^{-1}\| \frac{\|\mathbf{R}\|}{\|\mathbf{b}\|}$$

This last result relates the *relative residual* $\|\mathbf{R}\|/\|\mathbf{b}\|$ to the *relative error* $\|\mathbf{E}\|/\|\mathbf{x}\|$.

Let $K = \|\mathbf{A}\| \cdot \|\mathbf{A}^{-1}\|$. We call K the **condition number** for matrix $\mathbf{A}$. If K is small, then the relative error cannot be much larger than the relative residual. If K is large, then the relative error may be very large, even though the relative residual is small.

EXAMPLE 4.25   Given the matrix

$$\mathbf{A} = \begin{bmatrix} 2 & 3 & -1 \\ 4 & 4 & -3 \\ -2 & 3 & -1 \end{bmatrix}$$

the inverse is

$$\begin{bmatrix} 0.75 & 0.25 & -0.25 \\ 0 & 0.60 & 0.10 \\ 1 & 0.40 & -0.20 \end{bmatrix}$$

Then $\|\mathbf{A}\| = 11$ and $\|\mathbf{A}^{-1}\| = 1.60$, so $K = 17.6$. Values of K from 10 to 1000 are fairly common.   ■

EXAMPLE 4.26   Given the matrix

$$\mathbf{A} = \begin{bmatrix} 33 & 16 & 72 \\ -24 & -10 & -57 \\ -8 & -4 & -17 \end{bmatrix}$$

we can use the method of Section 4.6 to find the inverse

$$\begin{bmatrix} -9.66572 & -2.66647 & -31.9969 \\ 7.99920 & 2.49983 & 25.4974 \\ 2.66640 & 0.666610 & 8.99910 \end{bmatrix}$$

So $\|\mathbf{A}\| = 121$ and $\|\mathbf{A}^{-1}\| = 44.329$. Then $K = 5360$. This is a fairly large value of K.
                                                                                    ■

Note that given two matrices $\mathbf{A}$ and $\mathbf{B}$ and any nonzero vector $\mathbf{x}$, we have

$$\|(\mathbf{AB})\mathbf{x}\| = \|\mathbf{A}(\mathbf{Bx})\| \le \|\mathbf{A}\| \cdot \|\mathbf{Bx}\|$$

$$\le \|\mathbf{A}\| \cdot \|\mathbf{B}\| \cdot \|\mathbf{x}\|$$

So

$$\frac{\|(\mathbf{AB})\mathbf{x}\|}{\|\mathbf{x}\|} \le \|\mathbf{A}\| \cdot \|\mathbf{B}\|$$

It follows that

$$\|\mathbf{AB}\| \le \|\mathbf{A}\| \cdot \|\mathbf{B}\|$$

Applying this fact to $\mathbf{I} = \mathbf{AA}^{-1}$, we see that

$$1 = \|\mathbf{I}\| \le \|\mathbf{A}\| \cdot \|\mathbf{A}^{-1}\| = K$$

So the condition number of a nonsingular matrix cannot be less than 1. The value for the identity, of course, is 1.

Let us now consider an application of the preceding results to an error estimation problem.

EXAMPLE 4.27   The system $\mathbf{Ax} = \mathbf{b}$ with

$$\mathbf{A} = \begin{bmatrix} 33 & 16 & 72 \\ -24 & -10 & -57 \\ -8 & -4 & -17 \end{bmatrix} \quad \text{and} \quad \mathbf{b} = \begin{bmatrix} 359 \\ -281 \\ -85 \end{bmatrix}$$

is solved by Gauss elimination, arriving at the approximate solution

$$\hat{\mathbf{x}} = \begin{bmatrix} -0.999328 \\ 1.999447 \\ 4.999814 \end{bmatrix}$$

Estimate the error.

Solution   We find that

$$\mathbf{R} = \mathbf{A}\hat{\mathbf{x}} - \mathbf{b} = \begin{bmatrix} -0.000060 \\ -0.000000 \\ -0.000002 \end{bmatrix}$$

Then $\|\mathbf{R}\| = 0.00006$, $\|\mathbf{b}\| = 359$, and $\|\mathbf{R}\|/\|\mathbf{b}\| = 1.67 \times 10^{-7}$. In Example 4.26, we found that the condition number of the matrix $\mathbf{A}$ is 5360. So it follows that

$$\frac{\|\mathbf{E}\|}{\|\mathbf{x}\|} \leq (5360)(1.67 \times 10^{-7}) = 8.958 \times 10^{-4}$$

But $\|\mathbf{x}\|$ is approximately $\|\hat{\mathbf{x}}\| = 4.999814 < 5$. Hence

$$\|\mathbf{E}\| \leq (8.958 \times 10^{-4})(5) = 44.79 \times 10^{-4}$$

The estimates of x_1, x_2, and x_3 should not be in error by more than 4.479×10^{-3} in this case. The exact solution is

$$\mathbf{x} = \begin{bmatrix} -1 \\ 2 \\ 5 \end{bmatrix} \quad \text{and} \quad \mathbf{E} = \begin{bmatrix} 0.000672 \\ -0.000553 \\ -0.000186 \end{bmatrix}$$

So the true $\|\mathbf{E}\| = 6.72 \times 10^{-4}$. This agrees with the theoretical bound, which is 44.79×10^{-4}. In fact, the theoretical bound is very "safe." This is usually the case with theoretical error bounds.   ∎

■ *Exercise Set 4.8*

1. Find the norm of each matrix.

a. $\begin{bmatrix} 2 & 3 & -1 \\ 4 & -2 & 3 \\ 5 & 1 & 6 \end{bmatrix}$ **b.** $\begin{bmatrix} 1 & 0 & -1 & 0 \\ 1 & 2 & -1 & 3 \\ 0 & 1 & 0 & 2 \\ 1 & 1 & 1 & 0 \end{bmatrix}$ **c.** $\begin{bmatrix} 20 & 13 & -10 \\ 33 & 12 & 4 \\ 2 & 6 & -30 \end{bmatrix}$

2. Given

$$A = \begin{bmatrix} 2 & 1 & 3 \\ 5 & 2 & 1 \\ 3 & -2 & 4 \end{bmatrix} \quad \text{and} \quad x = \begin{bmatrix} 2 \\ 1 \\ 3 \end{bmatrix}$$

verify that $\|Ax\| < \|A\| \cdot \|x\|$.

3. Find the condition number for each matrix.

a. $\begin{bmatrix} 11 & 10 \\ 9 & 8 \end{bmatrix}$ **b.** $\begin{bmatrix} 2 & 1 & 3 \\ 3 & -1 & 2 \\ 4 & 1 & 1 \end{bmatrix}$ **c.** $\begin{bmatrix} 1 & 0 & 0 \\ 1 & 1 & 0 \\ 3 & 1 & 1 \end{bmatrix}$

4. Explain why the matrix

$$\begin{bmatrix} 100 & 101 & 102 \\ 99 & 100 & 101 \\ 1 & 1 & 1 \end{bmatrix}$$

does not have a condition number.

5. Suppose that $Ax = rx$ for some real number r and vector $x \neq 0$. Show that $|r| \leq \|A\|$ must follow.

6. Consider the matrices

$$A = \begin{bmatrix} 2 & 1 & 3 \\ 3 & 4 & -1 \\ 4 & 1 & 5 \end{bmatrix} \quad \text{and} \quad b = \begin{bmatrix} 4 \\ -2 \\ 8 \end{bmatrix}$$

a. Find the exact solution of $Ax = b$.
b. Find $\|E\|$ and $\|R\|$ for the approximate solution $x_1 = 0.99$, $x_2 = -0.99$, $x_3 = 1.02$.
c. Find K, the condition number of A.
d. Verify the inequality

$$\frac{1}{K}\frac{\|R\|}{\|b\|} \leq \frac{\|E\|}{\|x\|} \leq K\frac{\|R\|}{\|b\|}$$

7. Consider the system $Ax = b$ with

$$A = \begin{bmatrix} 1 & -2 & 1 \\ 3 & 1 & 2 \\ 5 & 4 & -1 \end{bmatrix} \quad \text{and} \quad b = \begin{bmatrix} 2 \\ 1 \\ 3 \end{bmatrix}$$

and approximate solution

$$\hat{x} = \begin{bmatrix} 1.107 \\ -0.821 \\ -0.750 \end{bmatrix}$$

a. Use the method of Example 4.27 to find a bound for the error in $\hat{x}$.
b. Find the exact solution of the system by hand computation.
c. Find the exact error, and compare with the error bound in part (a).

4.9   Ill-Conditioned Systems

The methods described in this chapter are quite adequate to find approximate solutions for most small or medium-sized systems you might encounter. It is even possible to do some error analysis and estimate errors in the solutions found. In the case of large systems, the methods described present storage problems and due to excessive round-off error may produce inaccurate results. That is not surprising. What is surprising, however, is that even in the case of some small or medium-sized systems, results from Gauss elimination can be very inaccurate.

EXAMPLE 4.28   Solve the system whose augmented matrix is

$$\begin{bmatrix} 1.99 & 4 & 6 & 8 & -4.01 \\ 4 & 6 & 8 & 10 & -4.00 \\ 5 & 7 & 9 & 11 & -4.00 \\ 7 & 8.99 & 10 & 8 & 0.01 \end{bmatrix}$$

Solution   The approximate solution obtained using the Gauss algorithm with trivial pivoting is

$$X(1) = 0.999809$$
$$X(2) = -0.999551$$
$$X(3) = 0.999679$$
$$X(4) = -0.999963$$

The exact solution is 1, −1, 1, −1. The machine solution is very inaccurate. We would expect to get a more accurate solution using scaled partial pivoting. The resulting solution with scaled partial pivoting turns out to be $X(1) = 1.000164$, $X(2) = -1.000384$, $X(3) = 1.000274$, and $X(4) = -1.000054$. Using scaled partial pivoting has produced a slightly better approximation, but there is still a surprising amount of error for such a small system.

An approximate inverse for the **A** matrix in this system is

$$\begin{bmatrix} -100.000 & 300.000 & -200.000 & 0.000 \\ 232.558 & -703.987 & 471.096 & -0.332 \\ -165.116 & 502.473 & -337.193 & 0.664 \\ 32.558 & -99.487 & 67.096 & -0.332 \end{bmatrix}$$

So the norm of $\mathbf{A}^{-1}$ is approximately 1408. The norm of **A** is easily seen to be 33.99. The condition number of **A** is approximately $1408(33.99) = 47,858$. It is usually difficult to find an accurate solution for the system $\mathbf{Ax} = \mathbf{b}$ by the Gauss algorithm if **A** has a large condition number. We say that the system $\mathbf{Ax} = \mathbf{b}$ is *ill-conditioned* in this case.   ∎

It is often believed that small shifts in the parameters or data of a problem cause small (hopefully negligible) shifts in the solution. But as pointed out in Section 3.5, this is not always the case. Some so-called ill-conditioned problems do not obey that rule; small shifts in the parameters or data for the problem may cause sizable changes in the solution.

EXAMPLE 4.29 Solve the system with

$$
A = \begin{bmatrix} \frac{1}{5} & \frac{1}{6} & \frac{1}{7} & \frac{1}{8} \\ \frac{1}{6} & \frac{1}{7} & \frac{1}{8} & \frac{1}{9} \\ \frac{1}{7} & \frac{1}{8} & \frac{1}{9} & \frac{1}{10} \\ \frac{1}{8} & \frac{1}{9} & \frac{1}{10} & \frac{1}{11} \end{bmatrix} \quad \text{and} \quad b = \begin{bmatrix} 2250 \\ 2222 \\ 1920 \\ 1770 \end{bmatrix}
$$

Solution We cannot enter the coefficients exactly, so for the machine solution, coefficients such as $\frac{1}{6}$ and $\frac{1}{7}$ are replaced by six-decimal equivalents. The machine used is a personal computer using single-precision arithmetic. The machine output follows.

INPUT MATRIX IS

0.200000	0.166667	0.142857	0.125000	2550
0.166667	0.142857	0.125000	0.111111	2222
0.142857	0.125000	0.111111	0.100000	1970
0.125000	0.111111	0.100000	0.090909	1770

STAGE NUMBER 3

0.200000	0.166667	0.142857	0.125000	2550
0.000000	0.003968	0.005952	0.006944	96.995850
0.000000	0.000000	0.000141	0.000297	3.057251
0.000000	0.000000	0.000000	0.000005	0.037770

The solution is

X(4) = 8037.461
X(3) = 4788.707
X(2) = 3195.668
X(1) = 1643.024

The exact solution (as you can verify) is $x_1 = 1680$, $x_2 = 3024$, $x_3 = 5040$, and $x_4 = 7920$. The machine solution is very inaccurate.

Changing to double-precision arithmetic sometimes helps in such cases. In this case, the new solution is X(4) = 7915.696729, X(3) = 5048.961983, X(2) = 3018.077820, and X(1) = 1681.224309. There is some improvement, but we still have an undesirably large amount of error. The matrix A is a notorious example of a system matrix leading to an ill-conditioned linear system. Another example similar to this one is found in Fox (1965) together with a lengthy discussion of ill-conditioned systems. ∎

It is important to note that if the system in Example 4.29 were solved using rational arithmetic, with no decimal approximations at any stage, then the Gauss elimination method would produce the exact solution. Our difficulty arises from the fact that machines do not do exact arithmetic. In fact, the initial data sometimes cannot be handled in their exact form by the machine. In ill-conditioned systems, this accumulation of error snowballs and becomes excessive.

It is important to note that even in systems with integer coefficients, ill conditioning may occur. The following example is due to T. S. Wilson.

EXAMPLE 4.30   Solve the system with

$$
A = \begin{bmatrix} 10 & 7 & 8 & 7 \\ 7 & 5 & 6 & 5 \\ 8 & 6 & 10 & 9 \\ 7 & 5 & 9 & 10 \end{bmatrix} \quad \text{and} \quad b = \begin{bmatrix} 32 \\ 23 \\ 33 \\ 31 \end{bmatrix}
$$

Solution   There is no rounding error in the input, of course, and very little during the forward elimination. At that point, the machine produces the coefficient matrix

10.000000	7.000000	8.000000	7.000000	32.000000
0.000000	0.100000	0.400000	0.100000	0.600000
0.000000	0.000000	1.999998	3.000000	4.999995
0.000000	0.000000	0.000000	0.499995	0.499998

Clearly, the element $A(3, 3)$ should be 2.000000, and $A(4, 4)$ should be 0.5000000, but these are normal-sized errors. However, the element $A(4, 5)$ should also be 0.500000, and the fact that the approximate values for 0.5 differ as much as they do introduces an error in $X(4)$ that propagates and adds to the other errors produced. The exact solution to the system is $X(1) = X(2) = X(3) = X(4) = 1$. But the machine produces $X(4) = 1.000007$, $X(3) = 1.000043$, $X(2) = 0.999989$, and $X(1) = 0.999974$. These errors may not seem excessive, but they are in view of the fact that the system is very small and there is no error in the input data. If the earlier **b** vector is replaced by vector

$$
b = \begin{bmatrix} 32.01 \\ 22.99 \\ 32.99 \\ 31.01 \end{bmatrix}
$$

the machine solution becomes $X(4) = 0.8900145$, $X(3) = 1.189976$, $X(2) = 0.180103$, and $X(1) = 1.499937$. A relatively small change in the data has produced a large shift in the solution. Hence, the system is ill-conditioned.   ∎

Let **x** be the exact solution of the system $A\mathbf{x} = \mathbf{b}$. Let $\hat{\mathbf{x}}$ be the exact solution of system $\hat{A}\hat{\mathbf{x}} = \mathbf{b}$, where $\hat{A} = A + \Delta A$. Then $\hat{\mathbf{x}} = \mathbf{x} + \Delta\mathbf{x}$, where

$$
(A + \Delta A)(\mathbf{x} + \Delta\mathbf{x}) = \mathbf{b}
$$

Expanding the left side and using $\mathbf{b} = \mathbf{Ax}$, we find that

$$\Delta\mathbf{A}\ \mathbf{x} + \Delta\mathbf{A}\ \Delta\mathbf{x} + \mathbf{A}\ \Delta\mathbf{x} = 0$$

It follows that

$$\Delta\mathbf{A}\ \hat{\mathbf{x}} = -\mathbf{A}\ \Delta\mathbf{x}$$

so

$$\Delta\mathbf{x} = -\mathbf{A}^{-1}(\Delta\mathbf{A})\hat{\mathbf{x}}$$

Then

$$\|\Delta\mathbf{x}\| \leq \|\mathbf{A}^{-1}\| \cdot \|\Delta\mathbf{A}\| \cdot \|\hat{\mathbf{x}}\|$$

and

$$\frac{\|\Delta\mathbf{x}\|}{\|\hat{\mathbf{x}}\|} \leq \|\mathbf{A}^{-1}\| \cdot \|\mathbf{A}\| \left(\frac{\|\Delta\mathbf{A}\|}{\|\mathbf{A}\|}\right)$$

so

$$\frac{\|\Delta\mathbf{x}\|}{\|\hat{\mathbf{x}}\|} \leq k\left(\frac{\|\Delta\mathbf{A}\|}{\|\mathbf{A}\|}\right)$$

where K is the condition number of matrix $\mathbf{A}$. Of course,

$$\frac{\|\Delta\mathbf{x}\|}{\|\hat{\mathbf{x}}\|} \cong \frac{\|\Delta\mathbf{x}\|}{\|\mathbf{x}\|}$$

which is a measure of the relative error in the solution produced by an error matrix $\Delta\mathbf{A}$ if the coefficient shifts are errors in the input data. If the condition number of $\mathbf{A}$ is large, then small errors or shifts in the matrix coefficients can produce large shifts in the solution or approximate solution.

The equation $\Delta\mathbf{x} = -\mathbf{A}^{-1}(\Delta\mathbf{A})\hat{\mathbf{x}}$ derived here suggests that the shift in the solution could be estimated from matrix $\Delta\mathbf{A}$ and the approximate solution $\hat{\mathbf{x}}$, but this cannot be done in practice because in the case of ill-conditioned systems, neither $\mathbf{A}^{-1}$ nor $\mathbf{x}$ is known with usable precision. Note that $\mathbf{x}$ refers to the exact solution of the system $\mathbf{Ax} = \mathbf{b}$ in this case.

It is well to note in this connection that if system $\mathbf{Ax} = \mathbf{b}$ is ill-conditioned, then the numerical algorithms for estimating the inverse of $\mathbf{A}$ also produce inaccurate results. The following is an example cited in Lanczos (1956).

EXAMPLE 4.31   Find the inverse of the matrix

$$\mathbf{A} = \begin{bmatrix} 33 & 16 & 72 \\ -24 & -10 & -57 \\ -8 & -4 & -17 \end{bmatrix}$$

The machine result for **A** is

$$\begin{bmatrix} -9.666560 & -2.666638 & -31.999650 \\ 7.999911 & 2.499976 & 25.499700 \\ 2.666638 & 0.666659 & 8.999904 \end{bmatrix}$$

The exact inverse using rational arithmetic is

$$\begin{bmatrix} -\frac{29}{3} & -\frac{8}{3} & -32 \\ 8 & \frac{5}{2} & \frac{51}{2} \\ \frac{8}{3} & \frac{2}{3} & 9 \end{bmatrix}$$

The machine results are quite inaccurate for such a small system, compared to the accuracy attained with the same machine and same software on other 3×3 systems. ∎

▪ *Exercise Set 4.9*

1. Solve the system

$$\frac{x}{5} + \frac{y}{6} = \frac{11}{30}$$

$$\frac{x}{6} - \frac{y}{7} = \frac{1}{42}$$

by hand using rational arithmetic. Then solve by machine, using the Gauss algorithm with trivial pivoting. Compare results.

2. Use a machine program to solve the system with the given augmented matrix. Compare the results with the solution of the system in Example 4.8. What do you conclude about the effect on the solution when small changes in matrix **A** or vector **b** are introduced?

a.
$$\begin{bmatrix} 0.2 & 0.32 & 0.12 & 0.3 & 0.94 \\ 0.1 & 0.15 & 0.24 & 0.32 & 0.81 \\ 0.2 & 0.25 & 0.46 & 0.36 & 1.26 \\ 0.6 & 0.4 & 0.32 & 0.21 & 1.52 \end{bmatrix}$$

b.
$$\begin{bmatrix} 0.2 & 0.32 & 0.12 & 0.3 & 0.95 \\ 0.1 & 0.15 & 0.25 & 0.32 & 0.80 \\ 0.2 & 0.24 & 0.46 & 0.36 & 1.25 \\ 0.6 & 0.4 & 0.32 & 0.21 & 1.53 \end{bmatrix}$$

3. It was shown in this section that if $\mathbf{A}$ is replaced by matrix $\mathbf{A} + \Delta\mathbf{A}$, the solution changes to $\mathbf{x} + \Delta\mathbf{x}$, where

$$\frac{\|\Delta\mathbf{x}\|}{\|\mathbf{x}\|} \leq K \frac{\|\Delta\mathbf{A}\|}{\|\mathbf{A}\|}$$

a. Use this inequality and the results of exercise 2a to find a lower bound for the condition number K of matrix $\mathbf{A}$.

b. Find $\mathbf{A}^{-1}$ and compute $K = \|\mathbf{A}\| \cdot \|\mathbf{A}^{-1}\|$ directly. Compare with the results in part (a).

4. Let $\mathbf{A}$ be the matrix in Example 4.30, and choose

$$\Delta\mathbf{A} = \begin{bmatrix} 0 & 0 & 0 & 0 \\ 0 & 0 & 0.001 & 0 \\ 0 & 0.001 & 0 & 0 \\ 0 & 0 & 0 & 0 \end{bmatrix}$$

a. Solve the system $(\mathbf{A} + \Delta\mathbf{A})(\mathbf{x} + \Delta\mathbf{x}) = \mathbf{b}$ for $\mathbf{x} + \Delta\mathbf{x}$, using the same $\mathbf{b}$ as in Example 4.30.

b. Use the result of exercise 4a and the inequality in exercise 3 to find a lower bound for condition number K.

c. Compute $\mathbf{A}^{-1}$, and determine $\|\mathbf{A}\| \cdot \|\mathbf{A}^{-1}\| = K$. Compare with the results in part (b).

d. Do you think a usable estimate of K can be determined by the method used in part (b)?

5. Use the result of exercise 4 in Exercise Set 4.1 to find the inverse of matrix $\mathbf{A}$ in Example 4.31. Now find the condition number. How does this explain the difficulty in computing the inverse of $\mathbf{A}$ by machine?

6. Let

$$\mathbf{A} = \begin{bmatrix} 1 & \frac{1}{2} & \frac{1}{3} \\ \frac{1}{2} & \frac{1}{3} & \frac{1}{4} \\ \frac{1}{3} & \frac{1}{4} & \frac{1}{5} \end{bmatrix}$$

This is the Hilbert matrix of order 3.

a. Use the method of exercise 5 to find the condition number of $\mathbf{A}$.

b. Suppose that the elements a_{13}, a_{22}, and a_{31} are replaced by the decimal 0.33333 at machine input when you are solving system $\mathbf{A}\mathbf{x} = \mathbf{b}$ for some $\mathbf{b}$. Find a bound for the relative error $\|\Delta\mathbf{x}\|/\|\mathbf{x}\|$.

4.10 The Method of Iterative Improvement (Optional)

Let $\hat{\mathbf{x}}$ be an approximate solution for the system $\mathbf{A}\mathbf{x} = \mathbf{b}$. The error in $\hat{\mathbf{x}}$ is defined to be $\mathbf{E} = \hat{\mathbf{x}} - \mathbf{x}$. So

$$\mathbf{A}\mathbf{E} = \mathbf{A}\hat{\mathbf{x}} - \mathbf{A}\mathbf{x} = \mathbf{A}\hat{\mathbf{x}} - \mathbf{b} = \mathbf{R}$$

where $\mathbf{R}$ is the residual for $\hat{\mathbf{x}}$. Then error vector $\mathbf{E}$ is the solution of the system $\mathbf{A}\mathbf{E} = \mathbf{R}$. The preceding facts suggest the following procedure.

Suppose we solve the system $\mathbf{A}\mathbf{x} = \mathbf{b}$ by the Gauss algorithm and obtain approximate solution $\mathbf{x}^{(0)}$. Let $\mathbf{R}^{(0)} = \mathbf{A}\mathbf{x}^{(0)} - \mathbf{b}$. Let $\mathbf{E}^{(0)}$ be an approximate solution of $\mathbf{A}\mathbf{E} = \mathbf{R}^{(0)}$. Then $\mathbf{x}^{(1)} = \mathbf{x}^{(0)} - \mathbf{E}^{(0)}$ should be a better approximation to $\mathbf{x}$. Now let us define $\mathbf{R}^{(1)} = \mathbf{A}\mathbf{x}^{(1)} - \mathbf{b}$ and let $\mathbf{E}^{(1)}$ be an approximation to the solution of system

$\mathbf{AE} = \mathbf{R}^{(1)}$. Then $\mathbf{x}^{(2)} = \mathbf{x}^{(1)} - \mathbf{E}^{(1)}$ should be an even better approximation to the exact solution $\mathbf{x}$ than $\mathbf{x}^{(1)}$ was.

In this manner, we could in theory generate a sequence of approximate solutions $\mathbf{x}^{(n)}$ by repeating the procedure described by

$$\mathbf{x}^{(n)} = \mathbf{x}^{(n-1)} - \mathbf{E}^{(n-1)}$$

with $\mathbf{E}^{(n-1)}$ an approximate solution of $\mathbf{AE} = \mathbf{R}^{(n-1)}$ and $\mathbf{R}^{(n-1)} = \mathbf{Ax}^{(n-1)} - \mathbf{b}$ for $n = 1, 2, 3, \ldots$. This procedure is called the **method of iterative improvement**. We will assume that $\mathbf{x}^{(0)}$ is obtained by use of the Gauss algorithm. Since the linear system $\mathbf{AE} = \mathbf{R}$ must be solved repeatedly, the matrices $\mathbf{L}$ and $\mathbf{U}$ should be saved while $\mathbf{x}^{(0)}$ is being computed. Then solutions $\mathbf{E}^{(k)}$ should be obtained by the forward and back substitution algorithm.

Note that vectors $\mathbf{b}$ and $\mathbf{Ax}^{(n-1)}$ are very close together, so to avoid loss of significant digits, we must compute residuals $\mathbf{R}^{(n-1)}$ by use of double-precision machine arithmetic. The rest of the computation can be done in single precision for added speed of execution. In the examples given in this section, such a procedure is used.

EXAMPLE 4.32   Solve the system with augmented matrix

$$\begin{bmatrix} 100 & 101 & 102 & 101 \\ 101 & 102 & 103 & 101 \\ 200 & 202 & 203 & 201 \end{bmatrix}$$

Solution   The approximate solution from the Gauss algorithm is

$$\mathbf{x}^{(0)} = \begin{bmatrix} -100.054800 \\ 99.054200 \\ 1.000000 \end{bmatrix}$$

with

$$\mathbf{R}^{(0)} = \begin{bmatrix} -0.00073242187500 \\ -0.00128173828125 \\ -0.00146484375000 \end{bmatrix}$$

The corresponding error vector is

$$\mathbf{E}^{(0)} = \begin{bmatrix} -0.05477862 \\ 0.05422901 \\ 0.00000000 \end{bmatrix}$$

yielding

$$\mathbf{x}^{(1)} = \begin{bmatrix} -99.99997 \\ 98.99997 \\ 1.00000 \end{bmatrix}$$

The new residual vector is

$$\mathbf{R}^{(1)} = \begin{bmatrix} -0.000030517578125 \\ -0.000030517578125 \\ -0.000061035156250 \end{bmatrix}$$

The error vector and new vector **x** are

$$\mathbf{E}^{(1)} = \begin{bmatrix} 3.053449E - 05 \\ -3.053432E - 05 \\ 0.0000000 \end{bmatrix} \quad \text{and} \quad \mathbf{x}^{(2)} = \begin{bmatrix} -100 \\ 99 \\ 1 \end{bmatrix}$$

In this example, the method of residual correction works very well. The corrected solution $\mathbf{x}^{(2)}$ after two steps of the iterative improvement method is in fact the exact solution, even though the initial Gauss estimate contained a sizable amount of error.   ∎

EXAMPLE 4.33   Solve the system with augmented matrix

$$\begin{bmatrix} 3.333 & 15920 & -10.333 & 15913 \\ 2.222 & 16.71 & 9.612 & 28.544 \\ 1.5611 & 5.1791 & 1.6852 & 8.4254 \end{bmatrix}$$

Solution   The solution obtained from the Gauss algorithm is

$$\mathbf{x}^{(0)} = \begin{bmatrix} 0.999709 \\ 1.000000 \\ 1.000106 \end{bmatrix}$$

with

$$\mathbf{R}^{(0)} = \begin{bmatrix} -1.673965487043461D - 04 \\ 3.748124476601333D - 04 \\ -2.740567608157107D - 04 \end{bmatrix}$$

The solution of $\mathbf{AE} = \mathbf{R}^{(0)}$ is

$$\mathbf{E}^{(0)} = \begin{bmatrix} -2.902498\mathrm{E} - 04 \\ 1.189766\mathrm{E} - 07 \\ 1.058843\mathrm{E} - 04 \end{bmatrix} \quad \text{and} \quad \mathbf{x}^{(1)} = \begin{bmatrix} 0.9999996 \\ 1 \\ 1 \end{bmatrix}$$

The new residual is

$$\mathbf{R}^{(1)} = \begin{bmatrix} -3.893971552315634\mathrm{D} - 06 \\ 3.049375152386347\mathrm{D} - 07 \\ 8.190869493773789\mathrm{D} - 08 \end{bmatrix}$$

The corresponding error and new $\mathbf{x}$ are

$$\mathbf{E}^{(1)} = \begin{bmatrix} 2.472800\mathrm{E} - 08 \\ -2.326298\mathrm{E} - 10 \\ 2.641274\mathrm{E} - 08 \end{bmatrix} \quad \text{and} \quad \mathbf{x}^{(2)} = \begin{bmatrix} 0.9999996 \\ 1 \\ 1 \end{bmatrix}$$

The exact solution is $x_1 = x_2 = x_3 = 1$, so the improvement in accuracy is apparent. The second iteration provided no improvement in our estimate of $\mathbf{x}$ owing to the small size of the error. The method of iterative improvement works very well in this example also.   ∎

The systems in Examples 4.32 and 4.33 have system matrices $\mathbf{A}$ with large condition numbers. The matrix in Example 4.32 has a condition number 122,210, while the condition number for the matrix in Example 4.33 is approximately 15,997. For systems that are well-conditioned, there is little or no value in applying the iterative improvement method. For systems with large condition numbers, it is advisable to try to improve the accuracy of the solution by using iterative improvement. As seen in the preceding examples, this can lead to a dramatic improvement with only one or two steps of the improvement process.

It is true, however, that for some systems, the method of iterative improvement provides only a small improvement in accuracy or none at all. The following example illustrates that fact.

EXAMPLE 4.34   Solve the system with augmented matrix

$$\begin{bmatrix} 0.2 & 0.32 & 0.12 & 0.3 & 0.94 \\ 0.1 & 0.15 & 0.24 & 0.32 & 0.81 \\ 0.2 & 0.24 & 0.46 & 0.36 & 1.26 \\ 0.6 & 0.4 & 0.32 & 0.21 & 1.53 \end{bmatrix}$$

Solution   Gauss elimination produces the approximate solution

$$\mathbf{x}^{(0)} = \begin{bmatrix} 0.999995 \\ 1.000003 \\ 1.000000 \\ 1.000000 \end{bmatrix}$$

with

$$\mathbf{R}^{(0)} = \begin{bmatrix} 3.844495921612179D - 08 \\ 4.261732966526210D - 08 \\ -5.602839348028965D - 08 \\ -1.545548538928188D - 08 \end{bmatrix}$$

The error vector and new vector **x** are

$$\mathbf{E}^{(0)} = \begin{bmatrix} -4.904515E - 06 \\ 3.196784E - 06 \\ 5.126385E - 07 \\ -2.171311E - 07 \end{bmatrix} \quad \text{and} \quad \mathbf{x}^{(1)} = \begin{bmatrix} 1.000000 \\ 0.9999998 \\ 0.9999998 \\ 1.0000000 \end{bmatrix}$$

The small residual suggests that the solution $\mathbf{x}^{(0)}$ is quite accurate. If the condition number of the coefficient matrix is small (not too far from 1), then the size of the residual is a good indicator of the size of the error. There is some improvement in the new solution estimate $\mathbf{x}^{(1)}$. Can we do better? If we continue, we find that the new residual is

$$\mathbf{R}^{(1)} = \begin{bmatrix} 5.066400188269427D - 09 \\ 2.086160577619012D - 09 \\ 3.552713678800501D - 15 \\ 1.370906943520822D - 08 \end{bmatrix}$$

The solution of $\mathbf{AE} = \mathbf{R}^{(1)}$ is then

$$\mathbf{E}^{(1)} = \begin{bmatrix} 4.2671365E - 08 \\ -2.1867800E - 08 \\ -2.3803120E - 08 \\ 2.1287320E - 08 \end{bmatrix} \quad \text{and} \quad \mathbf{x}^{(2)} = \begin{bmatrix} 1 \\ 1.9999998 \\ 0.9999998 \\ 1 \end{bmatrix}$$

The exact solution to the system is $x_1 = x_2 = x_3 = x_4 = 1$, so all of the approximate solutions have about the same amount of error. Solution vectors $\mathbf{x}^{(k)}$ and error corrections $\mathbf{E}^{(k)}$ were computed using single-precision arithmetic. All residuals were computed using double-precision arithmetic.   ∎

It is important to note that the same machine using double-precision arithmetic for all computations produces the exact solution for this system using only the Gauss algorithm with no residual correction. So the errors produced in Example 4.34 are entirely due to round-off error in the single-precision arithmetic computations involved in the Gauss elimination stage of the algorithm. The error estimates contained in error vectors $\mathbf{E}^{(0)}$ and $\mathbf{E}^{(1)}$ are clearly much too small as a result of this.

When we are using this procedure, a good working rule is to stop the procedure when $\|\mathbf{E}^{(k)}\| < 10^{-t}$ if t-digit arithmetic is being used. In Example 4.34, that would mean to stop after the computation of $\mathbf{E}^{(1)}$. If no significant improvement is obtained after doing this, then the only way to obtain a more accurate solution is to return to the Gauss algorithm with a pivoting strategy and use double- or even multiple-precision arithmetic if that is available.

Ideally, we would like to have $\lim_n \mathbf{x}^{(n)} = \mathbf{x}$ for the sequence of approximations to the solution of the system. As the following argument shows, under certain conditions this should be the case.

Vector $\mathbf{x}^{(0)}$ is an approximate solution of the system $\mathbf{Ax} = \mathbf{b}$. So $\mathbf{x}^{(0)}$ is not equal to $\mathbf{A}^{-1}\mathbf{b}$. It is true, however, that $\mathbf{x}^{(0)} = \mathbf{Cb}$, where $\mathbf{C}$ is an approximation to $\mathbf{A}^{-1}$ that represents the actual operations used in the numerical calculation leading to $\mathbf{x}^{(0)}$.

Vector $\mathbf{x}^{(0)}$ is an approximate solution of the system $\mathbf{Ax} = \mathbf{b}$. So $\mathbf{x}^{(0)}$ is not equal to $\mathbf{A}^{-1}\mathbf{b}$. It is true, however, that $\mathbf{x}^{(0)} = \mathbf{Cb}$, where $\mathbf{C}$ is an approximation to $\mathbf{A}^{-1}$ that represents the actual operations used in the numerical calculation leading to $\mathbf{x}^{(0)}$.

$$\mathbf{x}^{(n+1)} = \mathbf{x}^{(n)} - \mathbf{E}^{(n)} = \mathbf{x}^{(n)} - \mathbf{CR}^{(n)}$$

Then

$$\mathbf{x}^{(n+1)} - \mathbf{x} = \mathbf{x}^{(n)} - \mathbf{x} - \mathbf{CR}^{(n)} = \mathbf{x}^{(n)} - \mathbf{x} - \mathbf{C}(\mathbf{Ax}^{(n)} - \mathbf{b})$$

Or

$$\mathbf{x}^{(n+1)} - \mathbf{x} = \mathbf{x}^{(n)} - \mathbf{x} - \mathbf{C}(\mathbf{Ax}^{(n)} - \mathbf{Ax})$$
$$= \mathbf{x}^{(n)} - \mathbf{x} - \mathbf{CA}(\mathbf{x}^{(n)} - \mathbf{x})$$

It follows that

$$\mathbf{x}^{(n+1)} - \mathbf{x} = (\mathbf{I} - \mathbf{CA})(\mathbf{x}^{(n)} - \mathbf{x})$$

But $\mathbf{e}^{(n)} = \mathbf{x}^{(n)} - \mathbf{x}$ is the error of the nth iteration. So we have shown that $\mathbf{e}^{(n+1)} = \mathbf{He}^{(n)}$, where $\mathbf{H} = \mathbf{I} - \mathbf{CA}$. If $\|\mathbf{H}\| < 1$, then $\|\mathbf{e}^{(n+1)}\| < \|\mathbf{e}^{(n)}\|$, of course. In fact, if $\|\mathbf{H}\| < 1$, it easily follows that $\|\mathbf{e}^{(n)}\|$ approaches zero as n approaches infinity. So this theoretical argument gives a sufficient condition for obtaining $\lim_n \mathbf{x}^{(n)} = \mathbf{x}$ in the iterative improvement method. $\|\mathbf{H}\| < 1$ will do. Note that

$$\mathbf{H} = \mathbf{I} - \mathbf{CA} = (\mathbf{A}^{-1} - \mathbf{C})\mathbf{A}$$

so

$$\|\mathbf{H}\| \le \|\mathbf{A}^{-1} - \mathbf{C}\| \cdot \|\mathbf{A}\|$$

If

$$\|\mathbf{A}^{-1} - \mathbf{C}\| < \frac{1}{\|\mathbf{A}\|}$$

then $\|\mathbf{H}\| < 1$ follows. If $\mathbf{C}$ is sufficiently close to $\mathbf{A}^{-1}$, then $\lim_n \mathbf{x}^{(n)} = \mathbf{x}$ follows. The argument just given tells us how close will suffice. The condition $\|\mathbf{A}^{-1} - \mathbf{C}\| < 1/\|\mathbf{A}\|$ is, of course, only a sufficient condition, not a necessary one.

The method of iterative improvement worked quite well in both Examples 4.32 and 4.33. This indicates that matrix $\mathbf{C}$ in these cases is a reasonably good approximation to $\mathbf{A}^{-1}$—at least good enough to permit an accurate computation of the error vectors. Less satisfactory results were obtained from this method in Example 4.34. The iterations did converge, however. There is a minimal amount of error resulting from random round-off errors in the machine computations, which cannot be overcome by the iterative improvements. Clearly, the heart of the problem lies in the computation of error vectors $\mathbf{E}^{(n)}$. If this cannot be done accurately, then the method will fail to improve the estimate of the solution. In some systems, such as the one in Example 4.34, accurate computation of the error vector may not be possible if the computation of $\mathbf{L}$ and $\mathbf{U}$ is accomplished using single-precision arithmetic. In these cases, we are forced to use higher-precision arithmetic to obtain an accurate numerical solution.

In conclusion, it is interesting to note that the coefficient matrix in Example 4.32 has a condition number of 122,210, the coefficient matrix in Example 4.33 has a condition number of approximately 15,997, and the coefficient matrix in Example 4.34 has a condition number of approximately 29.4. The iterative improvement method seems to work best in the case of systems with large condition numbers, as these examples illustrate.

There are several subroutines in the IMSL MATH/LIBRARY that can be used to solve a linear system. The subroutine LSLRG can be used to obtain the solution without iterative improvement. The subroutine LSARG will solve the system with iterative improvement. If the IMSL software is available, these routines should be used to find solutions of some of the linear systems in the exercises in this chapter. Compare the IMSL solutions with those found by use of your own programs. The real advantage in using the professionally written IMSL software comes in solving larger, more complex problems than those in this chapter, but the linear systems studied here can be used to learn how to use these software tools.

■ *Exercise Set 4.10*

1. Apply the method of residual correction to the system with the given augmented matrix. Use a machine program or calculator to find the corrected solution $\mathbf{x}^{(1)}$. In which cases does the residual correction provide an improved estimate?

a. $\begin{bmatrix} 1.05 & 0.99 & 3.09 \\ 0.99 & 0.98 & 2.96 \end{bmatrix}$ **b.** $\begin{bmatrix} 3 & 15000 & 1000 & 31497 \\ 2 & 20 & 2000 & 3038 \\ 1 & 100 & 900 & 1549 \end{bmatrix}$

c. $\begin{bmatrix} 1 & 0.99 & 0.99 & 0.298 \\ 1 & 1.01 & 1.02 & 0.303 \\ 1 & 1.02 & 1.01 & 0.303 \end{bmatrix}$ **d.** The system in Example 4.13

2. Given the system with augmented matrix

$$\begin{bmatrix} 3 & 4 & 5 & 11.99 \\ 2 & -3 & 1 & 0.01 \\ 1 & -2 & 4 & 2.98 \end{bmatrix}$$

it is easy to see that 1, 1, 1 is an approximate solution.

a. Find the residual $\mathbf{R}$, and estimate the error $\mathbf{E}$ by solving system $\mathbf{AE} = \mathbf{R}$. Then compute $\mathbf{x}^{(1)} = \mathbf{x}^{(0)} - \mathbf{E}$.

b. Solve the system by hand (with calculator) and compare your solution with the solution $\mathbf{x}^{(1)}$ found in part (a).

3. Given the system with augmented matrix

$$\begin{bmatrix} 1 & 0.5 & 0.3333 & 1 \\ 0.5 & 0.3333 & 0.25 & 0 \\ 0.3333 & 0.25 & 0.2 & 0 \end{bmatrix}$$

a student obtained the solution

$$\mathbf{x} = \begin{bmatrix} 9.06175 \\ -36.32324 \\ 30.30265 \end{bmatrix}$$

a. Estimate the error $\mathbf{E}$ for this solution by solving the system $\mathbf{AE} = \mathbf{R}$.

b. How much confidence do you have in $\mathbf{E}$ as an error estimate? Give reasons for your conclusion.

4. Given the system $\mathbf{Ax} = \mathbf{b}$ with

$$\mathbf{A} = \begin{bmatrix} 1 & 1 \\ 1 & -1 \end{bmatrix} \quad \text{and} \quad \mathbf{b} = \begin{bmatrix} 3 \\ 1 \end{bmatrix}$$

a. Verify that

$$\mathbf{C} = \begin{bmatrix} 0.49 & 0.49 \\ 0.49 & -0.49 \end{bmatrix}$$

is an approximation to the inverse of $\mathbf{A}$.

b. Find the corresponding approximate solution $\mathbf{x}^{(0)} = \mathbf{Cb}$.

c. Compute the sequence of vectors $\mathbf{x}^{(k)}$ described by iterations

$$\mathbf{R}^{(k)} = \mathbf{Ax}^{(k)} - \mathbf{b}, \quad \mathbf{E}^{(k)} = \mathbf{CR}^{(k)}, \quad \text{and} \quad \mathbf{x}^{(k+1)} = \mathbf{x}^{(k)} - \mathbf{E}$$

Verify that this sequence converges to the solution of the system.

d. Find matrix $\mathbf{H} = \mathbf{I} - \mathbf{CA}$ in this case. Is $\|\mathbf{H}\| < 1$?
e. Clearly, the exact solution of the system is $x_1 = 2$ and $x_2 = 1$. Use this fact to compute the true errors $\mathbf{e}^{(1)}$ and $\mathbf{e}^{(2)}$ in estimates $\mathbf{x}^{(1)}$ and $\mathbf{x}^{(2)}$. Then verify the theoretical claim that $\|\mathbf{e}^{(2)}\| \leq \|\mathbf{H}\| \cdot \|\mathbf{e}^{(1)}\|$.

4.11   Iterative Methods

In Section 2.5, we found that we can sometimes approximate a root of an equation $f(x) = 0$ by using a fixed-point iteration scheme of the form

$$x_{n+1} = g(x_n)$$

for some properly selected function g. Similarly, a linear system $\mathbf{Ax} = \mathbf{b}$ can sometimes be solved by using a properly selected iteration scheme.

For example, suppose that we want to solve the system

$$10x + 2y - 3z = 5$$

$$2x + 8y + z = 6$$

$$3x - y + 15z = 12$$

Solving the first equation for x, the second equation for y, and the third equation for z, we obtain

$$x = \frac{5 - 2y + 3z}{10}$$

$$y = \frac{6 - 2x - z}{8}$$

$$z = \frac{12 - 3x + y}{15}$$

We select initial estimates $x^{(0)} = 0.5$, $y^{(0)} = 1$, and $z^{(0)} = 1$. Substituting in the equations for x, y, and z, we obtain new estimates

$$x^{(1)} = \frac{5 - 2(1) + 3(1)}{10} = 0.6$$

$$y^{(1)} = \frac{6 - 2(0.5) - 1}{8} = 0.5$$

$$z^{(1)} = \frac{12 - 3(0.5) + 1}{15} = 0.766667$$

Substituting these values in the equations for x, y, and z yields new estimates

$$x^{(2)} = \frac{5 - 2(0.5) + 3(0.766667)}{10} = 0.630000$$

$$y^{(2)} = \frac{6 - 2(0.6) - 0.766667}{8} = 0.504167$$

$$z^{(2)} = \frac{12 - 3(0.6) + 0.5}{15} = 0.713333$$

Continuing this iteration process, we eventually arrive at estimates $x^{(8)} = 0.611831$, $y^{(8)} = 0.508104$, and $z^{(8)} = 0.711507$, with very little change in further estimates. We conclude that our iterations have "converged on" the solution. Using the Gauss algorithm yields the solution $x = 0.611832$, $y = 0.508104$, and $z = 0.711507$, which confirms our conclusion.

The success of this initial effort is to some extent misleading, since a little experimentation shows that a similar approach used on other systems sometimes leads to nonconvergent iterations. We need to develop some theoretical base for developing such iterative methods in order to understand what is happening here. A variety of iterative methods have been developed for linear systems. The introductory discussion in this section will help you understand how such methods are developed and analyzed.

In analogy to the procedure followed in Section 2.5 for the equations $f(x) = 0$, we note that for a system $\mathbf{Ax} = \mathbf{b}$ there are various choices for matrices $\mathbf{G}$ and $\mathbf{F}$ such that the system $\mathbf{x} = \mathbf{Gx} + \mathbf{F}$ is equivalent to the system $\mathbf{Ax} = \mathbf{b}$. For example, if $\mathbf{C}$ is any nonsingular matrix, then the equation

$$\mathbf{x} = \mathbf{x} + \mathbf{C}(\mathbf{Ax} - \mathbf{b})$$

is equivalent to $\mathbf{Ax} = \mathbf{b}$. There is a well-known and useful general procedure for finding equivalents. First we decompose $\mathbf{A}$ into a sum $\mathbf{A} = \mathbf{A}_1 + \mathbf{A}_2$, where $\mathbf{A}_1$ has an easily determined inverse (diagonal or triangular matrix, for example). Then $\mathbf{Ax} = \mathbf{b}$ is equivalent to

$$(\mathbf{A}_1 + \mathbf{A}_2)\mathbf{x} = \mathbf{b}$$

This is equivalent to $\mathbf{A}_1\mathbf{x} = -\mathbf{A}_2\mathbf{x} + \mathbf{b}$, which is equivalent to $\mathbf{x} = -\mathbf{A}_1^{-1}\mathbf{A}_2\mathbf{x} + \mathbf{A}_1^{-1}\mathbf{b}$. So we can always choose $\mathbf{G} = -\mathbf{A}_1^{-1}\mathbf{A}_2$ and $\mathbf{F} = \mathbf{A}_1^{-1}\mathbf{b}$ for matrices $\mathbf{A}_1$ and $\mathbf{A}_2$ with the described properties.

Given the equivalent system $\mathbf{x} = \mathbf{Gx} + \mathbf{F}$, let

$$\mathbf{x}^{(n+1)} = \mathbf{Gx}^{(n)} + \mathbf{F}$$

with $\mathbf{x}^{(0)}$ as an initial estimate of the solution of system $\mathbf{Ax} = \mathbf{b}$. We need to investigate convergence of this iterative algorithm. Let

$$\mathbf{e}^{(k)} = \mathbf{x}^{(k)} - \mathbf{x}$$

Then

$$\mathbf{e}^{(k+1)} = \mathbf{x}^{(k+1)} - \mathbf{x} = (\mathbf{Gx}^{(k)} + \mathbf{F}) - (\mathbf{Gx} + \mathbf{F})$$

So

$$\mathbf{e}^{(k+1)} = \mathbf{G}(\mathbf{x}^{(k)} - \mathbf{x}) = \mathbf{Ge}^{(k)}$$

It follows that if $\|\mathbf{G}\| < 1$, then

$$\|\mathbf{e}^{(k)}\| < \|\mathbf{G}\|^k \cdot \|\mathbf{e}^0\| \to 0 \quad \text{as } k \to \infty$$

So, for every choice of the initial vector, $\|\mathbf{G}\| < 1$ is a sufficient condition for convergence of the iterations to the solution of the system.

Now consider the case where

$$\mathbf{A}_1 = \begin{bmatrix} a_{11} & 0 & \cdots & 0 \\ 0 & a_{22} & \cdots & 0 \\ 0 & 0 & \cdots & 0 \\ 0 & 0 & \cdots & a_{nn} \end{bmatrix} \quad \text{and} \quad \mathbf{A}_2 = \mathbf{A} - \mathbf{A}_1$$

Assume that $a_{ii} \neq 0$ for every i. Then

$$\mathbf{A}_1^{-1} = \begin{bmatrix} 1/a_{11} & 0 & 0 & \cdots & 0 \\ 0 & 1/a_{22} & 0 & \cdots & 0 \\ 0 & 0 & 1/a_{33} & \cdots & 0 \\ \vdots & \vdots & \vdots & & \vdots \\ 0 & 0 & \cdots & & 1/a_{nn} \end{bmatrix}$$

and

$$\mathbf{G} = \begin{bmatrix} 0 & -a_{12}/a_{11} & -a_{13}/a_{11} & \cdots & -a_{1n}/a_{11} \\ -a_{21}/a_{22} & 0 & -a_{23}/a_{22} & \cdots & -a_{2n}/a_{22} \\ \vdots & \vdots & & & \vdots \\ -a_{n-1,1}/a_{n-1,n-1} & & & \cdots & -a_{n-1,n}/a_{n-1,n-1} \\ -a_{n1}/a_{nn} & & & \cdots & 0 \end{bmatrix}$$

while

$$\mathbf{F} = \begin{bmatrix} b_1/a_{11} \\ b_2/a_{22} \\ \vdots \\ b_n/a_{nn} \end{bmatrix}$$

The iterations are represented by the formula

$$x_i^{(k+1)} = -\frac{1}{a_{ii}} \left\{ \sum_{\substack{j=1 \\ j \neq i}}^{n} a_{ij} x_j^{(k)} + b_i \right\} \qquad i = 1, 2, \ldots, n$$

Traditionally, this iterative algorithm for estimating the solution of a linear system has been called the **Jacobi method**. This is the algorithm used to compute the solution of the system discussed at the beginning of this section.

In this case, the norm of **G** is the maximum of the sums

$$\sum_{\substack{j=1 \\ j \neq i}}^{n} \frac{|a_{ij}|}{|a_{ii}|} \qquad i = 1, 2, \ldots, n$$

If

$$|a_{ii}| > \sum_{\substack{j=1 \\ j \neq i}}^{n} |a_{ij}|$$

for every i, then $\|\mathbf{G}\| < 1$, so the iterations will converge. A matrix with this property is said to be *strictly diagonally dominant*. We now have a well-known result due to Jacobi, who proved that the Jacobi iterations will converge to the solution of the system if the matrix **A** is strictly diagonally dominant. The Jacobi iterations do not always converge, so this result is important. It is also important to note that strict diagonal dominance is a sufficient condition for convergence, but not a necessary condition.

EXAMPLE 4.35   Solve the system with augmented matrix

$$\begin{bmatrix} 10 & 3 & 1 & 14 \\ 2 & -10 & 3 & -5 \\ 1 & 3 & 10 & 14 \end{bmatrix}$$

Solution   We use initial estimates $x_1 = x_2 = x_3 = 0$. Here is a summary of the machine output:

Iteration No.	x_1	x_2	x_3
1	1.4	0.5	1.4
5	1.01159	0.9953	1.01159
10	0.999954	1.00013	0.999954
15	1	0.999999	1
16	1	1	1

The iterations converge to the solution as predicted by the strictly diagonal dominant coefficient matrix for the system. Several iterations were required, however.   ■

EXAMPLE 4.36   Solve the system with matrix

$$\begin{bmatrix} 10 & 9 & 0 & 1 \\ 5 & 20 & 13 & -2 \\ 2 & 10 & 15 & 7 \end{bmatrix}$$

Solution   We use initial estimates $x_1 = x_2 = x_3 = 0$ again. Here is a summary of results:

Iteration No.	x_1	x_2	x_3
1	0.1	−0.1	0.466667
5	0.724468	−0.705878	0.857642
10	0.910754	−0.958386	0.943891
15	0.992477	−0.978194	0.998682
20	0.992073	−1.001	0.99414

The iterations are converging on the solution 1, −1, 1. The convergence is very slow. After iteration 15, there is very little improvement. The problem lies with the limitations of machine arithmetic.   ■

Convergence of the Jacobi iterations is often slow. It seems reasonable to try to speed up the iterations by making use of the improved estimates of the x_j's as they are calculated. The iterations are then calculated by the equations

$$x_i^{(k+1)} = \frac{b_i - \sum_{j=1, j \neq i}^{n} a_{ij} z_j}{a_{ii}} \qquad i = 1, 2, \ldots, n$$

where

$$z_j = \begin{cases} x_j^{(k)} & \text{if } j > i \\ x_j^{(k+1)} & \text{if } j < i \end{cases}$$

In this case, the iterative algorithm is known as the **Gauss-Seidel iteration method**.

EXAMPLE 4.37   Apply the Gauss-Seidel iterations to the system in Example 4.36.

Solution   We use initial estimates $x_1^{(0)} = 0.5$, $x_2^{(0)} = -0.5$, and $x_3^{(0)} = 0.5$. Here is a summary of the machine results:

Iteration No.	x_1	x_2	x_3
1	0.55	−0.6925	0.855
5	0.961933	−0.978677	0.99086
10	0.998767	−0.99931	0.999705
15	0.99996	−0.999978	0.99999
20	0.999999	−1	1

The improvement is clear. The Gauss-Seidel iterations converge to the solution even though the Jacobi iterations cannot.   ■

In the Gauss-Seidel iteration method, the matrices A_1 and A_2 can be identified as

$$
A_1 = \begin{bmatrix} a_{11} & 0 & \cdots & 0 \\ a_{21} & a_{22} & \cdots & 0 \\ \vdots & \vdots & & \vdots \\ a_{n-1,1} & & & \\ a_{n1} & a_{n2} & \cdots & a_{nn} \end{bmatrix} \quad \text{and} \quad A_2 = \begin{bmatrix} 0 & a_{12} & \cdots & a_{1n} \\ 0 & 0 & a_{23} & a_{2n} \\ \vdots & \vdots & & \vdots \\ 0 & 0 & \cdots & a_{n-1,n} \\ 0 & 0 & \cdots & 0 \end{bmatrix}
$$

The Gauss-Seidel iteration equations correspond to solving the system

$$
A_1 x^{(k+1)} = -A_2 x^{(k)} + b
$$

from the top down. We determine $x_1^{(k+1)}$ from the first equation; we substitute the result in the second equation and solve for $x_2^{(k+1)}$; and so on. We never need to determine A_1^{-1} explicitly, although that could be done if desired, since A_1 is a lower triangular matrix and easy to invert.

The Gauss-Seidel iterations will also converge whenever matrix A is strictly diagonally dominant, but we expect the Gauss-Seidel iterations to converge faster than the Jacobi iterations. That is frequently the case, but not always.

A detailed investigation of the convergence of the Jacobi and Gauss-Seidel iterations usually employs some theory of eigenvalue problems (see Chapter 6). It can be shown that convergence of iteration algorithms is determined by the maximum size of the absolute values of the eigenvalues of matrix G. A good discussion of some convergence theory of this type can be found in Ortega (1972).

One of the principal reasons for the study of iterative algorithms for solving linear systems is that for large, sparse systems, the Gauss algorithm, or similar so-called **direct methods**, require a large amount of storage, whereas iterative algorithms, such as the Jacobi algorithm, require less storage, are more convenient to use, and produce a final solution that is less affected by round-off error. The usefulness of iterative methods is illustrated in Chapter 11 in connection with the development of algorithms for solving elliptic partial differential equations.

■ *Exercise Set 4.11*

1. Use the Jacobi iteration method to estimate the solution of the system with the given augmented matrix.

a. $\begin{bmatrix} 10 & 1 & -1 & 12 \\ 2 & 15 & 3 & 14 \\ 3 & 5 & 25 & -17 \end{bmatrix}$ b. $\begin{bmatrix} 12 & 5 & 4 & 2 \\ 15 & 20 & -1 & 5 \\ 3 & 5 & 9 & 7 \end{bmatrix}$

2. Estimate solutions for the systems in exercise 1 by using the Gauss-Seidel iteration method.

3. Use the Gauss-Seidel method to estimate the solution of the system with augmented matrix

$$\begin{bmatrix} 10 & 1 & 0 & 10 \\ 1 & 20 & 1 & 5 \\ 0 & 1 & 100 & 1 \end{bmatrix}$$

using the third iteration as your final estimate. Begin with initial estimates $x_1 = 1$, $x_2 = 0$, $x_3 = 0$. Then find the exact solution by hand calculation. Compare results. What does this calculation indicate concerning rate of convergence for the Gauss-Seidel iterations?

4. Repeat exercise 3 using the Jacobi iterations. Compare results with the results of exercise 3.

5. Given the system with augmented matrix

$$\begin{bmatrix} 10 & 9 & 0 & 1 \\ 5 & 20 & 13 & -2 \\ 2 & 10 & 15 & 7 \end{bmatrix}$$

a. Beginning with initial estimates $x_1 = 0.5$, $x_2 = -0.5$, $x_3 = 0.5$ and using a computer program that executes the Gauss-Seidel algorithm, estimate the solution as accurately as you can.
b. Repeat using a computer program that executes the Jacobi algorithm.
c. Which algorithm performs better?

4.12 Applications

In this section, we will consider some applied problems that require the solution of a linear system. Examples will be selected from a variety of disciplines.

Consider the problem of finding the center and radius of a sphere passing through four noncoplanar points (x_i, y_i, z_i), where $i = 1, 2, 3, 4$. If the sphere has radius r and center (x_0, y_0, z_0), then

$$(x_i - x_0)^2 + (y_i - y_0)^2 + (z_i - z_0)^2 = r^2$$

for $i = 1, 2, 3, 4$. It would seem that we must solve a nonlinear system of four equations for the unknowns x_0, y_0, z_0, and r. However, when we expand and simplify the system, we obtain

$$x_i x_0 + y_i y_0 + z_i z_0 + w = \frac{x_i^2 + y_i^2 + z_i^2}{2}$$

for $i = 1, 2, 3, 4$, where $w = -(x_0^2 + y_0^2 + z_0^2 - r^2)/2$. We solve the linear system for x_0, y_0, z_0, and w. Then we compute

$$r = \sqrt{x_0^2 + y_0^2 + z_0^2 + 2w}$$

EXAMPLE 4.38   Find the center and radius of the sphere passing through points $(1, 2, 3)$, $(2, 1, 1)$, $(1, -2, 1)$ and $(-1, 3, 2)$.

Solution   The system we must solve is

$$x_0 + 2y_0 + 3z_0 + w = 7$$

$$2x_0 + y_0 + z_0 + w = 3$$

$$x_0 - 2y_0 + z_0 + w = 3$$

$$-x_0 + 3y_0 + 2z_0 + w = 7$$

Using the Gauss algorithm, we obtain $x_0 = -0.666667$, $y_0 = 0.222222$, $z_0 = 1.555555$, and $w = 2.555556$. So the center is at $(-0.666667, 0.222222, 1.555555)$, and $r = \sqrt{2.9135789 + 2(2.555556)} = 2.832788$. ∎

Note that the procedure illustrated in Example 4.38 has a number of applications outside of mathematics. Consider the problem of locating the center of a sound source known to be equally distant from four observation points in space. The source is at the center of a sphere passing through the four points. Similar reasoning applies in the case of a source of electromagnetic radiation.

Let us now consider a heat flow problem. A window is to be constructed with two glass panes separated by an air space as shown in Figure 4.1. The panes of glass have thicknesses x_1 and x_3 as shown, while the air space is of width x_2. Assume that the heat conductivities of the glass panes are k_1 and k_3, while the conductivity of air is k_2. The temperatures at the boundaries are T_0, T_1, T_2, and T_3 as shown, with

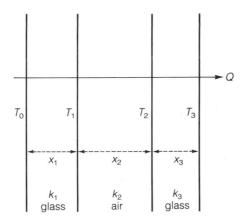

Figure 4.1

$T_0 > T_3$, so the heat flow is from left to right. T_0 might be the inside room temperature, while T_3 would correspond to the outside temperature. Let Q denote the rate of heat flow from left to right through the system. The problem is to find Q from the values given and the temperatures T_0 and T_3.

The flows through the left pane, air space, and right pane, respectively, are

$$k_1 \frac{T_0 - T_1}{x_1}, \quad k_2 \frac{T_1 - T_2}{x_2}, \quad \text{and} \quad k_3 \frac{T_2 - T_3}{x_3}$$

But these are all equal after equilibrium is reached, so we have

$$k_1 \frac{T_0 - T_1}{x_1} = k_2 \frac{T_1 - T_2}{x_2}$$

and

$$k_2 \frac{T_1 - T_2}{x_2} = k_3 \frac{T_2 - T_3}{x_3}$$

These simplify to the system

$$(k_2 x_1 + k_1 x_2)T_1 - k_2 x_1 T_2 = k_1 x_2 T_0$$
$$k_2 x_3 T_1 - (k_2 x_3 + k_3 x_2)T_2 = -k_3 x_2 T_3$$

We must solve this system for T_1 and T_2. Then the heat flow is

$$Q = k_1 \frac{T_0 - T_1}{x_1} \quad \text{or} \quad Q = k_2 \frac{T_1 - T_2}{x_2}$$

Since we only have a 2×2 system, the solution could be found using only a pocket calculator. But it is simple to write a PC program that accepts the thicknesses, conductivities, and inside-outside temperatures and then computes temperatures T_1 and T_2 and the heat flow Q. With such software, an engineer could experiment with various combinations of input parameters and get results quickly. It would be quite easy, in fact, to generate a table of values of Q for various values of T_3 with T_0 fixed.

Let us now consider the case where $T_0 = 65\,°F$ and $T_3 = 30\,°F$ while $x_1 = x_3 = 0.125$ in. and $x_2 = 0.25$ in. Assume conductivities $k_1 = k_3 = 5.39$ and $k_2 = 0.154$ (dimensions are Btu/ft$^2 \cdot$ hr $\cdot °F$/in.). Our system for the intermediate temperatures is

$$1.36675T_1 - 0.01925T_2 = 87.5875$$

$$0.01925T_1 - 1.36675T_2 = -40.425$$

Using the Gauss algorithm, we find that

$$T_1 = 64.5139 \quad \text{and} \quad T_2 = 30.48611$$

The corresponding heat flow calculated from $k_1(T_0 - T_1)/x_1$ is 20.96 Btu/hr/ft^2, and from $k_2(T_1 - T_2)/x_2$ we obtain the same value.

It is also worth noting a very general situation in which linear systems occur. If we want to find extreme values of a quadratic function of n variables

$$y = \sum_{i,\,j=1}^{n} a_{ij} x_i x_j + \sum_{i=1}^{n} b_i x_i$$

where $a_{ij} = a_{ji}$, we find that

$$\frac{\partial y}{\partial x_k} = 2 \sum_{j=1}^{n} a_{kj} x_j + b_k = 0$$

for $k = 1, 2, \ldots, n$. So we must solve a linear system. This is a common problem in applied mathematics.

Next we consider Leontief input/output models used in economic theory. The basic ideas were first introduced by W. Leontief (1966) and have been frequently used by economic modelers; see, for example, Searl (1973).

Suppose we have a collection of n industries, each producing a "good" (steel, electric power, microchips, etc.) used by these industries themselves and possibly other sectors of the economy. We consider production during a fixed time period, such as one year or a quarter. Industry i produces x_i units of its good. Industry k uses a_{ik} units of the output of industry i for each unit of its production. So industry k uses $a_{ik} x_k$ units of the output of industry i during the selected time period. Suppose that the rest of the economy requires d_i units of the output of industry i during this period. Then clearly, we must have

$$x_i = \sum_{k=1}^{n} a_{ik} x_k + d_i \qquad i = 1, 2, \ldots, n$$

In matrix form, $\mathbf{x} = \mathbf{A}\mathbf{x} + \mathbf{d}$, where $\mathbf{x}$ is a "production schedule" and $\mathbf{d}$ is the external demand vector. This equation is equivalent to

$$(\mathbf{I} - \mathbf{A})\mathbf{x} = \mathbf{d}$$

of course. It is usually assumed that $\mathbf{I} - \mathbf{A}$ is nonsingular, and there are economic reasons for believing that this is true. In economic theory, the conclusion is that

$$\mathbf{x} = (\mathbf{I} - \mathbf{A})^{-1}\mathbf{d}$$

so the production schedule is determined by matrix $\mathbf{A}$ when $\mathbf{d}$ is given. In the case of specific modeling problems, such as studies of the U. S. economy, as discussed in Searl (1973), the elements of $\mathbf{A}$ can be estimated, so we can explore production schedules required to produce a given demand vector $\mathbf{d}$. The system $(\mathbf{I} - \mathbf{A})\mathbf{x} = \mathbf{d}$ is quite large in such models, and there is no guarantee that the system will not be ill conditioned. The computation of $\mathbf{x}$ must be done by machine. From the discussion in this chapter, it is clear that accurate values of the solution will not be obtained unless the numerical work is performed carefully. It is also clear that in such numerical work, it would not be wise to compute the inverse of $\mathbf{I} - \mathbf{A}$ and use that to compute $\mathbf{x}$ from $\mathbf{d}$ as suggested by the theory. The forward and back substitution algorithm or an iterative method would be better.

Numerous other applications of linear systems occur in problems in electrical circuit theory, statics of structures, regression analysis in statistical methods, and equilibrium temperatures in plane regions, for example. You can consult texts on electrical circuit theory, engineering mechanics, or statistical methods for details of such applications.

■ *Exercise Set 4.12*

1. In a certain experiment, a sound source is to be placed at a point equally distant from four detectors. An origin and coordinate axes are selected for measurement purposes. With respect to this coordinate system, the detectors have locations (50, 60, 10), (−40, 25, 15), (−35, −30, 12), and (20, −32, 8). Find the coordinates for the sound source location and the distance to each detector.

2. Assume that

$$f(x_1, x_2, x_3) = \sum_{i,j=1}^{3} a_{ij} x_i x_j + \sum_{i=1}^{3} b_i x_i$$

Find extreme values of the function f for each of the following matrices $\mathbf{A}$ and $\mathbf{b}$.

a. $\mathbf{A} = \begin{bmatrix} 2 & 1 & -3 \\ 1 & 3 & 4 \\ -3 & 4 & 5 \end{bmatrix}$ and $\mathbf{b} = \begin{bmatrix} 2 \\ 1 \\ 3 \end{bmatrix}$

b. $\mathbf{A} = \begin{bmatrix} 1 & 4 & -2 \\ 4 & 1 & 5 \\ -2 & 5 & 1 \end{bmatrix}$ and $\mathbf{b} = \begin{bmatrix} 1 \\ 0 \\ 1 \end{bmatrix}$

3. Suppose for a certain three-industry Leontief input/output model we have

$$A = \begin{bmatrix} 0.12 & 0.4 & 0.4 \\ 0.40 & 0.01 & 0.2 \\ 0.48 & 0.15 & 0 \end{bmatrix} \quad \text{and} \quad d = \begin{bmatrix} 9749 \\ 7529 \\ 8878 \end{bmatrix}$$

Find production schedule x.

4. The outside wall of a building is constructed with three layers of construction materials with differing widths and heat conduction properties. The inside layer is 1 in. thick with heat conductivity $k_1 = 6.5$. The outside layer is 2 in. thick with conductivity $k_3 = 8.2$. The middle layer is 3 in. thick with conductivity 0.21. Assume an inside temperature $T_0 = 60°$ F and an outside temperature $T_3 = 25°$ F. Find the heat flow through the wall in Btu/hr/ft^2.

5 NONLINEAR SYSTEMS

In Chapter 4, we studied linear systems of equations. In this chapter, we turn to nonlinear systems. Although nonlinear systems are much more difficult to handle, it is still possible to develop enough theory and methods so that most of the nonlinear systems you will encounter can be understood and solved with reasonable accuracy. In the past, the computational difficulties involved in implementing these methods posed a serious problem, but with the computing tools available today, that problem is not a serious one. In fact, the primary difficulty with nonlinear systems is the lack of a body of theory such as we have for linear systems.

5.1 Introduction

We want to solve systems of the form

$$f_k(x_1, x_2, \ldots, x_n) = 0 \qquad k = 1, \ldots, n$$

In the case of $n = 2$, we have systems such as

$$
\begin{array}{ll}
x^2 + y^2 = 4 & x^2 + y - 11 = 0 \\
x + y = 1 & y^2 + x - 7 = 0
\end{array}
\quad \text{or}
$$

In these cases, we may interpret the desired solutions as coordinates of intersection points for the graphs of the equations involved. In general, we may do that when $n = 2$, although in many cases, the curves involved will be much more complicated than those involved in these two simple examples. When it is feasible to sketch the graphs, we can use this graphical method to determine the number of solutions and their approximate values.

The first fact that becomes clear from consideration of such two-dimensional examples is that a nonlinear system may have no solutions or it may have infinitely many, even in the case of $n = 2$. In general, there seems to be no way to determine the number of solutions analytically from some property, or properties, of the equations in the system. We can do so in special cases, of course. It is usually possible to determine the number of solutions in the case of $n = 2$ either from the graphs or from properties of the functions involved.

For example, if we want to solve the system

$$\frac{x^2}{16} - \frac{y^2}{9} = 1$$

$$x^2 + y^2 = 1$$

we know that points on the hyperbola are all at least three units away from the origin, while points on the circle are all one unit from the origin, so no solutions exist. Similarly, if we want to solve the system

$$y - \tan x = 0$$

$$xy = 1$$

we know that the graph of the hyperbola must cross the graph of the tangent function infinitely many times, so an infinite number of solutions exist.

Also note that in the case where $n = 2$, it is often a simple matter to eliminate one of the variables and convert the system to a single equation. For the system

$$x + y^2 = 7$$

$$x^2 + y = 11$$

we have $y = 11 - x^2$ from the second equation. Substituting in the first equation yields the polynomial equation

$$x^4 - 22x^2 + x + 114 = 0$$

We need real roots of this equation to complete the solution. For some systems with $n > 2$, such a reduction to a single equation, with one unknown, might also be possible, but this is not a feasible approach for systems in general.

In summary, we have a number of difficulties to consider in the general case.

1. We do not know how many solutions exist.
2. After a solution is found, there is no general method for using this information to simplify the problem (such as deflation of polynomials). We simply start from the beginning again to find the next solution (if there is one).
3. Finding initial estimates of solutions can be difficult.

Some theorems have been proved concerning existence and uniqueness of solutions for nonlinear systems, but all are too abstract to be of use in the practical work of finding solutions by machine.

■ *Exercise Set 5.1*

1. For each of the following systems, sketch graphs of the equations to determine how many solutions exist and their approximate locations.

   a. $x^2 + y^2 = 9$
   $x + 2y = 2$

   b. $16x^2 + 9y^2 = 144$
   $x^2 + y^2 = 10$

   c. $ye^x = \sin x$
   $x + 3y = 3$

   d. $4x^2 + 9y^2 = 36$
   $x^2 - y^2 = 1$

   e. $x^{2/3} + y^{2/3} = 9$
   $xy = 1$

   f. $\sqrt{x} + \sqrt{y} = 1$
   $e^x - e^{-x} = 2y$

2. For each of the following systems, determine how many solutions exist by using either graphs or analytical reasoning.

   a. $x^2 + y^2 = 1$
   $x^3 - y = 0$

   b. $|x| + |y| = 10$
   $xy = 1$

   c. $ye^x = 2$
   $16x^2 + 4y^2 = 64$

   d. $y \cot x = 1$
   $xe^y = 1$

   e. $(x^2 + 1)y = x$
   $10y - 2x = 1$

   f. $xy - x^2 = 1$
   $ye^{x^2} = 1$

3. Show that the system

$$x^2 + y^2 + z^2 = 1$$

$$z = 2\sqrt{x^2 + y^2}$$

$$y^2 - z^2 = 1$$

   has no solutions.

4. Suppose that $P(x)$ and $Q(x)$ are polynomials, and each has degree $n \geq 1$. Show that the nonlinear system

$$y = P(x)$$

$$y = Q(x)$$

   cannot have more than n solutions.

5. Consider the equation

$$x^4 + y^4 + 2x^2y^2 + 2xy - 9x^2 - 9y^2 + 25 = 0$$

   a. Show (by completing the square on properly grouped terms) that the given equation has the same solutions as the equation

$$(x^2 + y^2 - 5)^2 + (x + y)^2 = 0$$

   b. Explain why the solutions are solutions of the system

$$x^2 + y^2 = 5$$

$$x + y = 0$$

   and solve this system to find the solutions.

6. Use the method of exercise 5 to find the solutions (if any exist) for each of the following equations.

a. $2x^4 + 2y^4 - 10x^2 - 6y^2 + 17 = 0$

b. $x^4 + x^2y^2 + y^2 - 2xy - 2x^2y + 1 = 0$

7. Even two-variable nonlinear systems can be quite difficult to solve. Note that $(x, y) = (0, 1)$ is a solution of the system

$$2e^x + xy - 2 = 0$$

$$e^x + y + \cos xy = 3$$

Can you find other solutions or show that no other solutions exist?

5.2   Method of Steepest Descent

We begin the problem of approximating solutions of nonlinear systems by examining a method for finding an initial estimate of the location of a solution. This method is known as the **method of steepest descent**. To illustrate the procedure, we will consider the case of two equations with two unknowns.

Let us define the function $S(x, y)$ by

$$S = \frac{f(x, y)^2 + g(x, y)^2}{2}$$

Note that $S(\hat{x}, \hat{y}) = 0$ if and only if $(\hat{x}, \hat{y})$ is a solution of the system of equations

$$f(x, y) = 0$$

$$g(x, y) = 0$$

It is also true that $S(x, y) > 0$ if (x, y) is not a solution of that system. Hence, any solution $(\hat{x}, \hat{y})$ of the system is a minimum point for the function $S(x, y)$, and the minimum value of $S(x, y)$ is zero. It follows that we can find a solution for our nonlinear system by locating the point where $S(x, y)$ assumes its minimum value of zero.

From calculus we know that the gradient vector $\nabla S = (S_x, S_y)$ points in the direction of maximum rate of increase of S at each point (x, y). Then $-\nabla S$ must point in the direction of maximum rate of decrease at each point. Let t be a small real number. Let (x_0, y_0) be any point in the plane. Then we should have

$$S(x_0 - tS_x(x_0, y_0), y_0 - tS_y(x_0, y_0)) < S(x_0, y_0)$$

if t is sufficiently small. Hence, starting from any point (x_0, y_0) in the plane, if we move a *short* distance in the direction of $-\nabla S$, the value of $S(x, y)$ should decrease. This is not necessarily true if we move too far (t too large). We conclude that if we continue to take short steps in the direction of $-\nabla S(x, y)$ at each point in the plane starting from a point (x_0, y_0), we should eventually arrive at the point $(\hat{x}, \hat{y})$, where $S = 0$. We know the direction to go at each point after we arrive there. We do not yet know how far we should step. That is, we do not know the appropriate choice for t. Let

$$M = \|\nabla S(x_0, y_0)\|$$

and define

$$h(t) = S\left(x_0 - \frac{tS_x(x_0, y_0)}{M}, \; y_0 - \frac{tS_y(x_0, y_0)}{M}\right)$$

The ideal choice for t would be the value of t that minimizes $h(t)$. Let t_0 be an approximation to this optimum value of t. Suppose that a procedure has been found to obtain a good approximation t_0 at each point (x_0, y_0). Then we could move in the direction of $-\nabla S(x_0, y_0)$ a distance t_0 to a point (x_1, y_1). Next we would compute $-\nabla S(x_1, y_1)$ and a new value for t_0 appropriate for the point (x_1, y_1). Using this information we would move from (x_1, y_1) a distance t_0 in the direction of $-\nabla S(x_1, y_1)$ to a new point (x_2, y_2). By repeating this procedure as we arrive at each point, we should eventually arrive at the minimum point, since each move decreases the value of $S(x, y)$, and the value cannot continue to decrease forever. As a practical measure, we might arrange to stop the procedure when the value of $S(x, y)$ is sufficiently small.

A method often used in practice to determine the optimum value of t at each step is as follows. Compute $h(t)$ for t_1, t_2, and t_3. Next, find a quadratic function $q(t)$ such that $q(t_i) = h(t_i)$ for $i = 1, 2, 3$. Give t that value, which minimizes $q(t)$. In Algorithm 5.1, the procedure outlined in the previous paragraph is used. But a procedure simpler than quadratic interpolation is used to select an appropriate value of t at each step. Initially, t is set equal to 0.1. After computing the partial derivatives $S_x(x_0, y_0)$ and $S_y(x_0, y_0)$, we replace t by $t/\|\nabla S(x_0, y_0)\|$. The value of $\|\nabla S(x_0, y_0)\|$ is the rate of decrease of $S(x, y)$ in the direction in which we plan to move. If this number is large, we decrease the step size; if the rate of change is small, we increase the step size. If the value of S at the new point

$$(x_0 - tS_x(x_0, y_0), \; y_0 - tS_y(x_0, y_0))$$

is less than $S(x_0, y_0)$, then we move to that point and repeat the process with t initially set equal to the distance moved at the last step. Otherwise, we terminate the process and use our present location as the estimate of the solution.

Algorithm 5.1   (Steepest Descent Algorithm)

Given the system

$$f(x, y) = 0$$
$$g(x, y) = 0$$

compute the first partials of f and g as follows.

Step 1 Select a starting point (x_0, y_0). Set $t = 0.1$ and $n = 0$.
Step 2 Set

$$S = \frac{f(x_n, y_n)^2 + g(x_n, y_n)^2}{2}$$

Compute

$$S_x(x_n, y_n) = f(x_n, y_n)f_x(x_n, y_n) + g(x_n, y_n)g_x(x_n, y_n)$$

and

$$S_y(x_n, y_n) = f(x_n, y_n)f_y(x_n, y_n) + g(x_n, y_n)g_y(x_n, y_n)$$

Set

$$M = \sqrt{S_x(x_n, y_n)^2 + S_y(x_n, y_n)^2} \quad \text{and} \quad t = \frac{t}{M}$$

Step 3 Let

$$x_{n+1} = x_n - tS_x(x_n, y_n)$$
$$y_{n+1} = y_n - tS_y(x_n, y_n)$$

Step 4 Compute $S(x_{n+1}, y_{n+1})$. If $S(x_{n+1}, y_{n+1}) > S$, then go to step 7.
Step 5 Let $t = \sqrt{(x_{n+1} - x_n)^2 + (y_{n+1} - y_n)^2}$ and $n = n + 1$.
Step 6 Go to step 2.
Step 7 Print x_{n+1} and y_{n+1}.

The procedure in this algorithm is easily generalized to systems with more than two equations. However, the case of two equations is easier to understand and implement.

EXAMPLE 5.1   Find a solution of the system

$$f(x, y) = x^2 + y^2 - 4 = 0$$

$$g(x, y) = xy - 1 = 0$$

Solution   First we compute $f_x = 2x$, $f_y = 2y$, $g_x = y$, and $g_y = x$. Next we select $x_0 = 1$ and $y_0 = 0$. More precise estimates of a solution could be made by inspection of the graphs, but the intention here is to demonstrate that the method of steepest descent does not require precise estimates. Almost any starting point somewhere in the general area of an anticipated solution can be used. Let us set $t = 0.1$. Using these initial values for x_0, y_0, and t, we obtain

$$S_x = ff_x + gg_x = (-3)(2) + (-1)(0) = -6$$

$$S_y = ff_y + gg_y = (-3)(0) + (-1)(1) = -1$$

$$M = \sqrt{36 + 1} = \sqrt{37} = 6.082763$$

So

$$x_1 = x_0 - \frac{(0.1)S_x}{M} = 1 - \frac{(0.1)(-6)}{6.08276} = 1.09864$$

$$y_1 = y_0 - \frac{(0.1)S_y}{M} = 0 - \frac{(0.1)(-1)}{6.08276} = 0.01644$$

Now we should recompute S_x, S_y, and M and then compute the new estimates x_2 and y_2, and so on. Since the computations are difficult to do by hand, we turn to machine computation. Machine output for this system is as follows.

X	Y
1.098639	0.0164399
1.196877	3.513122E − 02
1.294647	5.613406E − 02
1.391862	7.956891E − 02
1.488404	0.1056396
1.584095	0.1346778
1.678647	0.1672336
1.771532	0.2042797
1.861587	0.2477554
1.945316	0.302431
2.002336	0.3845815
1.917617	0.4377094
1.998095	0.4970666
1.900755	0.4741552

ESTIMATES OF X AND Y ARE: 1.900755   0.474155

If we choose starting point $x = -1$ and $y = 0$, the iterations converge to another solution, as the following machine output shows.

X	Y
−1.098639	−0.0164399
−1.196877	−3.513122E − 02
−1.294647	−5.613406E − 02
−1.391862	−7.956891E − 02
−1.488404	−0.1056396
−1.584095	−0.1346778
−1.678647	−0.1672336
−1.771532	−0.2042797
−1.861587	−0.2477554
−1.945316	−0.302431
−2.002336	−0.3845815
−1.917617	−0.4377094
−1.998095	−0.4970666
−1.900755	−0.4741552

ESTIMATES OF X AND Y ARE: −1.900755   −0.474155   ■

   In general, this algorithm does not yield a precise value of the solution, but it will provide more than adequate approximations for the "refinement" methods discussed in Sections 5.3 and 5.4. Note that the length of the gradient vector ∇S approaches zero as we near the solution point. This slows down the convergence and eventually prevents the algorithm from producing more precise approximations. The iterations should be stopped when the value of $S(x_{n+1}, y_{n+1})$ is greater than or equal to $S(x_n, y_n)$. After this point, the numerical estimation procedure fails to produce new points closer to the desired minimum point.

   The extension of Algorithm 5.1 to three or more variables should be clear. We only need to compute more partials, compute another component S_z for the gradient of S, and so forth.

   Before developing algorithms to provide precise solutions from the initial estimates, such as those produced by the method of steepest descent, it is worth examining a vector formulation of our problem.

■ *Exercise Set 5.2*

1. For each system, use Algorithm 5.1 with the values of x_0 and y_0 given to compute two new estimates (x_1, y_1) and (x_2, y_2).

   a. $x^2 - y = 2$ **b.** $x^2 + y^2 = 4$
          $2x - y^2 = 2$ $y + 3x = 3$
          $x_0 = 1, \quad y_0 = 1$ $x_0 = 0, \quad y_0 = 2$

   c. $2x + 3y = 5$ **d.** $\sin x + \sin y = 0.5$
          $2x - 3y = -1$ $\cos x - \cos y = 1$
          $x_0 = 2, \quad y_0 = 2$ $x_0 = -1, \quad y_0 = 2$

2. Verify the estimates (x_2, y_2) and (x_3, y_3) given in Example 5.1.

3. Use a machine program that implements Algorithm 5.1 to find better estimates of a solution for each system in exercise 1. Use the indicated values of x_0 and y_0 in each case.

4. Using algebraic elimination, solve the systems in exercises 1a, 1b, and 1c. Compare these results with the results in exercise 3 for these systems.

5. Find an analytic solution of the system in exercise 1d. There are two solution points. Compare results with the machine solution obtained in exercise 3. (*Hint:* Add to the given equations the two trig identities $\sin^2 x + \cos^2 x = 1$ and $\sin^2 y + \cos^2 y = 1$. Solve this system of four equations in four unknowns by elimination and/or substitution methods.

6. Consider the system $x^2 + y^2 = 4$ and $xy = 1$.
   a. Eliminate y and solve the resulting equation for values of x. You should find four distinct values.
   b. Write down the complete set of solutions for the system.
   c. You should have one solution near point (1.93, 0.52). Use $x_0 = 1.93$ and $y_0 = 0.52$ in the steepest descent algorithm to compute new solution estimate $(x_1\, y_1)$. Is (x_1, y_1) closer to the actual solution than (x_0, y_0)? Use the Euclidean distance between points as your measure of closeness.

7. Write down function $S(x, y)$ for the system in exercise 6.
   a. Compute $S(x_0, y_0)$ and $S(x_1, y_1)$ for points (x_0, y_0) and (x_1, y_1) in exercise 6c. Which is smaller?
   b. Compute gradient vector $\nabla S(x_0, y_0)$. Make a sketch showing points (x_0, y_0), (x_1, y_1) and the solution point. Draw vector $\nabla S(x_0, y_0)$ attached to point (x_0, y_0). Is this sketch consistent with the theory of the method of steepest descent?
   c. Make a sketch of the graphs of the equations in the system for $1 \le x \le 2$ and $0 \le y \le 1$. Mark the intersection point within this rectangle as point P. Sketch gradient vectors $\nabla S(1, 0)$ and $\nabla S(2, 0)$.
   d. Suppose that $x_0 = 1$ and $y_0 = 0$ were used with Algorithm 5.1 for this system. Does your sketch in part (c) indicate that point (x_1, y_1) would indeed be closer than point (x_0, y_0) to the solution point represented by the intersection point P of the graphs?
   e. Repeat part (d) for $x_0 = 2$ and $y_0 = 0$.

5.3   Vector Formulation

Any two functions $f(x, y)$ and $g(x, y)$ can be used to define a mapping $\mathbf{T} : R^2 \to R^2$, where $\mathbf{T}(x, y) = (f(x, y),\ g(x, y))$. Briefly, we have $\mathbf{T}(\mathbf{v}) = (F(\mathbf{v}),\ g(\mathbf{v}))$ for vector $\mathbf{v} = (x, y)$.

So the problem of solving the system

$$f(x,\ y) = 0$$

$$g(x,\ y) = 0$$

is equivalent to solving a single vector equation

$$\mathbf{T}(\mathbf{v}) = 0$$

Similarly, the problem of solving a system of three equations can be converted to a single vector equation where $\mathbf{v} = (x, y, z)$. The extension to n equations is clear. The mathematical ideas involved can be illustrated by considering the case of two equations and two unknowns.

It is convenient to use standard matrix notation in the following, so we will treat the solution as a column matrix and adopt the usual conventions for working with vectors as column matrices in R^2. All vector norms are Euclidean norms in the following.

If **T** is a linear transformation, then our system is a homogeneous linear system, so the following treatment is an extension of some of the theory in Chapter 4 to nonlinear systems.

Suppose the mapping $\mathbf{T} : R^2 \rightarrow R^2$ is given and **P** is a point in R^2. We will say that mapping **T** is *continuous at point **P*** iff for every $\varepsilon > 0$ there exists a $\delta > 0$ such that $\|\mathbf{P} - \mathbf{v}\| < \delta$ implies that $\|\mathbf{T}(\mathbf{P}) - \mathbf{T}(\mathbf{v})\| < \varepsilon$. Loosely speaking, **T** is continuous at point **P** in R^2 iff **v** close to **P** implies **T(v)** close to **T(P)**.

In our discussion of systems, we will assume that **T** is continuous at **P** if **T(P)** = 0.

More generally, if **A** is a fixed vector in R^2, and for every $\varepsilon > 0$ there exists a $\delta > 0$ such that $0 < \|\mathbf{P} - \mathbf{v}\| < \delta$ implies that $\|\mathbf{T}(\mathbf{v}) - \mathbf{A}\| < \varepsilon$, then we say that $\lim \mathbf{T}(\mathbf{v}) = \mathbf{A}$ as $\mathbf{v} \rightarrow \mathbf{P}$.

If $\{\mathbf{x}^n\}$ is a sequence of vectors in R^2 and **A** is a fixed vector in R^2, then we say that the sequence $\{\mathbf{x}^n\}$ converges to **A**, or $\lim \mathbf{x}^n = \mathbf{A}$ iff for every $\varepsilon > 0$ there exists a positive integer N such that $n \geq N$ implies that $\|\mathbf{x}^n - \mathbf{A}\| < \varepsilon$.

Now suppose that the equation **T(v)** = 0 is equivalent to the equation $\mathbf{v} = \mathbf{G}(\mathbf{v})$ for some function $\mathbf{G} : R^2 \rightarrow R^2$. Suppose that **R** is a fixed point of **G** so that $\mathbf{R} = \mathbf{G}(\mathbf{R})$. Then clearly, $\mathbf{v} = \mathbf{R}$ is a root of the vector equation **T(v)** = 0. Now let us define a sequence of vectors as follows. We choose $\mathbf{x}^0$ to be an approximation to a fixed point **R** of function **G**. Then we let

$$\mathbf{x}^{n+1} = \mathbf{G}(\mathbf{x}^n) \qquad n = 0, 1, 2, \ldots$$

Now

$$\|\mathbf{x}^{n+1} - \mathbf{R}\| = \|\mathbf{G}(\mathbf{x}^n) - \mathbf{G}(\mathbf{R})\|$$

Under certain conditions, we can show that $\|\mathbf{x}^{n+1} - \mathbf{R}\|$ approaches zero as n approaches infinity.

Definition 5.1

Mapping $\mathbf{G} : R^2 \rightarrow R^2$ is a **contraction mapping** on a region $D \leq R^2$ if there exists a constant L such that $0 \leq L < 1$ and

$$\|\mathbf{G}(\mathbf{v}) - \mathbf{G}(\mathbf{u})\| \leq L\|\mathbf{v} - \mathbf{u}\|$$

for all pairs **v** and **u** in D.

Assume now that **G** is a contraction mapping on region D, and **R** in D is a fixed point of **G**. Then

$$\|\mathbf{x}^{n+1} - \mathbf{R}\| \leq L\|\mathbf{x}^n - \mathbf{R}\|$$

So

$$\|\mathbf{x}^{n+1} - \mathbf{R}\| \leq L^2\|\mathbf{x}^{n-1} - \mathbf{R}\| \leq L^3\|\mathbf{x}^{n-2} - \mathbf{R}\|$$

By continuing this reasoning, we finally arrive at

$$\|\mathbf{x}^{n+1} - \mathbf{R}\| \leq L^{n+1}\|\mathbf{x}^0 - \mathbf{R}\|$$

Since $0 \leq L < 1$ and $\|\mathbf{x}^0 - \mathbf{R}\|$ is a fixed number, the term on the right has a limit of zero as n approaches infinity. It follows that the sequence $\{\mathbf{x}^n\}$ of vectors converges to vector $\mathbf{R}$.

The preceding theoretical result is a vector analogy of the theory developed in Chapter 3 for fixed-point iterations. In the earlier theory, the constant L was a bound on $|g'(x)|$ in a neighborhood of a fixed point. The present result for vector fixed points is a generalization of the earlier theory. This result also suggests an iterative procedure for solving a nonlinear system. For the system

$$f(x, y) = 0$$
$$g(x, y) = 0$$

suppose that the first equation can be solved for x and the second equation can be solved for y, yielding equations

$$x = F(x, y)$$
$$y = G(x, y)$$

The procedure would be as follows:

1. Choose estimates x_0 and y_0 of a solution.
2. Compute new estimates with the iterations

$$x_{n+1} = F(x_n, y_n) \qquad n = 0, 1, 2, \ldots$$
$$y_{n+1} = G(x_n, y_n) \qquad n = 0, 1, 2, \ldots$$

This is a nonlinear extension of the Jacobi iterative method for linear systems. The Jacobi iterations do not always converge to a solution, so it is not surprising that the present iterative method for nonlinear systems does not always converge to a solution. Consider the following examples.

EXAMPLE 5.2 Solve the system

$$x^2 + y^2 = 4$$
$$x + y = 1$$

with initial estimates $x_0 = -0.75$ and $y_0 = 1.8$.

Solution We choose iteration equations

$$x_{n+1} = 1 - y_n$$
$$y_{n+1} = \sqrt{4 - x_n^2}$$

The machine results are as follows.

Iterations	Xn	Yn
1	−0.800000	1.854050
2	−0.854050	1.833030
3	−0.833030	1.808480
4	−0.808480	1.818258
5	−0.818258	1.829306
10	−0.824189	1.823299
15	−0.822962	1.822755
20	−0.822851	1.822868
25	−0.822874	1.822878
30	−0.822876	1.822876

The iterations have converged on a solution in 30 iterations. The operation of the algorithm is satisfactory. ∎

EXAMPLE 5.3　Estimate the solution of the system

$$x^2 y = 1$$
$$y + 20x - 10x^2 = 10$$

using $x_0 = 2$ and $y_0 = 1$ with iteration equations

$$x_{n+1} = \sqrt{2x_n - 1 + \frac{y_n}{10}}$$

$$y_{n+1} = \frac{1}{x_n^2}$$

Solution　Here is a sample of the machine results.

Iterations	Xn	Yn
1	1.760682	0.250000
2	1.595733	0.322581
3	1.491216	0.392717
4	1.421866	0.449696
5	1.374300	0.494633
10	1.278308	0.603311
15	1.258270	0.629605
20	1.253794	0.635673
25	1.252780	0.637058
30	1.252549	0.637374
39	1.252486	0.637460
40	1.252485	0.637462

The iterations are convergent in this case, and the approximate solution is $x = 1.252485$ and $y = 0.637462$, with some doubt about the last digit in each estimate. ∎

It is also true that convergence of the iterations can sometimes be accelerated by using x_{n+1} rather than x_n in the computation of y_{n+1}. The iterations are then computed using iteration equations

$$x_{n+1} = F(x_n, y_n)$$
$$y_{n+1} = G(x_{n+1}, y_n)$$

EXAMPLE 5.4 Solve the system in Example 5.2, using the same initial estimates but with the revised iteration procedure.

Solution

Iterations	Xn	Yn
1	−0.800000	1.833030
2	−0.833030	1.818258
3	−0.818258	1.824953
4	−0.824953	1.821936
5	−0.821936	1.823299
10	−0.822893	1.822868
15	−0.822875	1.822876
16	−0.822876	1.822875
17	−0.822875	1.822876

We now have convergence in only 17 iterations rather than 30 as in Example 5.2. The improvement is clear. ∎

Using the revised iterations does not improve the situation in the case of the system in Example 5.3, however, as the next example illustrates.

EXAMPLE 5.5 Estimate the solution of the system in Example 5.3 using the revised iterations

$$x_{n+1} = \sqrt{2x_n - 1 + \frac{y_n}{10}}$$

$$y_{n+1} = \frac{1}{x_{n+1}^2}$$

with $x_0 = 2$ and $y_0 = 1$.

Solution The machine results are as follows.

Iterations	Xn	Yn
1	1.760682	0.322581
2	1.598005	0.391601
3	1.495049	0.447393
4	1.426477	0.491440
5	1.379166	0.525735
10	1.281449	0.608972
15	1.259568	0.630314
20	1.254243	0.635678
25	1.252921	0.637020
30	1.252591	0.637355
39	1.252490	0.637458
40	1.252488	0.637460

The iterations are convergent in this case, and the approximate solution is $x = 1.25249$ and $y = 0.63746$. Use of the revised iterations does not produce any significant improvement in the rate of convergence for this system. ∎

■ *Exercise Set 5.3*

1. Given the system

$$x^2 - y = 2$$
$$2x - y^2 = 2$$

do the following.
a. Show that this system has two solutions.
b. Show that the iteration scheme

$$x_{n+1} = \sqrt{2 + y_n}, \quad y_{n+1} = \sqrt{2x_n - 2} \qquad x_0 = 2, \quad y_0 = 1.5$$

converges to one of these solutions. Find that solution.
c. Show that the iteration scheme

$$x_{n+1} = \sqrt{2 + y_n}, \quad y_{n+1} = -\sqrt{2x_n - 2} \qquad x_0 = 1.2, \quad y_0 = -1$$

converges to one of these solutions. Find that solution.

2. Given that the system

$$x^2 + y^2 = 4$$
$$xy = 1$$

has four solutions, do the following.

a. Locate one of these solutions using iterations

$$x_{n+1} = \sqrt{4 - y_n^2}, \quad y_{n+1} = \frac{1}{x_n} \qquad x_0 = 1, \quad y_0 = 0.5$$

b. Find the remaining three solutions.

3. Given the system

$$x^2 + y = 11$$
$$y^2 + x = 7$$

which of the following iteration schemes converges to a solution of the system? Find the solution for each convergent iteration scheme.

a. $x_{n+1} = 7 - y_n^2$, $y_{n+1} = 11 - x_n^2$ $x_0 = -2.8$, $y_0 = 3$

b. $x_{n+1} = -\sqrt{11 - y_n}$, $y_{n+1} = \sqrt{7 - x_n}$ $x_0 = -2.8$, $y_0 = 3$

c. $x_{n+1} = \dfrac{11 - y_n}{x_n}$, $y_{n+1} = \dfrac{7 - x_n}{y_n}$ $x_0 = -3$, $y_0 = 3$

d. $x_{n+1} = \sqrt{11 - y_n}$, $y_{n+1} = \sqrt{7 - x_n}$ $x_0 = 3.6$, $y_0 = 1.8$

e. $x_{n+1} = \sqrt{11 - y_n}$, $y_{n+1} = -\sqrt{7 - x_n}$ $x_0 = 3.6$, $y_0 = -1.8$

4. Repeat exercises 1b and 1c using x_{n+1} rather than x_n in the computation of y_{n+1}. Compare the rate of convergence with the results of exercise 1.

5. Given the system

$$x^2 + y - 11 = 0$$
$$y^2 + x - 7 = 0$$

do the following.
a. Sketch graphs of the equations, and show that the system has four solutions.
b. Eliminate y to obtain a polynomial equation for x. Use the Newton algorithm for polynomials to find all real zeros of this polynomial. Compute corresponding values of y to obtain the four solutions of the system.

6. Let $T(x, y) = (7 - y^2, 11 - x^2)$.
a. Explain why the solutions of the system in exercise 5 are fixed points of **T**.
b. Suppose that (x_0, y_0) is one of the fixed points of **T**. Let $x_1 = x_0 + h$ and $y_1 = y_0 + k$. Show that

$$\|T(x_1, y_1) - T(x_0, y_0)\| \cong 2\sqrt{k^2 y_0^2 + h^2 x_0^2}$$

if h and k are small. The Euclidean norm is used here.
c. Use the result of part (b) to show that

$$\frac{\|T(x_1, y_1) - T(x_0, y_0)\|}{\|(x_1, y_1) - (x_0, y_0)\|} \cong \frac{2\sqrt{k^2 y_0^2 + h^2 x_0^2}}{\sqrt{k^2 + h^2}}$$

d. Let (x_0, y_0) be the fixed point $(-2.805119, 3.131313)$. Use the result in part (c) to show that **T** cannot be a contraction mapping on any neighborhood of point (x_0, y_0).
e. Explain why you would not expect iterations.

$$x_{n+1} = 7 - y_n^2, \quad y_{n+1} = 11 - x_n^2 \qquad x_0 = -2.8, \quad y_0 = 3.1$$

to converge to a fixed point of **T**.

7. Use fixed-point iterations to find all solutions of the system

$$x^2 y = 1$$
$$y + 20x - 10x^2 = 10$$

8. Use iterations

$$x_{n+1} = 3 - y_n - z_n \qquad\qquad x_0 = 0$$

$$y_{n+1} = \sqrt{z_n - x_n^2} \qquad\qquad y_0 = 1$$

$$z_{n+1} = \sqrt{4 - x_n^2 - y_n^2} \qquad z_0 = 1$$

to find a solution of the system

$$x + y + z = 3$$

$$x^2 + y^2 + z^2 = 4$$

$$x^2 + y^2 = z$$

Then do the following.

a. Explain why a second solution can be found by simply interchanging x and y in the solution computed from the iteration procedure.

b. Show that this system has exactly two solutions.

5.4   Newton's Method for Systems

Suppose that a transformation $\mathbf{T} : R^2 \to R^2$ is given, where

$$\mathbf{T}(x, y) = (f(x, y), g(x, y))$$

for some functions f and g. Suppose that functions f and g have continuous first partial derivatives in a neighborhood of a point $\mathbf{v}^0 = (x_0, y_0)$ in the domain of $\mathbf{T}$. Consider $\mathbf{T}(\mathbf{v}^0 + \mathbf{h}) - \mathbf{T}(\mathbf{v}^0)$, where $\mathbf{h} = (h_1, h_2)$ and $\|\mathbf{h}\|$ is sufficiently small so that point $\mathbf{v}^0 + \mathbf{h}$ is still in the domain of $\mathbf{T}$. Clearly,

$$\mathbf{T}(\mathbf{v}^0 + \mathbf{h}) - \mathbf{T}(\mathbf{v}^0) = (f(x_0 + h_1, y_0 + h_2)$$

$$- f(x_0, y_0), g(x_0 + h_1, y_0 + h_2) - g(x_0, y_0))$$

We know from calculus that under the conditions given,

$$f(x_0 + h_1, y_0 + h_2) - f(x_0, y_0) \cong f_x(x_0, y_0)h_1 + f_y(x_0, y_0)h_2$$

and

$$g(x_0 + h_1, y_0 + h_2) - g(x_0, y_0) \cong g_x(x_0, y_0)h_1 + g_y(x_0, y_0)h_2$$

Combining results, we find that

$$\mathbf{T}(\mathbf{v}^0 + \mathbf{h}) - \mathbf{T}(\mathbf{v}^0) \cong \begin{bmatrix} f_x(x_0, y_0) & f_y(x_0, y_0) \\ g_x(x_0, y_0) & g_y(x_0, y_0) \end{bmatrix} \begin{bmatrix} h_1 \\ h_2 \end{bmatrix}$$

We define the matrix

$$\mathbf{DT} = \begin{bmatrix} f_x(x_0, y_0) & f_y(x_0, y_0) \\ g_x(x_0, y_0) & g_y(x_0, y_0) \end{bmatrix}$$

to be the *derivative* of transformation $\mathbf{T}$ at point $\mathbf{v}^0$. In summary, we have

$$\mathbf{T}(\mathbf{v}^0 + \mathbf{h}) - \mathbf{T}(\mathbf{v}^0) \cong \mathbf{DTh}$$

Now consider the nonlinear system $\mathbf{T}(\mathbf{v}) = 0$. Suppose that $\mathbf{v}^0 = (x^0, y^0)$ is an approximation of solution $\mathbf{R}$. Now

$$\mathbf{T}(\mathbf{R}) - \mathbf{T}(\mathbf{v}^0) \cong \mathbf{DT}(\mathbf{R} - \mathbf{v}^0)$$

so

$$-\mathbf{T}(\mathbf{v}^0) \cong \mathbf{DT}(\mathbf{R} - \mathbf{v}^0)$$

If $(\mathbf{DT})^{-1}$ exists, then we find

$$-\mathbf{DT}^{-1}\mathbf{T}(\mathbf{v}^0) \cong \mathbf{R} - \mathbf{v}^0$$

and

$$\mathbf{R} \cong \mathbf{v}^0 - \mathbf{DT}^{-1}\mathbf{T}(\mathbf{v}^0)$$

We expect that the vector $\mathbf{v}^0 - \mathbf{DT}^{-1}\mathbf{T}(\mathbf{v}^0)$ will be a better approximation to the root $\mathbf{R}$ than the initial estimate $\mathbf{v}^0$.

This leads to the Newton algorithm for systems.

Algorithm 5.2

Step 1 Choose estimate $\mathbf{v}^0$ of a root $\mathbf{R}$ of the system.
Step 2 Compute vector estimates

$$\mathbf{v}^{(n+1)} = \mathbf{v}^{(n)} - \mathbf{DT}^{-1}(\mathbf{v}^{(n)})\mathbf{T}(\mathbf{v}^{(n)})$$

for $n = 0, 1, 2, \ldots$ until satisfied.

The inverse of $\mathbf{DT}$ exists iff $J = \det(\mathbf{DT}) \neq 0$. We will assume that is true in the following. In this case, the matrix $\mathbf{DT}^{-1}$ is

$$\begin{bmatrix} \dfrac{g_y(x_0, y_0)}{J} & \dfrac{-f_y(x_0, y_0)}{J} \\[2mm] \dfrac{-g_x(x_0, y_0)}{J} & \dfrac{f_x(x_0, y_0)}{J} \end{bmatrix}$$

Let

$$\mathbf{T}(\mathbf{v}^0) = \begin{bmatrix} w_1^0 \\ w_2^0 \end{bmatrix} = \mathbf{w}^0$$

Then the Newton iterations assume the explicit form

$$\mathbf{v}^{n+1} = \mathbf{v}^n - \begin{bmatrix} g_y/J & -f_y/J \\ -g_x/J & f_x/J \end{bmatrix}\begin{bmatrix} w_1^{(n)} \\ w_2^{(n)} \end{bmatrix}$$

So

$$x^{(n+1)} = x^{(n)} - \frac{w_1^{(n)}g_y - w_2^{(n)}f_y}{J}$$

$$y^{(n+1)} = y^{(n)} - \frac{-w_1^{(n)}g_x + w_2^{(n)}f_x}{J}$$

This Newton algorithm is easily programmed for machine use. The following example illustrates the results for a system of two equations.

EXAMPLE 5.6   Solve the system

$$x^2 + y^2 = 4$$

$$xy = 1$$

by use of Newton iterations.

Solution   We must supply the functions f and g and their first partials in the machine code. We must also supply initial estimates of a solution. In this case, it is easy to see that there is a solution with $1 < x < 2$ and $0 < y < 1$. The resulting machine output is as follows.

```
ENTER INITIAL VALUES X, Y    ? 1.5, .5
NUMBER OF ITERATIONS WAS   5
FINAL ESTIMATES ARE:

X  =    1.93185165          Y  =     0.51763809
F  =    2.25697238676D – 12    G  = – 1.1270151478D – 12
H1 = –7.977404226D – 13     H2 =   7.971398708D – 13
```

In this example, the convergence is rapid. Values F and G are values of the functions f and g for the final values of x and y. The values of F and G suggest that the approximate solution is very close to the exact solution. But these values can be small, even when the estimates are some distance away from the true solution. A better check of accuracy is given by the values of increments h_1 and h_2, labeled here as H1 and H2. Just as in the case of the Newton method for a single equation, these increments provide estimates of the error in the estimates of x and y. The error is very small and quite acceptable.   ∎

Note that the Newton iterations can be organized in the form

$$\mathbf{v}^{(n+1)} = \mathbf{v}^{(n)} - \mathbf{h}^{(n)}$$

where $\mathbf{DT}(\mathbf{v}^{(n)})\mathbf{h}^{(n)} = \mathbf{T}(\mathbf{v}^{(n)})$. In this case, the "correction vector" $\mathbf{h}^{(n)}$ is regarded as the solution of a linear system and could be computed by an algorithm such as the Gauss algorithm. This form is best when solving systems of three or more equations. In the case of n equations with n unknowns, the ith row of matrix $\mathbf{DT}$ contains the n first partials of the ith function in the system. Since an $n \times n$ system must be solved repeatedly during execution of the algorithm, the computations proceed very slowly when n is much larger than 3.

It is very difficult to write programs to solve large nonlinear systems. It would be wiser in that case to employ professionally written software, such as the IMSL subroutines NEQNF or DNEQNF. Both of these subroutines use an algorithm (Levenberg-Marquardt) that employs both steepest descent and the Newton method. The partials are approximated by finite differences. The routine DNEQNF provides double-precision results. You might want to try these routines on the systems used in the examples in this chapter to become familiar with their implementation. The IMSL documentation also contains some sample systems for user practice.

■ *Exercise Set 5.4*

1. Use Algorithm 5.2 to find all the solutions for the systems in exercise 1 of Exercise Set 5.2.

2. Use Algorithm 5.2 to find all the solutions for the systems in exercise 1 of Exercise Set 5.1.

3. Use Algorithm 5.2 to find all the solutions for the systems in exercise 2 of Exercise Set 5.1.

4. Find derivative **DT** for each transformation **T**. Vector $\mathbf{x} = (x, y)^t$ in each case.
   a. $\mathbf{T}(\mathbf{x}) = (xe^x + y^2, x, y + \sin y)^t$
   b. $\mathbf{T}(\mathbf{x}) = (2x + 3y, 5x - 2y)^t$
   c. $\mathbf{T}(\mathbf{x}) = (e^x + \ln y, xy \sin x)^t$

5. Compute $\mathbf{DT}^{-1}$ for each matrix **DT** from exercise 4.

6. Find derivative **DT** for the transformation

$$\mathbf{T}(\mathbf{x}) = (x + y + z, x^2 + y^2 + z^2, xyz)^t$$

   Now let $\mathbf{x} = (1, 1, 1)^t$. Find $\mathbf{DT}^{-1}(\mathbf{x})$.

7. Write a machine program for solving 3×3 nonlinear systems using the Newton algorithm. Then solve each of the following systems.
   a. The system in exercise 8 of Exercise Set 5.3.
   b. $x^2 + y^2 + z^2 = 4$ **c.** $x + y^2 + z^2 = 2.78$
   $\qquad 3x - 2y = 1$ $e^x \sin \pi y + \cos \pi z = 1.36$
   $\qquad 3z - y = 2$ $zxe^{-y} - 1.42x + 1 = 0$
   d. $\qquad x^2 + y^2 = \sqrt{z}$
   $\qquad x^2 + y^2 + z^2 = 4$
   $\qquad 10x + 9y - z = 0$

8. Find local maximum or minimum points for the function

$$f(x, y) = xy^3 + \sin y + \tfrac{1}{3}x^3 - \tfrac{1}{2}x^2$$

   by solving the system

$$f_x(x, y) = 0$$

$$f_y(x, y) = 0$$

5.5 Bairstow's Method (Optional)

In this section, we will examine an algorithm that can be used to determine all the roots of a polynomial equation. The algorithm requires us to solve a nonlinear system at one point in the process, so we employ the Newton method as described in Section 5.4. The basic idea of the algorithm is to determine the quadratic factors of the polynomial and then solve the resulting quadratic equations. We must begin by discussing the algebra involved in synthetic division by a quadratic expression.

Suppose that

$$P(x) = a_n x^n + \cdots + a_1 x + a_0$$

is to be divided by $x^2 + px + q$. If the quotient is

$$Q(x) = b_{n-2} x^{n-2} + \cdots + b_1 x + b_0$$

with remainder $Sx + R$, then, assuming $n > 2$,

$$P(x) = (x^2 + px + q) \cdot Q(x) + (Sx + R)$$

We substitute the assumed form for $Q(x)$ and carry out the indicated multiplication. After collecting powers of x and matching up coefficients of powers of x on each side of the equation, we find that

$$a_n = b_{n-2}$$
$$a_{n-1} = b_{n-3} + pb_{n-2}$$
$$a_{n-2} = b_{n-4} + pb_{n-3} + qb_{n-2}$$
$$\vdots \qquad\qquad \vdots \qquad \vdots$$
$$a_2 = b_0 + pb_1 + qb_2$$
$$a_1 = S + pb_0 + qb_1$$
$$a_0 = R + qb_0$$

Alternatively, we can write

$$b_{n-2} = a_n$$
$$b_{n-3} = a_{n-1} - pb_{n-2}$$
$$b_{n-4} = a_{n-2} - pb_{n-3} - qb_{n-2}$$
$$\vdots \qquad\qquad \vdots$$
$$b_0 = a_2 - pb_1 - qb_2$$
$$S = a_1 - pb_0 - qb_1$$
$$R = a_0 - qb_0$$

This set of equations permits us to compute the coefficients recursively. If we define $b_n = b_{n-1} = 0$ and $b_{-1} = S$, the set of equations can be written compactly as

$$b_k = a_{k+2} - pb_{k+1} - qb_{k+2} \qquad k = n-2, \ldots, 1, 0, -1$$

Then

$$S = b_{-1} \quad \text{and} \quad R = a_0 - qb_0$$

EXAMPLE 5.7 Divide $x^4 + 3x^3 + x^2 - x + 7$ by $x^2 + 2x + 3$.

Solution We have $a_4 = 1$, $a_3 = 3$, $a_2 = 1$, $a_1 = -1$, and $a_0 = 7$, and we have $p = 2$ and $q = 3$. The quotient is $b_2 x^2 + b_1 x + b_0$ with remainder $Sx + R$, where

$$b_2 = a_4 = 1$$
$$b_1 = a_3 - pb_2 = 3 - (2)(1) = 1$$
$$b_0 = a_2 - pb_1 - qb_2 = 1 - (2)(1) - 3(1) = -4$$
$$S = a_1 - pb_0 - qb_1 = -1 - 2(-4) - 3(1) = 4$$
$$R = a_0 - qb_0 = 7 - 3(-4) = 19$$

Hence, $Q(x) = x^2 + x - 4$ with remainder $4x + 19$.   ■

Now suppose that $x^2 + px + q$ is a factor of polynomial $P(x)$. Then we must have $R = S = 0$ in the preceding division algorithm. In general, if $P(x)$ is divided by a quadratic of the assumed form, then R and S will be nonlinear functions of p and q if we treat p and q as variables. Let us call these functions $R(p, q)$ and $S(p, q)$.

If $x^2 + px + q$ is a factor of $P(x)$, then (p, q) is a solution of the nonlinear system

$$R(p, q) = 0$$
$$S(p, q) = 0$$

The idea now is to develop a method for solving this system. If (p_0, q_0) is a solution, then we have found a quadratic factor $x^2 + p_0 x + q_0$ of $P(x)$. The solutions of the quadratic equation

$$x^2 + p_0 x + q_0 = 0$$

are roots of the polynomial equation $P(x) = 0$.

We could solve the system beginning with some approximation (p_0, q_0) of a solution by using the Newton method if we knew the values of the partial derivatives $R_p(p_0, q_0)$, $R_q(p_0, q_0)$, $S_p(p_0, q_0)$, and $S_q(p_0, q_0)$. We can compute these partials by a recursive algorithm that we will now derive.

Clearly,

$$\frac{\partial R}{\partial p} = -q \frac{\partial b_0}{\partial p} \quad \text{and} \quad \frac{\partial S}{\partial p} = \frac{\partial b_{-1}}{\partial p}$$

and

$$\frac{\partial b_k}{\partial p} = -b_{k+1} - p \frac{\partial b_{k+1}}{\partial p} - q \frac{\partial b_{k+2}}{\partial p} \quad k = n-2, \ldots, 1, 0, -1$$

If we let

$$c_{k+1} = -\frac{\partial b_k}{\partial p}$$

then

$$c_{k+1} = b_{k+1} - p c_{k+2} - q c_{k+3} \quad k = n-2, \ldots, 1, 0, -1$$

while $c_{n+1} = c_n = 0$, since $b_n = b_{n-1} = 0$. In summary, we have $c_n = c_{n-1} = 0$ and

$$c_j = b_j - p c_{j+1} - q c_{j+2} \quad j = n-1, \ldots, 1, 0$$

By definition of the c_j's, we find that

$$c_0 = -\frac{\partial b_{-1}}{\partial p} = -\frac{\partial S}{\partial p} \quad \text{so} \quad \frac{\partial S}{\partial p} = -c_0$$

and

$$c_1 = -\frac{\partial b_0}{\partial p} \quad \text{so} \quad \frac{\partial R}{\partial p} = q c_1$$

Thus, partials $S_p(p_0, q_0)$ and $R_p(p_0, q_0)$ can be computed from c_0 and c_1, which can be computed by a recursive algorithm from coefficients b_k.

Similarly, we find that

$$\frac{\partial S}{\partial q} = \frac{\partial b_{-1}}{\partial q} \quad \text{and} \quad \frac{\partial R}{\partial q} = -b_0 - q \frac{\partial b_0}{\partial q}$$

Now

$$\frac{\partial b_k}{\partial q} = -p\,\frac{\partial b_{k+1}}{\partial q} - q\,\frac{\partial b_{k+2}}{\partial q} - b_{k+2} \qquad k = n-2,\ldots,0,\,-1$$

Let

$$d_{k+2} = -\frac{\partial b_k}{\partial q}$$

Then

$$d_{n+2} = d_{n+1} = 0$$

and

$$d_j = b_j - pd_{j+1} - qd_{j+2} \qquad j = n,\ldots,2,\,1$$

Comparing the algorithms for computing the c_j's and the b_j's, we see that the form is the same. In fact, it is clear that if we set $c_{n+2} = c_{n+1} = 0$, then

$$d_{k+2} = c_k \qquad k = n-2,\ldots,1,\,0$$

Now

$$\frac{\partial S}{\partial q} = \frac{\partial b_{-1}}{\partial q} = -d_1$$

and

$$\frac{\partial R}{\partial q} = -b_0 - q\,\frac{\partial b_0}{\partial q} = -b_0 + qd_2$$

Hence, to find the desired partials of R and S, we need only compute the d_k sequence. In summary,

$$\frac{\partial S}{\partial p} = -c_0 = -d_2 \qquad \frac{\partial R}{\partial p} = qc_1 = qd_3$$

$$\frac{\partial S}{\partial q} = -d_1 \qquad \frac{\partial R}{\partial q} = -b_0 + qc_0 = -b_0 + qd_2$$

We now have all the computational tools necessary to formulate an algorithm for finding all roots of a polynomial equation. This algorithm is known as **Bairstow's algorithm**.

Algorithm 5.3   Bairstow's Algorithm

Given coefficients $a_n, a_{n-1}, \ldots, a_1, a_0$ of $P(x)$.

Step 1 Choose approximate values for p and q.
Step 2 Set $B_{n+2} = B_{n+1} = D_n = D_{n-1} = 0$.
Step 3 Compute

$$B_k = a_k - pB_{k+1} - qB_{k+2} \qquad k = n, \ldots, 2, 1$$

Set $B_0 = a_0 - qB_2$.
Step 4 Compute

$$D_k = -B_{k+2} - pD_{k+1} - qD_{k+2} \qquad k = n - 2, \ldots, 2, 1$$

Set $D_0 = -B_2 - qD_2$.
Step 5 Set

$$J = pD_0 D_1 - qD_1^2 - D_0^2$$

Step 6 Compute

$$DP = \frac{-(B_0 D_1 - B_1 D_0)}{J} \quad \text{and} \quad DQ = \frac{B_0 D_0 + (qB_1 - pB_0)D_1}{J}$$

Step 7 Replace p by $p + DP$ and q by $q + DQ$.
Step 8 If $|DP|$ and $|DQ|$ are not sufficiently small, return to step 3. Otherwise, proceed to step 9.
Step 9 Solve the quadratic equation $x^2 + px + q = 0$ for roots r_1 and r_2.

In step 1, it is common practice to simply choose $p = q = 1$ when implementing this algorithm on a computer. Results are usually satisfactory. Also note that in step 5, the value of J could be zero, so in machine programs, this should be tested before proceeding to step 6. Usually, in this case, the values of p and q can be shifted slightly to produce a nonzero J, and then we can proceed with the computations. In step 8, it is assumed that some tolerance level has been set for maximum allowable values of the increments DP and DQ. The tolerance level selected depends on machine precision level and degree of accuracy desired. A choice must be made by the user of the algorithm. It is also wise to count the number of returns to step 3 and place a limit on this count to avoid an excessive number of iterations.

Let us now look at some sample output from Bairstow's algorithm.

EXAMPLE 5.8   Find all roots of the polynomial equation

$$x^4 - 26x^2 + 24x + 21 = 0$$

Solution   BAIRSTOW METHOD FOR POLYNOMIAL EQUATIONS
ENTER DEGREE OF POLYNOMIAL   4
ENTER COEFFICIENTS

A(0) = ?	21
A(1) = ?	24
A(2) = ?	−26
A(3) = ?	0
A(4) = ?	1

REAL ROOTS: 1.5857864376269   −0.55051025721682
REAL ROOTS: 4.4142135623731   −5.4494897427832
ALL ROOTS FOUND

Note that the only input required is the degree of the polynomial and the coefficients. Bairstow's algorithm does the rest. Roots are located in pairs, since the method is based on determining a quadratic factor and then computing the roots of the corresponding quadratic equation. After the first two real roots were found, the deflation procedure with synthetic division for quadratic factors was employed. The polynomial degree was reduced by 2 at this step. When the degree of the deflated polynomial reaches 2, as it did in this case after the first factor was found, the algorithm proceeds immediately to solve the resulting quadratic equation. In this case, two more real roots were found, and the solution of the problem was finished. The exact roots in this case are easily verified to be $3 + \sqrt{2}$, $3 - \sqrt{2}$, $-3 + \sqrt{6}$, and $-3 - \sqrt{6}$. Using this information, we find that all of the displayed digits of the root estimates are correct.   ∎

EXAMPLE 5.9   Find the roots of the equation

$$x^4 - 3x^2 + 18x - 20 = 0$$

Solution   BAIRSTOW METHOD FOR POLYNOMIAL EQUATIONS
ENTER DEGREE OF POLYNOMIAL   4
ENTER COEFFICIENTS

A(0) = ?	−20
A(1) = ?	18
A(2) = ?	−3
A(3) = ?	0
A(4) = ?	1

REAL ROOTS: 1.236067977500   −3.236067977500
COMPLEX ROOT PAIR: 1 + 2I   and   1 − 2I
ALL ROOTS FOUND

In this example, we have both real and complex roots. It is interesting to note the accuracy of the results. Again, the real roots are accurate through all displayed digits, and the complex roots are exactly correct. It is unusual to obtain exact results after a polynomial deflation using approximations to the earlier roots found.   ■

EXAMPLE 5.10   Find all roots of the equation

$$x^6 + x^5 + x^4 + x^3 + x^2 + x + 1 = 0$$

Solution   BAIRSTOW METHOD FOR POLYNOMIAL EQUATIONS
ENTER DEGREE OF POLYNOMIAL   6
ENTER COEFFICIENTS

A(0) = ?	1
A(1) = ?	1
A(2) = ?	1
A(3) = ?	1
A(4) = ?	1
A(5) = ?	1
A(6) = ?	1

COMPLEX ROOTS: −0.22252093395631 + 0.97492791218182 I
              AND   −0.22252093395631 − 0.97492791218182 I
COMPLEX ROOTS: −0.90096886790242 + 0.43388373911756 I
              AND   −0.90096886790242 − 0.43388373911756 I
COMPLEX ROOTS: 0.62348980185873 + 0.78183148246803 I
              AND   0.62348980185873 − 0.78183148246803 I
ALL ROOTS FOUND

Even when all roots are complex, Bairstow's algorithm is usable on machines that only do real arithmetic, since we only need to determine quadratic factors with real coefficients. All arithmetic operations are real. Again, the results are found to be correct in all digits displayed.   ■

■ *Exercise Set 5.5*

1. Use the quadratic synthetic division algorithm to find the quotient and remainder when $P(x)$ is divided by the quadratic $x^2 + px + q$.
   a. $P(x) = x^4 - 26x^2 + 24x + 21$ $p = 6,\ q = 3$
   b. $P(x) = x^4 - 3x^2 + 18x - 20$ $p = -2,\ q = 5$
   c. $P(x) = x^4 - 3x^2 + 18x - 20$ $p = 1,\ q = 1$
   d. $P(x) = x^5 + 2x - 1$ $p = 1,\ q = 1$

2. Verify that the complex roots found in Example 5.10 are the complex seventh roots of unity.

3. Write a machine program that executes Algorithm 5.3. Use this program to find all roots of each of the following polynomial equations.
   a. $x^{10} - 1 = 0$
   b. $x^7 + x^6 + x^5 + x^4 + x^3 + x^2 + x + 1 = 0$
   c. $x^4 + 8x^2 + x^3 + 5x + 15 = 0$
   d. $3x^{12} - 37x^5 + 123x^3 - 41x + 5 = 0$

e. $x^7 - 9.86x^5 + 11.05x^4 + 0.051x^3 - 2.86 = 0$
f. $x^3 + 2x^2 + 5x - 8 = 0$
g. $x^5 - 2x^4 + x - 2 = 0$

4. Use the program written in exercise 3 to find all roots of each of the equations in exercise 3 of Exercise Set 3.3.

5. Leonardo of Pisa (1225) claimed in one of his writings that 1.3688081075 is a root of the third-degree polynomial equation

$$x^3 + 2x^2 + 10x - 20 = 0$$

It is uncertain today how he obtained this result, although some historians think that he used geometrical arguments. Verify that his solution is correct, and determine the remaining roots.

6. Find all roots of the equation

$$\sum_{k=0}^{19} a_k x^k = 0$$

where

$$a_{2k} = 0 \quad \text{and} \quad a_{2k+1} = \frac{(-1)^k}{(2k+1)!} \qquad k = 0, 1, \dots, 9$$

7. If the roots of a polynomial equation $P(x) = 0$ are known, then the factors of $P(x)$ can be determined. Use the roots of $P(x) = 0$ to factor each of the following polynomials.
a. $P(x) = x^4 + x^3 - x^2 - 2x - 2$
b. $P(x) = x^4 - 6x^3 + 6x^2 + 6x + 1$
c. $P(x) = x^5 - 2x^4 + x^3 + x^2 - 2$

5.6 Applied Problems

There are numerous situations in applied mathematics where it becomes necessary to solve a nonlinear system of equations. We will examine two of these situations in this section.

First, let us consider an example from astronomy. A binary or double star system is a pair of stars that are orbiting around each other. If one star is much more massive than the other, we can say that the smaller star is in orbit around the larger one, as shown in Figure 5.1. The mean distance between the stars will be denoted by a, and this distance is measured in AU (astronomical units), where 1 AU is the mean distance from the earth to our sun. Hence, 1 AU is approximately

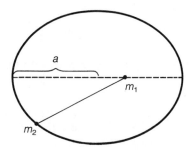

Figure 5.1

92,900,000 miles. As seen from the earth, the stars appear very close in the sky, with some angular separation α that will be measured in seconds of arc. Typical values for some well-known systems range from a few seconds of arc to a few thousandths of a second of arc. Let us assume that the stars have masses μ_1 and μ_2 measured in multiples of the sun's mass. For example, a star with a mass of 2 solar masses has twice the mass of our sun. Each binary system has an orbital period T that is the length of time for each member to complete one complete revolution about the other member. The period T is measured in earth years.

If a star is observed from the earth, its apparent direction changes as the earth revolves around the sun. The angular difference between positions measured at six-month intervals is called the parallax of the star and is denoted by p. Using the diameter of the earth's orbit and simple trigonometry, it is possible to compute the distance of a star from its parallax. For nearby stars, the parallax can be measured directly, even though the values are very small fractions of a second of arc. For the more distant stars, however, usable values are not obtainable by direct measurement, so indirect methods are needed.

In the case of binary systems, a method known as the method of *dynamical parallaxes* is used. The method rests on the fact that certain theoretical relations exist between visual magnitudes (apparent brightness as seen from earth) m_1 and m_2 of the component stars in the system, the parallax p, and the individual masses μ_1 and μ_2. These relations provide us with a nonlinear system to solve for the three unknowns μ_1, μ_2, and p after the observable values of m_1, m_2, α, and T are given. Details of these relationships are supplied in most introductory astronomy texts. They are as follows:

$$p^3 = \frac{\alpha^3}{T^2(\mu_1 + \mu_2)} \tag{1}$$

$$M_1 = m_1 + 5 + 5 \log p \quad \text{and} \quad M_2 = m_2 + 5 + 5 \log p \tag{2}$$

$$0.4(M_0 - M_1) = 3.5 \log \mu_1 \quad \text{and} \quad 0.4(M_0 - M_2) = 3.5 \log \mu_2 \tag{3}$$

The usual textbook procedure for determining p, μ_1, and μ_2 is to begin by assuming that the sum of the masses is 2. We substitute that value in Equation 1 and solve for p (we know α and T). We substitute this value of p in Equations 2 to determine M_1 and M_2, the absolute magnitudes of the stars. In Equations 3, the number M_0 is the absolute magnitude of the sun, which is about 4.8. We use this value of M_0 and the values of M_1 and M_2 found from Equations 2 to compute μ_1 and μ_2 from Equations 3. Now we substitute these values in Equations 1 to recompute p. We continue this iterative procedure until the new value of p agrees well with the preceding one.

Clearly, this procedure is nothing more than the iteration procedure described at the end of Section 5.3 for solving a nonlinear system. In this case, we have five equations and five unknowns.

We can improve on this procedure by substituting the values of the absolute magnitudes M_1 and M_2 from Equations 2 into Equations 3. We obtain the equations

$$M_0 - m_1 - 5 = 5 \log p + 8.75 \log \mu_1$$
$$M_0 - m_2 - 5 = 5 \log p + 8.75 \log \mu_2$$

Substituting $M_0 = 4.8$ and combining these equations with Equation 1, we obtain the system of three equations

$$p^3 = \frac{\alpha^3}{T^2(\mu_1 + \mu_2)}$$

$$-m_1 - 0.2 = 5 \log p + 8.75 \log \mu_1$$

$$-m_2 - 0.2 = 5 \log p + 8.75 \log \mu_2$$

to solve for the three unknowns p, μ_1, and μ_2. We assume that α, T, m_1, and m_2 are given. The procedure is to choose an initial estimate of the sum of the masses and substitute in the first equation. We solve for p, then compute μ_1 and μ_2 from the next two equations. Now we return to the first equation with the new values of μ_1 and μ_2 to solve for p again. Each time we solve for p, we compare the new value with the previous one. After the difference becomes smaller than 0.0001 second of arc, we stop the iterations. This procedure can easily be converted to a machine program. The following examples illustrate some results from such a program.

EXAMPLE 5.11   Given values $T = 88$ years, $\alpha = 4.6$ seconds of arc, $m_1 = 4.2$, and $m_2 = 6.0$ for a certain double star system, find the dynamical parallax.

Solution   The machine results are as follows.

```
ENTER ALPHA    ?   4.6
ENTER VISUAL MAGNITUDES    ?   4.2, 6.0
ENTER PERIOD IN YEARS    ?   88
```

PARALLAX	MASS #1	MASS #2
0.184541	0.825123	0.513812
0.210952	0.764406	0.476003
0.216395	0.753358	0.469123
0.217448	0.751272	0.467824
0.217649	0.750875	0.467577
0.217687	0.750799	0.467530

The iterations have "converged" on a parallax of about 0.2177 and masses of about 0.751 and 0.468 solar masses. The distance to this system in light-years is given by $3.258/p$, so in this case, the distance is about 15.0 light-years. This particular system is sufficiently close to allow direct measurement of the parallax, so it is interesting to note that the accepted value is 0.199 and the corresponding correct distance is 16.4 light-years. It is only expected in the case of dynamical parallaxes that the estimated values are somewhere in the neighborhood of the true values. For stars at great distances, even such approximations are better than nothing. ■

EXAMPLE 5.12   A certain binary system has an angular separation of 2 seconds of arc, with visual magnitudes of 2.91 and 6.44. The period is 320 years. Find the dynamical parallax.

Solution

PARALLAX	MASS #1	MASS #2
3.393022E − 02	3.049553	1.204502
2.638329E − 02	3.521025	1.390723
2.514884E − 02	3.618772	1.42933
2.492034E − 02	3.637695	1.436805
2.487705E − 02	3.641311	1.438233

The final value of the parallax is 0.0249, with masses about 3.64 and 1.44 solar masses. For this system, the directly measured parallax is 0.021 second, with a corresponding distance of about 155 light-years. For stars at distances greater than 100 light-years, however, direct measurement of parallaxes is difficult, and the values obtained may contain much error. In this case, it is interesting to note that the dynamical parallax gives a distance of about 131 light-years, so there is at least approximate agreement. It is difficult to say in this particular case which value is more precise. ■

The next application is taken from the Bohr theory of the hydrogen spectrum in atomic physics. A certain constant known as the Rydberg constant, usually denoted by R_∞, plays an important part in that theory, but it cannot be measured directly. The value must be inferred from measurements of other constants. Approximate values R_H and R_{He} can be determined from measurements of the spectrum of hydrogen and helium, respectively. If we knew the exact mass M_H of the nucleus of a hydrogen atom and the exact mass m of an electron, then theoretically, we could determine R_∞ from the equation

$$R_\infty = \left(1 + \frac{m}{M_H}\right) R_H$$

Suppose, for the moment, that m and M_H are unknown (because even our best values are inferred from other measured variables). The atomic weight A_H of hydrogen can be measured. Also, $A_H = M_H + m$. Similar equations hold for the helium atom. So we have a system of four equations:

$$R_\infty = \left(1 + \frac{m}{M_H}\right) R_H \qquad R_\infty = \left(1 + \frac{m}{M_{He}}\right) R_{He}$$

$$A_H = M_H + m \qquad A_{He} = M_{He} + m$$

In this system, the numbers A_H, A_{He}, R_H, and R_{He} can be taken as known because their values can be determined from laboratory measurements. The number R_∞ and masses m, M_H, and M_{He} will be regarded as unknown. Their values can be determined by solving the system of four equations.

The work is simplified if we first make some algebraic changes in the system. Let $x_1 = R_\infty$, $x_2 = m$, $x_3 = M_H$, and $x_4 = M_{He}$. Then our system becomes

$$x_1 = \left(1 + \frac{x_2}{x_3}\right)R_H \tag{1}$$

$$x_1 = \left(1 + \frac{x_2}{x_4}\right)R_{He} \tag{2}$$

$$x_3 + x_2 = A_H \tag{3}$$

$$2x_2 + x_4 = A_{He} \tag{4}$$

Equations 1 and 2 can be written in the equivalent forms

$$x_3 x_1 = (x_3 + x_2)R_H = A_H R_H$$

and

$$x_4 x_1 = (x_4 + x_2)R_{He} = (A_{He} - x_2)R_{He}$$

Now, using $x_3 = A_H - x_2$ from Equation 3 and $x_4 = A_{He} - 2x_2$ from Equation 4, we can eliminate the variables x_3 and x_4 from our system. The result is the following system of two equations with two unknowns:

$$A_H x_1 - x_2 x_1 = A_H R_H$$
$$A_{He} x_1 + R_{He} x_2 - 2x_2 x_1 = A_{He} R_{He}$$

Solving the second equation for x_1 and the first equation for x_2, we obtain

$$x_1 = \frac{(A_{He} - x_2)R_{He}}{A_{He} - 2x_2} \quad \text{and} \quad x_2 = A_H\left(1 - \frac{R_H}{x_1}\right)$$

These two equations can be used to solve for x_1 and x_2 iteratively. Since $x_1 = R_\infty$ and R_∞ is approximated by the measured number R_H, we have an initial estimate of x_1. Similarly, since x_2 is the mass of the electron and we know that the value is somewhere around $\frac{1}{2000}$ of a unit of atomic mass, an initial estimate of 0.0005 for x_2 seems reasonable. We need values of A_H, A_{He}, R_H, and R_{He} to perform the calculations. Suppose laboratory measurements produced the following values: $A_H = 1.00812$, $A_{He} = 4.00388$, $R_H = 109677.68$, and $R_{He} = 109722.34$. What are the corresponding values for Rydberg constant R_∞ and masses m, M_H, and M_{He}? A machine implementation of the iterations suggested here produced the following results.

```
ENTER B1, B2, A1, A2 ? 109677.68, 109722.34, 1.00812, 4.00388
ENTER ESTIMATES OF X1, X2   ?   109700, .0005
    109736.0454245943              2.051161203281595D – 04
    109727.9615787803              5.361898327421165D – 04
    109737.0376844038              4.619594173690497D – 04
    109735.0024592504              5.453005663704973D – 04
    109737.2874899085              5.266133533002925D – 04
    109736.7751111532              5.475942052249215D – 04
    109737.3503791225              5.428896866651015D – 04
    109737.2213862266              5.481716333877225D – 04
    109737.3662116304              5.469872621569454D – 04
    109737.333737355               5.483170021853407D – 04
    109737.3701975002              5.480188341937248D – 04
    109737.3620220271              5.483535991033088D – 04
    109737.3712009519              5.482785346513505D – 04
    109737.3691427581              5.483628124592398D – 04
    109737.3714535731              5.483439148158865D – 04
    109737.3709354182              5.483651319421784D – 04
    109737.3715171711              5.483603744181861D – 04

RINFINITY =                        109737.37
ELECTRON MASS =                    0.00054837
MASS OF HYDROGEN NUCLEUS =         1.00757163
MASS OF HELIUM NUCLEUS =           4.00278327
HYDROGEN MASS/ELECTRON MASS =      1837.41
```

The value of one atomic mass unit is 1.6599×10^{-24} g. So the computed mass of the electron in this case is equivalent to

$$0.00054837 \times 1.6599 \times 10^{-24} = 9.103 \times 10^{-28} \text{ g}$$

The accepted value is 9.11×10^{-28} g, so the computed value is quite satisfactory.

■ *Exercise Set 5.6*

1. Use the dynamical parallax method with the procedure illustrated in this section to estimate the distance in light-years to the binary system with the given values of T, α, m_1, and m_2.
   a. $T = 40.2$, $\alpha = 4.26$, $m_1 = 0.5$, $m_2 = 13.5$
   b. $T = 420$, $\alpha = 2.0$, $m_1 = 2.0$, $m_2 = 2.8$
   c. $T = 620$, $\alpha = 4.3$, $m_1 = 2.1$, $m_2 = 3.4$

2. Suppose that laboratory measurements produced values $A_H = 1.00815$, $A_{He} = 4.00393$, $R_H = 109677.89$, and $R_{He} = 109722.05$. Use the method described in this section to find corresponding values of the Rydberg constant R_∞ and masses m, M_H, and M_{He}.

3. If the system of differential equations

$$\frac{dx}{dt} = f(x, y)$$

$$\frac{dy}{dt} = g(x, y)$$

describes a physical or biological system, the steady-state solutions (if any exist) are found by solving the nonlinear system

$$f(x, y) = 0$$

$$g(x, y) = 0$$

Find the steady-state solutions, if any exist, for each of the following systems.

a. $x'(t) = -4y + 2xy - 8$
$y'(t) = 4y^2 - x^2$

b. $x'(t) = -x^2 - y^2 + 1$
$y'(t) = 2xy + y^2 + 2$

4. An engineer has the following heat flow problem. A wall consists of three layers of different materials with thicknesses $x_1 = 2$ in., $x_2 = 3.5$ in., and $x_3 = 6.0$ in. Inside and outside wall temperatures are known to be $T_1 = 1000°$ F and $T_4 = 185°$ F. The engineer needs to determine interior temperatures T_2 and T_3 and the heat flow Q. She has the following mathematical facts to offer. Heat conductivities k_1, k_2, and k_3 are temperature dependent with

$$k_i = 0.5a_i(T_i + T_{i+1}) + b_i \qquad i = 1, 2, 3$$

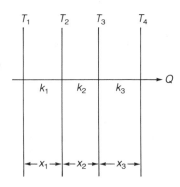

Values of the coefficients are $a_1 = 0.0021$, $a_2 = 0.00015$, $a_3 = 0.00052$, $b_1 = 0.0596$, $b_2 = 0.0257$, and $b_3 = 1.9978$. The heat flow is

$$Q = \frac{T_1 - T_4}{\dfrac{x_1}{k_1} + \dfrac{x_2}{k_2} + \dfrac{x_3}{k_4}}$$

Also,

$$Q = \frac{k_1(T_1 - T_2)}{x_1} = \frac{k_2(T_2 - T_3)}{x_2} = \frac{k_3(T_3 - T_4)}{x_3}$$

Develop an iterative procedure for solving this problem using the preceding equations. Begin by estimating temperatures T_2 and T_3, and show how better estimates can be found. Then repeat the process until satisfactory results are obtained.

6 EIGENVALUE PROBLEMS

It is assumed that you are familiar with the definitions and simpler theorems concerning eigenvalues of an $n \times n$ matrix. Nevertheless, we will review some of this material before proceeding to numerical methods. Most linear algebra courses contain some material on eigenvalues, but the amount included varies widely. In the opening section, we will review and collect the material that is most relevant to the numerical problem of estimating eigenvalues.

6.1 General Theory

There are a number of places in the study of both pure and applied mathematics where we must consider the following problem. Given an $n \times n$ matrix $\mathbf{A}$, we need to find a nonzero vector $\mathbf{x}$ and a number λ such that $\mathbf{A}\mathbf{x} = \lambda\mathbf{x}$. If $\mathbf{x} \neq \mathbf{0}$ satisfies this equation for some number λ, then we say that $\mathbf{x}$ is an **eigenvector** of $\mathbf{A}$ and λ is an **eigenvalue** of $\mathbf{A}$. If an eigenvector $\mathbf{x}$ exists, then $\mathbf{x}$ must be a nontrivial solution of the system

$$(\mathbf{A} - \lambda\mathbf{I})\mathbf{x} = 0$$

It follows that matrix $\mathbf{A} - \lambda\mathbf{I}$ is singular and

$$\det(\mathbf{A} - \lambda\mathbf{I}) = \mathbf{0}$$

Since

$$\mathbf{A} - \lambda\mathbf{I} = \begin{bmatrix} a_{11} - \lambda & a_{12} & a_{13} & \cdots & a_{1n} \\ a_{21} & a_{22} - \lambda & a_{23} & & a_{2n} \\ a_{31} & a_{32} & a_{33} - \lambda & \cdots & a_{3n} \\ \vdots & & & \ddots & \vdots \\ a_{n1} & a_{n2} & & \cdots & a_{nn} - \lambda \end{bmatrix}$$

it is clear that $\det(\mathbf{A} - \lambda\mathbf{I}) = P(\lambda)$ is a polynomial of degree n. So any eigenvalue of $\mathbf{A}$ is a solution of a polynomial equation $P(\lambda) = 0$. We call this polynomial equation the **characteristic equation** for matrix **A**. Some writers call the solutions **characteristic values** for **A**. This is an alternate term for eigenvalues.

EXAMPLE 6.1   Find the eigenvalues of the matrix

$$\mathbf{A} = \begin{bmatrix} 1 & 2 \\ 2 & 1 \end{bmatrix}$$

Solution   Since

$$\mathbf{A} - \lambda\mathbf{I} = \begin{bmatrix} 1 - \lambda & 2 \\ 2 & 1 - \lambda \end{bmatrix}$$

then

$$P(\lambda) = (1 - \lambda)^2 - 4 = \lambda^2 - 2\lambda - 3 = 0$$

The eigenvalues are clearly 3 and -1.   ■

EXAMPLE 6.2   Find the eigenvalues of the matrix

$$\mathbf{A} = \begin{bmatrix} 3 & 1 & 2 \\ 4 & 1 & -6 \\ 1 & 0 & 1 \end{bmatrix}$$

Solution   In this case,

$$P(\lambda) = \det(\mathbf{A} - \lambda\mathbf{I}) = -\lambda^3 + 5\lambda^2 - \lambda - 9$$

We can solve the cubic equation $P(\lambda) = 0$ using the methods of Chapter 3. We find the values to be approximately 1.85956464, 4.27307281, and -1.13263749. The eigenvalues are irrational roots of a polynomial equation of degree 3.   ■

The situation illustrated in Example 6.2 is a common one and the case that concerns us most. That is, we are most interested in the case where the roots are irrational real roots or possibly complex roots of a polynomial equation of degree 3 or more. Although in theory we could compute polynomial $P(\lambda)$ and then use standard root-finding methods from Chapter 3 or Bairstow's method from Chapter 5, it is not feasible to do this when the matrix is larger than 3×3. Even in the case of 3×3 matrices, the computations in this procedure would sometimes be quite difficult to carry out.

There are a number of definitions and well-known theorems that will be useful in the development of numerical methods for the eigenvalue problem. Proofs of the theorems can be found in any standard linear algebra text. All matrices in the following are $n \times n$ with $n \geq 2$. The entries in some matrices may be complex numbers.

Definition 6.1

If there exists a nonsingular matrix $\mathbf{P}$ such that $\mathbf{A} = \mathbf{P}^{-1}\mathbf{BP}$, then matrix $\mathbf{A}$ is *similar* to matrix $\mathbf{B}$.

Theorem 6.1

Suppose that matrix $\mathbf{A}$ is similar to $\mathbf{B}$ with $\mathbf{A} = \mathbf{P}^{-1}\mathbf{BP}$. If λ is an eigenvalue of $\mathbf{A}$ with associated eigenvector $\mathbf{x}$, then λ is also an eigenvalue of $\mathbf{B}$ and $\mathbf{Px}$ is an associated eigenvector of $\mathbf{B}$.

Theorem 6.2

If $\mathbf{A}$ is triangular or diagonal, then the diagonal entries of $\mathbf{A}$ are the eigenvalues of $\mathbf{A}$.

Theorem 6.3

If $\lambda_1, \lambda_2, \ldots, \lambda_k$ are distinct real eigenvalues of real matrix $\mathbf{A}$ with associated eigenvectors $\mathbf{x}^{(1)}, \mathbf{x}^{(2)}, \ldots, \mathbf{x}^{(k)}$, then $\{\mathbf{x}^{(1)}, \mathbf{x}^{(2)}, \ldots, \mathbf{x}^{(k)}\}$ is a linearly independent set of vectors in R^n.

Definition 6.2

Real matrix $\mathbf{P}$ is *orthogonal* if and only if $\mathbf{P}^{-1} = \mathbf{P}^t$.

Theorem 6.4

If $\mathbf{A}$ is a real symmetric matrix, then there exists an orthogonal matrix $\mathbf{P}$ such that $\mathbf{P}^{-1}\mathbf{AP} = \mathbf{D}$ is a diagonal matrix. (Note that in this case, the diagonal entries of $\mathbf{D}$ are the eigenvalues of matrix $\mathbf{A}$.)

Theorem 6.5

The eigenvalues of a real symmetric matrix are all real numbers.

All of the preceding definitions and theorems are contained in virtually every introduction to linear algebra. We will now list some definitions and theorems at a slightly more difficult level that are useful in the development and study of numerical algorithms. Proofs can be found in various linear algebra texts.

Definition 6.3

If $\mathbf{A}$ is a matrix with complex entries, then the **Hermitian transpose** of $\mathbf{A}$ is the transpose of the complex conjugate of $\mathbf{A}$ and is denoted by $\mathbf{A}^H$.

Definition 6.4

If $A = A^H$ (matrix A is Hermitian), then the eigenvalues of A are all real numbers.

Definition 6.5

Matrix U is a *unitary* matrix if and only if $U^{-1} = U^H$.

Theorem 6.6

If A is any $n \times n$ matrix, then there exists a unitary matrix U such that $U^H A U = T$ is upper triangular and the eigenvalues of A are the entries on the diagonal of T.

Theorem 6.7

If A is a real matrix and all the eigenvalues of A are real, then there exists an orthogonal matrix P such that $T = P^{-1}AP$ is upper triangular and the diagonal entries of T are the eigenvalues of A.

Theorem 6.8

If A is a real $n \times n$ matrix with n distinct real eigenvalues, then A is similar to a diagonal matrix.

These theorems and definitions are only a small part of the general theory of matrix eigenvalue problems, but these few results provide an adequate theoretical base for understanding the numerical methods to be discussed in the remainder of this chapter.

■ *Exercise Set 6.1*

1. Given that λ is an eigenvalue of A with associated eigenvector x, show that each of the following is true.
   a. λ^2 is an eigenvalue of A^2.
   b. If A is nonsingular, then λ^{-1} is an eigenvalue of A^{-1}.
   c. If A is nonsingular, then $\lambda \neq 0$.
   d. λ is also an eigenvalue of the transpose of A.

2. Suppose $n \times n$ matrix A has eigenvalues $\lambda_1, \lambda_2, \ldots, \lambda_n$ with characteristic polynomial

$$P(\lambda) = (-1)^n(\lambda^n + a_{n-1}\lambda^{n-1} + \cdots + a_1\lambda + a_0)$$

   a. Show that $P(0) = \lambda_1 \lambda_2 \ldots \lambda_n$.
   b. Show that $\det A = \lambda_1 \lambda_2 \ldots \lambda_n$.
   c. Show that $a_{n-1} = -(\lambda_1 + \lambda_2 + \cdots + \lambda_n)$.

3. Verify that the results in exercise 2 are true for the matrix

$$\begin{bmatrix} 1 & 2 \\ 2 & 1 \end{bmatrix}$$

4. Repeat exercise 3 for the matrix

$$\begin{bmatrix} 1 & 5 \\ 2 & 3 \end{bmatrix}$$

5. Verify the characteristic polynomial $P(\lambda)$ and the eigenvalues given in Example 6.2.

6. Find all eigenvalues of the matrix

$$\begin{bmatrix} \dfrac{1}{\sqrt{2}} & \dfrac{-1}{\sqrt{2}} & 0 \\ \dfrac{1}{\sqrt{2}} & \dfrac{1}{\sqrt{2}} & 0 \\ 0 & 0 & 1 \end{bmatrix}$$

7. Verify $\det \mathbf{A} = \lambda_1 \lambda_2 \lambda_3$ for the matrix in exercise 6.

8. Find all eigenvalues for each of the following matrices.

a. $\begin{bmatrix} 2 & 3 & 5 & 6 \\ 0 & 4 & -1 & 0 \\ 0 & 0 & 1 & 0 \\ 0 & 0 & 0 & 3 \end{bmatrix}$ **b.** $\begin{bmatrix} 1 & 0 & 0 & 0 \\ 4 & 1 & 0 & 0 \\ 0 & 3 & 1 & 0 \\ 0 & 0 & 2 & 1 \end{bmatrix}$

c. $\begin{bmatrix} 1 & -1 & 0 & 0 \\ 2 & 0 & 0 & 0 \\ 0 & 0 & 3 & 5 \\ 0 & 0 & 1 & 7 \end{bmatrix}$ **d.** $\begin{bmatrix} 0 & 0 & 0 & 4 \\ 0 & 0 & 3 & 0 \\ 0 & 2 & 0 & 0 \\ 1 & 0 & 0 & 0 \end{bmatrix}$

9. Verify that -2, -1, and -3 are eigenvalues of the matrix

$$\begin{bmatrix} -33 & -16 & -72 \\ 24 & 10 & 57 \\ 8 & 4 & 17 \end{bmatrix}$$

Find associated eigenvectors for each eigenvalue.

10. Let $\mathbf{P}$ be the matrix

$$\begin{bmatrix} \dfrac{1}{\sqrt{2}} & 0 & \dfrac{-1}{\sqrt{2}} \\ 0 & 1 & 0 \\ \dfrac{1}{\sqrt{2}} & 0 & \dfrac{1}{\sqrt{2}} \end{bmatrix}$$

a. Show that $\mathbf{P}^{-1} = \mathbf{P}^t$.
b. Compute matrix $\mathbf{B} = \mathbf{P}^{-1}\mathbf{AP}$ for the matrix $\mathbf{A}$ in exercise 9.
c. Verify that $\mathbf{B}$ and $\mathbf{A}$ have the same eigenvalues.

6.2   Numerical Estimation of Eigenvalues

Our primary concern in this text is the exploration of numerical algorithms for solving problems. So we are naturally most concerned with finding numerical values of eigenvalues of a matrix by use of a machine algorithm. Sometimes we can find the eigenvalues of a matrix by using the methods of Chapter 3 to estimate the roots of the characteristic polynomial equation

$$P(\lambda) = \det(\mathbf{A} - \lambda\mathbf{I}) = 0$$

We will begin by exploring this approach via some numerical examples. For small matrices, it is sometimes feasible to solve the characteristic polynomial equation directly, but as the matrix size grows, this method quickly becomes impractical.

EXAMPLE 6.3 Find the eigenvalues of the matrix

$$\mathbf{A} = \begin{bmatrix} 3 & 1 & 2 \\ 4 & 1 & -6 \\ 1 & 0 & 1 \end{bmatrix}$$

Solution The determinant of $\mathbf{A} - \lambda\mathbf{I}$ is

$$P(\lambda) = -\lambda^3 + 5\lambda^2 - \lambda - 9$$

We find $P(1) = -6$ and $P(2) = 1$, so there is a root in the interval $(1, 2)$. Also, $P(4) = 3$ and $P(5) = -14$, so there is a root in the interval $(4, 5)$. Finally, note that $P(-2) = 21$ and $P(-1) = -4$, so there is a root in the interval $(-2, -1)$. Using the Newton algorithm from Chapter 3, we find that precise values of the eigenvalues are 1.85956464, 4.27307281, and -1.13263749. ∎

EXAMPLE 6.4 Find all eigenvalues of the matrix

$$\mathbf{A} = \begin{bmatrix} 8 & -1 & -5 \\ -4 & 4 & -2 \\ 18 & -5 & -7 \end{bmatrix}$$

Solution Polynomial $P(\lambda)$ is the determinant of the matrix

$$\begin{bmatrix} 8 - \lambda & -1 & -5 \\ -4 & 4 - \lambda & -2 \\ 18 & -5 & -7 - \lambda \end{bmatrix}$$

Some computation yields

$$P(\lambda) = -\lambda^3 + 5\lambda^2 - 24\lambda + 20$$

The roots of $P(\lambda) = 0$ are easily found to be real number 1 and complex numbers $2 \pm 4i$. ∎

One advantage of the simple direct method used in Example 6.3 and 6.4 is that complex eigenvalues present no particular problem if $P(\lambda)$ is known. We have already developed methods for finding real and complex roots of a polynomial equation.

In the case of real eigenvalues, the present method can be implemented without actually determining $P(\lambda)$ explicitly. We can simply evaluate the determinant of $A - \lambda I$ directly for selected values of λ as needed in the numerical work. If the determinant is large, we should evaluate the determinant by first converting the matrix $A - \lambda I$ to triangular form using elementary row operations.

EXAMPLE 6.5   Find eigenvalues of the matrix

$$A = \begin{bmatrix} 2 & 3 & 1 \\ 4 & -1 & 2 \\ 0 & 1 & 3 \end{bmatrix}$$

Solution

$$P(\lambda) = \det \begin{bmatrix} 2-\lambda & 3 & 1 \\ 4 & -1-\lambda & 2 \\ 0 & 1 & 3-\lambda \end{bmatrix}$$

So

$$P(2) = \det \begin{bmatrix} 0 & 3 & 1 \\ 4 & -3 & 2 \\ 0 & 1 & 1 \end{bmatrix} = -8$$

$$P(3) = \det \begin{bmatrix} -1 & 3 & 1 \\ 4 & -4 & 2 \\ 0 & 1 & 0 \end{bmatrix} = 6$$

We have located an eigenvalue in the interval (2, 3).

$$P(2.5) = \det \begin{bmatrix} -0.5 & 3 & 1 \\ 4 & -3.5 & 2 \\ 0 & 1 & 0.5 \end{bmatrix} = \det \begin{bmatrix} -0.5 & 3 & 1 \\ 0 & 20.5 & 10 \\ 0 & 1 & 0.5 \end{bmatrix}$$

$$= -\det \begin{bmatrix} -0.5 & 3 & 1 \\ 0 & 1 & 0.5 \\ 0 & 20.5 & 10 \end{bmatrix} = -\det \begin{bmatrix} -0.5 & 3 & 1 \\ 0 & 1 & 0.5 \\ 0 & 0 & -0.25 \end{bmatrix}$$

So $P(2.5) = -0.125$, and we know that the eigenvalue is in the interval (2.5, 3). Similarly, we find $P(2.6) = 1.2640$, so the eigenvalue lies in (2.5, 2.6).

Letting the value be denoted by $2.5 + h$, we can approximate h by solving the proportion

$$\frac{h}{0.1} = \frac{0.125}{0.125 + 1.2640}$$

We find that h is $0.0089993 \cong 0.009$, so the eigenvalue is approximated by 2.509. Similar calculations show that there are eigenvalues with approximate values 4.904 and -3.410. ∎

The calculations are tedious in this method, but they can be accomplished with the aid of a small calculator. If approximate values with two- or three-digit accuracy are sufficient, this method can be used with a pocket calculator. The procedure can be converted to a machine program, of course, if that is desired. We simply need to modify one of the algorithms given in Chapter 3 for finding roots of polynomial equations so that the input is matrix **A** and the polynomial evaluations are accomplished by evaluating determinants $P(\lambda) = \det(\mathbf{A} - \lambda\mathbf{I})$ for the various values of λ.

■ *Exercise Set 6.2*

1. Determine the characteristic polynomial $P(\lambda)$ for each matrix. Then solve $P(\lambda) = 0$ to determine the eigenvalues.

a.
$$\begin{bmatrix} 2 & -1 & 0 \\ -1 & 2 & -1 \\ 0 & -1 & 2 \end{bmatrix}$$
b.
$$\begin{bmatrix} 2 & 0 & 0 \\ 0 & 1 & 4 \\ 0 & 4 & 3 \end{bmatrix}$$

c.
$$\begin{bmatrix} 1 & 3 & 0 & 0 \\ 3 & 5 & 0 & 0 \\ 0 & 0 & 2 & 6 \\ 0 & 0 & 6 & 1 \end{bmatrix}$$
d.
$$\begin{bmatrix} 1 & 2 & 0 & 0 \\ 0 & 3 & 5 & 0 \\ 0 & 0 & 4 & 3 \\ 0 & 0 & 3 & 2 \end{bmatrix}$$

2. Estimate all eigenvalues of each matrix with absolute error less than 0.01. Do not construct $P(\lambda)$ explicitly. Use $P(\lambda)$ in determinant form with the numerical procedure illustrated in Example 6.5.

a.
$$\begin{bmatrix} 2 & -1 & 0 \\ -1 & 2 & -1 \\ 0 & -1 & 2 \end{bmatrix}$$
b.
$$\begin{bmatrix} 2 & 0.5 & -1.5 \\ 0.5 & 0 & 1 \\ -1.5 & 1 & -1 \end{bmatrix}$$

c.
$$\begin{bmatrix} 3 & 1 & 2 \\ 4 & 1 & -6 \\ 1 & 0 & 1 \end{bmatrix}$$
d.
$$\begin{bmatrix} 1 & -1 & 2 & 2 \\ -1 & 2 & 1 & -1 \\ 2 & 1 & 3 & 2 \\ 2 & -1 & 2 & 1 \end{bmatrix}$$

e.
$$\begin{bmatrix} 5 & 7 & 6 & 5 \\ 7 & 10 & 8 & 7 \\ 6 & 8 & 10 & 9 \\ 5 & 7 & 9 & 10 \end{bmatrix}$$

3. Given that the matrix

$$\begin{bmatrix} 10 & 7 & 8 & 7 \\ 7 & 5 & 6 & 5 \\ 8 & 6 & 10 & 5 \\ 7 & 5 & 9 & 10 \end{bmatrix}$$

has an eigenvalue in the interval (0.8, 0.9), find the value with an error less than 0.001.

4. Consider the matrix

$$\mathbf{A} = \begin{bmatrix} 1 & 1 & 1 \\ 1 & 2 & 1 \\ 1 & 1 & 2 \end{bmatrix}$$

   a. Compute det $\mathbf{A}$.
   b. Given that two of the eigenvalues have approximate values 1.000000 and 0.267949, estimate the third eigenvalue by using this information with the value of det $\mathbf{A}$ from part (a).

5. Consider the matrix

$$\mathbf{A} = \begin{bmatrix} 3 & 3 & 4 \\ 4 & 7 & 8 \\ 5 & 1 & 3 \end{bmatrix}$$

   a. Show that $\mathbf{A} - \mathbf{I}$ is a singular matrix.
   b. Use the result in part (a) to find one eigenvalue.
   c. Show that $\det(\mathbf{A} + \mathbf{I}) = 24$ and $\det \mathbf{A} = -1$.
   d. Use the information in parts (a), (b), and (c) to determine $P(\lambda)$.
   e. Solve $P(\lambda) = 0$ to find the remaining eigenvalues.

6. Suppose that matrix $\mathbf{A}$ has an eigenvalue λ_1. Let $\mathbf{B} = \mathbf{A} + c\mathbf{I}$. Show that $\lambda_1 + c$ is an eigenvalue of $\mathbf{B}$.

7. Combine the results of exercise 1 with the result in exercise 6 to find eigenvalues of each matrix.

   a. $\begin{bmatrix} 7 & -1 & 0 \\ -1 & 7 & -1 \\ 0 & -1 & 7 \end{bmatrix}$ b. $\begin{bmatrix} -1 & 0 & 0 \\ 0 & -2 & 4 \\ 0 & 4 & 0 \end{bmatrix}$

   c. $\begin{bmatrix} -1 & 3 & 0 & 0 \\ 3 & 3 & 0 & 0 \\ 0 & 0 & 0 & 6 \\ 0 & 0 & 6 & -1 \end{bmatrix}$ d. $\begin{bmatrix} 5 & 2 & 0 & 0 \\ 0 & 7 & 5 & 0 \\ 0 & 0 & 8 & 3 \\ 0 & 0 & 3 & 6 \end{bmatrix}$

8. Let $\mathbf{A}$ be the matrix

$$\begin{bmatrix} \dfrac{1}{\sqrt{2}} & 0 & \dfrac{-1}{\sqrt{2}} \\ 0 & 1 & 0 \\ \dfrac{1}{\sqrt{2}} & 0 & \dfrac{1}{\sqrt{2}} \end{bmatrix}$$

   a. Verify that $\mathbf{A}^{-1} = \mathbf{A}^t$.
   b. Verify that $\|\mathbf{A}\| = 1$.
   c. Use the results in parts (a) and (b) to show that $|\lambda| = 1$ for every eigenvalue λ of $\mathbf{A}$.
   d. Determine all eigenvalues of $\mathbf{A}$.

6.3   The Power Method

In this section, we begin our investigation of some numerical methods that are more sophisticated than merely trying to determine the zeros of the characteristic polynomial directly. These methods are also better suited for machine estimation of eigenvalues.

Suppose that A is a real $n \times n$ matrix with eigenvalues $\lambda_1, \lambda_2, \lambda_3, \ldots, \lambda_n$, where

$$|\lambda_1| > |\lambda_2| \geq |\lambda_3| \geq \cdots \geq |\lambda_n|$$

That is, we assume that A has a dominant eigenvalue λ_1. We also assume that the associated eigenvectors $x^1, x^2, x^3, \ldots, x^n$ form a basis for R^n. Suppose that v is any nonzero vector in R^n. Then v is a linear combination of the basis vectors, so we have

$$v = c_1 x^1 + c_2 x^2 + \cdots + c_n x^n$$

for some real numbers $c_1, c_2, \ldots, c_n$. It follows that

$$Av = c_1 \lambda_1 x^1 + c_2 \lambda_2 x^2 + \cdots + c_n \lambda_n x^n$$
$$A^2 v = c_1 \lambda_1^2 x^1 + c_2 \lambda_2^2 x^2 + \cdots + c_n \lambda_n^2 x^n$$
$$\vdots \qquad\qquad \vdots$$
$$A^m v = c_1 \lambda_1^m x^1 + c_2 \lambda_2^m x^2 + \cdots + c_n \lambda_n^m x^n$$

Then

$$A^m v = \lambda_1^m \left[c_1 x^1 + \left(\frac{\lambda_2}{\lambda_1} \right)^m c_2 x^2 + \cdots + \left(\frac{\lambda_n}{\lambda_1} \right)^m c_n x^n \right]$$

Since $|\lambda_j/\lambda_1| < 1$ for each j, for large m we have

$$A^m v \cong \lambda_1^m c_1 x^1$$

It follows that $A^{m+1} v \cong \lambda_1 A^m v$ for large m. If $(A^m v)_i$ is not zero, we may conclude that

$$\lambda_1 \cong \frac{(A^{m+1} v)_i}{(A^m v)_i}$$

if m is large. Also note that if we let

$$u^m = \frac{A^m v}{\| A^m v \|} \qquad m = 1, 2, 3, \ldots$$

then for large m, vector $\mathbf{u}^m$ is approximately

$$\frac{\lambda_1^m c_1 \mathbf{x}^1}{|\lambda_1^m| \cdot |c_1| \cdot \|\mathbf{x}^1\|} = \frac{\pm \mathbf{x}^1}{\|\mathbf{x}^1\|}$$

So the sequence of vectors $\mathbf{u}^1, \mathbf{u}^2, \ldots, \mathbf{u}^m, \ldots$ converges to an eigenvector of $\mathbf{A}$ associated with the dominant eigenvalue λ_1. In forming these conclusions, we have assumed implicitly that vector $\mathbf{v}$ has been selected so that $c_1 \neq 0$. If we start with a vector $\mathbf{v}$ for which $c_1 = 0$, the argument fails.

The preceding argument suggests the following algorithm for estimating the dominant eigenvalue of a matrix when one exists. In the application of this algorithm, we assume that $\mathbf{A}$ is a real $n \times n$ matrix with a dominant eigenvalue and n eigenvectors that form a basis for R^n.

Algorithm 6.1

Given an $n \times n$ matrix $\mathbf{A}$ satisfying the preceding conditions with dominant eigenvalue λ_1.

Step 1 Select initial vector $\mathbf{v} \neq 0$. Select $M =$ maximum number of iterations permitted. Select a value for TOL = maximum permissible value for the absolute value of the difference between successive estimates of the eigenvalue. Set $k = 0$. Set LAST $= 0$.
Step 2 Compute product $\mathbf{y} = \mathbf{A}\mathbf{v}$.
Step 3 Determine j for which

$$|y_j| = \max_{1 \leq i \leq M} |y_i|$$

Set $S = |y_j|$.
Step 4 Set $L = y_j/v_j$.
Step 5 Set $\mathbf{v} = \mathbf{y}/S$.
Step 6 Output vector $\mathbf{v}$ and number L (L is the eigenvalue estimate).
Step 7 Set $k = k + 1$. If $k \geq M$, then stop.
Step 8 If $|\text{LAST} - L| < \text{TOL}$, then stop.
Step 9 Otherwise, set LAST $= L$ and go to step 2.

We will now consider some numerical output from this algorithm.

EXAMPLE 6.6   Estimate the dominant eigenvalue of

$$\mathbf{A} = \begin{bmatrix} 1 & 2 & 3 \\ 2 & 4 & 5 \\ 3 & 5 & 6 \end{bmatrix}$$

Solution   The initial vector is $\mathbf{v} = \begin{bmatrix} 1 & 1 & 1 \end{bmatrix}^t$ for the following machine output.

EIGENVECTOR COMPONENTS			EIGENVALUE ESTIMATE
0.428571	0.785714	1.000000	14.000000
0.445860	0.802548	1.000000	11.214290
0.445006	0.801908	1.000000	11.350320
0.445044	0.801939	1.000000	11.344560
0.445042	0.801938	1.000000	11.344830
0.445042	0.801938	1.000000	11.344810

There is no change in the output for further iterations, so we see that for the given matrix, there is a dominant eigenvalue with approximate value 11.344810 and an associated eigenvector with components 0.445042, 0.801938, and 1.000000. The eigenvalues of the matrix $\mathbf{A}$ are, in fact, approximately 11.344810, -0.515730, and 0.170914, so it is easy to see why the convergence was so rapid.   ■

EXAMPLE 6.7   Estimate the dominant eigenvalue of matrix

$$\mathbf{A} = \begin{bmatrix} 3 & 1 & 2 \\ 4 & 1 & -6 \\ 1 & 0 & 1 \end{bmatrix}$$

Solution   The initial vector is the same as in Example 6.6. The machine output is as follows.

ITERATIONS	EIGENVECTOR COMPONENTS			EIGENVALUE ESTIMATE
1	1.	-0.166667	0.333333	6
2	1.	0.523810	0.380952	3.5
5	1.	0.638471	0.309524	4.256
10	1.	0.661694	0.305594	4.272607
15	1.	0.662021	0.305524	4.273066
17	1.	0.662025	0.305524	4.273072
20	1.	0.662026	0.305523	4.273073

No significant change in output occurs after iteration 20. In this case, the eigenvalues are -1.1326375, 1.8595646, and 4.2730728. There is a dominant eigenvalue, but the ratio of the absolute values of the two largest eigenvalues is not quite so large as in Example 6.6, resulting in a slower rate of convergence.   ■

EXAMPLE 6.8   Find the dominant eigenvalue of matrix

$$\mathbf{A} = \begin{bmatrix} 2 & 0.5 & -1.5 \\ 0.5 & 0 & 1 \\ -1.5 & 1 & -1 \end{bmatrix}$$

Solution   Applying Algorithm 6.1 with initial vector $\mathbf{v} = \begin{bmatrix} 1 & 1 & 1 \end{bmatrix}^t$, again we find, after 85 iterations, the eigenvalue estimates converge on 2.624015 with eigenvector components 1.000000, 0.036666, and -0.403788. The convergence is very slow because the eigenvalues are 2.624015, 2.189657, and −0.565641. The two largest eigenvalues are close together, so we should expect the convergence to be slow.   ∎

■ *Exercise Set 6.3*

1. Each of the following matrices has a dominant eigenvalue. Write a machine program that implements Algorithm 6.1, and use your program to find the dominant eigenvalue with an associated eigenvector for each matrix.

a.
$$\begin{bmatrix} 4 & -1 & 1 \\ -1 & 3 & -2 \\ 1 & -2 & 3 \end{bmatrix}$$
b.
$$\begin{bmatrix} 3 & -1 & 0 \\ -1 & 2 & -1 \\ 0 & -1 & 3 \end{bmatrix}$$

c.
$$\begin{bmatrix} 1 & 1 & -1 \\ 2 & 3 & -4 \\ 4 & 1 & -4 \end{bmatrix}$$
d.
$$\begin{bmatrix} 10 & 7 & 8 & 7 \\ 7 & 5 & 6 & 5 \\ 8 & 6 & 10 & 9 \\ 7 & 5 & 9 & 10 \end{bmatrix}$$

e.
$$\begin{bmatrix} 1 & 3 & 0 & 0 & 0 \\ 3 & 2 & 1 & 0 & 0 \\ 0 & 1 & 4 & 5 & 0 \\ 0 & 0 & 5 & 1 & 6 \\ 0 & 0 & 0 & 6 & 3 \end{bmatrix}$$

2. Use hand calculations with Algorithm 6.1 beginning with initial vector $v = [1 \ 1 \ 1]^t$ to find the first two iterates for the matrices in exercises 1a and 1b. Compare your results with the results obtained in exercises 1a and 1b.

3. The matrix
$$\begin{bmatrix} 4 & 2 & 0 \\ 2 & 3 & 1 \\ 0 & 1 & 2 \end{bmatrix}$$

has a dominant eigenvalue in interval $(5, 6)$. There is an eigenvector close to $v = [1 \ 0.8 \ 0.2]^t$. Use this vector v as the initial vector in Algorithm 6.1. Compute the first three iterates by hand to obtain an estimate of the eigenvalue.

4. Apply Algorithm 6.1 to the matrix
$$\begin{bmatrix} 4 & -1 & -1 & -1 \\ -1 & 4 & -1 & -1 \\ -1 & -1 & 4 & -1 \\ -1 & -1 & -1 & 4 \end{bmatrix}$$

using the indicated initial vector in each case.
a. $v = [1 \ 1 \ 1 \ 1]^t$ b. $v = [1 \ 0 \ 0 \ 0]^t$
c. $v = [0 \ 1 \ 0 \ 0]^t$ d. $v = [0 \ 0 \ 1 \ 0]^t$
e. $v = [0 \ 0 \ 0 \ 1]^t$

5. What conclusion do you draw about the eigenvalues and eigenvectors of the matrix in exercise 4 from examination of the numerical results obtained in exercise 4?

6. Using the determinant form of $P(\lambda)$ for the matrix

$$\begin{bmatrix} 2 & 3 & 1 \\ 4 & -1 & 2 \\ 0 & 1 & 3 \end{bmatrix}$$

it has been verified that there are three real eigenvalues in intervals $(-4, -3)$, $(2, 3)$, and $(4, 5)$. Use Algorithm 6.1 to find an accurate estimate of the dominant eigenvalue. Is it helpful to know in advance that the eigenvalue lies in the interval $(4, 5)$?

7. Consider the matrix

$$\mathbf{A} = \begin{bmatrix} 1 & 1 & 1 \\ 1 & 2 & 1 \\ 1 & 1 & 2 \end{bmatrix}$$

  a. Compute det $\mathbf{A}$.
  b. Explain why 1 must be an eigenvalue of $\mathbf{A}$.
  c. Show that $\det(\mathbf{A} - \lambda\mathbf{I}) = 0$ for $\lambda = 2 - \sqrt{3}$.
  d. Explain why $2 + \sqrt{3}$ must be an eigenvalue.
  e. Use Algorithm 6.1 to estimate the dominant eigenvalue. Compare numerical results with your result in part (d).

8. Let $\mathbf{A}$ be the matrix

$$\begin{bmatrix} 0 & 1 & 0 & 0 \\ 1 & 0 & 1 & 0 \\ 0 & 1 & 0 & 1 \\ 0 & 0 & 1 & 0 \end{bmatrix}$$

  a. Use Algorithm 6.1 with initial vector $[1 \ \ 1 \ \ 1 \ \ 1]^t$ to find an eigenvalue.
  b. Show that $P(\lambda) = \lambda^4 - 3\lambda^2 + 1$. Note that $P(\lambda)$ is an even function. What does that tell you about the eigenvalues? Does $\mathbf{A}$ have a dominant eigenvalue?
  c. Find all eigenvalues of $\mathbf{A}$.

6.4   The Householder Algorithm

Most of the algorithms for estimating eigenvalues of an $n \times n$ matrix $\mathbf{A}$ make use of Theorem 6.1, which states that similar matrices have the same set of eigenvalues. Since the eigenvalues of a diagonal matrix are the diagonal entries, an ideal algorithm would be one that could produce a diagonal matrix similar to the given matrix $\mathbf{A}$. Unfortunately, not all matrices are similar to a diagonal matrix. However, as stated in Theorems 6.4 and 6.5, if $\mathbf{A}$ is an $n \times n$ symmetric matrix, then $\mathbf{A}$ is similar to a diagonal matrix with real entries on the diagonal.

A number of methods have been developed for finding the eigenvalues of a symmetric matrix. Some of these methods do construct a diagonal matrix similar to

A that has the eigenvalues of **A** along the diagonal. The best-known algorithm of this type is one due to Jacobi. The Jacobi algorithm generates an infinite sequence of matrices

$$\mathbf{A}^{(k+1)} = \mathbf{R}^{(k)}\mathbf{A}^{(k)}(\mathbf{R}^{(k)})^t \qquad \mathbf{A}^{(0)} = \mathbf{A}$$

where $\mathbf{R}^{(k)}$ is a "rotation matrix." Matrices $\mathbf{R}^{(k)}$ are all orthogonal, so $\mathbf{A}^{(k+1)}$ is similar to $\mathbf{A}^{(k)}$ at each step in the iterative process. Rotation matrix $\mathbf{R}^{(k)}$ is selected at each step to "zero out" the largest (max absolute value) off-diagonal element in the matrix $\mathbf{A}^{(k+1)}$. Unfortunately, elements that are zeroed out at one step do not necessarily remain zero in successive steps of the algorithm. Nevertheless, using the Jacobi strategy for selecting the rotation matrices, it has been proved that the sequence of matrices $\mathbf{A}^{(k)}$ does converge to a diagonal matrix similar to the given matrix **A**. The numerical process is stopped when the maximum of the absolute values of off-diagonal elements is sufficiently small.

To avoid the difficulty of working with an infinite sequence of matrices, the method of Givens makes a different selection of the rotation matrices with the object being to produce a tridiagonal matrix similar to **A** rather than a diagonal matrix. It has been found that if **A** is $n \times n$, then in only $(n-1)(n-2)/2$ steps of the Givens procedure we arrive at the desired tridiagonal matrix. The number of steps is finite, and the computations are much simpler.

For details on the algorithms of Jacobi and Givens, consult Ralston (1965) or Fox (1965). In this section, we will explore a procedure developed by Householder that produces from a given symmetric matrix a tridiagonal matrix that is similar to the given matrix. This algorithm is more efficient than the method of Givens in that several zero elements are produced at each step, and zeros remain in all of these positions as the iterations proceed. In Section 6.5, we will investigate an algorithm for estimating the eigenvalues of a matrix in tridiagonal form. By combining that algorithm with the Householder algorithm, we will have a powerful procedure for determining eigenvalues of a symmetric matrix. Let us begin by developing the Householder algorithm.

Suppose that **u** is a column vector with $\|\mathbf{u}\| = 1$. We can define matrix **P** by

$$\mathbf{P} = \mathbf{I} - 2\mathbf{u}\mathbf{u}^t$$

where **I** is the $n \times n$ identity and $\mathbf{u}^t$ is the transpose of **u**. You can verify that **P** is a symmetric $n \times n$ matrix that is also orthogonal.

EXAMPLE 6.9   Suppose that $\mathbf{u}^t = [1/\sqrt{3} \quad 1/\sqrt{3} \quad 1/\sqrt{3}]$. Then $\mathbf{u}\mathbf{u}^t$ is the matrix

$$\begin{bmatrix} \frac{1}{3} & \frac{1}{3} & \frac{1}{3} \\ \frac{1}{3} & \frac{1}{3} & \frac{1}{3} \\ \frac{1}{3} & \frac{1}{3} & \frac{1}{3} \end{bmatrix}$$

and

$$
P = \begin{bmatrix} \frac{1}{3} & -\frac{2}{3} & -\frac{2}{3} \\ -\frac{2}{3} & \frac{1}{3} & -\frac{2}{3} \\ -\frac{2}{3} & -\frac{2}{3} & \frac{1}{3} \end{bmatrix}
$$

∎

In the case of an $n \times n$ matrix A, it is possible to select vector $\mathbf{u}$ so that in the matrix $B = PAP$, we have

$$
b_{1j} = b_{j1} = 0 \qquad j = 3, 4, \ldots, n
$$

For the moment, let us restrict our attention to the case of 4×4 matrices. In that case, we choose a vector $\mathbf{v}$ with $v_1 = 0$, so that

$$
\mathbf{v}^t = \begin{bmatrix} 0 & v_2 & v_3 & v_4 \end{bmatrix}
$$

where elements v_2, v_3, and v_4 are computed by the following procedure. Let

$$
S = a_{12}^2 + a_{13}^2 + a_{14}^2
$$

$$
Z = \frac{1 \pm a_{12}/\sqrt{S}}{2} \quad \text{and} \quad v_2 = \sqrt{Z}
$$

$$
v_j = \frac{\pm a_{1j}}{2v_2\sqrt{S}} \qquad j = 3, 4
$$

The algebraic sign in the formula for Z is chosen so as to make $|v_2|$ as large as possible. The sign in the formulas for v_3 and v_4 is chosen to be the same as selected in computing Z. In the case of a 3×3 matrix, $\mathbf{v}$ is selected similarly, with the first component being zero.

EXAMPLE 6.10   Use the Householder algorithm to find a tridiagonal matrix similar to matrix

$$
A = \begin{bmatrix} 3 & 1 & 0.5 \\ 1 & 1 & 0.25 \\ 0.5 & 0.25 & 2 \end{bmatrix}
$$

Solution   First we compute

$$
S = 1^2 + (0.5)^2 = 1.25
$$

Then

$$\sqrt{S} = 1.118034$$

Next

$$Z = \frac{1}{2}\left(1 + \frac{1}{\sqrt{S}}\right) = 0.9472136$$

So

$$v_2 = \sqrt{Z} = 0.9732490$$

and

$$v_3 = \frac{0.5}{2(v_2)\sqrt{S}} = 0.2297529$$

Then

$$\mathbf{v} = \begin{bmatrix} 0 \\ 0.9732490 \\ 0.2297529 \end{bmatrix} \quad \text{and} \quad \mathbf{vv}^t = \begin{bmatrix} 0 & 0 & 0 \\ 0 & 0.9472136 & 0.2236068 \\ 0 & 0.2236068 & 0.0527864 \end{bmatrix}$$

$$\mathbf{P} = \mathbf{I} - 2\mathbf{vv}^t = \begin{bmatrix} 1 & 0 & 0 \\ 0 & -0.8944272 & -0.4472136 \\ 0 & -0.4472136 & 0.8944272 \end{bmatrix}$$

So finally, we obtain

$$\mathbf{PAP} = \begin{bmatrix} 3 & -1.118034 & 0 \\ -1.118034 & 1.4 & -0.55 \\ 0 & -0.55 & 1.6 \end{bmatrix}$$

which is the desired tridiagonal matrix.   ∎

Note that for a 3 × 3 matrix, only one product **PAP** needs to be constructed. In the case of a 4 × 4 matrix, we need to form two such products, as the following example shows.

EXAMPLE 6.11   Use the Householder procedure to find a tridiagonal matrix similar to matrix

$$\mathbf{A} = \begin{bmatrix} 1 & -1 & 2 & 2 \\ -1 & 2 & 1 & -1 \\ 2 & 1 & 3 & 2 \\ 2 & -1 & 2 & 1 \end{bmatrix}$$

Solution   We find that

$$S = [(-1)^2 + 2^2 + 2^2] = 9 \quad \text{and} \quad \sqrt{S} = 3$$

$$Z = \frac{1 \pm -\frac{1}{3}}{2} = \frac{1 + \frac{1}{3}}{2} = \frac{2}{3}$$

$$v_2 = \sqrt{Z} = \frac{2}{\sqrt{6}}$$

$$v_3 = \frac{-a_{13}}{2v_2(3)} = \frac{-1}{\sqrt{6}}$$

$$v_4 = \frac{-a_{14}}{2v_2(3)} = \frac{-1}{\sqrt{6}}$$

So

$$\mathbf{v}^t = [0 \quad 2/\sqrt{6} \quad -1/\sqrt{6} \quad -1/\sqrt{6}]$$

Then matrix $\mathbf{P}$ is $\mathbf{I} - 2\mathbf{v}\mathbf{v}^t$, where

$$\mathbf{v}\mathbf{v}^t = \begin{bmatrix} 0 & 0 & 0 & 0 \\ 0 & \frac{2}{3} & -\frac{1}{3} & -\frac{1}{3} \\ 0 & -\frac{1}{3} & \frac{1}{6} & \frac{1}{6} \\ 0 & -\frac{1}{3} & \frac{1}{6} & \frac{1}{6} \end{bmatrix}$$

so

$$\mathbf{P} = \begin{bmatrix} 1 & 0 & 0 & 0 \\ 0 & -\frac{1}{3} & \frac{2}{3} & \frac{2}{3} \\ 0 & \frac{2}{3} & \frac{2}{3} & -\frac{1}{3} \\ 0 & \frac{2}{3} & -\frac{1}{3} & \frac{2}{3} \end{bmatrix}$$

Next we find that

$$\mathbf{PAP} = \begin{bmatrix} 1 & 3 & 0 & 0 \\ 3 & \frac{34}{9} & \frac{7}{9} & \frac{1}{9} \\ 0 & \frac{7}{9} & \frac{25}{9} & \frac{10}{9} \\ 0 & \frac{1}{9} & \frac{10}{9} & -\frac{5}{9} \end{bmatrix} = \mathbf{B}$$

This is the first step in converting the matrix to tridiagonal form. We now repeat the procedure for determination of vector $\mathbf{v}$ as if the matrix $\mathbf{A}$ were the 3×3 matrix

$$\begin{bmatrix} \frac{34}{9} & \frac{7}{9} & \frac{1}{9} \\ \frac{7}{9} & \frac{25}{9} & \frac{10}{9} \\ \frac{1}{9} & \frac{10}{9} & -\frac{5}{9} \end{bmatrix}$$

In this case, we assume that $\mathbf{v}^t$ is of the form $[0 \quad v_2 \quad v_3]$. We now find that

$$S = (\tfrac{7}{9})^2 + (\tfrac{1}{9})^2 = \tfrac{50}{81} \quad \text{and} \quad \sqrt{S} = 0.7856742$$

$$Z = \frac{1 \pm a_{12}/\sqrt{S}}{2} = \frac{1 + \tfrac{7}{9}(1/0.7856742)}{2}$$

so

$$Z = 0.99497475$$

Then

$$v_2 = \sqrt{Z} = 0.99748421$$

Next

$$v_3 = \frac{\pm a_{13}}{2v_2\sqrt{S}}$$

$$= \frac{\tfrac{1}{9}}{2v_2\sqrt{S}} = 0.07088902$$

Matrix $\mathbf{vv}^t$ is

$$\begin{bmatrix} 0 & 0 & 0 \\ 0 & v_2^2 & v_2 v_3 \\ 0 & v_3 v_2 & v_3 \end{bmatrix} = \begin{bmatrix} 0 & 0 & 0 \\ 0 & 0.994974747 & 0.070710678 \\ 0 & 0.070710678 & 0.005025253 \end{bmatrix}$$

and $\mathbf{I} - 2\mathbf{vv}^t$ with $\mathbf{I}$ as the 3×3 identity is

$$\begin{bmatrix} 1 & 0 & 0 \\ 0 & -0.98994950 & -0.14142136 \\ 0 & -0.14142136 & 0.98994950 \end{bmatrix}$$

Now we let $\mathbf{Q}$ be the 4×4 matrix

$$\begin{bmatrix} 1 & 0 & 0 & 0 \\ 0 & 1 & 0 & 0 \\ 0 & 0 & -0.98994950 & -0.14142136 \\ 0 & 0 & -0.14142136 & 0.098994950 \end{bmatrix}$$

and compute product $\mathbf{Q(PAP)Q}$ using $\mathbf{PAP}$ from above. The result is the matrix

$$\begin{bmatrix} 1 & 3 & 0 & 0 \\ 3 & 3.7777777 & -0.7856743 & 0 \\ 0 & -0.7856743 & 3.0222222 & -0.60000000 \\ 0 & 0 & -0.6000000 & -0.80000000 \end{bmatrix}$$

which is the desired tridiagonal matrix similar to the given matrix **A**. The computational procedure is easily generalized to matrices of any size n.   ∎

In the following algorithm, called the **Householder transformation**, we assume that the input is an $n \times n$ symmetric matrix **A** with n greater than 2.

Algorithm 6.2

Step 1 Set $k = 1$ and initialize **B** = **A**.

Step 2 Compute

$$s = \sum_{i=k+1}^{n} b_{ik}^2$$

If $s = 0$, then set $k = k + 1$ and recompute s.

Step 3 Let $s = \sqrt{s}$.

Step 4 Set

$$SG = \begin{cases} -1 \text{ if } b_{k+1,k} < 0 \\ +1 \text{ if } b_{k+1,k} \geq 0 \end{cases}$$

Step 5 Set

$$z = \frac{1 + SG \cdot b_{k+1,k}/s}{2}$$

Step 6 Set $v_i = 0$ for $i = 1, \ldots, k$. Set $v_{k+1} = \sqrt{z}$. Set

$$v_i = \frac{SGb_{kj}}{2v_{k+1}s} \qquad i = k+2, \ldots, n$$

Step 7 Set

$$P_{ij} = \delta_{ij} - 2v_i v_j \qquad i = 1, \ldots, n; \quad j = 1, \ldots, n$$

Step 8 Compute **A** = **PBP**.

Step 9 If $k = n - 2$, then output **A** and stop.

Step 10 Set $k = k + 1$, **B** = **A**, and go to step 2.

The following examples are samples of machine output from this algorithm.

EXAMPLE 6.12   For the matrix given in Example 6.11, using Algorithm 6.2 with machine computation, we obtain the following.

INPUT MATRIX A IS:

1.000000	−1.000000	2.000000	2.000000
−1.000000	2.000000	1.000000	−1.000000
2.000000	1.000000	3.000000	2.000000
2.000000	−1.000000	2.000000	1.000000

P MATRICES ARE:

K = 1

1.000000	0.000000	0.000000	0.000000
0.000000	−0.333333	0.666667	0.666667
0.000000	0.666667	0.666667	−0.333333
0.000000	0.666667	−0.333333	0.666667

K = 2

1.000000	0.000000	0.000000	0.000000
0.000000	1.000000	0.000000	0.000000
0.000000	0.000000	−0.989949	−0.141422
0.000000	0.000000	−0.141422	0.989949

AFTER HOUSEHOLDER TRANSFORMATION

1.000000	3.000000	−0.000000	0.000000
3.000000	3.777778	−0.785674	0.000000
0.000000	−0.785674	3.022223	−0.599999
0.000000	0.000000	−0.599999	−0.800000

Some slight rounding in the machine arithmetic is noticeable, but the results are essentially the same as in Example 6.11. ∎

EXAMPLE 6.13   Apply the Householder algorithm to the matrix

$$\mathbf{A} = \begin{bmatrix} 10 & 7 & 5 & 6 \\ 7 & 2 & 3 & 0 \\ 5 & 3 & -1 & 2 \\ 6 & 0 & 2 & 3 \end{bmatrix}$$

Solution   INPUT MATRIX IS:

10.000000	7.000000	5.000000	6.000000
7.000000	2.000000	3.000000	0.000000
5.000000	3.000000	−1.000000	2.000000
6.000000	0.000000	2.000000	3.000000

P MATRICES ARE:

K = 1

1.000000	0.000000	0.000000	0.000000
0.000000	−0.667424	−0.476731	−0.572078
0.000000	−0.476731	0.863698	−0.163562
0.000000	−0.572078	−0.163562	0.803726

K = 2

1.000000	0.000000	0.000000	0.000000
0.000000	1.000000	0.000000	0.000000
0.000000	0.000000	−0.974361	−0.224989
0.000000	0.000000	−0.224989	0.974361

AFTER HOUSEHOLDER TRANSFORMATION

10.000000	-10.488090	0.000001	0.000000
-10.488090	4.645453	0.565465	0.000000
0.000001	0.565465	-2.737723	-1.718651
0.000000	0.000000	-1.718651	2.092269

In this case, the results are quite satisfactory, although some evidence of machine rounding errors is seen in the elements in the first row, third column and third row, first column, since these elements should be zero. ∎

■ *Exercise Set 6.4*

1. Construct matrix $\mathbf{P} = \mathbf{I} - 2\mathbf{u}\mathbf{u}^t$ for each vector $\mathbf{u}$. Verify in each case that $\mathbf{P}$ is orthogonal.

a. $\begin{bmatrix} \frac{1}{\sqrt{2}} \\ \frac{1}{\sqrt{2}} \\ 0 \end{bmatrix}$ **b.** $\begin{bmatrix} \frac{1}{\sqrt{3}} \\ 0 \\ \frac{-1}{\sqrt{3}} \\ \frac{1}{\sqrt{3}} \end{bmatrix}$ **c.** $\begin{bmatrix} \frac{3}{5} \\ \frac{4}{5} \end{bmatrix}$

2. Use the Householder transformation illustrated in Example 6.10 to convert each matrix to a similar tridiagonal matrix.

a. $\begin{bmatrix} 1 & 2 & 2 \\ 2 & 1 & -1 \\ 2 & -1 & 1 \end{bmatrix}$ **b.** $\begin{bmatrix} 1 & 3 & 4 \\ 3 & 2 & 5 \\ 4 & 5 & 1 \end{bmatrix}$ **c.** $\begin{bmatrix} 1 & 0 & 1 \\ 0 & 1 & 0 \\ 1 & 0 & 1 \end{bmatrix}$

3. Use the Householder transformation illustrated in Example 6.11 to convert each matrix to a similar tridiagonal matrix.

a. $\begin{bmatrix} 1 & -1 & 1 & 1 \\ -1 & 1 & 0 & 1 \\ 1 & 0 & 2 & 1 \\ 1 & 1 & 1 & 3 \end{bmatrix}$ **b.** $\begin{bmatrix} 2 & 1 & 1 & 2 \\ 1 & 5 & -1 & 2 \\ 1 & -1 & 1 & 3 \\ 2 & 2 & 3 & 1 \end{bmatrix}$

4. Write a machine program that implements Algorithm 6.2. Use your program to find a tridiagonal matrix similar to each matrix.

a. $\begin{bmatrix} 10 & 7 & 8 & 5 \\ 7 & 6 & 5 & 6 \\ 8 & 5 & 10 & 7 \\ 5 & 6 & 7 & 6 \end{bmatrix}$ **b.** $\begin{bmatrix} 1 & 2 & -1 & 3 & 4 \\ 2 & 1 & 3 & -1 & 5 \\ -1 & 3 & 4 & 2 & 1 \\ 3 & -1 & 2 & 1 & 3 \\ 4 & 5 & 1 & 3 & 2 \end{bmatrix}$

5. Ignore the requirement $n > 2$ preceding Algorithm 6.2, and apply the algorithm to

$$A = \begin{bmatrix} 1 & 2 \\ 2 & 1 \end{bmatrix}$$

Matrix **A** is already in tridiagonal form. Does the Householder transformation transform **A** into **A**?

6. (Brief Jacobi) Let **A** be any 2×2 symmetric matrix. Let **T** be the rotation matrix

$$\begin{bmatrix} \cos \theta & -\sin \theta \\ \sin \theta & \cos \theta \end{bmatrix}$$

Define **B** = **T**^{t}**AT**, where **T**^{t} is the transpose of **T**.
a. Find the value of θ needed to produce $b_{12} = b_{21} = 0$.
b. Using your result from part (a), construct a matrix **T** such that **B** = **T**^{t}**AT** is diagonal for matrix

$$A = \begin{bmatrix} 4 & 1 \\ 1 & 2 \end{bmatrix}$$

Compute **B**.
c. What are the eigenvalues of **A**?

7. Let **A** be the matrix

$$\begin{bmatrix} 3 & \dfrac{-1}{\sqrt{2}} & 2\sqrt{2} & -2 \\[2ex] \dfrac{-1}{\sqrt{2}} & 1 & 0 & \dfrac{-1}{\sqrt{2}} \\[2ex] 2\sqrt{2} & 0 & 3 & -2\sqrt{2} \\[2ex] -2 & \dfrac{-1}{\sqrt{2}} & -2\sqrt{2} & 3 \end{bmatrix}$$

a. Use Algorithm 6.2 to find a tridiagonal matrix similar to matrix **A**.
b. Let **B** denote the matrix

$$\begin{bmatrix} \dfrac{1}{\sqrt{2}} & 0 & 0 & \dfrac{1}{\sqrt{2}} \\[2ex] 0 & 1 & 0 & 0 \\[2ex] 0 & 0 & 1 & 0 \\[2ex] \dfrac{1}{\sqrt{2}} & 0 & 0 & \dfrac{-1}{\sqrt{2}} \end{bmatrix}$$

Verify that **B** is orthogonal, so **BAB** is similar to **A**. Compute **C** = **BAB**, and note that **C** is tridiagonal. So we have two distinct tridiagonal matrices that are similar to **A**.
c. Matrix **C**, computed in part (b), has the block form

$$\begin{bmatrix} C_1 & \vdots & O \\ \cdots & \cdots & \cdots \\ O & \vdots & C_2 \end{bmatrix}$$

where C_1 and C_2 are 2×2 matrices, while the **O** blocks are 2×2 zero matrices. Show that the eigenvalues of C_1 are eigenvalues of **A** and that the eigenvalues of C_2 are also eigenvalues of **A**.
d. Find all eigenvalues of **A**.

6.5 The QR Algorithm

In this section, we will examine a powerful algorithm for computing eigenvalues of a symmetric matrix. The algorithm was introduced in a paper by J. G. F. Francis (1961/1962). As in the case of the Muller algorithm, we again have an algorithm made feasible by the development of modern computing machines. The basic idea is simple enough, but the computations required to implement the algorithm would be unmanageable using hand calculations or desk calculators. The starting point for the development of this algorithm is the following theorem.

Theorem 6.9   If $\mathbf{A}$ is an $n \times n$ real matrix, then there exists an upper triangular matrix $\mathbf{R}$ and an orthogonal matrix $\mathbf{Q}$ such that $\mathbf{A} = \mathbf{QR}$.

The usual proof of this theorem is a general description of a method for constructing $\mathbf{Q}$ and $\mathbf{R}$. The procedure for doing this can easily be converted to a machine algorithm and will be illustrated with numerical examples in the following discussion. We employ so-called *rotation matrices* to convert $\mathbf{A}$ to the triangular matrix $\mathbf{R}$. The matrix $\mathbf{Q}$ can then be constructed from these rotation matrices.

For 3×3 matrices, a rotation matrix is a matrix in one of the forms

$$\begin{bmatrix} c & -s & 0 \\ s & c & 0 \\ 0 & 0 & 1 \end{bmatrix}, \quad \begin{bmatrix} 1 & 0 & 0 \\ 0 & c & -s \\ 0 & s & c \end{bmatrix}, \quad \text{or} \quad \begin{bmatrix} c & 0 & -s \\ 0 & 1 & 0 \\ s & 0 & c \end{bmatrix}$$

where $c = \cos \theta$ and $s = \sin \theta$ for some angle θ. As transformations in three dimensions, these matrices represent rotations around the z, x, and y axes, respectively. In the general case of $n \times n$ matrices (with $n > 1$), a rotation matrix is a matrix $\mathbf{P}$ formed from the $n \times n$ identity matrix by setting

$$\mathbf{P}_{ii} = \mathbf{P}_{jj} = \cos \theta$$

and

$$\mathbf{P}_{ji} = -\mathbf{P}_{ij} = \sin \theta$$

for some θ and some choice of row i and column j. In the case of 4×4 matrices, the matrices

$$\begin{bmatrix} 1 & 0 & 0 & 0 \\ 0 & c & 0 & -s \\ 0 & 0 & 1 & 0 \\ 0 & s & 0 & c \end{bmatrix} \quad \text{and} \quad \begin{bmatrix} c & 0 & -s & 0 \\ 0 & 1 & 0 & 0 \\ s & 0 & c & 0 \\ 0 & 0 & 0 & 1 \end{bmatrix}$$

with $c = \cos \theta$ and $s = \sin \theta$ are possible rotation matrices. There are several other 4×4 rotation matrices, of course.

Note that every rotation matrix $\mathbf{P}$ is an orthogonal matrix, so $\mathbf{P}^{-1} = \mathbf{P}^t$.

Now consider the product $\mathbf{PA}$ of a 4×4 rotation matrix $\mathbf{P}$ and any 4×4 matrix $\mathbf{A}$. Suppose that $\mathbf{P}$ is the second 4×4 rotation matrix above. Then $\mathbf{PA}$ is

$$\begin{bmatrix} ca_{11} - sa_{31} & ca_{12} - sa_{32} & ca_{13} - sa_{33} & ca_{14} - sa_{34} \\ a_{21} & a_{22} & a_{23} & a_{24} \\ sa_{11} + ca_{31} & sa_{12} + ca_{32} & sa_{13} + ca_{33} & sa_{14} + ca_{34} \\ a_{41} & a_{42} & a_{43} & a_{44} \end{bmatrix}$$

Note that only rows 1 and 3 are changed. Also note that by proper choice of θ, we can cause the element in row 3 and column 1 of the product to be zero. We need

$$sa_{11} + ca_{31} = 0$$

That will be the case if we select

$$s = \frac{-a_{31}}{\sqrt{a_{11}^2 + a_{31}^2}} \quad \text{and} \quad c = \frac{a_{11}}{\sqrt{a_{11}^2 + a_{31}^2}}$$

In the general case of $n \times n$ matrices, suppose that $\mathbf{P}$ is a rotation matrix with

$$P_{ii} = P_{jj} = c \quad \text{and} \quad P_{ji} = -P_{ij} = s$$

If $\mathbf{A}$ is any $n \times n$ matrix, the element in row j and column i of the product $\mathbf{PA}$ will be zero if we select

$$s = \frac{-a_{ji}}{\sqrt{a_{ii}^2 + a_{ji}^2}} \quad \text{and} \quad c = \frac{a_{ii}}{\sqrt{a_{ii}^2 + a_{ji}^2}}$$

EXAMPLE 6.14   Given matrix

$$\mathbf{A} = \begin{bmatrix} 1 & -3 & 4 \\ 1 & 2 & -1 \\ 1 & -2 & 5 \end{bmatrix}$$

find a rotation matrix $\mathbf{P}$ so that product $\mathbf{PA}$ has a zero in the third row and first column.

Solution   We choose

$$s = \frac{-a_{31}}{\sqrt{a_{11}^2 + a_{31}^2}} = \frac{-1}{\sqrt{1^2 + 1^2}} = \frac{-1}{\sqrt{2}}$$

$$c = \frac{a_{11}}{\sqrt{a_{11}^2 + a_{31}^2}} = \frac{1}{\sqrt{1^2 + 1^2}} = \frac{1}{\sqrt{2}}$$

Then **P** is the matrix

$$\begin{bmatrix} \dfrac{1}{\sqrt{2}} & 0 & \dfrac{1}{\sqrt{2}} \\ 0 & 1 & 0 \\ \dfrac{-1}{\sqrt{2}} & 0 & \dfrac{1}{\sqrt{2}} \end{bmatrix}$$

and the product **PA** is

$$\begin{bmatrix} \dfrac{2}{\sqrt{2}} & \dfrac{-1}{\sqrt{2}} & \dfrac{9}{\sqrt{2}} \\ 1 & 2 & -1 \\ 0 & \dfrac{1}{\sqrt{2}} & \dfrac{1}{\sqrt{2}} \end{bmatrix}$$ ∎

Now, given an $n \times n$ matrix **A**, we can convert to upper triangular matrix **R** by successive multiplications on the left with properly chosen rotation matrices. Let **P**(2, 1) be selected so that **P**(2, 1)**A** has a zero in row 2 and column 1. Now we select **P**(3, 1) so that the product **P**(3, 1)**P**(2, 1)**A** has a zero in row 3 and column 1. We proceed in this manner until we arrive at

$$\mathbf{P}(n, 1) \ldots \mathbf{P}(3, 1)\mathbf{P}(2, 1)\mathbf{A}$$

which has all zeros in column 1 for rows 2 through n.

Now we select rotation matrix **P**(3, 2) so that

$$\mathbf{P}(3, 2)\mathbf{P}(n, 1) \ldots \mathbf{P}(2, 1)\mathbf{A}$$

has a zero in row 3 and column 2. By successive multiplication by rotation matrices, we can clearly produce zero elements for column 2 for rows 3 through n. Continuing in this way, we eventually arrive, in a finite number of steps, at a product **R** that is upper triangular. In fact, the number of steps is the number of rotation matrices employed, which is at most

$$(n - 1) + (n - 2) + \cdots + 2 + 1 = \frac{n(n - 1)}{2} = r$$

Actually, we may not need this many steps, since we do not need to multiply by a rotation matrix in cases where a zero already exists in the desired row and column. When r matrix multiplications are required, however, we may denote the rotation matrices by

$$\mathbf{P}^{(1)}, \mathbf{P}^{(2)}, \ldots, \mathbf{P}^{(n)}$$

Then

$$\mathbf{P}^{(n)}\mathbf{P}^{(n-1)} \ldots \mathbf{P}^{(3)}\mathbf{P}^{(2)}\mathbf{P}^{(1)}\mathbf{A} = \mathbf{R}$$

is upper triangular and each $P^{(k)}$ is orthogonal. It follows that $A = QR$, where Q is the inverse of the product $P^{(n)}P^{(n-1)} \ldots P^{(1)}$. Since each rotation matrix is orthogonal, Q exists, and in fact,

$$Q = (P^{(1)})^t (P^{(2)})^t \ldots (P^{(n)})^t$$

The factorization $A = QR$ is referred to as the **orthogonal factorization** of A. The procedure outlined yields a unique Q and R for each square matrix A. It is applicable to any such matrix.

EXAMPLE 6.15   Find the orthogonal factorization of

$$A = \begin{bmatrix} 1 & 1 & 1 \\ 1 & 2 & 1 \\ 1 & 1 & 2 \end{bmatrix}$$

Solution   The matrices used are

$$P(2, 1) = \begin{bmatrix} \dfrac{1}{\sqrt{2}} & \dfrac{1}{\sqrt{2}} & 0 \\ \dfrac{-1}{\sqrt{2}} & \dfrac{1}{\sqrt{2}} & 0 \\ 0 & 0 & 1 \end{bmatrix} \qquad P(2, 1)A = \begin{bmatrix} \dfrac{2}{\sqrt{2}} & \dfrac{3}{\sqrt{2}} & \dfrac{2}{\sqrt{2}} \\ 0 & \dfrac{1}{\sqrt{2}} & 0 \\ 1 & 1 & 2 \end{bmatrix} = A1$$

$$P(3, 1) = \begin{bmatrix} \dfrac{2}{\sqrt{6}} & 0 & \dfrac{1}{\sqrt{3}} \\ 0 & 1 & 0 \\ \dfrac{-1}{\sqrt{3}} & 0 & \dfrac{2}{\sqrt{6}} \end{bmatrix} \qquad P(3, 1)A1 = \begin{bmatrix} \dfrac{3}{\sqrt{3}} & \dfrac{4}{\sqrt{3}} & \dfrac{4}{\sqrt{3}} \\ 0 & 1 & 0 \\ 0 & \dfrac{-1}{\sqrt{6}} & \dfrac{2}{\sqrt{6}} \end{bmatrix} = A2$$

$$P(3, 2) = \begin{bmatrix} 1 & 0 & 0 \\ 0 & \dfrac{\sqrt{3}}{2} & \dfrac{-1}{2} \\ 0 & \dfrac{1}{2} & \dfrac{\sqrt{3}}{2} \end{bmatrix} \qquad P(3, 2)A = \begin{bmatrix} \dfrac{3}{\sqrt{3}} & \dfrac{4}{\sqrt{3}} & \dfrac{4}{\sqrt{3}} \\ 0 & \dfrac{2}{\sqrt{6}} & \dfrac{-1}{\sqrt{6}} \\ 0 & 0 & \dfrac{1}{\sqrt{2}} \end{bmatrix} = R$$

Matrix **Q** is

$$\mathbf{Q} = \mathbf{P}(2,\,1)^t \mathbf{P}(3,\,1)^t \mathbf{P}(3,\,2)^t = \begin{bmatrix} \dfrac{1}{\sqrt{3}} & \dfrac{-1}{\sqrt{6}} & \dfrac{-1}{\sqrt{2}} \\[2mm] \dfrac{1}{\sqrt{3}} & \dfrac{2}{\sqrt{6}} & 0 \\[2mm] \dfrac{1}{\sqrt{3}} & \dfrac{-1}{\sqrt{6}} & \dfrac{1}{\sqrt{2}} \end{bmatrix}$$

You can verify that $\mathbf{QR} = \mathbf{A}$ as desired.   ■

Such orthogonal factorizations should be computed by a machine algorithm for matrices of larger sizes. The machine program can be no more than a direct conversion of the preceding method to the desired programming language. We need to "zero out" the elements systematically from column 1 to column $n - 1$, working from the top down in each column, beginning in row $k + 1$ of column k. Since only rows i and j are affected when matrix **A** is multiplied by rotation matrix $\mathbf{P}(i, j)$, the arithmetic in the matrix multiplications can be simplified. That is, if

$$\mathbf{B} = \mathbf{P}(i, j)\mathbf{A}$$

then

$$b_{ik} = ca_{ik} - sa_{jk}$$

and

$$b_{jk} = sa_{ik} + ca_{jk} \qquad k = 1,\ldots,n$$

while

$$b_{lk} = a_{lk} \quad \text{if} \quad l \neq i \quad \text{and} \quad l \neq j$$

EXAMPLE 6.16   Find the orthogonal factorization of the matrix

$$\begin{bmatrix} 1 & 1 & 1 \\ 1 & 2 & 1 \\ 1 & 1 & 2 \end{bmatrix}$$

Solution   Machine results are as follows.

INPUT MATRIX

1.000000	1.000000	1.000000
1.000000	2.000000	1.000000
1.000000	1.000000	2.000000

R MATRIX IS

1.732051	2.309401	2.309401
0.000000	0.816497	−0.408248
0.000000	0.000000	0.707107

Q MATRIX IS

0.577350	−0.408248	−0.707107
0.577350	0.816497	0.000000
0.577350	−0.408248	0.707107

This matrix is the same matrix used in Example 6.15, so the results are simply decimal equivalents of the results in Example 6.15. ∎

EXAMPLE 6.17   Find the orthogonal factorization of the matrix

$$\begin{bmatrix} 1 & -1 & 1 & -1 \\ -1 & 2 & 2 & 0 \\ 1 & 2 & 3 & -1 \\ -1 & 0 & -1 & 4 \end{bmatrix}$$

Solution   INPUT MATRIX

1.000000	−1.000000	1.000000	−1.000000
−1.000000	2.000000	2.000000	0.000000
1.000000	2.000000	3.000000	−1.000000
−1.000000	0.000000	−1.000000	4.000000

R MATRIX IS

2.000000	−0.500000	1.500000	−3.000000
0.000000	2.958040	3.296101	−0.845154
0.000000	0.000000	1.373213	−0.520157
0.000000	0.000000	0.000000	−2.831104

Q MATRIX IS

0.5000000	−0.253546	0.790638	−0.246183
−0.5000000	0.591608	0.582575	0.246183
0.5000000	0.760639	−0.187256	−0.369274
−0.5000000	−0.084515	0.020806	−0.861640

∎

Given an $n \times n$ matrix $\mathbf{A}$, we proceed as follows. Let $\mathbf{A} = \mathbf{Q}^{(0)}\mathbf{R}^{(0)}$. Define $\mathbf{A}^{(1)} = \mathbf{R}^{(0)}\mathbf{Q}^{(0)}$. Let $\mathbf{A}^{(1)} = \mathbf{Q}^{(1)}\mathbf{R}^{(1)}$. Define $\mathbf{A}^{(2)} = \mathbf{R}^{(1)}\mathbf{Q}^{(1)}$. Continuing in this manner, we generate a sequence of matrices $\mathbf{A}^{(k)}$ where

$$\mathbf{A}^{(0)} = \mathbf{A}$$

$$\mathbf{A}^{(k)} = \mathbf{Q}^{(k)}\mathbf{R}^{(k)} \quad \text{and} \quad \mathbf{A}^{(k+1)} = \mathbf{R}^{(k)}\mathbf{Q}^{(k)}$$

Clearly,

$$\mathbf{A}^{(k+1)} = (\mathbf{Q}^{(k)})^{-1}\mathbf{A}^{(k)}\mathbf{Q}^{(k)} \qquad k = 0, 1, \dots$$

so all the matrices in the infinite sequence $\mathbf{A}^{(0)}, \mathbf{A}^{(1)}, \mathbf{A}^{(2)}, \dots$, are similar and have the same eigenvalues. It is not very difficult to write a machine program that generates this sequence of matrices for a given matrix **A**. For the matrix

$$\begin{bmatrix} 1 & 1 & 1 \\ 1 & 2 & 1 \\ 1 & 1 & 2 \end{bmatrix}$$

some of the results are as follows:

$$\mathbf{A}^{(1)} = \begin{bmatrix} 3.666666 & 0.235702 & 0.408248 \\ 0.235702 & 0.833333 & -0.288675 \\ 0.408248 & -0.288675 & 0.500000 \end{bmatrix}$$

$$\mathbf{A}^{(2)} = \begin{bmatrix} 3.731707 & 0.005322 & 0.034080 \\ 0.005322 & 0.982578 & -0.111553 \\ 0.034080 & -0.111553 & 0.285714 \end{bmatrix}$$

$$\mathbf{A}^{(5)} = \begin{bmatrix} 3.732050 & 0.000000 & 0.000013 \\ 0.000000 & 0.999993 & -0.002199 \\ 0.000013 & -0.002200 & 0.267956 \end{bmatrix}$$

$$\mathbf{A}^{(7)} = \begin{bmatrix} 3.732050 & 0.000000 & 0.000000 \\ 0.000000 & 1.000000 & -0.000158 \\ 0.000000 & -0.000158 & 0.267949 \end{bmatrix}$$

$$\mathbf{A}^{(10)} = \begin{bmatrix} 3.732050 & 0.000000 & 0.000000 \\ 0.000000 & 1.000000 & -0.000003 \\ 0.000000 & -0.000003 & 0.267949 \end{bmatrix}$$

These results seem to indicate that the sequence is converging on a diagonal matrix with diagonal elements 3.732050, 1, and 0.267949. Since all the matrices in the sequence are similar, these three numbers should be approximations to the eigenvalues of **A**. In fact, it is easy to verify that the eigenvalues of **A** are $2 \pm \sqrt{3}$ and 1, so the "approximations" are correct to six decimal places.

What has happened here is no accident. If we begin with the matrix

$$\begin{bmatrix} 25 & -41 & 10 & -6 \\ -41 & 68 & -17 & 10 \\ 10 & -17 & 5 & -3 \\ -6 & 10 & -3 & 2 \end{bmatrix}$$

we find that matrix $\mathbf{A}^{(6)}$ is

$$\begin{bmatrix} 98.521710 & 0.000001 & 0.000000 & 0.000004 \\ -0.000000 & 1.186089 & -0.000352 & 0.000000 \\ 0.000000 & -0.000352 & 0.259197 & -0.000001 \\ 0.000000 & 0.000000 & -0.000001 & 0.033016 \end{bmatrix}$$

and further iterations produce no change in the diagonal entries, although the off-diagonal entries continue to decrease toward zero. Apparently, matrix **A** has eigenvalues with approximate values 98.521710, 1.186089, 0.259197, and 0.033016.

We now have the essence of the **QR algorithm** for estimation of the eigenvalues of a real symmetric matrix. We generate the infinite sequence of matrices defined by

$$\mathbf{A}^{(0)} = \mathbf{A}$$

$$\mathbf{A}^{(k+1)} = \mathbf{R}^{(k)}\mathbf{Q}^{(k)}$$

where

$$\mathbf{A}^{(k)} = \mathbf{Q}^{(k)}\mathbf{R}^{(k)} \qquad k = 0, 1, 2, \ldots$$

The convergence theory for this sequence of matrices is too complicated to consider here, but as we have seen in the preceding examples, the sequence sometimes converges to a diagonal matrix. When this occurs, the elements along the diagonal of matrix $\mathbf{A}^{(k)}$ are approximations to the eigenvalues of **A**, and the approximations improve as we move out in the sequence. Since all the matrices in the sequence are similar, and since some matrices are not similar to any diagonal matrix, clearly we cannot expect convergence to a diagonal matrix for every square matrix **A**. As noted in Theorem 6.8, if **A** has n distinct real eigenvalues, then **A** is similar to a real diagonal matrix, so in this case, we might expect the QR algorithm to produce a sequence of matrices that converge to a diagonal matrix with the desired eigenvalues along the diagonal. In particular for symmetric matrices, convergence is usually obtained.

Repeated eigenvalues do not necessarily cause any particular problem. For example, consider the matrix

$$\begin{bmatrix} 4 & -1 & -1 & -1 \\ -1 & 4 & -1 & -1 \\ -1 & -1 & 4 & -1 \\ -1 & -1 & -1 & 4 \end{bmatrix}$$

with eigenvalues 1, 5, 5, and 5. On the sixth iteration, the result is

$$\begin{bmatrix} 5.000000 & 0.000000 & 0.000000 & -0.000148 \\ 0.000000 & 4.999999 & 0.000000 & -0.000209 \\ 0.000000 & 0.000000 & 5.000000 & -0.000362 \\ -0.000148 & -0.000209 & -0.000362 & 1.000000 \end{bmatrix}$$

The iterations are converging rapidly to a diagonal matrix with very good approximations to the eigenvalues obtained after only six iterations.

Now suppose that **A** is a tridiagonal matrix. The only nonzero entries in **A** are the elements along the diagonal, the subdiagonal, and the superdiagonal. Typical such matrices are the matrices

$$
\begin{bmatrix}
1 & 1 & 0 & 0 \\
2 & 3 & -2 & 0 \\
0 & 1 & 4 & -2 \\
0 & 0 & -3 & 5
\end{bmatrix}
\quad \text{and} \quad
\begin{bmatrix}
-1 & 2 & 0 & 0 & 0 \\
2 & 1 & 3 & 0 & 0 \\
0 & 3 & -2 & 4 & 0 \\
0 & 0 & 4 & 2 & 5 \\
0 & 0 & 0 & 5 & -1
\end{bmatrix}
$$

For such matrices, the orthogonal factorization requires much less computation, since we only need to zero out the elements along the subdiagonal to produce upper triangular matrix **R**. This requires at most three multiplications by rotation matrices in the 4×4 case and at most four in the 5×5 case. If **A** is an $n \times n$ matrix, we need at most $n - 1$ rotations. So it is clear that the computations in the QR algorithm can be carried out much faster if **A** is a tridiagonal matrix.

   In Section 6.4, we found that if **A** is a symmetric matrix, then by use of the Householder algorithm, we can find a symmetric tridiagonal matrix that is similar to **A**. This suggests that we should precede the QR algorithm with an application of the Householder algorithm if **A** is symmetric and not already in tridiagonal form. That is the usual procedure in practice. It is important to note, however, that the QR algorithm does not require this; it is simply an advisable computational procedure. The value of this preliminary transformation to tridiagonal form clearly increases as the size of matrix **A** increases. For matrices of size 3×3 or 4×4, the advantage is slight. It is quite simple to write a machine program that accepts as input an $n \times n$ symmetric matrix **A**, transforms to tridiagonal form with the Householder algorithm, and then proceeds with the QR algorithm to produce estimates of the eigenvalues.

   It should also be noted that we are primarily interested in the diagonal elements of the matrices generated by the QR algorithm, so it is enough to provide as output from a machine program only the values of the diagonal elements after each iteration. The following examples were generated using a machine program based on these ideas.

EXAMPLE 6.18   Estimate eigenvalues for the matrix

$$
\begin{bmatrix}
1 & 1 & 1 \\
1 & 2 & 1 \\
1 & 1 & 2
\end{bmatrix}
$$

Solution   AFTER HOUSEHOLDER TRANSFORMATION

1.000000	−1.414214	0.000000
−1.414214	3.000000	0.000000
0.000000	0.000000	1.000000

QR ITERATIONS

ITERATE	EIGENVALUE ESTIMATES		
1	3.666668	0.333334	1.000000
2	3.731708	0.268293	1.000000
3	3.732049	0.267951	1.000000
4	3.732051	0.267949	1.000000
5	3.732051	0.267949	1.000000

The eigenvalues are approximately 3.732051, 0.267949, and 1.000000.   ∎

EXAMPLE 6.19   Find eigenvalues for the matrix

$$\begin{bmatrix} 5 & 7 & 6 & 5 \\ 7 & 10 & 8 & 7 \\ 6 & 8 & 10 & 9 \\ 5 & 7 & 9 & 10 \end{bmatrix}$$

Solution   AFTER HOUSEHOLDER TRANSFORMATION

5.000000	−10.488090	0.000000	0.000000
−10.488090	25.472730	3.521901	0.000000
0.000000	3.521901	3.680573	0.185813
0.000000	0.000000	0.185813	0.846701

QR ITERATIONS

ITERATE	EIGENVALUE ESTIMATES			
1	29.829640	4.307070	0.836982	0.026321
2	30.281140	3.865141	0.843578	0.010154
3	30.288570	3.858158	0.843130	0.010152
4	30.288690	3.858059	0.843109	0.010152
5	30.288690	3.858057	0.843108	0.010152

The eigenvalue estimates do not change after the fifth iteration, so the best values seem to be 30.288690, 3.858057, 0.843108, and 0.0101052.   ∎

In this section, we have considered the basic idea of the QR algorithm for estimation of eigenvalues of an $n \times n$ symmetric matrix. By combining the Householder algorithm for transformation of the given matrix to tridiagonal form with the QR algorithm, we have a powerful algorithm for estimating all the eigenvalues. This is only the beginning of a long story, however, with many variations on the theme. In the case of some matrices, the convergence of the algorithm as presented in this section is quite slow, and in such cases, "acceleration techniques" need to be employed. It is also possible to apply the algorithm to nonsymmetric matrices. It is not practical to explore all of these variations and extensions in a beginning text such as this one. In the next section, however, we will explore a few of the more important of these so as to produce as our final result a practical and widely usable algorithm for estimating eigenvalues.

■ *Exercise Set 6.5*

1. Given matrix

$$A = \begin{bmatrix} 2 & 1 & 2 \\ 1 & 5 & -1 \\ 4 & 2 & 1 \end{bmatrix}$$

find a rotation matrix **P** such that **PA** has a zero in the specified row and column. Compute **PA** in each case.
a. Row 2, column 1
b. Row 3, column 2
c. Row 1, column 3

2. Find the orthogonal factorization of each matrix.

a. $\begin{bmatrix} 1 & 1 & 2 \\ 2 & 1 & -1 \\ 1 & 2 & 1 \end{bmatrix}$ **b.** $\begin{bmatrix} 4 & 1 & 1 \\ 1 & 3 & 2 \\ 1 & 2 & -2 \end{bmatrix}$ **c.** $\begin{bmatrix} 1 & -2 & 0 \\ 2 & 1 & 3 \\ 1 & 3 & 5 \end{bmatrix}$

3. Write a machine program to generate the matrices **Q** and **R** for orthogonal factorization of matrix **A**. Use a machine to find **Q** and **R** for each matrix.

a. $\begin{bmatrix} 1 & 2 & 1 & -1 \\ -1 & 1 & 5 & 2 \\ 1 & 3 & 2 & 0 \\ -1 & 2 & 1 & 5 \end{bmatrix}$ **b.** $\begin{bmatrix} 3 & 2 & -5 & 6 \\ 5 & 1 & 7 & 8 \\ -2 & 3 & 1 & -5 \\ 7 & 6 & 8 & 1 \end{bmatrix}$

4. Use the machine program from exercise 3 to verify the numerical results given in Examples 6.16 and 6.17.

5. Write a machine program to generate the sequence of matrices $A^{(k)}$ defined by

$$A^{(0)} = A, \quad A^{(k)} = Q^{(k)}R^{(k)}, \quad \text{and} \quad A^{(k+1)} = R^{(k)}Q^{(k)}$$

Use this program to verify the results given in this section for matrices

$$\begin{bmatrix} 1 & 1 & 1 \\ 1 & 2 & 1 \\ 1 & 1 & 2 \end{bmatrix} \quad \text{and} \quad \begin{bmatrix} 25 & -41 & 10 & -6 \\ -41 & 68 & -17 & 10 \\ 10 & -17 & 5 & -3 \\ -6 & 10 & -3 & 2 \end{bmatrix}$$

6. Use the program written in exercise 5 to generate sequence $A^{(k)}$ for each of the following matrices. Identify the eigenvalues of **A** in each case.

a. $\begin{bmatrix} 3 & 5 & 1 \\ 5 & -2 & 6 \\ 1 & 6 & -4 \end{bmatrix}$ **b.** $\begin{bmatrix} 2 & 3 & -1 & 2 \\ 3 & 4 & 5 & 1 \\ -1 & 5 & 2 & 3 \\ 2 & 1 & 3 & 4 \end{bmatrix}$

7. Write a program whose input is an $n \times n$ symmetric matrix **A**. The program should implement the Householder transformation to convert **A** to tridiagonal form and then continue with generation of the sequence of matrices $A^{(k)}$, as defined in this section, until the off-diagonal elements are acceptably small in absolute value. The output should be the eigenvalues of matrix **A** with the number of iterations required.

8. Use the program written in exercise 7 to find all eigenvalues of each of the following matrices.

a.
$$\begin{bmatrix} 1 & 3 & 1 \\ 3 & 4 & 2 \\ 1 & 2 & 5 \end{bmatrix}$$
b.
$$\begin{bmatrix} 2 & 1 & 2 & 1 \\ 1 & 3 & 0 & 4 \\ 2 & 0 & 5 & -2 \\ 1 & 4 & -2 & -1 \end{bmatrix}$$

c.
$$\begin{bmatrix} 15 & 16 & 20 \\ 16 & -11 & 12 \\ 20 & 12 & 10 \end{bmatrix}$$
d.
$$\begin{bmatrix} 10 & 7 & 5 & 8 \\ 7 & 6 & 4 & 3 \\ 5 & 4 & 2 & 1 \\ 8 & 3 & 1 & 9 \end{bmatrix}$$

6.6   Some Comments on the Eigenvalue Problem (Optional)

You are now at the point where you should be able to find all eigenvalues for a large number of symmetric matrices. Before leaving this problem, however, we will consider some difficulties and investigate some approaches to their resolution. We begin with a numerical example.

EXAMPLE 6.20   Find all eigenvalues for

$$\begin{bmatrix} 10 & 7 & 5 & 6 \\ 7 & 2 & 3 & 0 \\ 5 & 3 & -1 & 2 \\ 6 & 0 & 2 & 8 \end{bmatrix}$$

Solution   Some of the machine output from the QR algorithm preceded by the Householder transformation is as follows.

AFTER HOUSEHOLDER TRANSFORMATION

10.000000	−10.488090	0.000000	0.000000
−10.488090	6.281817	2.427613	0.000000
0.000000	2.427613	5.808956	−0.477125
0.000000	0.000000	−0.477125	−3.090774

QR ITERATIONS

ITERATE	EIGENVALUE ESTIMATES			
1	18.528570	−0.973973	4.546378	−3.100978
5	18.797590	5.913739	−2.824515	−3.068819
6	18.979610	6.015050	−2.937663	−3.056992
10	18.979610	6.047313	−3.008579	−3.017987
14	18.979610	6.047529	−3.072653	−2.965960
20	18.979610	6.047529	−3.122584	−2.904452
30	18.979610	6.047529	−3.150396	−2.876640
50	18.979610	6.047529	−3.155859	−2.871179
70	18.979610	6.047529	−3.155986	−2.871050

After iteration 70, there is no significant change in the values, so we conclude that the eigenvalues are approximately 18.979610, 6.047529, −3.155986, and −2.871050. The convergence is extremely slow, however. We do have a good estimate of the largest eigenvalue after only 6 iterations, but even after 30 iterations, the estimates of the two smallest eigenvalues are not very accurate. ∎

It is typical that the eigenvalues are produced in order from largest to smallest (in absolute value), and the largest eigenvalue is estimated most rapidly. Suppose we have

$$|\lambda_1| > |\lambda_2| > |\lambda_3| > \cdots > |\lambda_n|$$

The convergence theory for the QR iterations tells us that the convergence will be most rapid if the ratios of absolute values $|\lambda_i|/|\lambda_{i+1}|$ are large. So we can expect some difficulty if absolute values of two or more of the eigenvalues are close together, as in the case of the two smallest eigenvalues in Example 6.20.

Given matrix $\mathbf{A}$, let $\mathbf{B}$ denote the "shifted matrix" $\mathbf{A} - s\mathbf{I}$, where s is some real number. If c is an eigenvalue of $\mathbf{B}$, and $\mathbf{v}$ is an associated eigenvector, then

$$\mathbf{Bv} = c\mathbf{v}$$

and

$$(\mathbf{A} - s\mathbf{I})\mathbf{v} = c\mathbf{v}$$

so

$$\mathbf{Av} = (s + c)\mathbf{v}$$

It follows that $c + s$ is an eigenvalue for matrix $\mathbf{A}$. So the eigenvalues of $\mathbf{B}$ are merely the eigenvalues of $\mathbf{A}$ shifted by amount s. However, the ratios of absolute values of eigenvalues for $\mathbf{B}$ are not the same as the ratios of absolute values for $\mathbf{A}$. If the shift s is appropriately chosen, it is possible to cause the QR algorithm to converge more rapidly to the eigenvalues for $\mathbf{B}$ than to the eigenvalues for $\mathbf{A}$. We can, of course, recover the eigenvalues for $\mathbf{A}$ by adding the shift s to each eigenvalue obtained for $\mathbf{B}$.

EXAMPLE 6.21 Given the matrix in Example 6.20, let $\mathbf{B}$ denote the shifted matrix $\mathbf{A} - s\mathbf{I}$, where $s = -2.88$. Compute the eigenvalues of $\mathbf{A}$ by first estimating eigenvalues for $\mathbf{B}$ using the QR algorithm and then adding back the shift.

Solution The machine results are as follows.

INPUT MATRIX:

12.880000	7.000000	5.000000	6.000000
7.000000	4.880000	3.000000	0.000000
5.000000	3.000000	1.880000	2.000000
6.000000	0.000000	2.000000	10.880000

AFTER HOUSEHOLDER TRANSFORMATION

12.880000	−10.488090	0.000000	0.000000
−10.488090	9.161818	2.427614	0.000000
0.000000	2.427614	8.688958	−0.477125
0.000000	0.000000	−0.477125	−0.210774

QR ITERATIONS

ITERATE	EIGENVALUE ESTIMATES			
1	21.668150	9.092740	−0.248095	0.007210
3	21.854210	8.932821	−0.275986	0.008952
5	21.859460	8.927578	−0.275986	0.008952
8	21.859610	8.927428	−0.275986	0.008952

Further iterations produce no change in results. In only three iterations, we have converged on the two smallest eigenvalues; after only eight iterations, we have converged on all the eigenvalues. We obtain the eigenvalues for $\mathbf{A}$ now by adding the shift $s = -2.88$ to these results and obtain 18.979610, 6.047428, −3.155986, and −2.871048. These results are essentially the same as the results produced in 70 iterations without shifting. The acceleration of the convergence is impressive.   ■

In Example 6.21 we "cheated" in that we knew the values of the eigenvalues for $\mathbf{A}$, so we could select the shift to produce large ratios for the absolute values of the ratios of eigenvalues for the shifted matrix. We need a procedure for estimating the shift in advance. One common technique is the following.

After transforming the matrix to tridiagonal form, we compute the eigenvalues of the 2×2 matrix in the lower right corner. For the matrix in Example 6.20, this is the matrix

$$\begin{bmatrix} 5.808956 & -0.477125 \\ -0.477125 & -3.090774 \end{bmatrix}$$

The eigenvalues are 5.834462 and −3.116280. Now we select for the shift the eigenvalue closest to a_{nn}. In this case, it is the value −3.116280. We subtract this number from each of the diagonal elements of matrix $\mathbf{A}$. Now using this matrix as the new $\mathbf{A}$ matrix, we compute matrix $\mathbf{A}^{(1)}$ in the QR algorithm. Again for matrix $\mathbf{A}^{(1)}$, we compute the shift and subtract that number from the diagonal elements of $\mathbf{A}^{(1)}$ to create a new $\mathbf{A}^{(1)}$. Now we compute $\mathbf{A}^{(2)}$, and so forth. The shifts computed at each stage must be accumulated as we proceed, since it is the sum of all of these shifts that must be added to the eigenvalue estimates obtained from the shifted matrices in order to recover the eigenvalues of $\mathbf{A}$.

EXAMPLE 6.22   Estimate the eigenvalues of the matrix in Example 6.20 using the acceleration-by-shifting technique described.

Solution   The machine output is as follows after the Householder transformation is completed as before.

INITIAL SHIFT IS -3.116279

ITERATE	EIGENVALUE ESTIMATES				SHIFT TOTAL
1	18.781060	6.129170	-2.755516	-3.154714	-3.155627
2	18.945110	6.081805	-2.870937	-3.155986	-3.155986
3	18.973630	6.053400	-2.871049	-3.155986	-3.155986
7	18.979600	6.047430	-2.871049	-3.155986	-3.155986
9	18.979600	6.047424	-2.871049	-3.155986	-3.155986

No further change in the estimates occurs after the ninth iteration. The values differ only slightly from those obtained in Example 6.20. The value of acceleration by shifting is clearly demonstrated. ∎

In the examples presented here, it is clear that the iterations frequently produce an accurate estimate of one of the eigenvalues (usually the largest or the smallest) after a few iterations. Can we use this value to deflate the matrix and produce a matrix of size $n - 1$ whose eigenvalues are the remaining eigenvalues of A? The answer is yes. There are various deflation techniques in the literature. You might want to consult Ralston (1962) or Fox (1965). Deflation techniques in general will not be discussed in this text, but a way to achieve the desired results is built into the QR algorithm. We will examine that case.

If at some stage in the iterations of the QR algorithm we arrive at a matrix $A^{(k)}$, where $a_{21}^{(k)}$ is extremely small, it is safe to assume that element $a_{11}^{(k)}$ is a good estimate of an eigenvalue and that the remaining eigenvalues are the eigenvalues of the $(n - 1) \times (n - 1)$ matrix formed by deleting row 1 and column 1 of the present matrix. Similarly, if at any point we find that $a_{n,n-1}^{(k)}$ is extremely small, we conclude that $a_{nn}^{(k)}$ is a good estimate of an eigenvalue and that the remaining eigenvalues are eigenvalues of the $(n - 1) \times (n - 1)$ matrix formed by deleting the nth row and nth column of the present matrix. It is quite common for one of these two instances to occur, so any machine algorithm should test for these cases after matrix $A^{(k)}$ is computed. It is also true that if some intermediate subdiagonal element is reduced to zero or becomes approximately zero, then we can separate the matrix into two smaller matrices before proceeding with the computations. This is of great value when working with large matrices.

In Sections 6.5 and 6.6, you have been presented with the pieces for assembling a variety of machine algorithms capable of approximating the eigenvalues of a real symmetric matrix. Most *QR algorithms* in the current literature consist of some combination of these pieces. Usually, the matrix is converted to tridiagonal form by use of the Householder algorithm, or it is assumed to already be in that form. As the iterations of the QR algorithm proceed, acceleration by shifting is usually employed, but it does not have to be included. Also, it is wise to test for zero elements along the subdiagonal to reduce the size of the matrices as we proceed, but this is an optional feature.

You should construct some such machine algorithm from the pieces presented here and run it on a machine to see firsthand the results obtained for a variety of matrices. There is no substitute for hands-on experience when trying to learn material as difficult as that presented in these last two sections.

The IMSL MATH/LIBRARY contains software for finding both real and complex eigenvalues for real or complex matrices. The subroutine EVLRG can be used to find all eigenvalues of a real matrix, for example. If the matrix is symmetric, you might prefer to use the routine EVLSF. This library contains a large number of subroutines for a variety of types of matrices. The EISPACK package, which is also distributed by IMSL, is especially designed for eigenvalue problems and contains numerous subroutines for this purpose. Writing good software capable of solving the eigenvalue problem for large matrices efficiently is quite difficult. This is an appropriate place to turn to professionally written software to do the job after the fundamental ideas of the methods are understood.

■ *Exercise Set 6.6*

1. Given that the eigenvalues of the matrix

$$\begin{bmatrix} -33 & -16 & -72 \\ 24 & 10 & 57 \\ 8 & 4 & 17 \end{bmatrix}$$

are -2, -1, and -3, determine the eigenvalues of each of the following matrices.

a. $\begin{bmatrix} -30 & -16 & -72 \\ 24 & 13 & 57 \\ 8 & 4 & 20 \end{bmatrix}$ **b.** $\begin{bmatrix} -3 & -16 & -72 \\ 24 & 40 & 57 \\ 8 & 4 & 47 \end{bmatrix}$

c. $\begin{bmatrix} -43 & -16 & -72 \\ 24 & 0 & 57 \\ 8 & 4 & 7 \end{bmatrix}$ **d.** $\begin{bmatrix} -20 & -16 & -72 \\ 24 & 23 & 57 \\ 8 & 4 & 30 \end{bmatrix}$

2. Given the matrix

$$A = \begin{bmatrix} 5 & 1 & 2 \\ 4 & 3 & -6 \\ 1 & 0 & 3 \end{bmatrix}$$

a. Find $P(\lambda)$.
b. Solve $P(\lambda) = 0$ for the eigenvalues of A.
c. Compare the results in part (b) with the eigenvalues found in Example 6.3. From examination of the results in Example 6.3, what would you expect to find as the eigenvalues for A?

3. Given the matrix

$$A = \begin{bmatrix} 6 & -1 & -5 \\ -4 & 2 & -2 \\ 18 & -5 & -9 \end{bmatrix}$$

a. Find $P(\lambda)$.
b. Solve $P(\lambda) = 0$ for the eigenvalues of A.
c. Compare the results in part (b) with the eigenvalues found in Example 6.4. From examination of the results in Example 6.3, what would you expect to find as the eigenvalues for A?

4. For a certain 3×3 matrix **A**, the Householder transformation produces the tridiagonal matrix

$$\begin{bmatrix} 1.37561 & 2.56125 & 0 \\ 2.56125 & 3.17019 & 4.01215 \\ 0 & 4.01215 & 5.00275 \end{bmatrix}$$

Find the shifted matrix that should be used in the computation of $\mathbf{A}^{(1)}$ if the technique of acceleration by shifting is being used.

5. Write a machine program that employs the QR algorithm with acceleration by shifting. Use this program to find the eigenvalues of each matrix in exercise 8 of Exercise Set 6.5. Compare the rate of convergence with the rate obtained in doing exercise 8 without shifting.

6. The spectral radius of matrix **A** is the maximum value of $|\lambda|$ for all eigenvalues λ of **A**. Find the spectral radius of each of the following matrices.

a. $\begin{bmatrix} 2 & 1 & 3 \\ 1 & 4 & -1 \\ 3 & -1 & 5 \end{bmatrix}$ b. $\begin{bmatrix} 1 & 2 & -1 & 2 \\ 2 & 3 & 0 & -5 \\ -1 & 0 & 1 & -3 \\ 2 & -5 & -3 & 4 \end{bmatrix}$

7. The *Gershgorin circle theorem* states the following: If λ is an eigenvalue of matrix **A**, then

$$|a_{ii} - \lambda| \le \sum_{\substack{j=1 \\ j \ne i}}^{n} |a_{ij}| \quad \text{for some } i$$

a. The truth in this theorem depends on two facts:
  i. If λ is an eigenvalue of **A**, then $\mathbf{A} - \lambda \mathbf{I}$ must be a singular matrix.
  ii. If a matrix is diagonally dominant, it must be nonsingular.
  Use this information to write a proof of the theorem.
b. Verify the truth of the theorem for each of the matrices in exercise 8 of Exercise Set 6.5.

8. If $\|\mathbf{A}\|$ is the matrix norm defined by Definition 4.5, where $\|\mathbf{x}\|$ is the Euclidean norm for vectors, then it can be shown that $\|\mathbf{A}\|$ is the square root of the spectral radius (see exercise 6) of the matrix $\mathbf{A}^t\mathbf{A}$, where $\mathbf{A}^t$ is the transpose of **A**. Find $\|\mathbf{A}\|$ for each of the following matrices.

a. $\begin{bmatrix} 2 & 3 & 1 \\ 3 & 5 & 4 \\ 1 & 4 & 3 \end{bmatrix}$ b. $\begin{bmatrix} 5 & 7 & -2 \\ 1 & 3 & -1 \\ -2 & 5 & 4 \end{bmatrix}$ c. $\begin{bmatrix} 2 & 5 & 7 & 8 \\ 5 & 4 & -1 & 0 \\ 7 & 3 & 1 & 2 \\ 3 & 6 & 2 & -6 \end{bmatrix}$

6.7 Some Applied Problems

The literature of applied mathematics and sciences contains numerous problems involving eigenvalues and eigenvectors. Eigenvalues arise in connection with vibration problems in engineering, quantum mechanical problems in physics and chemistry, and a variety of geometric and statistical problems. Most notable of these for people working in the physical sciences are the problems from quantum mechanics in which eigenvalues of the Hamiltonian matrix represent energy levels for the states of an atomic system. For details, see Schiff (1955). Unfortunately, most problems such as these are too difficult to discuss here. However, there are some problems in quantum chemistry for which the mathematics is simple enough to present here. So let us begin with such a problem for our first application. A detailed explanation of the theory on which the following application is based can be found in the discussion of molecular orbital calculations given in Roberts (1961).

In quantum mechanics, the state of a physical system such as a molecule is represented by a wave function ψ. In quantum chemistry, each state of an atom is called an *atomic orbital*. In the LCAO (linear combination of atomic orbitals) method, the chemist assumes that a molecular orbital is a linear combination of the atomic orbitals for the atoms involved. For a diatomic molecule, such as the hydrogen molecule, the assumed orbital can be represented by

$$\psi = c_1\psi_1 + c_2\psi_2$$

where ψ_1 and ψ_2 are orbitals (states) for the hydrogen atom. In general, if a molecule is composed of n atoms, the orbital is

$$\psi = c_1\psi_1 + c_2\psi_2 + \cdots + c_n\psi_n$$

If H is the Hamiltonian energy operator, then the energy of the molecule is given by

$$E = \frac{\int \psi H\psi \, d\tau}{\int \psi^2 \, d\tau}$$

The minimum value of E is desired. It is shown in Roberts (1961) that if we define

$$H_{ij} = \int \psi_i H\psi_j \, d\tau$$

$$S_{ij} = \int \psi_i\psi_j \, d\tau$$

then the desired value of E is a solution of a determinant equation represented by

$$\begin{vmatrix} H_{11} - ES_{11} & H_{12} - ES_{12} & \cdots & H_{1n} - ES_{1n} \\ H_{12} - ES_{12} & H_{22} - ES_{22} & \cdots & H_{2n} - ES_{2n} \\ \vdots & \vdots & & \vdots \\ H_{n1} - ES_{n1} & H_{n2} - ES_{n2} & \cdots & H_{nn} - ES_{nn} \end{vmatrix} = 0$$

This is often called the *secular equation* for the problem.

For the butadiene molecule, Roberts (1961) shows that the secular equation is

$$\begin{vmatrix} \alpha - E & \beta & 0 & 0 \\ \beta & \alpha - E & \beta & 0 \\ 0 & \beta & \alpha - E & \beta \\ 0 & 0 & \beta & \alpha - E \end{vmatrix} = 0$$

where $\alpha = H_{11}$ and $\beta = H_{12} = H_{23} = H_{34}$. If we let $x = (\alpha - E)/\beta$, the equation becomes

$$
\begin{vmatrix}
x & 1 & 0 & 0 \\
1 & x & 1 & 0 \\
0 & 1 & x & 1 \\
0 & 0 & 1 & x
\end{vmatrix} = 0
$$

Each possible value of x yields a possible value of the energy E.

Let

$$
\mathbf{A} = \begin{bmatrix}
0 & 1 & 0 & 0 \\
1 & 0 & 1 & 0 \\
0 & 1 & 0 & 1 \\
0 & 0 & 1 & 0
\end{bmatrix}
$$

Then

$$
\mathbf{A} - \lambda\mathbf{I} = \begin{bmatrix}
-\lambda & 1 & 0 & 0 \\
1 & -\lambda & 1 & 0 \\
0 & 1 & -\lambda & 1 \\
0 & 0 & 1 & -\lambda
\end{bmatrix}
$$

So we see that the values of x are the negatives of the eigenvalues of the matrix $\mathbf{A}$.

Using a machine program that implements the QR algorithm, we obtain for the eigenvalues of $\mathbf{A}$ the values -1.618034, -0.618032, 1.618032, and 0.618034. We may assume that the eigenvalues are approximated by ± 1.61803 and ± 0.61803. The energy levels of butadiene are then $\alpha \pm 1.61803\beta$ and $\alpha \pm 0.61803\beta$. Roberts obtained comparable values by constructing the characteristic polynomial equation and approximating the roots.

Let us now turn to a vibration problem. Vibrating structures are frequently modeled as a system of n masses interconnected with springs. So the mathematical analysis of such a system has wide application. Consider the case of three equal masses connected together by springs, as shown in Figure 6.1. If x_1, x_2, and x_3 represent the displacements of the masses from equilibrium during their motion, we find that

$$
m_1 x_1'' = -k_1 x_1 + k_2(x_2 - x_1)
$$

$$
m_2 x_2'' = -k_2(x_2 - x_1) + k_3(x_3 - x_2)
$$

$$
m_3 x_3'' = -k_3(x_3 - x_2) - k_4 x_3
$$

$$m_1 = m_2 = m_3 = m$$

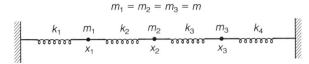

Figure 6.1

or

$$m_1 x_1'' = -(k_1 + k_2)x_1 + k_2 x_2$$

$$m_2 x_2'' = k_2 x_1 - (k_2 + k_3)x_2 + k_3 x_3$$

$$m_3 x_3'' = k_3 x_2 - (k_3 + k_4)x_3$$

In matrix form, we have

$$\mathbf{X}'' = \mathbf{AX}$$

where **A** is the symmetric matrix

$$\begin{bmatrix} -(k_1 + k_2)/m_1 & k_2/m_1 & 0 \\ k_2/m_2 & -(k_2 + k_3)/m_2 & k_3/m_2 \\ 0 & k_3/m_3 & -(k_3 + k_4)/m_3 \end{bmatrix}$$

and **X** is the column matrix for which

$$\mathbf{X}^t = \begin{bmatrix} x_1 & x_2 & x_3 \end{bmatrix}$$

For real symmetric matrix **A**, an orthogonal matrix **P** exists such that $\mathbf{P}^{-1}\mathbf{AP} = \mathbf{B}$ is diagonal with

$$\mathbf{B} = \begin{bmatrix} \lambda_1 & 0 & 0 \\ 0 & \lambda_2 & 0 \\ 0 & 0 & \lambda_3 \end{bmatrix}$$

where λ_1, λ_2, and λ_3 are real eigenvalues of **A**. Furthermore, the columns of **P** are the eigenvectors of **A**.

Let $\mathbf{X} = \mathbf{PY}$; so $\mathbf{Y} = \mathbf{P}^{-1}\mathbf{X}$. Then

$$\mathbf{PY}'' = \mathbf{APY} \Rightarrow \mathbf{Y}'' = \mathbf{P}^{-1}\mathbf{APY} = \mathbf{BY}$$

Matrix equation $\mathbf{Y}'' = \mathbf{BY}$ represents the system

$$y_i'' = \lambda_i y_i \qquad i = 1, 2, 3$$

If all the eigenvalues λ_i are negative, we can write

$$\lambda_i = -\omega_i^2 \qquad i = 1, 2, 3$$

for positive numbers ω_i^2. Then

$$y_i'' + \omega_i^2 y_i = 0 \qquad i = 1, 2, 3$$

Each such equation has a general solution

$$y_i = A_i \cos \omega_i t + B_i \sin \omega_i t$$

which represents an oscillatory solution with frequency f_i, where $f_i = \omega_i/2\pi$. The equation $\mathbf{X} = \mathbf{PY}$ states that solution functions $x_i(t)$ are linear combinations of the functions $y_i(t)$.

In many such problems, the primary interest is in the possible vibration frequencies for the structure. The analysis shows that these are determined by the eigenvalues of matrix $\mathbf{A}$. If all the eigenvalues are negative, then the vibration frequencies (fundamental modes) are given by

$$f_i = \frac{\sqrt{-\lambda_i}}{2\pi} \qquad i = 1, 2, 3$$

For example, suppose that we have a spring-mass system as shown in Figure 6.1, with $m = 2$ units and spring constants $k_1 = 0.5$, $k_2 = 1.0$, $k_3 = 0.75$, and $k_4 = 0.4$. Then

$$\mathbf{A} = \begin{bmatrix} -0.75 & 0.5 & 0 \\ 0.5 & -0.875 & 0.375 \\ 0 & 0.375 & -0.395 \end{bmatrix}$$

The QR algorithm gives eigenvalue estimates -1.399735, -0.548647, and -0.071618. The corresponding vibration frequencies are 0.188297, 0.117887, and 0.042592 oscillations per unit time.

■ *Exercise Set 6.7*

1. To determine the energy levels for a certain molecule by the LCAO method, it is necessary to solve the secular equation

$$\begin{vmatrix} x & 1 & 0 & 0 & 0 & 0 \\ 1 & x & 1 & 0 & 0 & 0 \\ 0 & 1 & x & 1 & 0 & 0 \\ 0 & 0 & 1 & x & 1 & 0 \\ 0 & 0 & 0 & 1 & x & 1 \\ 0 & 0 & 0 & 0 & 1 & x \end{vmatrix} = 0$$

where $x = (\alpha - E)/\beta$ for certain constants α and β. Possible values of E are desired. Find these values using the eigenvalue method demonstrated in this section.

2. Let $P_n(x)$ be the determinant of the $n \times n$ tridiagonal matrix with diagonal elements x and 1s above and below the diagonal. Then $P_6(x)$ is the determinant in exercise 1. Clearly, $P_1(x) = x$ and $P_2(x) = x^2 - 1$.
   a. Show that $P_n(x) = xP_{n-1}(x) - P_{n-2}(x)$ for $n \geq 3$.
   b. Use the result in part (a) to construct polynomials $P_3(x)$, $P_4(x)$, $P_5(x)$, and $P_6(x)$.
   c. Solve the equation $P_6(x) = 0$ using one of the methods in Chapter 3. Compare results with the results in exercise 1.

3. Verify the numerical values of the energy levels of butadiene obtained in this section.

4. Find the possible vibration frequencies for a spring-mass system of five equal masses, with $m = 4$ units and spring constants $k_1 = k_6 = 0.1$ and $k_2 = k_3 = k_4 = k_5 = 0.2$.

5. In calculating the quantum mechanical energy levels of the asymmetric top, physicist D. M. Dennison (1931) found that he needed the eigenvalues of the matrix

$$\begin{bmatrix} 36 & -4.062 & 0 & 0 \\ -4.062 & 16 & 8.216 & 0 \\ 0 & 8.216 & 4 & 14.49 \\ 0 & 0 & 14.49 & 0 \end{bmatrix}$$

He obtained approximate values 36.89, 22.34, −13.67, and 10.43. Use a machine with the QR algorithm to find the values as accurately as you can.

7 POLYNOMIAL INTERPOLATION

In this chapter, we will consider the following problem. Given values $y_k = f(x_k)$ for a function f at $n + 1$ distinct points $x_0 < x_1 < \cdots < x_n$, we want to estimate $f(x)$ for some x in interval (x_0, x_n). We call this the **interpolation problem**. This is a generalization of the familiar problem of trying to estimate the value of a trigonometric function from a table of values that does not include the value of x with which we are concerned. In that case, most students are taught to use the two closest points in the table and estimate the intermediate value by using a linear approximation of function f. If we actually have more than two points available, it is usually possible to make a better approximation of $f(x)$. In this chapter, we will examine some of the methods for doing this.

The interpolation problem is very old and has been studied by some of the world's greatest mathematicians. We have solution methods associated with Lagrange, Newton, and even Gauss. The fact that Carl Gauss studied the problem is no surprise to anyone familiar with his life and work. Gauss was director of an astronomical observatory most of his life and spent a great deal of time on calculations concerning the orbital motion of planets and comets. One of the most important applications of interpolation procedures in the past was the estimation of the celestial coordinates of the sun or a planet at a given intermediate time from previous computations of the location at a finite number of previous times. Today we are concerned with interpolation problems that arise in numerous places in the sciences and engineering. So it is important to have a general procedure for treating such problems.

The oldest method for handling interpolation problems is to find a polynomial P that approximates function f over interval $[x_0, x_n]$ and then estimate the desired value by substituting the polynomial value $P(x)$ as an approximation for $f(x)$.

7.1 Basic Principles and Theory

Before we can proceed to developing a usable method for making such polynomial approximations, we must agree on some criteria to be used in determining the approximating polynomial. We might agree to choose a polynomial $P_n(x)$ of degree *n* or less for which

$$P_n(x_k) = f(x_k) \qquad k = 0, 1, 2, \ldots, n$$

or we might agree to use a least squares fit in which we choose the polynomial $P_N(x)$ for which

$$\sum_{k=0}^{N} [P_N(x_k) - f(x_k)]^2$$

is minimized. There are a number of other possible criteria that could be used. It is important to note that some agreement must be made concerning a precise rule that defines the approximating polynomial. In this chapter, we will use the exact fit criterion:

$$f(x_k) = P_n(x_k) \qquad k = 0, 1, \ldots, n$$

If polynomial $P_n(x)$ satisfies this condition for a given function and a given set of points $\{x_0, x_1, \ldots, x_n\}$ with $x_0 < x_1 < \cdots < x_n$ and with the degree of $P_n(x)$ less than or equal to *n*, then we say that $P_n(x)$ is the *interpolating polynomial for f* on set $\{x_0, x_1, \ldots, x_n\}$. We will now show that $P_n(x)$ as defined actually exists and is uniquely determined by the stated criteria.

Theorem 7.1   Given $n + 1$ points $\{x_0, x_1, \ldots, x_n\}$ in the domain of a function f with $n \geq 1$ and $x_0 < x_1 < \cdots < x_n$, then there exists at most one polynomial $P_n(x)$ of degree less than or equal to *n* such that

$$P_n(x_k) = f(x_k) \qquad k = 0, 1, \ldots, n$$

Proof   Suppose that $P_n(x)$ and $Q_n(x)$ are polynomials of degree less than or equal to *n*, with

$$P_n(x_k) = f(x_k) \quad \text{and} \quad Q_n(x_k) = f(x_k) \qquad k = 0, 1, \ldots, n$$

Let

$$h(x) = P_n(x) - Q_n(x)$$

Then $h(x)$ is a polynomial of degree *n* or less. Also,

$$h(x_k) = 0 \qquad k = 0, 1, \ldots, n$$

Hence, $h(x)$ has $n + 1$ zeros. But this is impossible if the degree of $h(x)$ is greater than or equal to 1. Then $h(x)$ has degree zero. So $h(x)$ is a constant polynomial. Clearly, $h(x)$ is the zero polynomial. It follows that $P_n(x) = Q_n(x)$ for all *x*.   □

Theorem 7.2 The interpolating polynomial exists for any function f on any set $\{x_0, x_1, \ldots, x_n\}$ where $x_0, x_1, \ldots, x_n$ are distinct points in the domain of f.

Proof Let

$$q_k(x) = \prod_{\substack{j=0 \\ j \neq k}}^{n} (x - x_j) \qquad k = 0, 1, \ldots, n$$

Then each $q_k(x)$ is a polynomial of degree n. Furthermore,

$$q_k(x_i) = 0 \quad \text{if} \quad i \neq k$$

and

$$q_k(x_k) \neq 0 \quad \text{for} \quad k = 0, 1, \ldots, n$$

Let

$$L_k(x) = \frac{q_k(x)}{q_k(x_k)}$$

Then $L_k(x)$ is a polynomial of degree n for which

$$L_k(x_j) = 0 \quad \text{if} \quad j \neq k$$

and

$$L_k(x_k) = 1$$

Let

$$P_n(x) = \sum_{k=0}^{n} f(x_k) L_k(x)$$

Clearly, $P_n(x)$ is a polynomial of degree n or less. Also,

$$P_n(x_j) = \sum_{k=0}^{n} f(x_k) L_k(x_j) = f(x_j) \qquad j = 0, 1, \ldots, n$$

Hence, $P_n(x)$ is the desired polynomial. $\square$

EXAMPLE 7.1 Given $f(x) = \sin x$ with $x_0 = 0$, $x_1 = \pi/4$, and $x_2 = \pi/2$, find the interpolating polynomial.

Solution   We find that

$$q_0(x) = \left(x - \frac{\pi}{4}\right)\left(x - \frac{\pi}{2}\right)$$

$$q_1(x) = x\left(x - \frac{\pi}{2}\right)$$

$$q_2(x) = x\left(x - \frac{\pi}{4}\right)$$

and

$$L_0(x) = \frac{q_0(x)}{\left(-\dfrac{\pi}{4}\right)\left(-\dfrac{\pi}{2}\right)} = \frac{8}{\pi^2}\left(x - \frac{\pi}{4}\right)\left(x - \frac{\pi}{2}\right)$$

$$L_1(x) = \frac{q_1(x)}{\dfrac{\pi}{4}\left(-\dfrac{\pi}{4}\right)} = \frac{-16}{\pi^2}\,x\left(x - \frac{\pi}{2}\right)$$

$$L_2(x) = \frac{q_2(x)}{\dfrac{\pi}{2}\left(\dfrac{\pi}{4}\right)} = \frac{8}{\pi^2}\,x\left(x - \frac{\pi}{4}\right)$$

Also,

$$f(x_0) = 0, \quad f(x_1) = \frac{1}{\sqrt{2}}, \quad \text{and} \quad f(x_2) = 1$$

Hence,

$$P_2(x) = \frac{-16}{\pi^2\sqrt{2}}\,x\left(x - \frac{\pi}{2}\right) + \frac{8}{\pi^2}\,x\left(x - \frac{\pi}{4}\right)$$ ■

EXAMPLE 7.2   Given function f with values $f(a)$ and $f(b)$, where $a < b$, find a formula for $P_1(x)$ if $a < x < b$.

Solution   We find that

$$q_0(x) = x - b \quad \text{and} \quad q_1(x) = x - a$$

So

$$L_0(x) = \frac{x - b}{a - b} \quad \text{and} \quad L_1(x) = \frac{x - a}{b - a}$$

Then

$$P_1(x) = f(a)\frac{x-b}{a-b} + f(b)\frac{x-a}{b-a}$$

Note that some slight rearranging gives us

$$P_1(x) = \frac{bf(a) - af(b)}{b-a} + \frac{f(b) - f(a)}{b-a}x$$

which is a familiar form for the equation of the straight line through points $(a, f(a))$ and $(b, f(b))$. ∎

The formula for $P_n(x)$, derived in the proof of Theorem 7.2 and used in the examples in this section, is called the **Lagrange formula** for the interpolating polynomial. This is only one of a number of formulas that have been derived for constructing interpolating polynomials.

In this section, we have developed some basic theory for the interpolation problem. In the following sections, we will consider a variety of numerical examples and applications as well as some of the alternative formulas for the interpolating polynomial.

■ *Exercise Set 7.1*

1. Find an equation for the quadratic or linear function whose graph passes through the given points.
 a. $(-1, 1), (0, 0), (2, 5)$
 b. $(-1, 0), (3, 5), (6, 0)$
 c. $(-1, -1), (0, -1), (1, -1)$
 d. $(-1, 1), (0, 3), (1, 5)$

2. Show that $\sum_{k=0}^{n} L_k(x) = 1$. This is a useful fact for checking the computation of numbers $L_k(x)$ for a given x.

3. For a certain polynomial $P(x)$ of degree 5, it is known that $P(1) = -4$, $P(2) = 29$, $P(0) = -7$, $P(-1) = -10$, and $P(-2) = -43$. Find $P(0.5)$ using the Lagrange formula, but without actually constructing an algebraic formula for $P(x)$. That is, compute numerical values of the coefficients $L_k(0.5)$ rather than algebraic formulas.

4. The Lagrange polynomial can be used to interpolate in standard tables of function values to obtain higher-order estimates than those obtained by linear interpolation (the usual elementary method). Using the given table of values for $\sin x$, estimate the value of $\sin 0.25$ using the following methods.

x	$\sin x$
0.1	0.09983
0.2	0.19867
0.3	0.29552
0.4	0.38942

 a. Use linear interpolation.
 b. Use points 0.1, 0.2, and 0.3 (quadratic interpolation).
 c. Use points 0.2, 0.3, and 0.4.
 d. Use all four points.

5. Let $f(x) = \sqrt{x}$. Then $f(1) = 1$, $f(1.21) = 1.1$, and $f(1.44) = 1.2$. Use these values and the Lagrange interpolation polynomial of degree 2 to estimate $\sqrt{1.3}$.

6. The boiling point of water varies with atmospheric pressure. Given the following table of handbook values, estimate the boiling point of water at a pressure of 753 mm using linear interpolation and then quadratic interpolation.

Pressure (mm)	Boiling Point (°C)
750	99.630
755	99.815
760	100.000
765	100.184

7. If a projectile is fired vertically from the surface of a planet and the initial velocity is small, then the altitude h (feet) at time t (seconds) is closely approximated by a quadratic function of t. So we may assume that $h(t)$ is a quadratic function. Given that $h(1) = 695$, $h(3) = 2055$, and $h(4) = 2720$, estimate the altitude at time $t = 2$. Estimate the acceleration (it's a constant). Is this the planet earth?

8. An astronomy student consults an almanac to find the times of sunrise and sunset on May 4 for that year. She finds the table shown here. Since May 4 is not included, she must use interpolation to find the desired times. Use quadratic interpolation and find the times for her. You can do this two different ways. Do both for comparison.

Date	Sunrise	Sunset
May 1	5:18 A.M.	6:37 P.M.
May 3	5:16	6:39
May 5	5:14	6:40
May 7	5:12	6:41

9. Explain why it would not be wise to use the polynomial interpolation method of this section to estimate the desired number in each of the following situations.
   a. A financial analyst knows the total dollar volume of sales for the New York Stock Exchange on Monday, Tuesday, Wednesday, and Thursday of this week. He would like to estimate the sales volume for Friday (tomorrow).
   b. A woman knows how much her husband spent for lunch on Tuesday and Thursday. She would like to estimate how much he spent on Wednesday.
   c. An astronomy student knows how many satellites the earth, Mars, and Saturn each have. She needs to know how many Jupiter has to answer the next question on the exam.
   d. The track coach has observed that one of his runners was timed at 11 seconds for the 100 yd dash, 51 seconds for the 440 yd dash, and 9.4 minutes for the mile run. He would like to estimate the time that runner would require to run 880 yd.

7.2   The Lagrange Polynomial

As stated at the end of Section 7.1, the polynomial

$$P_n(x) = \sum_{j=0}^{n} f(x_j)L_j(x)$$

which was used as the interpolating polynomial in the proof of Theorem 7.2, is known as the Lagrange polynomial. It has been known by mathematicians for a long time. We are concerned only with its use in connection with the interpolation problem. Let us examine its use for the numerical estimation of values of a function from tabulated values $f(x_0), f(x_1), \ldots, f(x_n)$.

EXAMPLE 7.3   Given the following table of values for a function f, estimate $f(2.5)$.

x	$f(x)$
2	0.30103
3	0.47712

Solution   We need to compute

$$P_1(2.5) = f(2)L_0(2.5) + f(3)L_1(2.5)$$

We find that

$$L_0(2.5) = \frac{2.5 - 3}{2 - 3} = 0.5 \quad \text{and} \quad L_1(2.5) = \frac{2.5 - 2}{3 - 2} = 0.5$$

Hence,

$$P(2.5) = 0.30103(0.5) + 0.47712(0.5)$$
$$= 0.150515 + 0.23856 = 0.389075$$

So $f(2.5)$ is approximately 0.38908 after rounding.   ∎

Example 7.3 is an example of the method of linear interpolation used by high school students working with trigonometric tables or tables of logarithms.

EXAMPLE 7.4   Given the following table of values for function f, estimate $f(2.12)$.

x	$f(x)$
2.0	0.69315
2.1	0.74194
2.2	0.78846

Solution   We find that

$$L_0(2.12) = \frac{0.02(-0.08)}{(-0.1)(-0.2)} = -0.08$$

$$L_1(2.12) = \frac{(0.12)(-0.08)}{(0.1)(-0.1)} = 0.96$$

$$L_2(2.12) = \frac{(0.12)(0.02)}{(0.2)(0.1)} = 0.12$$

Therefore,

$$P_2(2.12) = 0.69315(-0.08) + 0.74194(0.96) + 0.78846(0.12) = 0.75142$$

Hence, $f(2.12)$ is approximately 0.75142. How accurate is this estimate? In this case, the "unknown" function is actually $f(x) = \ln x$. So we can verify that $f(2.12) = 0.75142$ to six decimal places. Hence, our estimate is correct to five decimal places, since rounding our previous estimate gives the approximate value 0.75142 for $f(2.12)$. ∎

EXAMPLE 7.5   Using only the values for $f(x) = \sin x$ at $x = 0$, $\pi/2$, and π, estimate the integral

$$\int_0^\pi \sin x \, dx$$

Solution   The interpolating polynomial is $P_2(x) = L_1(x)$, where

$$L_1(x) = \frac{x(x - \pi)}{\frac{\pi}{2}(-\pi/2)} = \frac{4}{\pi^2}(\pi x - x^2)$$

The integral is estimated by

$$\int_0^\pi f(x) \, dx \cong \int_0^\pi \frac{4}{\pi^2}(\pi x - x^2) \, dx = \tfrac{2}{3}\pi$$

We conclude that

$$\int_0^\pi \sin x \, dx \cong \tfrac{2}{3}\pi \cong 2.094$$

The true value is 2.00000, so the absolute error is 0.094 and the relative error is $0.094/2 = 0.047$. ∎

We will now derive an error estimation formula that will show the analytical dependence of the interpolation error on the function f. Frequently, we find ourselves working only with a table of values, and we do not know much, if anything, about the precise analytical form of f. Even in such cases, however, the formula we are about to derive can sometimes be used to bound the error if we can find bounds for derivatives of f. This can sometimes be done even when the exact analytical form of f and its derivatives is not known.

Let

$$f(x) = P_n(x) + R_n(x)$$

so $|R_n(x)|$ is the absolute error in the approximation of $f(x)$ by $P_n(x)$. We assume that f has $n + 1$ derivatives on interval (x_0, x_n) and that f is class C^{n+1} on the interval $[x_0, x_n]$. Since

$$R_n(x) = f(x) - P_n(x)$$

and

$$f(x_k) = P(x_k) \qquad k = 0, 1, \ldots, n$$

then

$$R_n(x_k) = 0 \qquad k = 0, 1, \ldots, n$$

We shall assume that

$$R_n(x) = C(x)(x - x_0)(x - x_1) \ldots (x - x_n)$$

for some function $C(x)$. We can now determine the function $C(x)$ using some elementary calculus. The argument depends primarily on the use of Rolle's theorem.

Given point x, let us consider the set of $n + 2$ points $\{x, x_0, x_1, \ldots, x_n\}$ as a set of fixed points in the following. We assume that

$$x \neq x_i \qquad i = 0, 1, \ldots, n \quad \text{and} \quad x_0 < x < x_n$$

We define a function F by

$$F(t) = f(t) - P_n(t) - C(x)(t - x_0)(t - x_1) \ldots (t - x_n)$$

Now

$$F(x_k) = 0 \qquad k = 0, 1, 2, \ldots, n$$

Also,

$$F(x) = 0$$

Hence, $F(t)$ has $n + 2$ distinct zeros in interval $[x_0, x_n]$. Under the assumed hypotheses for f, we may apply Rolle's theorem to function F. We conclude that $F'(t) = 0$ for some point in each of the subintervals between the points $x, x_0, x_1, \ldots, x_n$. So $F'(t)$ has at least $n + 1$ distinct zeros in $[x_0, x_n]$. Continuing this argument and applying Rolle's theorem to each of the derivatives of F from $F'(t)$ to $F^{(n+1)}(t)$, we find that $F^{(n+1)}(t)$ has at least one zero. We call this $t = \xi$ in $[x_0, x_n]$.

Now we have

$$F^{(n+1)}(t) = f^{(n+1)}(t) - P_n^{(n+1)}(t) - C(x) \frac{d^{n+1}}{dt^{n+1}} (t - x_0) \ldots (t - x_n)$$

But $P_n(t)$ is a polynomial of degree n or less. So $P_n^{(n+1)}(t) = 0$. It is also clear that

$$\psi(t) = (t - x_0)(t - x_1) \ldots (t - x_n)$$

is a polynomial of degree $n + 1$ and that the coefficient of t^{n+1} is 1. So

$$\psi^{(n+1)}(t) = (n + 1)!$$

Then

$$F^{(n+1)}(t) = f^{(n+1)}(t) - (n + 1)!C(x)$$

Setting $t = \xi$, we find that

$$0 = f^{(n+1)}(\xi) - (n + 1)!C(x)$$

We conclude that

$$C(x) = \frac{f^{n+1}(\xi)}{(n + 1)!} \qquad x_0 < \xi < x_n$$

We have shown that

$$R_n(x) = \frac{f^{n+1}(\xi)}{(n + 1)!} (x - x_0)(x - x_1) \ldots (x - x_n)$$

In summary, we find that

$$f(x) = \sum_{k=0}^{n} f(x_k)L_k(x) + \frac{f^{n+1}(\xi)}{(n + 1)!} (x - x_0) \ldots (x - x_n) \qquad x_0 < \xi < x_n$$

This formula is known as the **Lagrange interpolation formula** with error term.

We will now look at a typical application of this formula to estimate the value of a function from a table of values and then determine the accuracy of the estimate obtained.

EXAMPLE 7.6   Given the following table of common logarithms, estimate the value of log 7 by use of the Lagrange formula. Estimate the error.

x	$\log x$
5	0.69897
6	0.77815
8	0.90309
9	0.95424
10	1.00000

Solution   We find that

$$L_0(x) = \frac{(7 - 6)(7 - 8)(7 - 9)(7 - 10)}{(5 - 6)(5 - 8)(5 - 9)(5 - 10)} = -0.1$$

Similar calculations give $L_1 = 0.5$, $L_2 = 1$, $L_3 = -0.5$, and $L_4 = 0.1$. So,

$$\log 7 \cong -0.1 \log 5 + 0.5 \log 6 + \log 8 - 0.5 \log 9 + 0.1 \log 10$$

Substituting values of the logarithms yields $\log 7 \cong 0.845148$ as our estimate. The error is

$$E_4 = \frac{f^5(\xi)}{5!}(x-5)(x-6)(x-8)(x-9)(x-10)$$

where

$$f(x) = \log x \cong 0.43429 \ln x \quad \text{and} \quad 5 < \xi < 10$$

Then

$$f^5(x) = \frac{24}{x^5}(0.43429)$$

So

$$E = \frac{1}{5!}\frac{24(0.43429)}{\xi^5}(2)(1)(-1)(-2)(-3)$$

and

$$|E_4| = \frac{1}{5}\frac{0.43429}{\xi^5}(12) < \frac{12(0.43429)}{5^6} < 0.000334$$

We conclude that $\log 7 = 0.845148$ with absolute error not more than 0.00034. Or we may use the error bound to make an interval estimate:

$$0.845148 - 0.000334 \le \log 7 \le 0.845148 + 0.000334$$

$$0.844814 \le \log 7 \le 0.845482$$

The estimate of $\log 7$ is reliable only to three decimal places. We conclude that $\log 7 = 0.845$ to three decimal places.

Actually, it can be verified that $\log 7 = 0.84510$ to five decimal places. So the true absolute error in our estimate is 0.000048, which is less than 5×10^{-5}. We do, in fact, have four correct decimals. Error bounds obtained by the Lagrange formula are typically much larger than the true error. ∎

■ *Exercise Set 7.2*

1. Let $y_j = f(x_j)$ for $j = 0, 1, \ldots, n$, and write down explicitly the Lagrange polynomials $P_1(x)$ and $P_2(x)$.

2. We want to find a formula for the quadratic function whose graph passes through the points $(-h, y_0)$, $(0, y_1)$, and (h, y_2) for positive constant h. The formula can be written in several different forms.

a. Find numbers a, b, and c for the formula $Q(x) = ax^2 + bx + c$.

b. Find the Lagrange formula for $Q(x)$.

c. Find constants A_0, A_1, and A_2 such that

$$Q(x) = A_0 + A_1(x + h) + A_2 x(x + h)$$

d. Find constants B_0, B_1, and B_2 such that

$$Q(x) = B_0 + B_1 x + B_2(x - h)(x + h)$$

3. Show by use of the error term that the Lagrange polynomial gives exact values (zero error) of $f(x)$ when we interpolate f on set $\{x_0, x_1, \ldots, x_n\}$ if f is a polynomial of degree less than or equal to n.

4. Suppose that f'' is continuous on the interval $[a, b]$ containing points x_0 and $x_1 = x_0 + h$. Suppose that $|f''(x)| < M$ on (a, b).

a. Find a bound for the absolute error when values of $f(x)$ are estimated by linear interpolation on the given set.

b. Apply the result in part (a) to the problem of estimating values of $\sin x$ from a table of values where the spacing is $h = 0.1$. What is the error bound?

c. What is the maximum allowable spacing in a table of values of $\sin x$ if we want to be able to interpolate using linear interpolation with an error less than 10^{-5}?

d. Repeat parts (b) and (c) for $f(x) = \ln x$ on interval $[1, 2]$.

e. Repeat parts (b) and (c) for $f(x) = e^x$ on interval $[0, 1]$.

5. Given the table of values of $\sqrt{x}$ shown, estimate $\sqrt{x}$ for each of the following values of x. Use quadratic interpolation and find an error bound for each estimate. Compare your error bounds with the true errors computed by using accurate values of $\sqrt{x}$ (from tables or calculator).

a. $x = 1.6$ **b.** $x = 1.2$ **c.** $x = 1.9$ **d.** $x = 1.45$

x	$\sqrt{x}$
1.0	1.000000
1.5	1.224745
2.0	1.414214

6. Given the table of values of the Bessel function $J_0(x)$ as shown, estimate $J_0(x)$ for the following values of x using quadratic interpolation. Find a table giving more precise values of $J_0(x)$, and determine the error in each estimate. Which estimate is most accurate?

a. $x = 1.1$ **b.** $x = 1.3$ **c.** $x = 1.01$ **d.** $x = 1.39$

x	$J_0(x)$
1.0	0.7652
1.2	0.6711
1.4	0.5669

7. Suppose that f''' is continuous on $[a, b]$ and points x_0, $x_1 = x_0 + h$, and $x_2 = x_0 + 2h$ lie in (a, b). Assuming that $|f'''(x)|$ is bounded by constant M, find a bound for the error in estimates of $f(x)$ obtained by quadratic interpolation of f on the given set.

8. We want to interpolate $\sin x$ on the set of points $x_0 = 0$, $x_1 = 0.1$, and $x_2 = 0.2$ using quadratic interpolation.

   a. Find a formula for the error bound as a function of x for $0 \le x \le 0.2$.
   b. Sketch the graph of the error bound formula in part (a).
   c. For which values of x would you expect the error to be largest in this case?

7.3   The Newton Divided Difference Form

The Lagrange polynomial form for the interpolating polynomial is an important one, but many other forms for the interpolating polynomial have been found. Some of these are more convenient to use than the Lagrange form. Note that each time interpolation point x is changed, we must make a fresh computation of the polynomials $L_k(x)$. Also note that these polynomials are not written in a form that is convenient to use in machine computation.

   The Lagrange polynomial also suffers from another serious fault. If we increase the number of node points by adding a point x_{n+1} to the set $\{x_0, x_1, \ldots, x_n\}$, the new interpolation polynomial $P_{n+1}(x)$ cannot be computed in any simple way from the polynomial $P_n(x)$. We must begin afresh as if the polynomial $P_n(x)$ had not been computed. In practice, when a set of points $x_0, x_1, \ldots, x_n$ is available to use for interpolation of a function f, we usually do not know in advance how many points we should use for the desired interpolation. How many points give the best result? We sometimes begin with two or three points near interpolation point x and add other points one at a time until we are satisfied with the resulting estimate.

   This leads us to consider the problem of constructing $P_n(x)$ from $P_{n-1}(x)$. We proceed as follows. We define function $C_n(x)$ by

$$P_n(x) = P_{n-1}(x) + C_n(x)$$

Since

$$C_n(x) = P_n(x) - P_{n-1}(x)$$

$C_n(x)$ is a polynomial of degree n or less. Note that $C_n(x_k) = 0$ for $k = 0, 1, \ldots, n-1$. Hence,

$$C_n(x) = A_n(x - x_0)(x - x_1) \ldots (x - x_{n-1})$$

for some constant A_n. Since

$$P_n(x) = P_{n-1}(x) + A_n(x - x_0)(x - x_1) \ldots (x - x_{n-1})$$

we have

$$P_n(x_n) = P_{n-1}(x_n) + A_n(x_n - x_0)(x_n - x_1) \ldots (x_n - x_{n-1})$$

and constant A_n is a function of the numbers $x_0, x_1, \ldots, x_n$. The numbers $A_1, A_2, \ldots, A_n$ are called **divided differences**. They can be computed if we know f and points $x_0, x_1, \ldots, x_n$. Clearly,

$$A_1 = \frac{P_1(x_1) - P_0(x_1)}{x_1 - x_0}$$

But

$$P_1(x_1) = f(x_1) \quad \text{and} \quad P_0(x_1) = f(x_0)$$

so

$$A_1 = \frac{f(x_1) - f(x_0)}{x_1 - x_0}$$

In general, A_k is determined by points $x_0, x_1, \ldots, x_k$ and function f, so we write

$$A_k = f[x_0, x_1, \ldots, x_k]$$

We generate the sequence of polynomials

$$P_0(x) = f(x_0)$$
$$P_1(x) = f(x_0) + f[x_0, x_1](x - x_0)$$
$$P_2(x) = f(x_0) + f[x_0, x_1](x - x_0) + f[x_0, x_1, x_2](x - x_0)(x - x_1)$$

and so forth, where

$$P_k(x) = P_{k-1}(x) + f[x_0, x_1, \ldots, x_k](x - x_0) \ldots (x - x_{k-1})$$

We have shown that

$$f[x_0, x_1] = \frac{f(x_1) - f(x_0)}{x_1 - x_0}$$

Now observe that setting $x = x_2$ in $P_2(x)$ yields

$$f(x_2) = f(x_0) + f[x_0, x_1](x_2 - x_0) + f[x_0, x_1, x_2](x_2 - x_0)(x_2 - x_1)$$

It follows that

$$f(x_2) - f(x_1) = f(x_0) - f(x_1) + f[x_0, x_1](x_2 - x_0)$$
$$+ f[x_0, x_1, x_2](x_2 - x_0)(x_2 - x_1)$$
$$= -f[x_0, x_1](x_1 - x_0) + f[x_0, x_1](x_2 - x_0)$$
$$+ f[x_0, x_1, x_2](x_2 - x_0)(x_2 - x_1)$$

So

$$f(x_2) - f(x_1) = f[x_0, x_1](x_2 - x_1) + f[x_0, x_1, x_2](x_2 - x_0)(x_2 - x_1)$$

and

$$\frac{f(x_2) - f(x_1)}{x_2 - x_1} = f[x_0, x_1] + f[x_0, x_1, x_2](x_2 - x_0)$$

It follows that

$$\frac{f[x_1, x_2] - f[x_0, x_1]}{x_2 - x_0} = f[x_0, x_1, x_2]$$

In general, we find that the divided difference $f[x_0, x_1, \ldots, x_k]$ is

$$\frac{f[x_1, x_2, \ldots, x_k] - f[x_0, x_1, \ldots, x_{k-1}]}{x_k - x_0} \qquad k = 1, 2, \ldots, n$$

A proof of this fact will be given in Section 7.5.

It is important to note that if $z_0, z_1, z_2, \ldots, z_k$ is any reordering of the interpolation points $x_0, x_1, \ldots, x_k$, then

$$f[z_0, z_1, z_2, \ldots, z_k] = f[x_0, x_1, x_2, \ldots, x_k]$$

Hence,

$$f[x_0, x_1] = f[x_1, x_0]$$

and

$$f[x_2, x_1, x_0] = f[x_0, x_1, x_2]$$

for example.

We can prove this as follows. First note that $f[x_0, x_1, \ldots, x_k]$ is the coefficient of x^k in the interpolation polynomial $P_k(x)$. According to the Lagrange formula,

$$P_k(x) = \sum_{j=0}^{k} f(x_j) L_j(x)$$

Since

$$L_j(x) = \frac{(x - x_0)(x - x_1) \ldots (x - x_{j-1})(x - x_{j+1}) \ldots (x - x_k)}{(x_j - x_0)(x_j - x_1) \ldots (x_j - x_{j-1})(x_j - x_{j+1}) \ldots (x_j - x_k)}$$

the coefficient of x^k in $P_k(x)$ is also

$$\sum_{j=0}^{k} \frac{f(x_j)}{(x_j - x_0) \ldots (x_j - x_{j-1})(x_j - x_{j+1}) \ldots (x_j - x_k)}$$

Hence, this sum also equals $f[x_0, x_1, \ldots, x_k]$. The formula

$$f[x_0, x_1, \ldots, x_k] = \sum_{j=0}^{k} \frac{f(x_j)}{(x_j - x_0) \ldots (x_j - x_{j-1})(x_j - x_{j+1}) \ldots (x_j - x_k)}$$

is a well-known formula for the divided differences and has various theoretical uses. It is, of course, clear from the formula that rearrangement of points $x_0, x_1, \ldots, x_k$ merely leads to rearrangement of the terms in the sum, so no effect on the value of $f[x_0, x_1, \ldots, x_k]$ is produced.

Now since $f[x_0, x_1, \ldots, x_k]$ is the coefficient of x^k in $P_k(x)$, it follows that

$$f[x_0, x_1, \ldots, x_k] = \frac{1}{k!} \cdot \frac{d^k P_k(x)}{dx^k}$$

This is an interesting formula for $f[x_0, x_1, \ldots, x_k]$ that shows the relationship between divided differences and derivatives of the interpolation polynomial.

The polynomial defined by

$$P_n(x) = f(x_0) + f[x_0, x_1](x - x_0) + \cdots + f[x_0, x_1, \ldots, x_n](x - x_0) \ldots (x - x_{n-1})$$

with

$$f[x_0, x_1] = \frac{f(x_1) - f(x_0)}{x_1 - x_0}$$

and

$$f[x_0, x_1, \ldots, x_k] = \frac{f[x_1, \ldots, x_k] - f[x_0, x_1, \ldots, x_{k-1}]}{x_k - x_0} \qquad k = 2, 3, \ldots, n$$

is called the **Newton divided difference** form of the interpolation polynomial. We will now examine some applications of this divided difference polynomial to the interpolation problem.

EXAMPLE 7.7   Estimate $f(2.17)$ from the given table of values of $f(x)$.

x	$f(x)$			
2.1	1.449138			
		$> f[x_0, x_1] = 0.34102$		
2.2	1.483240		$> f[x_0, x_1, x_2] = -0.03835$	
		$> f[x_1, x_2] = 0.33335$		
2.3	1.516575			

Solution   We first compute the needed divided differences by extending the table as shown. Substitution in the formula for $P_2(x)$ with $x = 2.17$ yields

$$P_2(2.17) = 1.449138 + 0.34102(0.07) - 0.03835(0.07)(-0.03)$$

Thus

$$P_2(2.17) = 1.473090$$

and

$$f(2.17) = 1.47309 \quad \text{(approximately)}$$

The function in this example is $f(x) = \sqrt{x}$. So we can compare our estimate with the true value. In fact, $f(2.17) = 1.473092$, so our error is about 0.000002. ∎

Divided differences are usually listed in the triangular array shown in Example 7.7 when the work is done by hand. The arrangement displays the dependence of each divided difference on the numbers in the preceding column. Divided difference $f[x_0, x_1, x_2]$ is computed from $f[x_0, x_1]$ and $f[x_1, x_2]$, for example, so its value is placed to the right of and between the values of $f[x_0, x_1]$ and $f[x_1, x_2]$.

Now let us consider the error estimation problem. We could use the Lagrange error formula, of course, since we have the same interpolation polynomial (just written in a different form). However, there is an error estimation procedure that is better suited for use with the Newton divided difference formula.

Given $P_n(x)$, which interpolates $f(x)$ on the set of points $\{x_0, x_1, x_2, \ldots, x_n\}$, let $P_{n+1}(x)$ interpolate $f(x)$ on the set of points $\{x_0, x_1, \ldots, x_n, \bar{x}\}$, where $\bar{x}$ is another point in the domain of f distinct from $x_0, x_1, \ldots, x_n$. Then

$$P_{n+1}(\bar{x}) = f(\bar{x})$$

Also,

$$P_{n+1}(x) = P_n(x) + f[x_0, x_1, \ldots, x_n, \bar{x}](x - x_0) \ldots (x - x_n)$$

So

$$f(\bar{x}) = P_n(\bar{x}) + f[x_0, x_1, \ldots, x_n, \bar{x}](\bar{x} - x_0) \ldots (\bar{x} - x_n)$$

Then the error in estimation of $f(\bar{x})$ by $P_n(\bar{x})$ is

$$f[x_0, x_1, \ldots, x_n, \bar{x}](\bar{x} - x_0)(\bar{x} - x_1) \ldots (\bar{x} - x_n)$$

If we can estimate $f[x_0, x_1, \ldots, x_n, \bar{x}]$, we can estimate the error. The procedure is to estimate the error term by

$$f[x_0, x_1, \ldots, x_n, x_{n+1}](x - x_0)(x - x_1) \ldots (x - x_n)$$

where x_{n+1} is some point near x for which the value of $f(x_{n+1})$ is known, so that $f[x_0, x_1, \ldots, x_n, x_{n+1}]$ can be computed.

EXAMPLE 7.8   Estimate $f(0.12)$ from the following table of values of f and the needed divided differences.

x	$f(x)$	Divided Differences		
0.00	1.0000			
		1.0260		
0.05	1.0513		0.5200	
		1.0780		0.1333
0.10	1.1052		0.5400	
		1.1320		
0.15	1.1618			

Solution   Using these points and the formula for $P_3(x)$, we find that $f(0.12)$ is approximated by

$$1.0000 + 1.0260(0.12) + 0.5200(0.12)(0.07) + 0.1333(0.12)(0.07)(0.02) = 1.1275$$

The error is

$$f[0, 0.05, 0.1, 0.15, 0.12](0.12)(0.07)(0.02)(-0.03)$$

Given the added information that $f(0.20) = 1.2214$ from a table of values, we can approximate the error by

$$f[0, 0.05, 0.1, 0.15, 0.20](0.12)(0.07)(0.02)(-0.03)$$

We find by adding another row to our divided difference table that $f[0, 0.05, 0.1, 0.15, 0.2] = 1.3334$, as you should verify. Hence, our error estimate is

$$1.3334(0.12)(0.07)(0.02)(-0.03) = -0.0000007 \qquad \blacksquare$$

It is important to realize that our estimate in Example 7.8 is an estimate of the error involved in approximation of function $f(x)$ by polynomial $P_3(x)$ assuming all values used in the computations are known with unlimited accuracy and the arithmetic has been carried out with no errors. In our example, the values of f are given only to four digits, and only four-decimal arithmetic is used, so our estimate includes propagated error and round-off error in addition to the truncation error accounted for by our error estimation formula. Our computation does show that the polynomial approximation of f introduced very little truncation error relative to the round-off error, so we may assume that $f(0.12) = 1.1275$ as estimated, with perhaps some doubt about the last digit because of the limited accuracy of the arithmetic used.

■ *Exercise Set 7.3*

   1. Assume that we know that $\log 5 = 0.69897$, $\log 6 = 0.77815$, $\log 8 = 0.90309$, and $\log 9 = 0.95424$. We want to estimate $\log 7$ by interpolating on these data.
      a. Estimate $\log 7$ using the values of $\log 6$ and $\log 8$ with linear interpolation using the Newton divided difference formula.

b. Repeat part (a) using the Lagrange formula from Section 7.2.

c. Estimate log 7 using the values of log 6, log 8, and log 9 with quadratic interpolation employing the Newton divided difference formula.

d. Repeat part (c) using the Lagrange formula.

2. Make error estimates for the estimates of log 7 produced in exercise 1c and d using the corresponding error terms for the Newton and Lagrange formulas. To estimate the error in the Newton divided difference estimate, choose as your added data point the value log 10 = 1. Theoretically, the error should be the same for both. Explain any difference found in your calculations.

3. Use the Newton divided difference formula to construct interpolation polynomial $P_2(x)$ for sin x on the points 0, $\pi/2$, and π. Do not replace π with a decimal approximation in your formula. Now compare accurate values of sin x with values of $P_2(x)$ for $x = \pi/6$ and $x = \pi/3$.

4. Try to improve on your polynomial approximation to sin x obtained in exercise 3 by adding points $\pi/3$ and $\pi/6$ to the interpolation set and constructing the Newton divided difference polynomial $P_4(x)$. Compare values of $P_4(x)$ with accurate values of sin x for $x = 0.01$, $x = 0.3$, and $x = 1$.

5. Compute absolute errors and relative errors for each of the estimates obtained in exercise 4. Does inspection of the relative errors and absolute errors affect your opinion as to which estimate is "best"?

6. Estimate the integral of sin x from 0 to $\pi/2$ using the polynomials $P_2(x)$ and $P_4(x)$ obtained in exercises 3 and 4. Compare with the exact value.

7. Repeat exercise 5 from Exercise Set 7.2 using the Newton divided difference formula. Use $\sqrt{2.25} = 1.5$ as the added data point for making an error estimate.

8. Repeat exercise 6 from Exercise Set 7.2 using the Newton divided difference formula.

9. Use the Newton divided difference formula with $n = 2$ to estimate sin 12°13′ from the first three entries in the given table of values. Compute the error using the fact that the correct value to five decimal places is 0.21161.

x	sin x
12°00′	0.20791
12°30′	0.21644
13°00′	0.22495
13°30′	0.23345
14°00′	0.24192

10. Repeat exercise 9 using the first four entries and then using the full set of five entries. Which of these estimates has the smallest absolute error?

11. We would like to express the polynomial $19 - 34x + 19x^2 - 3x^3$ in the form

$$A_0 + A_1(x - 1) + A_2(x - 1)(x - 2) + A_3(x - 1)(x - 2)(x - 3)$$

The desired coefficients A_0, A_1, A_2, and A_3 are divided differences. Compute these coefficients by each of the following ways.

a. Construct a divided difference table using values of the given polynomial at $x = 1, 2, 3$, and 4.

b. Solve the system of equations resulting from substituting 1, 2, 3, and 4 into the desired form and equating to values of the polynomial at these points.

c. Same as in part (b) using values of the polynomial at points $-1, 0, 1$, and 2.

12. Consider the Newton divided difference polynomial

$$A_0 + A_1(x - 0.5) + A_2(x - 0.5)(x - 1) + A_3(x - 0.5)(x - 1)(x - 1.5)$$

where $A_0 = 2$, $A_1 = -3$, $A_2 = 4$, and $A_3 = -2$.

a. Show that we can compute the value of this polynomial using a nested multiplication algorithm.

b. Use the nested multiplication algorithm in part (a) to compute the value for $x = 0.75$.

c. Extend the result of part (a) to Newton divided difference polynomials of degree n. Write your result as an algorithm that could be converted to a machine program.

7.4   The Newton Forward Difference Formula

Suppose that a function f is continuous on $[a, b]$ and values $f(x_i)$ are known for equally spaced points

$$x_i = a + ih \qquad i = 0, 1, \ldots, n$$

where $h = (b - a)/n$. Given a point x, we define $s = (x - x_0)/h$ so that $x = x_0 + sh$. Then

$$f(x) = f(x_0 + sh) = F(s) = f_s$$

where values f_s are known for $s = 0, 1, \ldots, n$.

Let us define

$$\Delta^1 f_s = f_{s+1} - f_s \qquad s = 0, 1, \ldots, n$$

$$\Delta^2 f_s = \Delta^1 f_{s+1} - \Delta^1 f_s \qquad s = 0, 1, \ldots, n - 1$$

and in general

$$\Delta^{i+1} f_s = \Delta^i f_{s+1} - \Delta^i f_s$$

It is also conventional to set $\Delta^0 f_s = f_s$. In table form, we have

$$
\begin{array}{cccccccc}
\Delta^0 f_0 & & & & & & & \\
 & \Delta^1 f_0 & & & & & & \\
\Delta^0 f_1 & & \Delta^2 f_0 & & & & & \\
 & \Delta^1 f_1 & & \Delta^3 f_0 & & & & \\
\Delta^0 f_2 & \vdots & \Delta^2 f_1 & & \Delta^4 f_0 & & & \\
 & \Delta^1 f_2 & & \Delta^3 f_1 & & & & \\
\Delta^0 f_3 & & \Delta^2 f_2 & & & & & \\
 & \Delta^1 f_3 & & & & & & \\
\Delta^0 f_4 & & & & & & &
\end{array}
$$

·Differences $\Delta^1 f_0$, $\Delta^2 f_0$, and so on, are computed by the preceding formulas. So $\Delta^2 f_1$ is computed from $\Delta^1 f_1$ and $\Delta^1 f_2$, for example, as indicated by the table arrangement. The numbers $\Delta^j f_k$ are called **forward differences**. Such forward differences can be computed from any set of data pairs (x_i, y_i) if points x_i are equally spaced. We often assume in that case that $y_i = f(x_i)$ for some function f that may or may not be a known function.

There are two interesting and useful results concerning forward differences that we need at this time.

Theorem 7.3   Under the preceding assumptions on f,

$$f[x_k, x_{k+1}, \ldots, x_{k+i}] = \frac{1}{i! h^i} \Delta^i f_k$$

Proof   The proof is by induction on i. If $i = 0$, the claim becomes

$$f[x_k] = f(x_k) = \frac{1}{h^0} \Delta^0 f_k = f_k$$

which is, of course, true. The theorem is also clearly true for $i = 1$. Suppose that the theorem is true for some positive integer $i = n$. Then

$$f[x_k, x_{k+1}, \ldots, x_{k+n}, x_{k+n+1}] = \frac{f[x_{k+1}, \ldots, x_{k+n}, x_{k+n+1}] - f[x_k, \ldots, x_{k+n}]}{x_{k+n+1} - x_k}$$

$$= \frac{\dfrac{1}{n! h^n} \Delta^n f_{k+1} - \dfrac{1}{n! h^n} \Delta^n f_k}{(n+1)h} = \frac{\Delta^{n+1} f_k}{(n+1)! h^{n+1}}$$

By the induction principle, the theorem is then true for all positive integers i.   □

Theorem 7.4   If $P_n(x)$ interpolates f on equally spaced points $x_0, x_1, \ldots, x_n$ with $x_i = x_0 + ih$ for $i = 0, 1, \ldots, n$, then

$$\frac{d^n P_n(x)}{dx^n} = \frac{\Delta^n f_0}{h^n}$$

Proof   In Section 7.3, it was shown that

$$f[x_0, x_1, \ldots, x_n] = \frac{1}{n!} \cdot \frac{d^n P_n(x)}{dx^n}$$

Using $k = 0$ and $i = n$ in Theorem 7.3, we see that

$$f[x_0, x_1, \ldots, x_n] = \frac{1}{n!h^n} \Delta^n f_0$$

Equating these two results completes the proof.   □

As just noted, if we choose $k = 0$ and $i = n$ in Theorem 7.3, we obtain

$$f[x_0, x_1, \ldots, x_n] = \frac{1}{n!h^n} \Delta^n f_0$$

This result is used in the following argument.
The Newton divided difference polynomial is

$$P_n(x) = f(x_0) + f[x_0, x_1](x - x_0) + \cdots + f[x_0, x_1, \ldots, x_n](x - x_0) \ldots (x - x_{n-1})$$

Using Theorem 7.3, we see that

$$P_n(x) = f(x_0) + \frac{\Delta^1 f_0}{h} (x - x_0)$$

$$+ \frac{\Delta^2 f_0}{2!h^2} (x - x_0)(x - x_1) + \cdots + \frac{\Delta^n f_0}{n!h^n} (x - x_0) \ldots (x - x_{n-1})$$

Now using

$$x = x_0 + sh \quad \text{and} \quad x_k = x_0 + kh$$

we see that

$$x - x_k = (s - k)h$$

Hence,

$$P_n(x) = f_0 + s\Delta^1 f_0 + \frac{s(s - 1)}{2!} \Delta^2 f_0 + \cdots + \frac{s(s - 1) \ldots (s - n + 1)}{n!} \Delta^n f_0$$

Introducing the notation

$$\binom{s}{k} = 1 \quad \text{if} \quad k = 0$$

$$\binom{s}{k} = \frac{s(s - 1) \ldots (s - k + 1)}{k!} \quad k = 1, 2, \ldots$$

we see that

$$P_n(x) = f_0 + \binom{s}{1}\Delta^1 f_0 + \binom{s}{2}\Delta^2 f_0 + \cdots + \binom{s}{n}\Delta^n f_0$$

In numerical analysis literature, this formula is known as the **Newton forward difference** interpolating polynomial. Briefly, we have

$$P_n(x) = f_0 + \sum_{j=1}^{n} \binom{s}{j}\Delta^j f_0 \qquad x = x_0 + sh$$

The following example uses this form of the interpolation polynomial.

EXAMPLE 7.9 Using the following table of values for $f(x) = \sin x$, estimate the value of $\sin 30°12'$.

x	$f(x)$	Δf	$\Delta^2 f$	$\Delta^3 f$
30.0	0.50000			
		0.00754		
30.5	0.50754		−0.00004	
		0.00750		0.00000
31.0	0.51504		−0.00004	
		0.00746		
31.5	0.52250			

Solution First we compute the forward differences as indicated in the table. Letting $x_0 = 30.0$, $h = 0.5$, and $n = 3$, we see that $f_0 = 0.50000$, $\Delta^1 f_0 = 0.00754$, $\Delta^2 f_0 = -0.00004$, and $\Delta^3 f_0 = 0.00000$. Also, $s = (30.2 - 30.0)/0.5 = 0.4$. Then

$$\binom{s}{1} = 0.4, \quad \binom{s}{2} = -0.12, \quad \text{and} \quad \binom{s}{3} = 0.064$$

The forward difference formula then yields approximation

$$0.50000 + 0.4(0.00754) - 0.12(-0.00004) = 0.50302$$

for the value of $\sin 30.2°$. It is interesting to note that this result is in fact correct to five decimal places. ∎

EXAMPLE 7.10 Using the following table of values for a function f, estimate $f(1.09)$.

x	$f(x)$	Δf	$\Delta^2 f$	$\Delta^3 f$
1.00	0.1924			
		0.0490		
1.05	0.2414		0.0029	
		0.0519		0.0011
1.10	0.2933		0.0040	
		0.0559		
1.15	0.3492			

Solution We first compute forward differences as shown in the table. Then we find $s = (1.09 - 1.00)/0.05 = 1.8$. So our estimate of $f(1.09)$ is

$$P_3(1.09) = 0.1924 + \binom{1.8}{1}0.0490 + \binom{1.8}{2}0.0029 + \binom{1.8}{3}0.0011$$

$$= 0.1924 + 1.8(0.0490) + \frac{1.8(0.8)}{2}(0.0029) + \frac{(1.8)(0.8)(-0.2)}{6}(0.0011)$$

$$= 0.1924 + 0.0882 + 0.00209 + (-0.000053) = 0.28264$$

Since values of f are given to only four decimal places, the final estimate should be rounded to 0.2826. ∎

Now we once again consider the problem of error estimation. From the Newton divided difference form, we know that the error is

$$f[x_0, x_1, \ldots, x_n, x](x - x_0)(x - x_1) \ldots (x - x_n)$$

which can be approximated by

$$f[x_0, x_1, \ldots, x_n, x_{n+1}](x - x_0)(x - x_1) \ldots (x - x_n)$$

where x_{n+1} is an added point for which we know the value of f. Using the Newton forward difference formula, we replace the preceding as follows. We choose the next point

$$x_{n+1} = x_0 + (n + 1)h$$

Then

$$f[x_0, x_1, \ldots, x_n, x_{n+1}] = \frac{\Delta^{n+1} f_0}{(n + 1)!h^{n+1}}$$

Also,

$$(x - x_0)(x - x_1) \ldots (x - x_n) = s(s - 1)(s - 2) \ldots (s - n)h^{n+1}$$

Then the error is estimated by

$$\frac{\Delta^{n+1}f_0}{(n+1)!h^{n+1}} s(s-1)(s-2) \dots (s-n)h^{n+1} = \Delta^{n+1}f_0 \frac{s(s-1) \dots (s-n)}{(n+1)!}$$

$$= \binom{s}{n+1}\Delta^{n+1}f_0$$

That is, the error is approximated by the next term in the Newton forward difference polynomial resulting from adding point x_{n+1}.

EXAMPLE 7.11   Estimate the error in the approximation of sin 30.2° computed in Example 7.9.

Solution   The estimate in Example 7.9 was actually made using only the three points 30.0°, 30.5°, and 31.0°, so we may regard 31.5° as an added point for the purpose of error estimation. The computation of difference $\Delta^3 f_0$ gave 0.00000, as indicated in the table. Our error formula would then estimate the error as zero. This is, of course, not true. The zero error estimate arises from our restriction of the arithmetic to five decimals. We do see, however, that the difference is less than 0.00001. So the error should be of the order of 0.00001 or less. We would expect the estimate to be correct to five decimal places, with some doubt, possibly, about the last digit because of round-off error. That is a correct conclusion, as you can confirm by checking the value of sin 30.2° using precise tables.   ∎

There are many other forms for the interpolating polynomial available in the literature. Most of the others are primarily of historical interest. There are formulas derived by Bessel and by Gauss, among others. Many of the great mathematicians of the past spent some time studying the interpolation problem because interpolation methods were widely used in astronomical computations and other physical problems. Modern computing machines have made it unnecessary to have so many specialized formulas. Some acquaintance with the problem and the forms of the solution, as presented here, is sufficient. Most of the formulas developed in the distant past are not convenient for machine use. Even the formulas discussed in this text so far suffer from that fault. These formulas are useful as a starting point for developing better algorithms for machine use, however, and are needed to at least begin to do some error analysis. In the next section, we will turn to the development of an algorithm that is better suited for machine interpolation.

■ *Exercise Set 7.4*

1. Consider the following table of values.

x	$f(x)$
15.00	0.9655
15.50	0.9622
16.00	0.9589
16.50	0.9556
17.00	0.9524

    a. Construct a table of forward differences.

    b. Use these differences and the Newton forward difference formula to estimate $f(15.25)$ and $f(15.75)$.

2. Use the Newton forward difference formula to estimate $\log 7$ using the values given in exercise 1 of Exercise Set 7.3. Make an error estimate using the method described at the end of this section.

3. Repeat exercise 9 of Exercise Set 7.3 using the Newton forward difference polynomial.

4. Repeat exercise 10 of Exercise Set 7.3 using the Newton forward difference polynomial.

5. Estimate $J_0(1.3)$ and $J_0(1.1)$ using the table of values given in exercise 6 of Exercise Set 7.2.

6. The American Ephemeris and Nautical Almanac for 1979 gave the values in the following table for the declination (celestial latitude) of the moon on January 29.

Time	Declination		
(P.M.)	(Deg	Min	Sec)
8:00	−9	31	59
9:00	−9	21	39
10:00	−9	11	16
11:00	−9	00	49

Use this table of values and the Newton forward difference polynomial with $n = 3$ to estimate the declination of the moon at 9:16 P.M. and at 10:47 P.M. on that date.

7. Define the *forward difference operator* Δ by

$$\Delta f(x) = f(x + h) - f(x)$$

where constant $h > 0$ is assumed given. Higher-order difference operators are defined by $\Delta^2 f(x) = \Delta(\Delta f(x))$, $\Delta^3 f(x) = \Delta(\Delta^2 f(x))$, and so on.

    a. Show that Δ is a linear operator; that is,

$$\Delta(f(x) + g(x)) = \Delta f(x) + \Delta g(x)$$

and

$$\Delta(cf(x)) = c(\Delta f(x))$$

    b. Show that $\Delta^n(x^n) = n!h^n$ if n is a positive integer.

    c. Show that if $P_n(x)$ is a polynomial of degree $n \geq 1$ with

$$P_n(x) = a_n x^n + a_{n-1} x^{n-1} + \cdots + a_1 x + a_0$$

then $\Delta^n P_n(x) = n!a_n h^n$ is a constant.

    d. Verify the truth of part (c) for the polynomial $P_3(x) = x^3 + 5x - 1$ using values at points 0.5, 1, 1.5, 2, 2.5, and 3.

8. The shift operator E is defined by $Ef(x) = f(x + h)$. Powers E^n of E are defined as products of E, so that $E^2 f(x) = E(Ef(x))$, for example. Let I denote the identity operator, so that $If(x) = f(x)$.

    a. Show that $\Delta = E - I$ and $E = I + \Delta$.

b. Show that $E^n f(x) = f(x + nh)$.

c. Use the result in part (a) to show that

$$E^n f(x_0) = \sum_{k=0}^{n} \binom{n}{k} \Delta^k f_0$$

d. Combine results from parts (b) and (c) to show that

$$f(x_n) = f_0 + \sum_{k=1}^{n} \binom{n}{k} \Delta^k f_0$$

9. Prove that the following table contains values of a polynomial of degree 3. Determine the polynomial coefficients in the standard form

$$a_3 x^3 + a_2 x^2 + a_1 x + a_0$$

(*Hint:* Use exercise 7c.)

x	$f(x)$
1	-2
2	13
3	52
4	127
5	250
6	433
7	688

7.5   The Aitken Interpolation Algorithm (Optional)

In this section, we will discuss an algorithm that is particularly well suited for machine estimation of values of a function by interpolation. The theory to be discussed leads directly to an algorithm for machine use in making numerical estimates. No algebraic formula for the interpolating polynomial is employed directly.

Let us begin with a new look at the solution of the two-point interpolation problem provided by the Lagrange form. Given points x_0 and x_1, let $P_{0,1}(x)$ denote the interpolation polynomial for function f on this set of points. Then

$$P_{0,1}(x) = f(x_0) \frac{x - x_1}{x_0 - x_1} + f(x_1) \frac{x - x_0}{x_1 - x_0}$$

So

$$P_{0,1}(x) = \frac{(x - x_0)f(x_1) - (x - x_1)f(x_0)}{x_1 - x_0}$$

This simple formula for computing $P_{0,1}(x)$ from the values of $f(x_0) = P_0(x)$ and $f(x_1) = P_1(x)$ is a special case of a more general iterative formula for computing $P_{0,1,\ldots,k,k+1}(x)$ from $P_{0,1,\ldots,k}(x)$ and $P_{0,1,\ldots,k-1,k+1}(x)$. The following theorem due to Aitken explains how to do this.

Theorem 7.5   Let $P_{0, 1, \ldots, k}(x)$ be the interpolation polynomial for function f over set $\{x_0, x_1, \ldots, x_k\}$. Then $P_{0, 1, \ldots, k, k+1}(x)$ is equal to

$$\frac{(x - x_k)P_{0, 1, \ldots, k-1, k+1}(x) - (x - x_{k+1})P_{0, 1, \ldots, k}(x)}{x_{k+1} - x_k}$$

Proof   Let $P(x)$ denote the polynomial defined by the preceding expression. Since $P_{0, 1, \ldots, k-1, k+1}(x)$ and $P_{0, 1, \ldots, k}(x)$ are polynomials of degree k or less, $P(x)$ is a polynomial of degree $k + 1$ or less. Now if x_i is one of the points $x_0, x_1, \ldots, x_{k-1}$, then

$$P(x_i) = \frac{(x_i - x_k)f(x_i) - (x_i - x_{k+1})f(x_i)}{x_{k+1} - x_k} = f(x_i)$$

Also,

$$P(x_k) = \frac{-(x_k - x_{k+1})f(x_k)}{x_{k+1} - x_k} = f(x_k)$$

and

$$P(x_{k+1}) = \frac{(x_{k+1} - x_k)f(x_{k+1})}{x_{k+1} - x_k} = f(x_{k+1})$$

So $P(x)$ interpolates f on set $\{x_0, x_1, \ldots, x_k, x_{k+1}\}$. By uniqueness of the interpolation polynomial, we conclude that $P(x) = P_{0, 1, \ldots, k, k+1}(x)$.   □

We can use this theorem to develop a procedure for computing the value of $P_{0, 1, \ldots, n}(x)$ systematically. Suppose that values of x_k and $f(x_k) = P_k$ have been given for $k = 0, 1, \ldots, n$, where $n \geq 1$. We compute values of

$$P_{0, 1} = \frac{(x - x_0)P_1 - (x - x_1)P_0}{x_1 - x_0}$$

and

$$P_{0, 2} = \frac{(x - x_0)P_2 - (x - x_2)P_0}{x_2 - x_0}$$

By Theorem 7.5, we see that

$$P_{0, 1, 2} = \frac{(x - x_1)P_{0, 2} - (x - x_2)P_{0, 1}}{x_2 - x_1}$$

Now if we compute $P_{0,3}$, we can also compute

$$P_{0,1,3} = \frac{(x - x_1)P_{0,3} - (x - x_3)P_{0,1}}{x_3 - x_1}$$

Then it follows from Theorem 7.5 that

$$P_{0,1,2,3} = \frac{(x - x_2)P_{0,1,3} - (x - x_3)P_{0,1,2}}{x_3 - x_2}$$

Continuing in this way, we can compute row by row the array of estimates

$$
\begin{array}{llll}
P_0 & & & \\
P_1 & P_{0,1} & & \\
P_2 & P_{0,2} & P_{0,1,2} & \\
P_3 & P_{0,3} & P_{0,1,3} & P_{0,1,2,3} \\
\text{etc.}
\end{array}
$$

Let us consider a numerical example.

EXAMPLE 7.12   Estimate the value of $f(1.08)$ from the given table of values.

x	$f(x)$
1.05	0.97350
1.10	0.95135
1.15	0.93304

Solution   First we compute

$$P_{0,1} = \frac{(1.08 - 1.05)0.95135 - (1.08 - 1.10)0.97350}{1.10 - 1.05} = 0.960210$$

Next we compute

$$P_{0,2} = \frac{(1.08 - 1.05)0.93304 - (1.08 - 1.15)0.97350}{1.15 - 1.05} = 0.961362$$

Next we compute

$$P_{0,1,2} = \frac{(1.08 - 1.10)0.961362 - (1.08 - 1.15)0.960210}{1.15 - 1.10} = 0.959749$$

Since the original data were given to only five-place accuracy, we round this last result and conclude that the interpolated value of $f(1.08)$ is 0.95975 if we use the three data points given.   ∎

The process of computing the interpolated values of a function f by repeated use of Theorem 7.5 can be organized into a systematic programmable algorithm. We call this procedure the **Aitken interpolation algorithm**. The computations can be viewed as filling in the following array.

x_k	$f(x_k)$			
x_0	P_0			
x_1	P_1	$P_{0,1}$		
x_2	P_2	$P_{0,2}$	$P_{0,1,2}$	
x_3	P_3	$P_{0,3}$	$P_{0,1,3}$	$P_{0,1,2,3}$

The computation proceeds from top to bottom of the array, and each row is computed in full before moving to the next row. The value of $P_{0,1}$ is computed from values of x_0, x_1, P_0, P_1, and interpolation point x. Then we compute values of $P_{0,2}$ and $P_{0,1,2}$, and so on, using Theorem 7.5 repeatedly.

The numbers $P_{0,1}, P_{0,1,2}, P_{0,1,2,3}$, and so on, are the results of interpolating over two points, three points, four points, and so forth. As the computation proceeds, the values along the top of the array form a sequence of estimates of interpolated value $f(x)$. If they approach a common value, the interpolation is proceeding well, and the last value computed may be taken to be the desired estimate. Usually, the values appear to converge on the desired value for a few steps, but eventually begin to diverge with a worsening of the estimate of $f(x)$.

When we do the calculations by hand or with a pocket calculator, we can see that the computations follow a simple pattern that is easy to memorize. Note, for example, that

$$P_{0,1,2} = \frac{(x - x_1)P_{0,2} - (x - x_2)P_{0,1}}{x_2 - x_1}$$

so if we insert a column of differences $x - x_i$ in the array, as shown in the following listing, we can interpret the preceding formula as describing the computation of the value of a 2×2 determinant divided by the difference $x_2 - x_1$. This is indicated by the dotted lines in the following array.

x_i	$x - x_i$	$f(x_i)$				
x_0	$x - x_0$	P_0				
x_1	$x - x_1$	P_1	$P_{0,1}$			
x_2	$x - x_2$	P_2	$P_{0,2}$	$P_{0,1,2}$		
x_3	$x - x_3$	P_3	$P_{0,3}$	$P_{0,1,3}$	$P_{0,1,2,3}$	
x_4	$x - x_4$	P_4	$P_{0,4}$	$P_{0,1,4}$	$P_{0,1,2,4}$	$P_{0,1,2,3,4}$

Try identifying the pattern of the computations with regard to this array for several elements of the array, and verify the numbers in the following example to test your understanding of the process.

EXAMPLE 7.13   Verify all numbers in the following array. The interpolation point is $x = 4.5$.

x_i	$x - x_i$	$f(x_i)$				
4.0	0.5	0.60206				
4.2	0.3	0.62325	0.65504			
4.4	0.1	0.64345	0.65380	0.65318		
4.6	−0.1	0.66276	0.65264	0.65324	0.65321	
4.8	−0.3	0.68124	0.65155	0.65330	0.65321	0.65321

Solution   Notice in this example that 0.65318 is the interpolated value using points 4.0, 4.2, and 4.4, while 0.65321 is the interpolated value using all five of the points given. Notice also how the estimates along the top row have "converged" on the value 0.65321.   ∎

EXAMPLE 7.14   Given the values of a function f as shown in columns 1 and 3 of the following array, estimate the value of $f(0.4142)$.

x_i	$x - x_i$	$f(x_i)$			
0.30	0.1142	1.608049			
0.35	0.0642	1.622528	1.641119		
0.40	0.0142	1.640000	1.644537	1.645508	
0.45	−0.0358	1.660886	1.648276	1.645714	1.645567

Solution   We compute the differences as shown in column 2 and then fill in the interpolated values as indicated using the Aitken algorithm. Our final estimate is 1.645567 using all of the points given. In this case, there is a sizable difference between the results of interpolating on the first three points and the result of using all four points. This indicates that the result could possibly be improved by using a fifth point. Assume that we find from a table that $f(0.5) = 1.685750$. Then we can add another row to the array as follows:

0.50	−0.0858	1.685750	1.652416	1.645954	1.645571	1.645563

Our new estimate is 1.645563. The results of interpolating on points 0.30 through 0.45 and the results using points 0.30 through 0.50 differ only slightly in this case, so we may assume that $f(0.4142)$ is approximately 1.645563, with some doubt only about the last digit.   ∎

When developing a programmable algorithm, it is best to denote the elements of the arrays with standard two-dimensional array notation, as follows:

$$
\begin{array}{llll}
A(1, 1) & & & \\
A(2, 1) & A(2, 2) & & \\
A(3, 1) & A(3, 2) & A(3, 3) & \\
A(4, 1) & A(4, 2) & A(4, 3) & A(4, 4)
\end{array}
$$

We let

$$A(i, 1) = f(x_i) \qquad i = 1, 2, \ldots, n + 1$$

and

$$D(i) = x - x_i$$

where the points have been renumbered as $x_1, x_2, x_3, \ldots, x_{n+1}$. Elements of the array are computed from the iteration formula

$$A(i, j) = \frac{D(j - 1)A(i, j - 1) - D(i)A(j - 1, j - 1)}{x_i - x_{j-1}}$$

$$i = 2, \ldots, n + 1 \quad \text{and} \quad j = 2, \ldots, i$$

Note that $A(k, k)$ is the result of interpolation on points $x_1, x_2, \ldots, x_k$. A computer program can easily be written that incorporates the preceding formula into a nested loop and computes the array $A(i, j)$.

Note that the Aitken algorithm produces a sequence of estimates of $f(x)$ using an increasing number of points. Rather than print out the entire array of estimates, the computer output can be restricted to printing out this sequence. These are the numbers $A(i, i)$ for $i = 1, 2, \ldots, n + 1$. This is a practical form for computer output from the algorithm. Usually, there is no need for the other elements of the array. It is not enough to print out only the final estimate $A(n + 1, n + 1)$, however. We get some idea of the accuracy of the result by inspecting the behavior of the sequence of estimates.

The difference of the last two estimates is related to the accuracy of the estimate of $f(x)$, as the following argument shows. This fact has already been assumed in our commentary on the accuracy of estimates of $f(x)$ in preceding examples. We will now justify those remarks.

Recall that

$$P_{0, 1, \ldots, n} = P_{0, 1, \ldots, n-1} + f[x_0, x_1, \ldots, x_n](x - x_0) \ldots (x - x_{n-1})$$

In Section 7.3, it was shown that

$$f(x) = P_{0, 1, \ldots, n} + E_n$$

where error E_n is equal to

$$f[x_0, x_1, \ldots, x_n, x](x - x_0)(x - x_1) \ldots (x - x_n)$$

So

$$E_{n-1} = f[x_0, x_1, \ldots, x_{n-1}, x](x - x_0)(x - x_1) \ldots (x - x_{n-1})$$

But

$$P_{0, 1,...,n} - P_{0, 1,...,n-1} = f[x_0,...,x_n](x - x_0) ... (x - x_{n-1})$$

Hence, error E_{n-1} is approximated by the difference

$$P_{0, 1,...,n} - P_{0, 1,...,n-1}$$

Of course, the accuracy of this error estimate depends on how accurately divided difference $f[x_0, x_1,...,x_n]$ approximates divided difference $f[x_0, x_1,..., x_{n-1}, x]$.

Theorem 7.5 is a special case of the following general result that can be used to produce a large number of other so-called iterative interpolation algorithms. The following theorem is sometimes called **Aitken's lemma**.

Theorem 7.6   Suppose that f is continuous on interval I. Let W be a set of $n + 1$ points $\{x_0, x_1,..., x_n\}$ in I. Suppose that values $y_k = f(x_k)$ are known. Let $S \subset W$ with $S \neq W$. Suppose $T \subset W$ also and $T \neq W$. Assume that point $x_i \in S$ and $x_j \in T$, while $x_i \notin T$ and $x_j \notin S$. Assume that $S \cap T = S \cup T - \{x_i, x_j\}$. Let $P_S, P_T,$ and $P_{S \cup T}$ denote interpolation polynomials over sets $S, T,$ and $S \cup T$, respectively. Then

$$P_{S \cup T}(x) = \frac{(x - x_j)P_S(x) - (x - x_i)P_T(x)}{x_i - x_j}$$

The proof of Theorem 7.6 is a straightforward generalization of the proof of Theorem 7.5, so it will be left as an exercise.

Theorem 7.5 follows if we choose sets $W, S,$ and T as follows. Let

$$W = \{x_0, x_1,..., x_k, x_{k+1}\}$$
$$S = \{x_0, x_1,..., x_{k-1}, x_{k+1}\}$$
$$T = \{x_0, x_1,..., x_k\}$$

You can easily verify that in this case the conclusion of Theorem 7.6 is the same as the conclusion of Theorem 7.5.

Many other iteration formulas similar to the Aitken formula in Theorem 7.5 can be derived using Theorem 7.6 with other choices of sets $S, T,$ and W. If we choose

$$S = \{x_1, x_2,..., x_n\} \quad \text{and} \quad T = \{x_0, x_1,..., x_{n-1}\}$$

we obtain

$$P_{0, 1, 2,...,n}(x) = \frac{(x - x_0)P_{1, 2,...,n}(x) - (x - x_n)P_{0, 1,...,n-1}(x)}{x_n - x_0}$$

We can use this result to prove the iteration formula for divided differences introduced in Section 7.3. We have

$$P_{1,2,\ldots,n}(x) = f(x_1) + f[x_1, x_2](x - x_1) + f[x_1, x_2, x_3](x - x_1)(x - x_2) + \cdots$$
$$+ f[x_1, x_2, \ldots, x_n](x - x_1)(x - x_2) \ldots (x - x_{n-1})$$

and

$$P_{0,1,\ldots,n-1}(x) = f(x_0) + f[x_0, x_1](x - x_0) + f[x_0, x_1, x_2](x - x_0)(x - x_1) + \cdots$$
$$+ f[x_0, x_1, \ldots, x_{n-1}](x - x_0)(x - x_1) \ldots (x - x_{n-2})$$

Also,

$$P_{0,1,\ldots,n}(x) = P_{0,1,\ldots,n-1}(x) + f[x_0, x_1, \ldots, x_n](x - x_0)(x - x_1) \ldots (x - x_{n-1})$$

We already found that

$$(x - x_0)P_{1,2,\ldots,n}(x) - (x - x_n)P_{0,1,\ldots,n-1}(x) = (x_n - x_0)P_{0,1,2,\ldots,n}(x)$$

Equating coefficients of x^n on each side of this equation, we find

$$f[x_1, x_2, \ldots, x_n] - f[x_0, x_1, \ldots, x_{n-1}] = (x_n - x_0)f[x_0, x_1, \ldots, x_n]$$

It follows that

$$f[x_0, x_1, \ldots, x_n] = \frac{f[x_1, x_2, \ldots, x_n] - f[x_0, x_1, \ldots, x_{n-1}]}{x_n - x_0}$$

as claimed in Section 7.3.

■ *Exercise Set 7.5*

Since the Aitken algorithm is designed specifically for machine use, it is advisable to write a machine program for computation of array $A(i, j)$ before attempting these exercises. The computations can be done with the aid of a small calculator if no machine is available.

1. Given the following table of values of $f(x)$, verify each polynomial interpolation value for $x = 104.35$.

x	$f(x)$
100	4.641589
103	4.687548
105	4.717694
110	4.791420

    a. $P_{0,1} = 4.708230$
    b. $P_{0,3} = 4.706766$

    c. $P_{0,1,2} = 4.707940$
    d. $P_{0,1,2,3} = 4.707938$

2. Use the Aitken algorithm to estimate $f(1.75)$ from the given table of values. Make an estimate that employs all the data. Use rational arithmetic to avoid round-off error.

x	$f(x)$
1.0	$\frac{3}{4}$
1.5	$\frac{1}{4}$
2.0	$-\frac{1}{4}$
2.5	$-\frac{3}{4}$

3. Repeat exercise 6 of Exercise Set 7.1 using the Aitken algorithm. Now use the method described in this section to make an error estimate for the value obtained from quadratic interpolation.

4. Using the values of $\log x$ given in exercise 1 of Exercise Set 7.3, make a sequence of estimates of $\log 7$ with the Aitken algorithm. Using values of $\log 5$, $\log 6$, $\log 8$, and $\log 9$, we obtain a four-point estimate. Add the fact that $\log 10 = 1$ to make a five-point estimate. By inspecting the difference between the four-point and five-point estimates, what can you conclude about the value of $\log 7$? How many correct digits are there?

5. Let $f(x) = 1/(1 + 25x^2)$. Construct a table of values of this function for $0 \le x \le 1$ with spacing $h = 0.1$.
    a. Use the Aitken algorithm with the tabulated values at 0.1, 0.2, 0.3, 0.4, 0.5, 0.6, and 0.7 to estimate $f(0.25)$.
    b. Repeat part (a) using values at 0.2, 0.4, 0.6, and 0.8.
    c. Application of the Aitken algorithm in part (a) requires computation of a sequence of estimates of $f(0.25)$ using two, three, four,..., seven points. Compare these estimates with the value of $f(0.25)$ computed directly. Which estimate is most accurate? Does interpolation over an increasing number of points necessarily increase the accuracy of an interpolation?

6. It is sometimes desirable, or necessary, to estimate the value of x that yields a particular value of $f(x)$. If a table of values of $f(x)$ is available, we can do this by interpolating on x rather than the $f(x)$ values. The Aitken algorithm provides an efficient and simple procedure for doing this. We only need to interchange the values of $f(x)$ and x for each x and treat x as a function of the $f(x)$'s. We call this procedure *inverse interpolation.*
    a. Use the given table of values with inverse interpolation to estimate square roots for $N = 2, 2.1$, and 2.2.

x	x^2
1.2	1.44
1.3	1.69
1.4	1.96
1.5	2.25
1.6	2.56

    b. Compute values of $f(x) = x^3 - x - 1$ for $x = 1.2, 1.3, 1.4$, and 1.5. Then use inverse interpolation to find a root of $f(x) = 0$.
    c. Repeat part (b) for $f(x) = e^{-x} - \cos x$ with $x = 1.2, 1.3, 1.4$, and 1.5.

7. By use of Theorem 7.5 with the appropriate renumbering of points, if necessary, show each of the following.

a. $P_{1,2,3} = \dfrac{(x - x_1)P_{2,3} - (x - x_3)P_{1,2}}{x_3 - x_1}$

b. $P_{0,1,2} = \dfrac{(x - x_0)P_{1,2} - (x - x_2)P_{0,1}}{x_2 - x_0}$

c. $P_{0,1,2,3} = \dfrac{(x - x_0)P_{1,2,3} - (x - x_3)P_{0,1,2}}{x_3 - x_0}$

d. Using the formulas given in parts (a)–(c) and the data in exercise 1, compute $P_{0,1,2,3}$ and compare with the value obtained in exercise 1d.

7.6   Error Terms and Error Estimation (Optional)

If $P_n(x)$ interpolates function f over the set of points $\{x_0, x_1, \ldots, x_n\}$, then

$$f(x) = P_n(x) + R_n(x)$$

where

$$R_n(x) = \frac{f^{n+1}(\xi)}{(n + 1)!} (x - x_0)(x - x_1) \ldots (x - x_n)$$

Briefly,

$$R_n(x) = \frac{f^{n+1}(\xi)}{(n + 1)!} \psi_n(x) \qquad x_0 < \xi < x_n$$

where

$$\psi_n(x) = (x - x_0)(x - x_1) \ldots (x - x_n)$$

The total error produced in interpolation of $f(x)$ by use of $P_n(x)$ is a combination of computational error and theoretical error. The term $R_n(x)$ gives only the theoretical error, so machine results always contain some error in addition to this. In fact, the computational error can be very large compared to the theoretical error $R_n(x)$. In such cases, estimation of error by use of $R_n(x)$ alone is useless. To obtain accurate interpolation values, we must keep both kinds of error small.

EXAMPLE 7.15   In Example 7.8, we estimated $f(0.12)$ from a table of values using the Newton divided difference polynomial form. The estimated value was 1.1275. Our estimate of the absolute value of the truncation error was 7×10^{-7}. In this case, the function was $f(x) = \exp x$, so the true value is $\exp 0.12 = 1.12749685$, and the absolute value of the total error is 3.14×10^{-6}. Hence, our interpolated value includes a computational error amounting to 3.848×10^{-6} in absolute value. Most of the error is round-off error resulting from the limited accuracy of the tabulated values of the function. The truncation error alone provides a poor estimate of the total error.   ∎

As the preceding example shows, some error in the interpolated value originates in the inaccuracy of the data. More error is then produced by the arithmetic operations. Hence, the total computational error increases as the number of arithmetic operations increases. We can expect that this error will be less if we use the least number of interpolation points possible. Also, some reduction in this computational error can be achieved by proper choice of the algorithm. The Aitken algorithm is good in this respect. Direct computation using the Lagrange form of the interpolation polynomial would be a poor choice. Even though the value of $R_n(x)$ is only part (sometimes a small part) of the total error, it helps to be able to compute this part of the error.

Let us now examine more closely the truncation error computed from error term $R_n(x)$. We can readily see from the form of $R_n(x)$ that in order to have $|R_n(x)|$ small, we need to keep the differences $x - x_i$ as small as possible in absolute value. Hence, we should choose interpolation points $x_0, x_1, \ldots, x_n$ as close to x as possible, assuming we have the choice. Point x should be as near the center of the distribution of interpolation points as possible, and extrapolation should be avoided.

The form of $R_n(x)$ also tells us something about the effect on error resulting from increasing the number of points used from a set of data points. Using another point has the following effects on $R_n(x)$.

1. $R_n(x)$ is multiplied by $(x - x_{n+1})$.
2. $R_n(x)$ is divided by $n + 2$.
3. $f^{n+1}(\xi)$ is replaced by $f^{n+2}(\hat{\xi})$, where $\hat{\xi}$ may lie in an interval larger than the interval in which ξ must lie.

Effects 1 and 3 both may produce an increase in $|R_n(x)|$. In regard to effects 2 and 3, it is interesting to note that the number

$$\frac{|f^{n+1}(\xi)|}{(n + 1)!}$$

often increases in value as n increases and sometimes fails to exist above some value of n.

EXAMPLE 7.16   Consider $f(x) = \ln x$ on $(0, 5)$. Since

$$f^{n+1}(x) = \frac{(-1)^n n!}{x^{n+1}}$$

we have

$$\left|\frac{f^{n+1}(\xi)}{(n + 1)!}\right| = \frac{1}{(n + 1)|\xi|^{n+1}}$$

If $|\xi|$ is greater than 1, this term approaches zero as n increases, but if $|\xi|$ is less than 1, the term may grow as n increases. In that case, the truncation error increases as n increases.   ∎

Typically, the truncation error begins to decrease in absolute value as n increases, but eventually, the error begins to increase with further increase in n. Even when the truncation error does not increase, the computational error normally increases with increasing n, so the total error eventually increases with increasing n.

For each interpolation problem, we have an optimum value of n that produces the minimum total error for that particular problem. Hence, in general, it is not advisable to use a large number of points when interpolating function values from a table of values. Total error will be miminized if we interpolate over small intervals using only a few points.

In the preceding sections, we have found four different ways to estimate truncation error $R_n(x)$.

1. Direct use of the Lagrange error formula (Section 7.1)
2. Using estimate $f[x_0, \ldots, x_n, x_{n+1}]\psi_n(x)$ (Section 7.3)
3. Using estimate $\binom{S}{n+1}\Delta^{n+1}f_0$ (Section 7.4)
4. Using the difference $P_{n+1}(x) - P_n(x)$ (Section 7.5)

In most cases, if the data values are sufficiently accurate, these methods provide usable estimates of the error in our interpolated values. If the estimate of truncation error produced by these methods is very small compared to the probable errors in the data or arithmetic operations, then we know that most of the error is round-off error, and we have no simple reliable method to estimate the error. In such cases, if a sufficient number of decimals has been carried in doing the arithmetic, it is usually assumed that the interpolated value is roughly of the same order of accuracy as the data values. For example, if values of function f are provided that are accurate to five decimal places, and if six- or seven-decimal arithmetic is used in making our estimate while our estimate of truncation error is less than 10^{-7}, we may reasonably assume that the interpolated value is accurate to at least five decimal places. Sample computations seem to confirm this assumption.

■ Exercise Set 7.6

1. Using the exact values of $f(x) = x^4$ shown in the following table, a machine computation using the Lagrange polynomial estimated $f(1.6)$ to be 6.532000. Find the total error, the theoretical error, and the computational error.

x	$f(x)$
1	1.0
1.5	5.0625
2.0	16.0000
2.5	39.0625

2. Using the same data as in exercise 1 a machine computation using the Aitken algorithm gave estimate 6.532001 for the value of $f(1.6)$. Verify that the total error seems to be slightly less than the theoretical error. Can round-off error produced in machine arithmetic be negative?

3. Using the Aitken algorithm and the following table of values for $\sin x$, a machine estimated $\sin 0.28$ to be 0.2763360. The true value is 0.2763557. Make an error analysis. Con-

sider total error versus theoretical error, round-off error, and data errors. What is the major source of error?

x	sin x
0.20	0.1987
0.25	0.2474
0.30	0.2955
0.35	0.3429
0.40	0.3894

4. Using the following table of values for sin x with the Aitken algorithm, a machine estimated sin 0.28 to be 0.276354. The true value is 0.2763557. Make an error analysis. Consider total error versus theoretical error, round-off error, and data errors. What is the major source of error?

x	sin x
0.20	0.19867
0.25	0.24740
0.30	0.29552
0.35	0.34290
0.40	0.38942

5. Using the following table of values for sin x with the Aitken algorithm, a machine estimated sin 0.28 to be 0.2763555 using the first four data points. Using all five data points, the machine estimated sin 0.28 to be 0.2763556. The true value of sin 0.28 is 0.27635565 to eight correct decimals.

x	sin x
0.20	0.19866933
0.25	0.24740396
0.30	0.29552021
0.35	0.34289781
0.40	0.38941834

a. Estimate the error by the method described in Section 7.5 using only the machine output.

b. Compute the theoretical error using the Lagrange error term.

c. Using the known total error, estimate the round-off error and other computational error in the estimate.

6. Suppose that the interpolation points are equally spaced so that $x_k = x_0 + kh$ for $k = 0, 1, \ldots, n$. Define s to be $(x - x_0)/h$. Show that the Lagrange error term is

$$\frac{f^{n+1}(\xi)}{(n + 1)!} s(s - 1)(s - 2) \ldots (s - n)h^{n+1}$$

7. Suppose that $n = 2$ and the error is

$$\frac{f'''(\xi)}{3!} \phi(s)h^3$$

where $\phi(s) = s(s - 1)(s - 2)$.

a. Show that $\max |\phi(s)| = 2/3\sqrt{3}$ and that $|\phi(s)|$ assumes this maximum value when $s = 1 \pm 1/\sqrt{3}$.

b. Assuming that $|f'''(x)| \le M$ for $x_0 < x < x_2$, use the result in part (a) to find an error bound for the case $n = 2$.

c. Using the result in part (b), determine the maximum spacing h for a table of values of $\sin x$ on interval $[0, \pi/4]$ if we need to be able to interpolate values using quadratic interpolation with an absolute theoretical error less than 10^{-5}.

7.7   Splines

In the previous sections, we explored the consequences of using the exact fit criterion for construction of an interpolation polynomial. Given distinct points $x_0, x_1, \ldots, x_n$ and function values $f(x_k)$, we constructed a polynomial $P(x)$ of degree n or less such that $P(x_k) = f(x_k)$ for $k = 0, 1, \ldots, n$. If end points x_0 and x_n are widely separated and n is large, the result is a polynomial of high degree that may oscillate wildly between the node points $x_0, x_1, \ldots, x_n$. This is illustrated in Figure 7.1 and Table 7.1. The Y EST. column in the table contains estimated values of the Runge function $f(x) = 1/(1 + 25x^2)$, while the Y TRUE column contains values computed directly from the formula. The relative error at points such as $x = \pm 0.95$ or even ± 0.05 is extremely large. The error column shows the oscillatory behavior of the interpolation polynomial in this case.

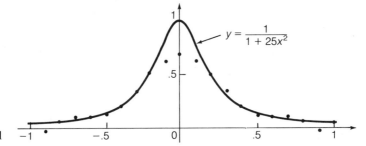

Figure 7.1

One way out of this difficulty is to use piecewise polynomial interpolation functions. For example, suppose $n \ge 4$. If $x_0 < x < x_2$, we might interpolate $f(x)$ by using the second-degree Lagrange polynomial fitted to the data points (x_0, y_0), (x_1, y_1), and (x_2, y_2). If $x_2 < x < x_4$, we could use the second-degree Lagrange polynomial for points (x_2, y_2), (x_3, y_3), (x_4, y_4), and so on. In this case, rather than use a single interpolation polynomial of degree $n = 4$ or larger, we are using a sequence of second-degree polynomials pieced together over the interval from x_0 to x_n. simpler version of this idea is to use linear approximations over each subinterval $[x_j, x_{j+1}]$. In that case, we effectively replace the function f by a function whose graph consists of straight-line segments joining data points as shown in Figure 7.2.

Piecewise polynomial functions such as these have the undesirable property of being nondifferentiable at the junction points. Can we find piecewise polynomial functions consisting of polynomials of low degree that fit the data points but are also differentiable over the entire interval from x_0 to x_n? The answer is yes. We use **spline functions,** defined as follows.

Table 7.1   INTERPOLATION OF RUNGE FUNCTION

X	Y EST.	Y TRUE	ERROR	%ERROR
−1	0.038463	0.038462	0.000001	0.003777
−0.95	−0.040931	0.042440	−0.083371	196.442600
−0.9	−0.028581	0.047059	−0.075639	160.733800
−0.85	0.015536	0.052459	−0.036923	70.385100
−0.8	0.058819	0.058824	−0.000004	0.007410
−0.75	0.087571	0.066390	0.021181	31.904170
−0.7	0.100100	0.075472	0.024628	32.632260
−0.65	0.101547	0.086486	0.015061	17.414090
−0.6	0.099999	0.100000	−0.000001	0.000991
−0.55	0.103723	0.116788	−0.013066	11.187460
−0.5	0.119395	0.137931	−0.018536	13.438870
−0.45	0.151121	0.164948	−0.013828	8.383005
−0.4	0.200000	0.200000	0.000000	0.000119
−0.35	0.264299	0.246154	0.018145	7.371313
−0.3	0.339848	0.307692	0.032156	10.450590
−0.25	0.420708	0.390244	0.030464	7.806376
−0.2	0.500002	0.500000	0.000002	0.000381
−0.15	0.570685	0.640000	−0.069315	10.830520
−0.1	0.626373	0.800000	−0.173627	21.703330
−0.05	0.661970	0.941177	−0.279206	29.665640
0	0.674208	1.000000	−0.325792	32.579210
0.05	0.661970	0.941177	−0.279207	29.665740
0.1	0.626371	0.800000	−0.173629	21.703630
0.15	0.570685	0.640000	−0.069315	10.830460
0.2	0.500000	0.500000	0.000000	0.000000
0.25	0.420710	0.390244	0.030466	7.806850
0.3	0.339846	0.307692	0.032154	10.449890
0.35	0.264300	0.246154	0.018146	7.371895
0.4	0.200000	0.200000	−0.000000	0.000015
0.45	0.151117	0.164948	−0.013831	8.385075
0.5	0.119399	0.137931	−0.018532	13.435610
0.55	0.103723	0.116788	−0.013065	11.186860
0.6	0.100000	0.100000	−0.000000	0.000007
0.65	0.101548	0.086486	0.015062	17.415120
0.7	0.100097	0.075472	0.024626	32.629080
0.75	0.087571	0.066390	0.021181	31.903390
0.8	0.058824	0.058824	−0.000000	0.000019
0.85	0.015532	0.052459	−0.036927	70.392520
0.9	−0.028579	0.047059	−0.075638	160.729800
0.95	−0.040936	0.042440	−0.083376	196.455300
1	0.038462	0.038462	0.000000	0.000000

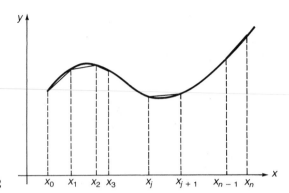

Figure 7.2

Suppose that an interval $[a, b]$ is given and that the points $x_0, x_1, \ldots, x_n$ form a partition of that interval. Suppose also that a sequence of polynomials

$$P_j(x) \qquad j = 0, 1, \ldots, n - 1$$

each of degree $k \geq 1$, is given for which

$$P_{j+1}(x_{j+1}) = P_j(x_{j+1}) \qquad j = 0, 1, \ldots, n - 2$$

If function g is defined by

$$g(x) = P_j(x) \qquad x_j < x < x_{j+1} \quad j = 0, 1, \ldots, n - 1$$

then we say that g is a *piecewise polynomial* of degree k on $[a, b]$. If derivatives $g^{(m)}(x)$ are continuous for $m \leq k - 1$, we say that g is a spline function of degree k on $[a, b]$.

If $k = 1$, then

$$P_j(x) = m_j x + c_j$$

for some constants m_j and c_j. Since $k - 1 = 0$, we only need continuity for the corresponding g in this case. That is, every first-degree piecewise polynomial function is a spline of degree 1. We call these *linear splines*. If $k = 2$, then each polynomial $P_j(x)$ is quadratic, and the piecewise polynomial function g must be differentiable on $[a, b]$ to be a spline (*quadratic spline*). The case of most interest to us is the case where $k = 3$. The function g is a cubic piecewise polynomial function and will be a *cubic spline* if the following conditions are satisfied:

1. $P_{j+1}(x_{j+1}) = P_j(x_{j+1})$
2. $P'_{j+1}(x_{j+1}) = P'_j(x_{j+1})$
3. $P''_{j+1}(x_{j+1}) = P''_j(x_{j+1})$
  for $j = 0, 1, \ldots, n - 2$

If these conditions are satisfied, then g will have continuous first and second derivatives on $[a, b]$.

Returning to the interpolation problem, we will now show how to construct a cubic spline $g(x)$ such that $g(x_k) = y_k$ for each data point (x_k, y_k). Let $P_j(x)$ have the form

$$a_j(x - x_j)^3 + b_j(x - x_j)^2 + c_j(x - x_j) + d_j \qquad j = 0, 1, \ldots, n - 1$$

Since $P_j(x_j) = d_j$, then $d_j = y_j$. Let

$$h_j = x_{j+1} - x_j \qquad j = 0, 1, 2, \ldots, n - 1$$

Then

$$P_j(x_{j+1}) = P_{j+1}(x_{j+1})$$

implies that

$$y_{j+1} = a_j h_j^3 + b_j h_j^2 + c_j h_j + d_j$$

so

$$y_{j+1} - y_j = a_j h_j^3 + b_j h_j^2 + c_j h_j \tag{1}$$

Differentiation of polynomial $P_j(x)$ gives

$$P_j'(x) = 3a_j(x - x_j)^2 + 2b_j(x - x_j) + c_j$$

and

$$P''(x) = 6a_j(x - x_j) + 2b_j$$

Now let

$$S_j = P_j''(x_j) = 2b_j$$

Then

$$b_j = \frac{S_j}{2} \tag{2}$$

Also, we find from

$$P_{j+1}''(x_{j+1}) = P_j''(x_{j+1})$$

that

$$S_{j+1} = 6a_j h_j + 2b_j$$

It follows that

$$S_{j+1} = 6a_j h_j + S_j$$

so

$$a_j = \frac{S_{j+1} - S_j}{6h_j} \qquad (3)$$

Substituting the results from Equations 2 and 3 into Equation 1 yields

$$y_{j+1} - y_j = \tfrac{1}{6}(S_{j+1} - S_j)h_j^2 + \tfrac{1}{2}S_j h_j^2 + c_j h_j$$

We solve this equation for c_j and obtain

$$c_j = \frac{y_{j+1} - y_j}{h_j} - \tfrac{1}{6}(S_{j+1} - S_j)h_j - \tfrac{1}{2}S_j h_j$$

which simplifies to

$$c_j = \frac{y_{j+1} - y_j}{h_j} - \tfrac{1}{6}(S_{j+1} + 2S_j)h_j \qquad (4)$$

If we knew the values of the numbers $S_0, S_1, \ldots, S_n$, we could compute coefficients a_j, b_j, and c_j. Then the polynomials $P_j(x)$ would be determined. We use the condition

$$P'_{j+1}(x_{j+1}) = P'_j(x_{j+1})$$

to find the relation between the numbers S_j.

From the formula for $P'_j(x)$, we find that

$$P'_j(x_{j+1}) = 3a_j h_j^2 + 2b_j h_j + c_j$$

and

$$P'_{j+1}(x_{j+1}) = c_{j+1}$$

It follows that

$$3a_j h_j^2 + 2b_j h_j + c_j = c_{j+1} \qquad (5)$$

Using Equations 2, 3, and 4 in Equation 5, we find that

$$\frac{S_{j+1} - S_j}{2}h_j + S_j h_j + \frac{y_{j+1} - y_j}{h_j} - \frac{S_{j+1} + 2S_j}{6}h_j$$

$$= \frac{y_{j+2} - y_{j+1}}{h_{j+1}} - \frac{S_{j+2} + 2S_{j+1}}{6}h_{j+1}$$

This equation is equivalent to

$$h_j S_j + 2(h_j + h_{j+1})S_{j+1} + h_{j+1}S_{j+2} = 6\left(\frac{y_{j+2} - y_{j+1}}{h_{j+1}} - \frac{y_{j+1} - y_j}{h_j}\right)$$

$$j = 0, 1, 2, \ldots, n - 2 \qquad \textbf{(6)}$$

Let

$$\hat{b}_j = 6\left(\frac{y_{j+2} - y_{j+1}}{h_{j+1}} - \frac{y_{j+1} - y_j}{h_j}\right)$$

in the following.

We have $n - 1$ equations to solve for the $n + 1$ numbers $S_0, S_1, \ldots, S_n$. We are free to specify two more conditions on these numbers. In practice, the added conditions are usually assumptions about end point values, such as the following:

1. $S_0 = S_n = 0$
2. $S_0 = S_1$ and $S_n = S_{n-1}$
3. $P_0'(x_0) = f'(x_0) = y_0'$ and $P_{n-1}'(x_n) = f'(x_n) = y_n'$

Case 1 Choose $S_0 = S_n = 0$. The numbers $S_1, \ldots, S_{n-1}$ satisfy the linear system of equations

$$\begin{bmatrix} 2(h_0 + h_1) & h_1 & 0 & 0 & \cdots & 0 \\ h_1 & 2(h_1 + h_2) & h_2 & 0 & \cdots & 0 \\ 0 & h_2 & 2(h_2 + h_3) & h_3 & \cdots & 0 \\ \vdots & \vdots & & \ddots & & \vdots \\ 0 & 0 & & \cdots & h_{n-2} & 2(h_{n-2} + h_{n-1}) \end{bmatrix} \begin{bmatrix} S_1 \\ S_2 \\ S_3 \\ \vdots \\ S_{n-1} \end{bmatrix} = \begin{bmatrix} \hat{b}_0 \\ \hat{b}_1 \\ \hat{b}_2 \\ \vdots \\ \hat{b}_{n-2} \end{bmatrix}$$

The system matrix is diagonally dominant, so the system has a unique solution.

Case 2 If we choose $S_0 = S_1$ and $S_n = S_{n-1}$, then the first equation ($j = 0$)

$$h_0 S_0 + 2(h_0 + h_1)S_1 + h_1 S_2 = \hat{b}_0$$

becomes

$$(3h_0 + 2h_1)S_1 + h_1 S_2 = \hat{b}_0$$

while the last equation ($j = n - 2$)

$$h_{n-2} S_{n-2} + 2(h_{n-2} + h_{n-1})S_{n-1} + h_{n-1}S_n = \hat{b}_{n-2}$$

becomes

$$h_{n-2} S_{n-2} + (2h_{n-2} + 3h_{n-1})S_{n-1} = \hat{b}_{n-2}$$

The system matrix is again diagonally dominant, so it is still true that a unique solution exists.

Case 3 Suppose we choose the conditions

$$P_0'(x_0) = y_0' \quad \text{and} \quad P_{n-1}'(x_n) = y_n'$$

These conditions imply certain restrictions on the values of the S_j's, which we can derive as follows. Recall that

$$P_j'(x) = 3a_j(x - x_j)^2 + 2b_j(x - x_j) + c_j$$

Then

$$P_0'(x_0) = c_0 = \frac{y_1 - y_0}{h_0} - \frac{S_1 + 2S_0}{6} h_0$$

so

$$\frac{h_0}{6} (S_1 + 2S_0) = \frac{y_1 - y_0}{h_0} - y_0' \tag{7}$$

Now $P_{n-1}'(x_n) = y_n'$ is equivalent to

$$3a_{n-1}(x_n - x_{n-1})^2 + 2b_{n-1}(x_n - x_{n-1}) + c_{n-1} = y_n'$$

Replacing a_{n-1}, b_{n-1}, and c_{n-1} by their equivalents in Equations 2, 3, and 4, we find after simplification that

$$\tfrac{1}{6}h_{n-1} S_{n-1} + \tfrac{1}{3}h_{n-1}S_n = y_n' - \frac{y_n - y_{n-1}}{h_{n-1}} \tag{8}$$

So we must add Equations 7 and 8 to our set of linear equations. The result is $(n + 1)$ equations for the $(n + 1)$ unknowns $S_0, S_1, S_2, \ldots, S_n$.

In all three cases that we consider, we need to solve a tridiagonal linear system. A modified form of the Gauss algorithm can be used. The following algorithm produces the numbers $S_1, S_2, \ldots, S_{n-1}$ for case 1. These numbers are then used to compute the coefficients for the appropriate polynomial $P_j(x)$, depending on the interpolation point x selected. The output is the interpolated value of $f(x)$ using cubic spline interpolation.

The assumed input for this algorithm is a point x and data points $(x_0, y_0), \ldots, (x_n, y_n)$.

Algorithm 7.1

Step 1 Set $S_0 = S_n = 0$.

Step 2 Compute $h_j = x_{j+1} - x_j$ for $j = 0, \ldots, n-1$.

Step 3 For $j = 1, \ldots, n-1$, set

$$a_{jj} = 2(h_{j-1} + h_j)$$

$$a_{j,j+1} = a_{j+1,j} = h_j$$

and

$$a_{jn} = 6\left(\frac{y_{j+1} - y_j}{h_j} - \frac{y_j - y_{j-1}}{h_{j-1}}\right)$$

Step 4 For $j = 1, \ldots, n-2$ and $i = j, \ldots, n$, compute

$$m_{j+1,j} = \frac{a_{j+1,j}}{a_{jj}}$$

and

$$a_{j+1,j} = a_{j+1,j} - m_{j+1,j}a_{ji}$$

Step 5 Set

$$S_{n-1} = \frac{a_{n-1,n}}{a_{n-1,n-1}}$$

Step 6 For $k = n-2, \ldots, 1$, set

$$S_k = \frac{a_{kn} - \sum_{j=k+1}^{n-1} a_{kj}S_j}{a_{kk}}$$

Step 7 For $i = 0, \ldots, n-1$, compute

$$c_{2i} = \frac{S_i}{2}$$

$$c_{3i} = \frac{S_{i+1} - S_i}{6h_i}$$

$$c_{1i} = (y_{i+1} - y_i)h_i - \frac{h_i(2S_i + S_{i+1})}{6}$$

$$c_{0i} = y_i$$

Step 8 For $i = 0, \ldots, n-1$, if $x < x_{i+1}$, then set $k = i$ and go to step 10.

Step 9 Output message X NOT IN INTERPOLATION INTERVAL. Stop.

Step 10 Compute

$$y = c_{3k}(x - x_k)^3 + c_{2k}(x - x_k)^2 + c_{1k}(x - x_k) + c_{0k}$$

Step 11 Output y.

This algorithm is easily converted to a machine program. Output from such a program is illustrated in the following examples.

EXAMPLE 7.17   Given the data points (2, 0.30103), (2.1, 0.32222), (2.3, 0.36173), (2.5, 0.39795), and (2.7, 0.43136), estimate values of the function at the points $x = 2.02$, 2.2, 2.35, 2.58, and 2.66.

Solution   In summary, the machine results are given in the following table.

x	Estimated $f(x)$
2.02	0.305306
2.2	0.342453
2.35	0.371078
2.58	0.411538
2.66	0.424790

The function in this case is the common logarithm function, so we can compare estimates with correct values. The true value of log 2.2 is 0.34242, as compared with the estimate of 0.342453; the value of log 2.66 is 0.42488, rather than 0.42479; and finally, log 2.35 was estimated to be 0.371078, while the true value is 0.37107. All of the interpolated values are acceptable.   ∎

EXAMPLE 7.18   Cubic spline estimates for values of the Bessel function $J_0(x)$ at $x = 0.6$, 0.8, 1.1, and 1.6 were computed using values from the table shown on the left. The estimates are summarized on the right.

x	$J_0(x)$
0.5	0.9385
0.7	0.8812
1.0	0.7652
1.3	0.6201
1.5	0.5118
1.7	0.3980

x	$J_0(x)$ Estimate	True
0.6	0.9111271	0.9120
0.8	0.8467071	0.8463
1.1	0.7195117	0.7196
1.6	0.4552738	0.4554

The values listed as true on the right are handbook values. The estimates compare very well with these values over the entire range of values of x. The errors range from 9×10^{-4} down to 1×10^{-4}.   ∎

EXAMPLE 7.19   Construct a table of values over the interval $[-1, 1]$ for a function whose values are known at -1, -0.7, -0.3, 0, 0.3, 0.7, and 1. These values are given in the following table:

x	0	± 0.3	± 0.7	± 1
$f(x)$	1	0.917431	0.671141	0.5000

Solution The graph of the function is a bell-shaped curve, which indicates that fitting a Lagrange interpolation polynomial to the entire set of points would probably not produce satisfactory results. We decide to use cubic spline interpolation as in Algorithm 7.1. The resulting machine output is shown in the following table:

X	Y	YTRUE	ERROR
−1.00	0.5000	0.5000	0.0000
−0.900	0.5547	0.5525	0.0023
−0.800	0.6112	0.6098	0.0015
−0.700	0.6711	0.6711	0.0000
−0.600	0.7353	0.7353	−0.0000
−0.500	0.8005	0.8000	0.0005
−0.400	0.8626	0.8621	0.0005
−0.300	0.9174	0.9174	0.0000
−0.200	0.9610	0.9615	−0.0006
−0.100	0.9897	0.9901	−0.0004
0.000	1.0000	1.0000	0.0000
0.100	0.9897	0.9901	−0.0004
0.200	0.9610	0.9615	−0.0006
0.300	0.9174	0.9174	0.0000
0.400	0.8626	0.8621	0.0005
0.500	0.8005	0.8000	0.0005
0.600	0.7353	0.7353	−0.0000
0.700	0.6711	0.6711	−0.0000
0.800	0.6112	0.6098	0.0015
0.900	0.5547	0.5525	0.0023

The function in this case is $f(x) = 1/(1 + x^2)$, which is notoriously difficult to interpolate using polynomial interpolation. Using a cubic spline, however, we see that good estimates are obtained over the entire range from −1 to 1 with errors of the order of 10^{-3} or 10^{-4}. ∎

Algorithm 7.1 is designed to provide cubic spline interpolation for end point condition 1. The splines constructed in this case are called *natural splines* and are widely used. Algorithm 7.1 is easily modified to handle the alternate end point conditions 2 and 3 to provide other kinds of cubic spline interpolants. There are many variations on the end point conditions in the current literature about splines. Some of these are incorporated in commercial software. You might want to investigate the IMSL subroutines CSDEC, CSVAL, and BSINT. The IMSL documentation should be consulted for details concerning these subroutines. The routines CSDEC and CSVAL compute cubic splines. The routine BSINT computes B-spline interpolants. An introduction to B-splines can be found in Wheatley and Gerald (1984).

■ *Exercise Set 7.7*

1. Given data points (0, 1), (0.5, 0.25), and (1, 0.5) do the following.
   a. Write down the linear system that determines the numbers S_0, S_1, and S_2 as defined in this section.

b. Assuming $S_0 = S_2 = 0$, find S_1.

c. Compute coefficients for the cubic polynomials $P_0(x)$ and $P_1(x)$.

d. Verify that the function defined by

$$g(x) = \begin{cases} P_0(x) & 0 \le x \le 0.5 \\ P_1(x) & 0.5 \le x \le 1 \end{cases}$$

satisfies the conditions required for a cubic spline interpolant on $[0, 1]$.

2. Repeat exercise 1 with part (b) replaced by the condition $S_0 = S_1$ and $S_2 = S_1$.

3. Write down the linear system that determines the numbers S_0, S_1, and S_2 for the data in Example 7.17.

4. Given that $S_1 = -0.119634$, $S_2 = -0.071599$, and $S_3 = -0.0874742$ for the data in Example 7.17, construct the polynomials $P_0(x)$ and $P_1(x)$.

5. Use the result in exercise 4 to verify the estimates of $f(2.02)$ and $f(2.2)$ given in Example 7.17.

6. Repeat exercise 4 using the data in Example 7.18.

7. Given that $S_1 = -0.510833$, $S_2 = -0.300560$, $S_3 = -0.226929$, and $S_4 = -0.1495163$ for the data in Example 7.18, construct polynomial $P_2(x)$. Use your polynomial to estimate $f(1.1)$ and $f(1.2)$.

8. Convert Algorithm 7.1 to a machine program and then do the following: Use values of $f(x) = e^{-x^2}$ at the points -1, -0.8, -0.4, -0.2, 0, 0.2, 0.4, 0.8, and 1 to construct a cubic spline interpolant on the interval $[-1, 1]$. Use this spline function to compute a table of values at the points $-1 + 0.1j$ for $j = 0, \ldots, 20$. Compare your estimates with values for this function obtained from the formula $f(x) = e^{-x^2}$ using a pocket calculator.

9. Repeat exercise 8 using $f(x) = \sin \pi x$ on $[-1, 1]$.

10. Given the data points (0, 0.5), (0.25, 1), (0.5, 1.5), (0.75, 2), and (1, 2.5), do the following.

a. Write down the linear system for $S_0, S_1, \ldots, S_4$.

b. Assume that $S_0 = S_4 = 0$ and determine $S_1, S_2,$ and S_3.

c. Compute the necessary coefficients and write the polynomials $P_0(x)$, $P_1(x)$, $P_2(x)$, and $P_3(x)$.

d. What do you conclude about $f(x)$? (Use caution.)

11. The following table gives altitude and velocity measurements for a certain V-2 launch t seconds after lift-off.

Time (sec)	Altitude (ft)	Velocity (ft/sec)
10	1840	358
20	7740	853
25	12660	1141
30	19230	1436
35	27100	1738
40	36610	2119

Use cubic spline interpolation to estimate the altitude and velocity at time $t = 32$.

7.8 Applications

We will now consider some uses of the interpolation methods introduced in this chapter. We will begin with the interpolation of function values from table values. All students have had some experience with this problem, beginning with attempts to evaluate trigonometric functions from tables. The traditional high school approach to this problem uses simple linear interpolation. Using the methods introduced here, it is possible to obtain more accurate values with some idea of the accuracy.

A chemistry student has the following table of values for the vapor pressure of water at several temperatures:

Temperature (°C):	20	21	22	23	24
Vapor Press (mm):	17.535	18.650	19.827	21.068	22.377

He needs to know the vapor pressure at 22.6 °C. Using the Aitken algorithm, we obtain the following output:

Temperature	Pressure	Interpolated Values			
20.0	17.535				
21.0	18.650	20.424000			
22.0	19.827	20.514600	20.562960		
23.0	21.068	20.596930	20.564350	20.563790	
24.0	22.377	20.682300	20.566430	20.564000	20.563710

The result of using all of the points is 20.563710. Using the first four points yields 20.563790.

Apparently, the desired value is 20.564 mm to three decimal places. A more precise value is not obtainable, owing to the limited accuracy of the data values, and is probably not needed. Some experimentation using other combinations of three or more points from the data set yields various estimates of the vapor pressure. All of these round off to 20.563 or 20.564, so there might be some doubt about the last digit.

Note that in this application, we not only obtain a more accurate estimate than could be obtained by simple linear interpolation, but we also have some idea of the accuracy of the estimate.

One of the oldest uses of interpolation methods has been estimation of values from tables of astronomical data. This application is still very much in use. The *Astronomical Almanac*, published annually by the U. S. Government Printing Office, always contains an appendix on interpolation methods with some instruction for their use in the almanac tables. As an example of such an application, suppose we want to find the declination (celestial latitude) of the sun on January 19, 1981, at $16^h23^m14\overset{s}{.}8$ E.T. (ephemeris time). From Table C4 in the 1981 *Astronomical*

Almanac, we find the following $0^h0^m0^s.0$ values:

Date	Declination
Jan. 18	$-20°35'59''.7$
Jan. 19	$-20°23'41''.5$
Jan. 20	$-20°11'00''.3$
Jan. 21	$-19°57'56''.4$

We convert the declination values to degrees with a decimal part, and we convert the time $16^h23^m14^s.8$ to a decimal part of 24 hours. In this way, our interpolation point becomes 19.6828102 for computational purposes. Using the Aitken algorithm again, we obtain the following results:

Date	Declination			
18	-20.599170			
19	-20.394860	-20.255360		
20	-20.183420	-20.249360	-20.251260	
21	-19.656670	-20.070490	-20.192240	-20.269980

If we use all four points, the interpolated value is $-20.26998°$, which is equivalent to $-20°16'11''.9$. Using the first three points yields estimate $-20.251260°$, which is equivalent to $-20°15'4''.5$. These results indicate a declination of $-20°15'$ with a possible error as large as 1 minute of arc.

There are clearly many instances when we must estimate the value of a function from a table that does not include the desired value so we interpolate. We conclude this section with an application to a relatively new field where such situations are very common. This is the field of nuclear physics.

The range of an alpha particle in air is determined by the initial energy. Tables of corresponding values of the energy and the range are readily available. Suppose that we need to find the range of an alpha particle with initial energy 0.5 MeV (millions of electron volts). Our table has values as follows:

Energy (MeV)	0.2	0.4	0.6	0.8	1.0
Range (cm)	0.17	0.27	0.38	0.47	0.57

Interpolation with the Aitken algorithm gives us the following:

Energy	Range				
0.2	0.17				
0.4	0.27	0.320000			
0.6	0.38	0.327500	0.323750		
0.8	0.47	0.320000	0.320000	0.325625	
1.0	0.57	0.320000	0.320000	0.324688	0.327031

Using all five points, we get a range of 0.327031 cm, while interpolation on the first four points yields a range of 0.325625. In view of the accuracy of the data, it is reasonable to accept as the desired value the rounded estimate of 0.33 cm.

Alternately, suppose that an alpha particle is observed to have a range of 0.5 cm. We can use the same table to estimate the initial energy of this particle. We interchange the columns of the table and interpolate on 0.5 cm in the following results:

Range		Energy			
0.17	0.20				
0.27	0.40	0.860000			
0.38	0.60	0.828571	0.794286		
0.47	0.80	0.860000	0.860000	0.881905	
0.57	1.00	0.860000	0.860000	0.835790	0.868070

So we see that the estimate is 0.86807 MeV using all five points, while the estimate is 0.881905 MeV using the first four points. Rounding, we find an estimate of 0.87 or 0.88 MeV, but the data do not permit a more accurate determination of the energy. Hence, estimation of the energy from measurement of the range alone cannot provide highly accurate energy estimates.

■ *Exercise Set 7.8*

For each of the following interpolation problems, choose a method from those discussed in this chapter. Experiment with various choices of data points. Try to decide which estimate is best in each case.

1. A reference book used by physicists and chemists gives the following values for specific gravity and percent of ethyl alcohol by volume for ethyl alcohol and water solutions:

%	Specific Gravity
1.00	0.99849
1.50	0.99775
2.00	0.99701
2.50	0.99629

  a. Estimate the specific gravity of a 1.67% alcohol solution.
  b. Repeat part (a) for a 2.4% solution.
  c. Estimate the percent of alcohol for a solution with a specific gravity of 0.99742.
  d. Repeat part (c) for a specific gravity of 0.99680.

2. The optical index of refraction of an aqueous solution of sucrose varies with the percent of sucrose. A standard table gives the following values:

% Sucrose	Index of refraction
10	1.3479
20	1.3639
30	1.3811
40	1.3997

a. Find the index of refraction if the percent of sugar is 12.

b. Repeat part (a) for a 27% solution.

c. Find the percent sugar if the index of refraction is 1.3546.

d. Find the percent sugar if the index of refraction is 1.3872.

3. Values of the elliptic integral

$$E = \int_0^{\pi/2} (1 - k^2 \sin^2 \phi)^{1/2} \, d\phi$$

where $k = \sin \theta$, have been computed for various values of θ, as shown in the following table:

θ	E
10°	1.5589
20°	1.5238
30°	1.4675
40°	1.3931

a. Find values of E for $\theta = 15°, 22°$, and $36°$.

b. Find values of E corresponding to $k = 0.2, 0.57$, and 0.375.

c. For which value of k will the value of E be 1.5000?

4. U.S. Census Bureau statistics give the following death rates for children under one year:

Year	Rate
1960	27.0
1961	25.4
1962	25.3
1963	25.3
1964	24.6
1965	24.1
1966	23.3

a. Use interpolation to estimate the death rates in the years 1961, 1963, and 1965 from the rates given for the years 1960, 1962, 1964, and 1966.

b. Discuss reasons why we should not expect interpolated values in data of this kind to agree very well with the actual values.

c. Name some other situations where tabulated statistical data are available but where interpolation in such tables cannot be expected to yield usable intermediate values. State why in each case.

5. Just before "burnout," the values of altitude, range, and velocity for a V-2 rocket were measured at 5-second intervals. The results are shown in the following table:

Time (sec)	Altitude (mi)	Range (mi)	Velocity (ft/sec)
45	7.833	3.029	3541.5
50	9.392	5.045	4150.7
55	11.043	7.695	4846.7
60	12.785	11.070	5650.2
65	14.617	15.287	6592.0

a. Estimate the altitude, range, and velocity at 52 seconds.

b. Repeat for 46 seconds and 64 seconds.

c. Estimate the velocity at the instant when the altitude was exactly 10 miles.

d. Estimate the time at which the velocity was exactly 5000 ft/second.

e. Estimate the range at 48 seconds, 59 seconds, and 63 seconds.

6. Suppose that a reference book on nuclear physics contains the following table relating the range in air for protons and deuterons of various energies:

Energy (MeV)	Range (cm) Proton	Deuteron
1.0	2.30	1.72
2.0	7.20	4.61
3.0	14.10	8.78
4.0	23.10	14.40
5.0	33.90	20.80

a. What is the range of a proton with an energy of 1.4 MeV? 2.5 MeV?

b. What is the range of a deuteron with an energy of 1.6 MeV? 2.5 MeV?

c. A more detailed table gives a range of 3.91 cm for a proton with an energy of 1.4 MeV and a range of 10.4 cm for an energy of 2.5 MeV. The same table gives ranges of 3.30 cm and 6.51 cm for deuterons with energies of 1.6 MeV and 2.5 MeV, respectively. How well do these values agree with your interpolated values? Can you explain why some values agree better than others?

d. What is the energy of a proton with an observed range of 4.67 cm? 8.21 cm?

e. What is the energy of a deuteron with an observed range of 10.42 cm?

f. Find the range and energy of a deuteron that has the same energy as a proton with a range of 12.45 cm.

8 APPROXIMATION OF FUNCTIONS

In this chapter, we will consider the general problem of approximating a function f on an interval $[a, b]$. We will be concerned with procedures for finding another function g such that g is a good approximation to f on $[a, b]$. The polynomial interpolation problem studied in Chapter 7 can be viewed as a special case. In the interpolation problem, we assume that values of a function f are known for each x_k in a given set of $n + 1$ interpolation points $x_0, x_1, \ldots, x_n$. We restrict the approximation function to be a polynomial P of degree n or less such that $P(x_k) = f(x_k)$ for each interpolation point x_k. We expect the number $P(x)$ to be close to $f(x)$ if the number x lies between x_0 and x_n.

The restrictions in the polynomial interpolation problem are not advisable under some conditions. For example, if the data contain some possible error, the exact fit criterion $P(x_k) = y_k$ is not justified. If n is large, the attempt to find a polynomial as defined in the interpolation problem can lead to a polynomial that oscillates widely between the interpolation points. In this chapter, we will consider approximations without these narrow restrictions.

8.1 Taylor Polynomials

In this section, we will consider the following problem. Given a function f that is continuous on interval $[a, b]$, we want to find a polynomial P such that P is a good approximation to f on $[a, b]$. Initially, we will consider P to be a good approximation to f on $[a, b]$ if $\max | f(x) - P(x)|$ remains small over the interval $[a, b]$. Later we will consider other definitions of a good approximating polynomial.

We must consider the following existence question. Given a function f that is continuous on the interval $[a, b]$, is it necessarily true that some polynomial P must exist such that P is a good approximation to f (as previously defined) on $[a, b]$? The answer is yes, as the following theorem—called the **Weierstrass approximation theorem**—states.

Theorem 8.1   If the real-valued function f is continuous on $[a, b]$, then corresponding to each $\varepsilon > 0$ there exists a polynomial P of some degree n such that

$$|f(x) - P(x)| < \varepsilon \qquad \text{for all } x \text{ in } [a, b]$$

The proof of this theorem will not be given here, but it can be found in almost any real analysis text.

   Such theorems and their proofs are important, but results such as these do not tell us how to find the polynomial P, or even what degree might be required. We do know now, however, that polynomial approximations of the desired type do at least exist.

   In beginning calculus courses, students are taught how to approximate a function f in the neighborhood of a point in the domain of f by use of Taylor polynomials. These are the simplest polynomial approximations, and because they are so useful in numerical work, we will review their construction and use.

   Given a function f and a point c in the domain of f, suppose that derivatives $f'(c)$ through $f^n(c)$ exist. Then we can construct the polynomial

$$P_n(x) = f(c) + f'(c)(x - c) + \frac{f''(c)}{2!}(x - c)^2 + \cdots + \frac{f^n(c)}{n!}(x - c)^n$$

Polynomial $P_n(x)$ is called the nth-degree **Taylor polynomial** approximation of function f at point c.

EXAMPLE 8.1   The Taylor polynomial of degree 3 for $f(x) = e^x$ at $c = 0$ is

$$P_3(x) = 1 + x + \frac{x^2}{2} + \frac{x^3}{6}$$

∎

EXAMPLE 8.2   The second-degree Taylor polynomial approximation of $f(x) = \cos x$ at $x = \pi/3$ is

$$P_2(x) = \frac{1}{2} - \frac{\sqrt{3}}{2}\left(x - \frac{\pi}{3}\right) - \frac{1}{4}\left(x - \frac{\pi}{3}\right)^2$$

∎

Theorem 8.2   Suppose that function f has $n + 1$ derivatives on the open interval (a, b) with c in (a, b). If x is in (a, b), then

$$f(x) = P_n(x) + \frac{f^{n+1}(\xi)}{(n + 1)!}(x - c)^{n+1}$$

where $P_n(x)$ is the nth-degree Taylor polynomial approximation of f at c and ξ lies between x and c.

Because of Theorem 8.2, we can treat $P_n(x)$ as an approximating polynomial for f in the neighborhood of c and make estimates of the error involved in the approximation of $f(x)$ by $P_n(x)$.

Taylor polynomials provide approximating polynomials that can approximate a function f very well near point c. For this reason, they are very useful in developing numerical algorithms. By use of a Taylor polynomial, we can sometimes provide accurate starting values for other simpler computational schemes. It is clear from Theorem 8.2 that the error in a Taylor polynomial approximation grows as we move away from point c. In Table 8.1, we see some numerical results from approximation of $\sin x$ by the Taylor polynomial of degree 5 at point $c = 0$.

Table 8.1

x	$P(x)$	$\sin x$	Error
0.1	0.0996668	0.0998334	−0.0001667
0.2	0.1973360	0.1986693	−0.0013333
0.3	0.2910203	0.2955202	−0.0045000
0.4	0.3787520	0.3894183	−0.0106663
0.5	0.4585938	0.4794255	−0.0208318
0.6	0.5286480	0.5646425	−0.0359945
0.7	0.5870673	0.6442177	−0.0571504
0.8	0.6320640	0.7173561	−0.0852921
0.9	0.6619208	0.7833270	−0.1214062

Column 3 of Table 8.1 gives numerical values of $P_5(x) - \sin x$ for $x = 0.1, 0.2,$ $\ldots, 0.9$. The error growth is quite rapid. In the next section, we will study a method for constructing polynomial approximations that provide more uniform approximations over an interval.

■ *Exercise Set 8.1*

1. Find the Taylor polynomial $P_4(x)$ for $f(x) = \cos x$ at point $c = 0$. Use this polynomial to approximate $\cos 0.1$ and $\cos 0.8$. Compare your results with more accurate values from tables.

2. Use Theorem 8.2 to find a bound for the error in the estimates of $\cos 0.1$ and $\cos 0.8$ obtained in exercise 1. Compare with an estimate of the true error obtained by computing values of $\cos x - P_4(x)$ for $x = 0.1$ and $x = 0.8$ using table values for $\cos x$.

3. Use Theorem 8.2 to find an error bound for the approximation of $\sin x$ by Taylor polynomial $P_5(x)$. Use this result to compute bounds for the error at $x = 0.1$ and $x = 0.9$. Compare with the error values given in Table 8.1.

4. The Taylor polynomial of degree n for $f(x) = e^x$ at $c = 0$ is

$$P_n(x) = 1 + x + \frac{x^2}{2!} + \cdots + \frac{x^n}{n!}$$

We want to use such a polynomial to estimate $e^{0.5}$ with error less than 10^{-6}.
   a. What is the minimum degree n required to do this?

b. Estimate $e^{0.5}$ by computing $P_n(0.5)$ using the value of n found in part (a). Compare with approximation 1.64872127, which is correct through eight decimal places. Is your error less than 10^{-6}?

5. Repeat exercise 4 for estimation of e^3. Compare your estimate with the table value of 20.0855369. Is your error as small as required?

6. Repeat exercise 4 for estimation of e^{-1}. Compare your estimate with the table value of 0.36787944. Is your error as small as required?

7. Repeat exercise 1 for $f(x) = \tan x$ using polynomial $P_5(x)$.

8. Repeat exercise 1 for $f(x) = e^{-x^2}$ using polynomial $P_5(x)$.

9. Use Theorem 8.2 to find an error bound for approximation of $\sin x$ near $c = 0$ by Taylor polynomial $P_n(x)$. Show that for all x, the error has a limit of zero as n approaches infinity. Interpret this result.

10. We want to approximate $f(x) = \sin x$ on the interval $[0, \pi/4]$ using the Taylor polynomial of degree 9 with $c = 0$. Let $R_9(x) = \sin x - P_9(x)$. Use Theorem 8.2 to find a bound M such that $|R_9(x)| \le M$ when $0 \le x \le \pi/4$.

11. Let $F(x) = \displaystyle\int_0^x e^{-t^2/2}\, dt$ on interval $[0, 1]$.

   a. Find Taylor polynomial $P_5(x)$ for $c = 0$.
   b. Make a table of values for $P_5(x)/\sqrt{2\pi}$ for $x = 0.1, 0.2, \ldots, 0.9$.
   c. Values of $F(x)/\sqrt{2\pi}$ are tabulated in statistical tables as values of probabilities for the standard normal distribution. We find that $F(0.5)/\sqrt{2\pi} = 0.19146$, for example, from such tables. Compare your values with the values given in such a standard normal table.

8.2   Chebyshev Polynomial Approximations

The Taylor polynomial approximation of a function f is not well suited to approximation of f over an interval $[a, b]$ if the approximation is to be uniformly accurate over the entire interval. Taylor polynomial approximations yield very small error near point c, but the error increases (quite rapidly in some cases) as we move away from c. We need approximating polynomials that provide a more uniform, and hopefully small, error over the interval concerned. It is possible to produce such approximating polynomials. We will investigate some methods that employ Chebyshev polynomials.

The **Chebyshev polynomial** of degree n is defined to be

$$T_n(x) = \cos(n \arccos x) \qquad n \ge 0, \quad -1 \le x \le 1$$

Let $\theta = \arccos x$, so $x = \cos \theta$. Then

$$T_n(x) = T_n(\cos \theta) = \cos n\theta \qquad 0 \le \theta < \pi$$

Since

$$\cos(n + 1)\theta = \cos n\theta \cos \theta - \sin n\theta \sin \theta$$

and

$$\cos(n - 1)\theta = \cos n\theta \cos \theta + \sin n\theta \sin \theta$$

then

$$T_{n+1}(\cos \theta) + T_{n-1}(\cos \theta) = 2 \cos \theta \cos n\theta$$

Hence,

$$T_{n+1}(x) + T_{n-1}(x) = 2x T_n(x)$$

Fact 1

$$T_{n+1}(x) = 2x T_n(x) - T_{n-1}(x) \quad \text{if} \quad n \geq 1$$

Clearly,

$$T_0(x) = 1 \quad \text{and} \quad T_1(x) = x$$

Using the recursion relation, it is easy to verify that

$$T_2(x) = 2x^2 - 1$$
$$T_3(x) = 4x^3 - 3x$$
$$T_4(x) = 8x^4 - 8x^2 + 1$$

and so on.

Fact 2   In general, $T_n(x)$ is a polynomial of degree n, and the coefficient of x^n is 2^{n-1}. Then $T_n(x)$ has n zeros at the points where

$$T_n(x) = \cos(n \arccos x) = 0$$

These are the points $x_k = \cos \theta_k$, where

$$\theta_k = \frac{(2k - 1)\pi}{2n} \qquad k = 1, 2, \ldots, n$$

Fact 3 $T_n(x)$ has n real zeros in interval $[-1, 1]$.

EXAMPLE 8.3 $T_4(x)$ has zeros at the points $x_1 = \cos(\pi/8) = 0.9238795$, $x_2 = \cos(3\pi/8)$ $= 0.3826834$, $x_3 = \cos(5\pi/8) = -x_2$, and $x_4 = \cos(7\pi/8) = -x_1$. Also, $T_n(x)$ has extreme values at the points where the derivative is zero. These are the points

$$\bar{x}_k = \cos(k\pi/n) \qquad k = 0, 1, \ldots, n$$ ■

Fact 4 Polynomial $T'_n(x)$ has $n + 1$ real zeros at the points

$$\bar{x}_k = \cos(k\pi/n) \qquad k = 0, 1, \ldots, n$$

in interval $[-1, 1]$. The value of $T_n(x)$ at point $\bar{x}_k$ is

$$\cos(n \arccos \bar{x}_k) = \cos(k\pi) = (-1)^k$$

Fact 5 If $|x| \le 1$, then $|T_n(x)| \le 1$. Now consider the polynomial

$$T_n^*(x) = 2^{1-n} T_n(x)$$

Each $T_n^*(x)$ is a **monic polynomial** of degree n. That is, the coefficient of x^n in $T_n^*(x)$ is 1. Polynomial $T_n^*(x)$ also has zeros at the points x_k and extreme values at points $\bar{x}_k$. However, $|T_n^*(x)| \le 2^{1-n}$ for $n \ge 1$. So

$$\max_{-1 < x < 1} |T_n^*(x)|$$

has a limit of zero as n approaches infinity.

In relation to other monic polynomials, the polynomials $T_n^*(x)$ have a very important minimal property, as expressed in the following theorem.

Theorem 8.3   If $P_n(x)$ is any monic polynomial of degree n, then

$$\max_{-1 \le x \le 1} |P_n(x)| \ge \max_{-1 \le x \le 1} |T_n^*(x)| = \frac{1}{2^{n-1}}$$

Proof Suppose that $P_n(x)$ is a monic polynomial of degree $n \ge 1$ and

$$\max_{-1 \le x \le 1} |P_n(x)| < \frac{1}{2^{n-1}}$$

Let

$$Q(x) = T_n^*(x) - P_n(x)$$

Then $Q(x)$ is a polynomial of degree $n - 1$ or less. Also,

$$Q(\bar{x}_k) = T_n^*(\bar{x}_k) - P_n(\bar{x}_k) = \frac{(-1)^k}{2^{n-1}} - P_n(\bar{x}_k)$$

Since

$$|P_n(\bar{x}_k)| < \frac{1}{2^{n-1}}$$

the algebraic sign of $Q(\bar{x}_k)$ is determined by the term

$$\frac{(-1)^k}{2^{n-1}}$$

Hence, $Q(\bar{x}_k) < 0$ if k is odd, and $Q(\bar{x}_k) > 0$ if k is even. It follows that $Q(x) = 0$ for at least n points in interval $[-1, 1]$. But if $Q(x)$ is not a constant polynomial, then $Q(x)$ cannot have more than $n-1$ zeros. If $Q(x)$ is a constant, then $Q(x)$ must be zero. But then it follows that $T_n^*(x) = P_n(x)$. This is not possible because

$$\max |P_n(x)| < \max |T_n^*(x)|$$

on $[-1, 1]$ due to our choice of $P_n(x)$. It follows that no such polynomial as $P_n(x)$ can exist.   □

We see, then, that among all the monic polynomials $P_n(x)$ of degree n, the polynomial $T_n^*(x)$ has the smallest possible value for $\max |P_n(x)|$ on $[-1, 1]$.

In the remainder of this section, we will explore two ways in which the preceding information can be used in conjunction with the problem of approximating a function by a polynomial.

Suppose that f is continuous on $[-1, 1]$ and we want to approximate f with a polynomial of degree n or less. We choose $n + 1$ points $x_0, x_1, \ldots, x_n$ in $[-1, 1]$ and form the interpolation polynomial $P_n(x)$. For each choice of the interpolation set, we obtain an approximating polynomial. Which choice provides the best approximation over the interval $[-1, 1]$? We know that

$$|f(x) - P_n(x)| = \frac{|f^{n+1}(\xi)|}{(n+1)!} |\psi_n(x)|$$

where

$$\psi_n(x) = (x - x_0)(x - x_1) \ldots (x - x_n)$$

If $|f^{n+1}(x)| \le M$ on $[-1, 1]$, then

$$|f(x) - P_n(x)| \le \frac{M}{(n+1)!} \max_{-1 \le x \le 1} |\psi_n(x)|$$

Now, $\psi_n(x)$ is a monic polynomial of degree $n + 1$. Hence, $\max_{-1 \le x \le 1} |\psi_n(x)|$ will be minimized if $\psi_n(x) = T_{n+1}^*(x)$. This means that the points $x_0, x_1, \ldots, x_n$ should be the zeros of $T_{n+1}(x)$ in $[-1, 1]$. If $P_n(x)$ is the interpolation polynomial over that set of points, then

$$|f(x) - P_n(x)| \le \frac{M}{(n+1)!} \cdot \frac{1}{2^n}$$

and this is the best we can expect to achieve from using an interpolation polynomial of degree n or less to approximate f over the interval $[-1, 1]$.

EXAMPLE 8.4   Let $f(x) = \sin x$ on $[-1, 1]$ and choose $n = 3$. Then

$$|f(x) - P_3(x)| < \frac{1}{4!} \cdot \frac{1}{2^3} = 5.208 \times 10^{-3} < 5.21 \times 10^{-3}$$

Construct the polynomial $P_3(x)$.

Solution   First we form a table of values of $\sin x$ at the zeros of $T_4(x)$ in $[-1, 1]$:

x_k	$\sin x_k$
-0.9238795	-0.797945906
-0.3826834	-0.373411110
0.3826834	0.373411110
0.9238795	0.797945906

We must now construct Lagrange polynomials $L_0(x)$, $L_1(x)$, $L_2(x)$, and $L_3(x)$. Then polynomial $P_3(x)$ is

$$-0.797945906L_0(x) - 0.373411110L_1(x) + 0.37341111L_2(x) + 0.797945906L_3(x)$$

Explicitly, we have

$$L_0(x) = -0.765367x^3 + 0.7071068x^2 + 0.1120854x - 0.1035534$$
$$L_1(x) = 1.847759x^3 - 0.7071068x^2 - 1.577161x + 0.6035534$$
$$L_2(x) = -1.847759x^3 - 0.7071068x^2 + 1.577161x + 0.6035534$$
$$L_3(x) = 0.765367x^3 + 0.7071068x^2 - 0.1120854x - 0.1035534$$

and

$$P_3(x) = -0.1585047x^3 + 0.9989828x \qquad \blacksquare$$

It is worth comparing the values of $P_3(x)$ and $\sin x$ on $[-1, 1]$. The results are given in Table 8.2. The error column tabulates the value of $\sin x - P_3(x)$ for each x. Note that the error is of the order of 10^{-4} throughout most of the interval, and this is well within the given error bound. The Taylor column in Table 8.2 lists values of the Taylor polynomial of degree 3 for $\sin x$ with $c = 0$. Note that although the Taylor approximation is much better close to $x = 0$, our polynomial $P_3(x)$ is more accurate near the ends of the interval and gives a more uniformly accurate approximation. Figure 8.1 provides a graphic view of the approximation. The agreement with $\sin x$ on the interval $[-1, 1]$ is clearly very good.

Table 8.2

X	P(X)	SIN X	ERROR	TAYLOR
−1	−0.8404779	−0.841471	−9.93073E − 04	−8.137703E − 03
−0.9	−0.7835345	−0.7833269	2.075434E − 04	−4.826904E − 03
−0.8	−0.7180317	−0.717356	6.756783E − 04	−2.689421E − 03
−0.7	−0.6449207	−0.6442177	7.030368E − 04	−1.384378E − 03
−0.6	−0.5651525	−0.5646425	5.100966E − 04	−6.425381E − 04
−0.5	−0.4796782	−0.4794255	2.526641E − 04	−2.589226E − 04
−0.4	−0.3894487	−0.3894183	3.042817E − 05	−8.499622E − 05
−0.3	−0.2954151	−0.2955202	−1.050532E − 04	−2.023578E − 05
−0.2	−0.1985284	−0.1986693	−1.408458E − 04	−2.667308E − 06
−0.1	−9.973969E − 02	−9.983335E − 02	−9.366125E − 05	−8.940697E − 08
0	7.443001E − 08	7.450581E − 08	7.58007E − 11	0
0.1	9.973984E − 02	9.983349E − 02	9.36538E − 05	8.195639E − 08
0.2	0.1985286	0.1986694	1.408309E − 04	2.667308E − 06
0.3	0.2954153	0.2955203	1.050234E − 04	2.020598E − 05
0.4	0.3894488	0.3894184	−3.042817E − 05	8.499622E − 05
0.5	0.4796783	0.4794256	−2.526939E − 04	2.588928E − 04
0.6	0.5651526	0.5646426	−5.10037E − 04	6.425381E − 04
0.7	0.6449208	0.6442178	−7.030368E − 04	1.384378E − 03
0.8	0.7180318	0.7173561	−6.756783E − 04	2.689421E − 03
0.9	0.7835345	0.7833271	−2.074838E − 04	4.826963E − 03

Note that if we had used $n = 8$ in the construction of the approximation in Example 8.4, the error bound would be

$$|\sin x - P_8(x)| < \frac{1}{9!} \cdot \frac{1}{2^8} = 1.077 \times 10^{-8} < 1.1 \times 10^{-8}$$

So great accuracy is possible with a polynomial of reasonably low degree.

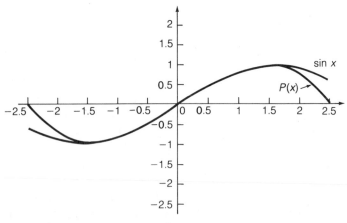

Figure 8.1 $P(x) = 0.9989828x - 0.1585047x^3$

To use this procedure to approximate a function on an interval $[a, b]$ other than $[-1, 1]$, we use the transformation

$$\hat{x} = \frac{(b - a)x + (a + b)}{2}$$

to convert the zeros of the selected Chebyshev polynomial to corresponding points in $[a, b]$.

Construction of an approximating polynomial by the procedure just described is laborious, even in the case of low-degree polynomials. It is important to know that results almost as accurate can be obtained with much less effort if we already have some kind of polynomial approximation, such as a Taylor polynomial. From such a polynomial, we can derive a polynomial of lower degree that may still have sufficient accuracy for the required application.

Suppose that a polynomial

$$P_n(x) = a_0 + a_1 x + a_2 x^2 + \cdots + a_n x^n$$

is given that approximates function f with

$$f(x) = P_n(x) + E_n(x) \qquad x \in [-1, 1]$$

Now

$$T_n^*(x) = 2^{1-n} T_n(x)$$

is a monic polynomial of degree n. So

$$Q_{n-1}(x) = P_n(x) - a_n T_n^*(x)$$

is a polynomial of degree $n - 1$ or less. But

$$f(x) = Q_{n-1}(x) + a_n T_n^*(x) + E_n(x)$$

So the error in approximating f by $Q_{n-1}(x)$ exceeds the error in approximating f by $P_n(x)$ by at most

$$|a_n T_n^*(x)| \le \frac{|a_n|}{2^{n-1}}$$

If $|a_n|$ is small or n is large, the error in approximating f by the polynomial $Q_{n-1}(x)$ will not differ significantly from the error involved in using approximation $P_n(x)$. We call the process of replacing a polynomial $P_n(x)$ by a lower-degree approximating polynomial **Chebyshev economization**. The lower-degree polynomial $Q_{n-1}(x)$ is called the economized polynomial.

EXAMPLE 8.5   Function e^x can be approximated on $[-1, 1]$ by Taylor polynomial

$$P_4(x) = 1 + x + \tfrac{1}{2}x^2 + \tfrac{1}{6}x^3 + \tfrac{1}{24}x^4$$

Form the economized polynomial $Q_3(x)$.

Solution   Let

$$Q_3(x) = P_4(x) - \tfrac{1}{24}T_4^*(x)$$

Since

$$T_4^*(x) = x^4 - x^2 + \tfrac{1}{8}$$

then

$$Q_3(x) = \tfrac{191}{192} + x + \tfrac{13}{24}x^2 + \tfrac{1}{6}x^3$$

and

$$|Q_3(x) - P_4(x)| = \tfrac{1}{24}|T_4^*(x)| < \tfrac{1}{24} \cdot \tfrac{1}{8} < 0.00521$$

For some purposes, this small increase in possible error would not be serious—a small cost to pay for the reduction in degree of the approximating polynomial.   ∎

As the following example shows, an economized polynomial also provides a more uniformly accurate approximation.

EXAMPLE 8.6   Let $f(x) = 1/(4 - x)$. Then f is approximated by the polynomial

$$P_4(x) = \tfrac{1}{4} + \tfrac{1}{16}x + \tfrac{1}{64}x^2 + \tfrac{1}{256}x^3 + \tfrac{1}{1024}x^4$$

on $(-4, 4)$. We are interested only in the interval $[-1, 1]$ in the following, however. Let

$$Q_3(x) = P_4(x) - \tfrac{1}{1024}T_4^*(x)$$

Then

$$Q_3(x) = \tfrac{2047}{8192} + \tfrac{1}{16}x + \tfrac{65}{1024}x^2 + \tfrac{1}{256}x^3$$

or

$$Q_3(x) = 0.24987793 + 0.0625x + 0.06347656x^2 + 0.0039062x^3$$

The following table shows the numerical results of approximating f by the economized polynomial $Q_3(x)$.

X	P(X)	1/(4-X)	ERROR
−1	0.2000732	0.2	−7.320941E − 05
−0.9	0.2042275	0.2040816	−1.458824E − 04
−0.8	0.2085029	0.2083333	−1.695752E − 04
−0.7	0.2129228	0.212766	−1.568645E − 04
−0.6	0.2175107	0.2173913	−1.19403E − 04
−0.5	0.22229	0.2222222	−6.778538E − 05
−0.4	0.2272842	0.2272727	−1.142919E − 05
−0.3	0.2325166	0.2325582	4.158914E − 05
−0.2	0.2380107	0.2380953	8.453429E − 05
−0.1	0.24379	0.2439025	1.124293E − 04
0	0.2498779	0.25	1.221001E − 04
0.1	0.2562978	0.2564103	1.124442E − 04
0.2	0.2630732	0.2631579	8.46982E − 05
0.3	0.2702275	0.2702703	4.276634E − 05
0.4	0.2777842	0.2777778	−6.377697E − 06
0.5	0.2857666	0.2857143	−5.230308E − 05
0.6	0.2941982	0.2941177	−8.055568E − 05
0.7	0.3031025	0.3030303	−7.221103E − 05
0.8	0.3125029	0.3125	−2.890825E − 06
0.9	0.3224228	0.3225807	1.578331E − 04

The values of $Q_3(x)$ are given in the P(X) column, and the error is tabulated in the ERROR column. Note that the size of the error does not vary greatly throughout the interval. In comparison, the following table shows the results of approximating f by the cubic resulting from dropping the last term of $P_4(x)$.

X	P(X)	1/(4-X)	ERROR
−1	0.1992188	0.2	7.81253E − 04
−0.9	0.2035586	0.2040816	5.230308E − 04
−0.8	0.208	0.2083333	3.333241E − 04
−0.7	0.2125664	0.212766	1.995564E − 04
−0.6	0.2172813	0.2173913	1.1006E − 04
−0.5	0.222168	0.2222222	5.425513E − 05
−0.4	0.22725	0.2272727	2.270937E − 05
−0.3	0.2325508	0.2325582	7.376075E − 06
−0.2	0.2380938	0.2380953	1.505017E − 06
−0.1	0.2439024	0.2439025	8.940697E − 08
0	0.25	0.25	0
0.1	0.2564102	0.2564103	1.192093E − 07
0.2	0.2631563	0.2631579	1.639128E − 06
0.3	0.2702617	0.2702703	8.553266E − 06
0.4	0.27775	0.2777778	2.777577E − 05
0.5	0.2856445	0.2857143	6.973744E − 05
0.6	0.2939688	0.2941177	1.488924E − 04
0.7	0.3027461	0.3030303	2.84195E − 04
0.8	0.312	0.3125	5.000234E − 04
0.9	0.3217539	0.3225807	8.267463E − 04

This Taylor polynomial of degree 3 gives more accurate results near $x = 0$, but the estimates are much less accurate near the ends of the interval. Approximation $Q_3(x)$ gives a more uniformly accurate approximation over the interval as a whole. In theory,

$$|Q_3(x) - P_4(x)| = \frac{1}{1024} T_4^*(x) < \frac{1}{1024} \cdot \frac{1}{8} < 1.221 \times 10^{-4}$$

The numerical results are consistent with this bound. For numerical comparison, the following table tabulates the results of approximation of f using the fourth-degree polynomial $P_4(x)$.

X	P(X)	1/(4-X)	ERROR
−1	0.2001953	0.2	−1.953095E − 04
−0.9	0.2041993	0.2040816	−1.176894E − 04
−0.8	0.2084	0.2083333	−6.66827E − 05
−0.7	0.2128009	0.212766	−3.492832E − 05
−0.6	0.2174078	0.2173913	−1.651049E − 05
−0.5	0.222229	0.2222222	−6.780029E − 06
−0.4	0.227275	0.2272727	−2.279878E − 06
−0.3	0.2325587	0.2325582	−5.364418E − 07
−0.2	0.2380953	0.2380953	−5.960465E − 08
−0.1	0.2439025	0.2439025	0
0	0.25	0.25	0
0.1	0.2564103	0.2564103	0
0.2	0.2631578	0.2631579	8.940697E − 08
0.3	0.2702696	0.2702703	6.556511E − 07
0.4	0.277775	0.2777778	2.771616E − 06
0.5	0.2857056	0.2857143	8.702278E − 06
0.6	0.2940953	0.2941177	2.232194E − 05
0.7	0.3029806	0.3030303	4.974008E − 05
0.8	0.3124	0.3125	1.000166E − 04
0.9	0.3223947	0.3225807	1.860261E − 04

Near the center of the interval, the error is much smaller than that obtained using $Q_3(x)$. But near the ends of the interval, $Q_3(x)$ actually provides an estimate that is equally accurate or better.   ■

■ *Exercise Set 8.2*

1. Use the recursion relation

$$T_{n+1}(x) = 2x T_n(x) - T_{n-1}(x)$$

to find polynomials $T_5(x)$ and $T_6(x)$.

2. Use the Newton algorithm (Algorithm 3.1) to find the zeros of $T_5(x)$ and $T_6(x)$. Compare your results with the theoretical values given before Example 8.3.

3. Find the extreme values of $T_4(x)$ on $[-1, 1]$ to show that $|T_4(x)| \leq 1$ on $[-1, 1]$.

4. Show that $\max_{-1 \leq x \leq 1} |P(x)| \geq 1/2^{n-1}$ for each of the following monic polynomials.
   a. $x^2 - 1$ **b.** $x^4 + 3x^2 - x + 2$ **c.** $x^3 + 0.4x^2 - 0.05x + 0.001$

5. Construct Chebyshev approximation polynomials $P_3(x)$ using the zeros of $T_4(x)$, as illustrated in Example 8.4, for each of the following functions.

   a. $f(x) = \cos \pi x$ **b.** $f(x) = xe^{-x}$ **c.** $f(x) = \cosh x = \dfrac{e^x + e^{-x}}{2}$

   d. $f(x) = \dfrac{x}{1 + x^2}$ **e.** $f(x) = \sqrt{1 + x}$

6. Use the transformation $\hat{x} = [(b - a)x + (a + b)]/2$ to transform the zeros of $T_4(x)$ to corresponding points in $[1, 2]$. Then find approximation polynomials $P_3(x)$ on $[1, 2]$ for each of the following functions.
   a. $f(x) = \ln x$ **b.** $f(x) = e^{-x}$
   c. $f(x) = e^{-x^2}$ **d.** $f(x) = \arctan x$

7. Find an error bound (as illustrated in Example 8.4) for the error $|f(x) - P_3(x)|$ for each function in exercise 5.

8. Compute values of $f(x)$ and $P_3(x)$ at points $x = 0, -0.5$, and 0.5 for each of the functions in exercise 5. Compare values of $|f(x) - P_3(x)|$ at these points with the error bounds found in exercise 7.

9. Use the method of Chebyshev economization, as illustrated in Examples 8.5 and 8.6, to derive a third-degree polynomial approximation $Q_3(x)$ from the fourth-degree Taylor polynomial at $c = 0$ for each of the following functions.

   a. $f(x) = \sin \pi x$ **b.** $f(x) = \dfrac{1}{1 + x^2}$

   c. $f(x) = \tan x$ **d.** $f(x) = \dfrac{\sin x}{x}$

10. It was shown in this section that

$$|\sin x - P_8(x)| \leq 1.1 \times 10^{-8}$$

on interval $[-1, 1]$ if $P_8(x)$ is the Chebyshev polynomial constructed by interpolating $\sin x$ on the zeros of $T_9(x)$. Construct polynomial $P_8(x)$, and verify the correctness of this error bound for several values of x in $[-1, 1]$.

11. Estimate each of the following integrals by integrating the Chebyshev polynomial $P_3(x)$ found in exercise 5 for the integrand function in each case. Compare your estimates with precise values determined by direct computation of the integral using the integration methods of calculus.

   a. $\displaystyle\int_0^1 \cos \pi x \, dx$ **b.** $\displaystyle\int_0^1 xe^{-x} \, dx$ **c.** $\displaystyle\int_0^1 \sqrt{1 + x} \, dx$

8.3 Least Squares Approximations

In this section, we will consider the approximation problem from a slightly more general viewpoint. Let $C[a, b]$ denote the set of all continuous functions on interval $[a, b]$. We want to approximate an element f of this set by another element P of this set. We might consider P a good approximation of f if P is close to f in some sense. We need to agree on what we mean by "P is close to f" in $C[a, b]$.

In Section 4.7, we considered a similar problem for vectors in R^n and introduced the concept of a norm (see Definition 4.3). That is the motivation for the following definitions.

Definition 8.1

Suppose that a rule has been given for assigning to each function f in $C[a, b]$ a real number $\|f\|$. We say that such a rule defines a *norm* on $C[a, b]$ if each of the following is satisfied for all functions f and g in $C[a, b]$:

> **i** $\|f\| \geq 0$ and $\|f\| = 0$ iff $f = 0$
> **ii** $\|cf\| = |c| \cdot \|f\|$ for each real number c
> **iii** $\|f + g\| \leq \|f\| + \|g\|$

Definition 8.2

If a norm has been defined on $C[a, b]$, then the *distance* between functions f and g is $\|f - g\|$.

A large number of possible norms for $C[a, b]$ are known, but we will only consider one type of norm in the following discussion. Suppose that $w(x)$ is a continuous function on $[a, b]$ with $w(x) \geq 0$. Assume further that $w(x)$ is not zero over any subinterval of $[a, b]$. We call $w(x)$ a **weight function**. Now we define

$$\|f\| = \sqrt{\int_a^b f(x)^2 w(x)\, dx}$$

Then $\|f\|$ satisfies the requirements in the definition of a norm. Each choice of $w(x)$ leads to a different norm, of course, so we have included several norms in this one rule. If we use $w(x) = 1$, then we have simply

$$\|f\| = \sqrt{\int_a^b f(x)^2\, dx}$$

and this is a possible norm for any choice of interval $[a, b]$. In all these cases, the distance from f to g is

$$\|f - g\| = \sqrt{\int_a^b (f(x) - g(x))^2 w(x)\, dx}$$

So we will now agree that polynomial $P(x)$ is a good approximation to function f if

$$\| f - P \|^2 = \int_a^b (f(x) - P(x))^2 w(x)\, dx$$

is small. If $P(x)$ is selected so that $\| f - g \|^2$ is minimized, we will call $P(x)$ a **least squares polynomial approximation** to f on $[a, b]$. The problem now is to develop a procedure for finding such polynomials. Developing the computational procedure is eased if we first introduce some other useful concepts.

Definition 8.3

If $w(x)$ is a weight function on $[a, b]$, then the **inner product** of functions f and g is

$$\langle f, g \rangle = \int_a^b f(x)g(x)w(x)\, dx$$

It follows that $\| f \|^2 = \langle f, f \rangle$, of course. It is easily verified that the following statements are true for inner products:

1. $\langle f, f \rangle \geq 0$ and $\langle f, f \rangle = 0$ iff $f = 0$
2. $\langle f, g \rangle = \langle g, f \rangle$
3. $\langle f, cg \rangle = c \langle f, g \rangle$ for any real number c
4. $\langle f, g + h \rangle = \langle f, g \rangle + \langle f, h \rangle$

All of these useful facts follow immediately from the definition of $\langle f, g \rangle$.
    Now suppose that

$$P(x) = \sum_{k=0}^{n} a_k x^k$$

Let

$$S = \| f - P \|^2 = \langle f - P, f - P \rangle$$

Then

$$S = \langle f, f \rangle - 2\langle f, P \rangle + \langle P, P \rangle$$

$$= \langle f, f \rangle - 2 \sum_{k=0}^{n} a_k \langle f, x^k \rangle + \sum_{k,\, j=0}^{n} a_k a_j \langle x^k, x^j \rangle$$

We need to have $\partial S/\partial a_i = 0$ for $i = 0, 1,\ldots,n$. Now

$$\frac{\partial S}{\partial a_i} = -2\langle f, x^i \rangle + \sum_{k=0}^{n} a_k \langle x^k, x^i \rangle + \sum_{j=0}^{n} a_j \langle x^i, x^j \rangle$$

$$= -2\langle f, x^i \rangle + 2 \sum_{k=0}^{n} a_k \langle x^k, x^i \rangle$$

So

$$\frac{\partial S}{\partial a_i} = 0 \quad \text{iff} \quad \sum_{k=0}^{n} a_k \langle x^k, x^i \rangle = \langle f, x^i \rangle$$

This is a system of $n + 1$ equations to be solved for each of the $n + 1$ unknowns $a_0, a_1,\ldots, a_n$.

EXAMPLE 8.7   Let $f(x) = \cos x$ on $[0, 1]$. We choose $w(x) = 1$ and want to determine a quadratic least squares polynomial approximation.

Solution   In this case,

$$\langle x^k, x^i \rangle = \int_0^1 x^{k+i}\, dx = \frac{1}{k+i+1}$$

and

$$\langle f, x^i \rangle = \int_0^1 x^i \cos x\, dx$$

So we must solve the sytem

$$\frac{a_0}{1} + \frac{a_1}{2} + \frac{a_2}{2} = \int_0^1 \cos x\, dx = 0.84147098$$

$$\frac{a_0}{2} + \frac{a_1}{3} + \frac{a_2}{4} = \int_0^1 x \cos x\, dx = 0.38177329$$

$$\frac{a_0}{3} + \frac{a_1}{4} + \frac{a_2}{5} = \int_0^1 x^2 \cos x\, dx = 0.23913363$$

The solution using the Gauss algorithm is $a_0 = 1.0034090$, $a_1 = -0.0365370$, and $a_2 = -0.4310094$, so the least squares quadratic desired is

$$P(x) = 1.0034090 - 0.0365370x - 0.4310094x^2$$

The numerical results from approximation of $f(x) = \cos x$ by this polynomial on $[0, 1]$ are shown in the following table.

X	P(X)	COS X	ERROR
0	1.003409	1	$-3.409028E - 03$
0.1	0.9954452	0.9950041	$-4.41134E - 04$
0.2	0.9788612	0.9800666	$1.205385E - 03$
0.3	0.953657	0.9553366	$1.67954E - 03$
0.4	0.9198327	0.921061	$1.228273E - 03$
0.5	0.8773882	0.8775826	$1.943708E - 04$
0.6	0.8263235	0.8253356	$-9.878874E - 04$
0.7	0.7666385	0.7648421	$-1.796365E - 03$
0.8	0.6983334	0.6967066	$-1.62673E - 03$
0.9	0.6214081	0.62161	$2.018809E - 04$

The maximum error occurs at $x = 0$, where the absolute error is approximately 3.4×10^{-3}. Throughout the remainder of the interval, the error remains below that maximum, with a minimum value of approximately 2×10^{-4}. The agreement is quite good considering the low degree of the approximating polynomial. ∎

The polynomials x^n, where n is a nonnegative integer, play a very special role in computing with polynomials, because every polynomial can be expressed as a unique finite linear combination of these particular polynomials. But there are other sets of polynomials with this property.

Definition 8.4

Set $\{\phi_k(x)\}$ will be called a **fundamental set of polynomials** if $\{\phi_k(x)\}$ is an infinite set of polynomials with the property that $\phi_k(x)$ has degree k for $k = 0, 1, 2, \ldots$.

If $\{\phi_k(x)\}$ is a fundamental set of polynomials, then for each polynomial $P_n(x)$ of degree n, there exist unique numbers $c_0, c_1, \ldots, c_n$ such that

$$P_n(x) = \sum_{k=0}^{n} c_k \phi_k(x)$$

By proper selection of such a fundamental set of polynomials, the computation required in the determination of least squares polynomial approximations can be simplified. Suppose now that $\{\phi_k(x)\}$ is any fundamental set of polynomials. Function f is given, and we want to select a polynomial $P_n(x)$ such that $\|f - P_n\|^2$ is minimized. Then

$$\|f - P_n\|^2 = \langle f, f \rangle - 2\langle f, P_n \rangle + \langle P_n, P_n \rangle$$

which becomes

$$\|f\|^2 - 2 \sum_{k=0}^{n} c_k \langle f, \phi_k \rangle + \sum_{k,j=0}^{n} c_k c_j \langle \phi_k, \phi_j \rangle$$

Now suppose that polynomials $\{\phi_k(x)\}$ have the additional property that $\langle \phi_k, \phi_j \rangle = 0$ if $j \neq k$. Then

$$\|f - P_n\|^2 = \|f\|^2 - 2 \sum_{k=0}^{n} c_k \langle f, \phi_k \rangle + \sum_{k=0}^{n} c_k^2 \langle \phi_k, \phi_k \rangle$$

This can be rearranged to the form

$$\|f - P_n\|^2 = \|f\|^2 + \sum_{k=0}^{n} \|\phi_k\|^2 \left(c_k - \frac{\langle f, \phi_k \rangle}{\|\phi_k\|^2} \right)^2 - \sum_{k=0}^{n} \frac{\langle f, \phi_k \rangle^2}{\|\phi_k\|^2}$$

This sum is minimized if

$$c_k = \frac{\langle f, \phi_k \rangle}{\|\phi_k\|^2}$$

Definition 8.5

If $\{\phi_k(x)\}$ is an infinite set of polynomials and for some weight function on interval $[a, b]$ it is true that

$$\langle \phi_k, \phi_j \rangle = 0 \quad \text{when } j \neq k$$

while

$$\langle \phi_k, \phi_k \rangle = \|\phi_k\|^2 \neq 0$$

then we say that $\{\phi_k(x)\}$ is an **orthogonal set of polynomials** on $[a, b]$ with respect to weight $w(x)$.

It can be shown that for each interval $[a, b]$ with specified weight function $w(x)$, there exists a unique set of orthogonal polynomials if we add the condition that $\|\phi_k(x)\| = 1$ for each k. There exists a procedure known as the Gram-Schmidt process for constructing the set of polynomials if $w(x)$ and interval $[a, b]$ are specified. In fact, there is a sizable body of literature concerning such orthogonal polynomial sets because they are useful in many areas of mathematics. In this section, it is sufficient to consider only two such sets: Legendre polynomials and Chebyshev polynomials.

For **Legendre polynomials**, the interval is $[-1, 1]$ and the weight function is $w(x) = 1$. The first two polynomials in this set are $P_0(x) = 1$ and $P_1(x) = x$. The remaining polynomials are determined by the recurrence relation

$$(n + 1)P_{n+1}(x) = (2n + 1)xP_n(x) - nP_{n-1}(x) \qquad n = 1, 2, 3, \ldots$$

We find that

$$P_2(x) = \tfrac{3}{2}(x^2 - \tfrac{1}{3})$$
$$P_3(x) = \tfrac{1}{2}(5x^3 - 3x)$$
$$P_4(x) = \tfrac{35}{8}(x^4 - \tfrac{6}{7}x^2 + \tfrac{3}{35})$$

For **Chebyshev polynomials**, the interval is $[-1, 1]$ and the weight function is $(1 - x^2)^{-1/2}$. The first two polynomials are $T_0(x) = 1$ and $T_1(x) = x$. The remaining polynomials are determined by the recurrence relation

$$T_{n+1}(x) = 2xT_n(x) - T_{n-1}(x) \qquad n = 1, 2, \ldots$$

These polynomials were discussed in Section 8.2. Some other polynomials in the set are listed there.

Now suppose we want to find a least squares polynomial approximation of degree n for a continuous function f on interval $[-1, 1]$. If $\phi_k(x)$ denotes the Legendre polynomial of degree k, then the desired polynomial is

$$P(x) = \sum_{k=0}^{n} c_k \phi_k(x)$$

where

$$c_k = \frac{\langle f, \phi_k \rangle}{\|\phi_k\|^2}$$

We do not need to solve a system of equations. We do need to compute some integrals. In this case, $\|\phi_k\|^2 = 2/(2k + 1)$, so

$$c_k = \frac{2k + 1}{2} \int_{-1}^{1} f(x)\phi_k(x) \, dx$$

EXAMPLE 8.8   Find a quadratic least squares approximation for $f(x) = e^x$ on $[-1, 1]$ using the Legendre polynomials as the orthogonal set.

Solution   The desired polynomial is $c_0 + c_1 x + c_2 \phi_2(x)$, where

$$\phi_2(x) = \tfrac{3}{2}(x^2 - \tfrac{1}{3})$$

So

$$c_0 = \frac{1}{2} \int_{-1}^{1} e^x \, dx = \frac{e - e^{-1}}{2} = 1.17520119$$

$$c_1 = \frac{3}{2} \int_{-1}^{1} xe^x \, dx = 3e^{-1} = 1.10363832$$

$$c_2 = \frac{5}{2} \int_{-1}^{1} \frac{3}{2}(x^2 - \frac{1}{3})e^x \, dx$$

$$= \frac{5}{2}(e - 7e^{-1}) = 0.35781435$$

and the desired polynomial is

$$P(x) = 0.35781435\phi_2(x) + 1.1036382x + 1.17520119$$

where $\phi_2(x)$ is the Legendre polynomial of degree 2. This polynomial can easily be converted to standard polynomial form if desired. ∎

Similarly, if we wish to express a least squares polynomial as a sum of Chebyshev polynomials

$$P(x) = \sum_{k=0}^{n} c_k T_k(x)$$

we need to compute coefficients

$$c_k = \frac{\langle f, T_k \rangle}{\| T_k \|^2}$$

In this case,

$$\| T_k \|^2 = \int_{-1}^{1} \frac{T_k(x)^2}{\sqrt{1 - x^2}} \, dx = \frac{\pi}{2}$$

So

$$c_k = \frac{2}{\pi} \int_{-1}^{1} f(x) T_k(x) \frac{1}{\sqrt{1 - x^2}} \, dx$$

Such integrals would need to be computed numerically for most choices of function f. Note that if we set $x = \cos \theta$, then

$$c_k = \frac{2}{\pi} \int_{0}^{\pi} f(\cos \theta) \cos k\theta \, d\theta$$

This is a useful alternate formula for the coefficients.

■ *Exercise Set 8.3*

1. Use the method illustrated in Example 8.7 to find the quadratic least squares polynomial approximation for each of the following functions on the indicated interval.
   a. $\sin x$ on $[0, 1]$ **b.** $\ln x$ on $[1, 2]$
   c. e^{-x} on $[-1, 1]$ **d.** $1/(1 + x^2)$ on $[0, 1]$

2. Given the fundamental set of polynomials for which $\phi_n(x) = x^n + n$, find numbers c_0, c_1, c_2, and c_3 such that

   a. $x^3 + 2x - 5 = \sum_{k=0}^{3} c_k \phi_k(x)$

   b. $x^4 + x^3 - x^2 + x - 1 = \sum_{k=0}^{4} c_k \phi_k(x)$

3. Given the fundamental set of polynomials for which

   $$\phi_0(x) = 1 \quad \text{and} \quad \phi_n(x) = (x - 1)(x - 2) \dots (x - n) \qquad n = 1, 2, 3, \dots$$

   find numbers c_0, c_1, c_2, and c_3 such that

   $$x^3 - 2x^2 + x - 7 = \sum_{k=0}^{3} c_k \phi_k(x)$$

4. Verify that Legendre polynomials $P_0(x)$, $P_1(x)$, $P_2(x)$, and $P_3(x)$ satisfy the orthogonality conditions

   $$\langle P_k, P_j \rangle = 0 \quad \text{if } k \neq j \quad \text{and} \quad \langle P_k, P_k \rangle \neq 0$$

   for values of j and k from 0 to 3.

5. Find a quadratic least squares Legendre polynomial approximation on $[-1, 1]$ for each of the following functions by using the procedure of Example 8.8.
   a. $f(x) = \sin x$ **b.** $f(x) = \cos x$
   c. $f(x) = \ln (x + 2)$ **d.** $f(x) = e^{-2x}$

6. Find cubic least squares Legendre polynomial approximations on $[-1, 1]$ for each of the functions in exercise 5.

7. A cubic polynomial approximation on $[-1, 1]$ for $f(x) = \sin x$ can be constructed by use of Taylor polynomials (see Section 8.1), by use of Chebyshev polynomials (see Section 8.2), or by use of Legendre least squares. In the exercises, you have been asked to construct each of these approximating polynomials.
   a. If you have not already done so, then construct these three polynomials now.
   b. Compute estimates of $\sin x$ at $x = 0.1$, 0.5, and 0.9 using each of the three approximating polynomials of part (a).
   c. Compare these three polynomials with respect to accuracy and ease of construction. Which do you prefer, and why?

8.4 Rational Approximations (Optional)

Polynomials are frequently used to approximate other functions because polynomials can be evaluated by machines using only arithmetic operations. It is also true that rational functions can be evaluated by machines using only arithmetic operations. Polynomials form a subset of the set of all rational functions, so it seems reasonable to believe that possibly better approximations to some functions might be obtainable if we extended our class of approximating functions to the set of all rational functions.

It is possible to develop methods for constructing rational function approximations for the exponential or log functions, trigonometric functions, and many others, as we will see, but the procedures are much more complicated than those used to construct polynomial approximations. We will explore one of the simpler methods in this section to demonstrate the problem, but no attempt will be made to explore the problem of rational function approximation in depth. If, after reading this section, you wish to study the problem in some depth, you should consult Ralston (1965) or Saff and Varga (1977).

Assume that we are given a function f with Maclaurin series expansion

$$f(x) = \sum_{k=0}^{\infty} c_k x^k$$

convergent on some nonzero interval $(-L, L)$. We want to approximate f with a rational function

$$R_{mn}(x) = \frac{A_m(x)}{B_n(x)} = \frac{a_0 + a_1 x + a_2 x^2 + \cdots + a_m x^m}{1 + b_1 x + b_2 x^2 + \cdots + b_n x^n}$$

We have assumed that $b_0 = 1$. If this is not so, then divide each coefficient by b_0. We *are* assuming that in all cases $b_0 \neq 0$, since f has no singularity at zero.

We must make a choice of m and n and then select $m + n + 1$ coefficients

$$a_0, a_1, \ldots, a_m, b_1, b_2, \ldots, b_n$$

to form the rational function $R_{mn}(x)$. To form a **Padé approximant**, we select integers m and n and then select the coefficients so that the $m + n + 1$ equations

$$f^{(j)}(0) = R_{mn}^{(j)}(0) \qquad j = 0, 1, 2, \ldots, m + n$$

are satisfied. This is an extension to rational approximations of the procedure used to form the Maclaurin series expansion of f.

Let

$$g(x) = f(x) - R_{mn}(x)$$

Then we need to satisfy

$$g^{(j)}(0) = 0 \qquad j = 0, 1, \ldots, m + n$$

This means that $g(x)$ must have a zero of order $m + n + 1$ at 0. Now

$$g(x) = f(x) - \frac{A_m(x)}{B_n(x)} = \frac{f(x)B_n(x) - A_m(x)}{B_n(x)}$$

If we carry out the indicated operations in the numerator of this expression for $g(x)$, we obtain an infinite series

$$f(x)B_n(x) - A_m(x) = \sum_{j=0}^{\infty} d_j x^j$$

where

$$d_j = \sum_{i=0}^{j} c_{j-i} b_i - a_j$$

Explicitly,

$$d_0 = c_0 b_0 - a_0$$
$$d_1 = c_1 b_0 + c_0 b_1 - a_1$$
$$d_2 = c_2 b_0 + c_1 b_1 + c_0 b_2 - a_2$$

and so on. Since the function $g(x)$ has a zero of order $m + n + 1$ at zero, then we must have

$$d_j = 0 \qquad j = 0, 1, \ldots, m + n$$

These conditions represent $m + n + 1$ linear equations for the unknown coefficients of $A_m(x)$ and $B_n(x)$. We solve this linear system, and the Padé rational approximant corresponding to our choice of m and n is determined.

EXAMPLE 8.9   Construct the Padé rational approximant for $f(x) = \exp(x)$ corresponding to the selection of $m = 2$ and $n = 2$.

Solution   The rational function is

$$R_{22} = \frac{a_0 + a_1 x + a_2 x^2}{1 + b_1 x + b_2 x^2}$$

The linear system determining the coefficients is

$$a_0 \qquad\qquad - b_0 \qquad\qquad = 0$$
$$a_1 \qquad - b_0 - b_1$$
$$a_2 - \frac{b_0}{2} - b_1 - b_2 = 0$$
$$\frac{b_0}{6} + \frac{b_1}{2} + b_2 = 0$$
$$\frac{b_0}{24} + \frac{b_1}{6} + \frac{b_2}{2} = 0$$

Now we set $b_0 = 1$ and eliminate b_1 from the last equation to obtain the new system

$$
\begin{aligned}
a_0 & & & = 1 \\
& a_1 & - b_1 & = 1 \\
& a_2 - b_1 & - b_2 & = \tfrac{1}{2} \\
& b_1 & + 2b_2 & = -\tfrac{1}{3} \\
& & b_2 & = \tfrac{1}{12}
\end{aligned}
$$

The complete solution is $a_0 = 1$, $a_1 = \tfrac{1}{2}$, $a_2 = \tfrac{1}{12}$, $b_1 = -\tfrac{1}{2}$, and $b_2 = \tfrac{1}{12}$. The rational approximant is

$$
R_{22} = \frac{1 + \dfrac{x}{2} + \dfrac{x^2}{12}}{1 - \dfrac{x}{2} + \dfrac{x^2}{12}} = \frac{12 + 6x + x^2}{12 - 6x + x^2}
$$

To evaluate the accuracy of this approximation, examine the numerical results in Table 8.3. The numbers in the RATIONAL column were computed from the preceding rational approximation. The numbers in the MACLAURIN column were computed using the terms through x^4 of the Maclaurin expansion for $\exp(x)$. The results from the rational approximation agree very well with the exact values and are clearly more accurate than the values obtained from the Maclaurin expansion.

Table 8.3

X	EXP(X)	RATIONAL	MACLAURIN
0.1	1.105171	1.105171	1.105171
0.2	1.221403	1.221402	1.2214
0.3	1.349859	1.349854	1.349837
0.4	1.491825	1.491803	1.491733
0.5	1.648721	1.648649	1.648438
0.6	1.822119	1.821918	1.8214
0.7	2.013753	2.013269	2.012171
0.8	2.225541	2.22449	2.2224
0.9	2.459603	2.45749	2.453838

In general, the Padé approximants have small error near $x = 0$, and the error increases as $|x|$ increases, just as in the case of a polynomial approximation obtained by truncating a Maclaurin series. The error is not uniform over any interval $[0, L]$. Rational function approximations with more uniform error over such intervals can be obtained by use of Chebyshev polynomial expansions for $f(x)$ and the numerator and denominator of the rational function. Ralston (1965) provides a detailed explanation of how to construct such rational approximations. A brief introduction to the method is given in Burden and Faires (1989).

■ *Exercise Set 8.4*

1. Find Padé approximants $R_{22}(x)$ for each of the following functions.
   a. $\sin x$ **b.** $\arcsin x$ on $(-1, 1)$ **c.** $\arctan x$
   d. $\ln (1 + x)$ **e.** $1 + x$

2. Find Padé approximants $R_{33}(x)$ for each of the functions in exercise 1.

3. Compute values of $\arctan x$ and the Padé approximant obtained in exercise 1c at points $x = 0.2, 0.5,$ and 0.7. Compute values of the error $|\arctan x - R_{22}(x)|$ at these points.

4. Repeat exercise 3 using the Padé approximant $R_{33}(x)$ from exercise 2.

5. The rational function $Q(x) = \dfrac{24x + 24x^2 + 8x^3}{24 + 36x + 18x^2 + 3x^3}$

   is an approximation to $\ln (1 + x)$ on $[0, 1]$.
   a. Compute the error in approximation $\ln (1 + x) \cong Q(x)$ for $x = 0.5$ and $x = 1$.
   b. Compute the error in approximation $\ln (1 + x) \cong R_{33}(x)$ for $x = 0.5$ and $x = 1$ using Padé approximant $R_{33}(x)$ from exercise 2.
   c. How would you decide whether $Q(x)$ or $R_{33}(x)$ is the better approximation to $\ln (1 + x)$ on $[0, 1]$? Comparing estimates at only two points is certainly not sufficient. What criteria would you choose? Which is better in your opinion?

6. The absolute error in approximation of $\ln (1 + x)$ by Padé approximant $R_{33}(x)$ on $[0, \infty)$ increases as x is increased.
   a. Verify this fact by computing the error for $x = 1, 2, 3,$ and 4.
   b. Show that a reasonably accurate value of $\ln 5$ can be obtained from $R_{33}(x)$ by using $\ln 5 = 2R_{33}(1) + R_{33}(0.25)$.
   c. Use $R_{33}(x)$ with properties of logarithms and properly selected values of x to estimate $\ln 10$ as accurately as you can.

7. Estimate π by using the approximation of $4 \arctan 1$ obtained from the Padé approximant $R_{33}(x)$ for $\arctan x$. Compare with a value of π correct to at least six decimals.

8. To construct Padé approximant $R_{33}(x)$ for a function, we must solve a 7×7 linear system

$$\sum_{i=0}^{j} c_{j-1} b_i - a_j = 0 \qquad j = 0, 1, \ldots, 6$$

   for unknowns $a_0, a_1, a_2, a_3, b_1, b_2,$ and b_3.
   a. Show that three of these equations contain only $b_1, b_2, b_3,$ and known values of the c_j's, so this 3×3 system can be solved separately for the b's.
   b. Show that after $b_1, b_2,$ and b_3 have been determined from the 3×3 system in part (b), the remaining four equations can be solved immediately (no elimination necessary) for the coefficients $a_0, a_1, a_2,$ and a_3.
   c. Generalize the results in parts (a) and (b) to the problem of determining coefficients for $R_{nn}(x)$.

9 NUMERICAL DIFFERENTIATION AND INTEGRATION

In this chapter, we will consider the problem of estimating the derivative or an integral of a function f using only the values

$$y_k = f(x_k) \qquad k = 0, 1, 2, \ldots, n$$

at $n + 1$ distinct points in the domain of f. The values y_k may be measured values of an unknown function or they may be tabulated values of some known function. In either case, we will construct an approximation polynomial P for f and use the derivative or integral of P as an estimate of the desired derivative or integral of f. A good understanding of Chapters 7 and 8 is needed to understand the methods described in this chapter.

9.1 Numerical Differentiation Formulas

Given the values $y_k = f(x_k)$ for $x_0, x_1, \ldots, x_n$ in the domain of f, we wish to estimate $f'(x_k)$ for some k where $0 \le k \le n$. Let $P_n(x)$ denote the interpolation polynomial for f on the given set of points. Then

$$f(x) = P_n(x) + R_n(x)$$

so

$$f'(x) = P'_n(x) + R'_n(x)$$

Now

$$R_n(x) = \frac{f^{n+1}(\xi)}{(n + 1)!}\, \psi_n(x) \qquad x_0 < \xi < x_n$$

with

$$\psi_n(x) = (x - x_0)(x - x_1) \ldots (x - x_n)$$

Note that ξ is a function of x in the following. We need to compute $R_n'(x)$. We find that

$$R_n'(x) = \frac{f^{n+1}(\xi)}{(n+1)!} \, \psi_n'(x) + \frac{d}{dx}\left(\frac{f^{n+1}(\xi)}{(n+1)!}\right)\psi_n(x)$$

But $\psi_n(x_k) = 0$, so

$$R_n'(x_k) = \frac{f^{n+1}(\xi)}{(n+1)!} \, \psi_n'(x_k)$$

Now

$$\psi_n'(x) = (x - x_1)(x - x_2) \ldots (x - x_n) + (x - x_0)(x - x_2) \ldots (x - x_n)$$
$$+ \cdots + (x - x_0)(x - x_1) \ldots (x - x_{n-1})$$

So

$$\psi_n'(x_k) = (x_k - x_0)(x_k - x_1) \ldots (x_k - x_{k-1})(x_k - x_{k+1}) \ldots (x_k - x_n)$$

In general, we have the following result:

$$f'(x_k) = P_n'(x_k) + \frac{f^{n+1}(\xi)}{(n+1)!}(x_k - x_0) \ldots (x_k - x_{k-1})(x_k - x_{k+1}) \ldots (x_k - x_n)$$

The value of $R_n'(x_k)$ is, of course, the error in the estimation of $f'(x_k)$ by $P_n'(x_k)$. We are now in a position to construct a variety of numerical differentiation formulas complete with error terms.

Case 1   Given $n = 1$ and $y_0 = f(x_0)$ and $y_1 = f(x_1)$, we want to estimate $f'(x_0)$. Using the Newton divided difference formula, we have

$$P_1(x) = f(x_0) + f[x_0, x_1](x - x_0)$$

Then

$$P_1'(x) = f[x_0, x_1]$$

and

$$f'(x_0) \cong P_1'(x_0) = f[x_0, x_1]$$

Explicitly, we have

$$f'(x_0) \cong \frac{f(x_1) - f(x_0)}{x_1 - x_0}$$

We also find that

$$R_1'(x_0) = \frac{f''(\xi)}{2!}(x_0 - x_1)$$

So the numerical differentiation formula, complete with error term, becomes

$$f'(x_0) = \frac{f(x_1) - f(x_0)}{x_1 - x_0} + \frac{f''(\xi)}{2}(x_0 - x_1) \qquad \textbf{(D1)}$$

Case 2   Given $n = 2$ and $y_k = f(x_k)$ for points $x_0, x_1,$ and $x_2,$ we wish to estimate $f'(x_0)$ and $f'(x_1)$. We can use the Newton divided difference formula again. We have

$$P_2(x) = f(x_0) + f[x_0, x_1](x - x_0) + f[x_0, x_1, x_2](x - x_0)(x - x_1)$$

Then

$$P_2'(x) = f[x_0, x_1] + f[x_0, x_1, x_2](2x - (x_0 + x_1))$$

So

$$P_2'(x_0) = f[x_0, x_1] + f[x_0, x_1, x_2](x_0 - x_1)$$

and

$$P_2'(x_1) = f[x_0, x_1] + f[x_0, x_1, x_2](x_1 - x_0)$$

Also,

$$R_2'(x_0) = \frac{f'''(\xi)}{3!}\psi_2'(x_0) = \frac{f'''(\xi)}{3!}(x_0 - x_1)(x_0 - x_2)$$

and

$$R_2'(x_1) = \frac{f'''(\xi)}{3!}\psi_2'(x_1) = \frac{f'''(\xi)}{3!}(x_1 - x_0)(x_1 - x_2)$$

Of course, $P_2'(x_0)$ is an estimate of $f'(x_0)$, and $P_2'(x_1)$ is an estimate of $f'(x_1)$, while $R_2'(x_0)$ and $R_2'(x_1)$ are the corresponding errors.

It is convenient to make a change in notation in these formulas when points $x_0,$ $x_1,$ and x_2 are equally spaced. If we relabel points, letting $a - h = x_0, a = x_1,$ and $a + h = x_2,$ then the formula for $P_2'(x_1)$ becomes

$$P_2'(a) = \frac{f(a + h) - f(a - h)}{2h} \quad \text{and} \quad R_2'(a) = -\tfrac{1}{6}f'''(\xi)h^2$$

So

$$f'(a) = \frac{f(a + h) - f(a - h)}{2h} - \tfrac{1}{6}f'''(\xi)h^2 \qquad \textbf{(D2)}$$

If we relabel as $a = x_0$, $a + h = x_1$, and $a + 2h = x_2$, then the formula for $P_2'(x_0)$ becomes

$$P_2'(a) = \frac{4f(a + h) - 3f(a) - f(a + 2h)}{2h} \quad \text{and} \quad R_2'(a) = \tfrac{1}{3}f'''(\xi)h^2$$

So

$$f'(a) = \frac{4f(a + h) - 3f(a) - f(a + 2h)}{2h} + \tfrac{1}{3}f'''(\xi)h^2 \qquad \textbf{(D3)}$$

Of course, Equation D1 can also be rewritten in the form

$$f'(a) = \frac{f(a + h) - f(a)}{h} - \tfrac{1}{2}f''(\xi)h \qquad \textbf{(D1)}$$

which we shall also refer to as Equation D1 in the following.

Higher-order formulas using four or more points are easily derived, but for a large number of numerical differentiation problems, Equations D1–D3 suffice.

In general, numerical estimation of derivatives cannot be expected to give very precise results. It is clear that even if $P(x_k) = f(x_k)$ for a set of points $x_0, x_1, \ldots, x_n$, the values of $P'(x_k)$ and $f'(x_k)$ may not agree well at all. Fortunately, in the case of "well-behaved" functions such as exponentials or the sine or cosine functions, satisfactory results can usually be obtained.

EXAMPLE 9.1   Estimate $f'(0.5)$ from the following table of values:

x	$f(x)$
0.49	0.470626
0.50	0.479426
0.51	0.488177
0.52	0.496880

Solution   Equation D2 gives estimate

$$\frac{0.488177 - 0.470626}{0.02} = 0.877550$$

Equation D3 gives estimate

$$\frac{4(0.488177) - 3(0.479426) - 0.496880}{0.02} = 0.877500$$

The true value is 0.877583 in this case because function f is the sine function. The estimates are close, with slightly better results obtained from the "symmetric difference" equation, Equation D2.   ■

EXAMPLE 9.2   Estimate $f'(1.25)$ from the following table of values:

x	$f(x)$
1.20	0.1823216
1.25	0.2231436
1.30	0.2622364
1.35	0.3001046

Solution   Equation D2 gives estimate

$$\frac{0.2622364 - 0.1823216}{0.1} = 0.799148$$

Equation D3 gives estimate

$$\frac{4(0.2622364) - 3(0.2231436) - 0.3001046}{0.1} = 0.794103$$

The function in this case is $\ln x$, so the true value of $f'(1.25)$ is 0.800000. Again, the results are close, with Equation D2 giving the better estimate.   ■

Numerical differentiation formulas can also be derived from Taylor formula expansions about a point a. Suppose that f is class C^3 on an interval $(a - c, a + c)$, where $c > 0$. If $|h| < c$, then

$$f(a + h) = f(a) + f'(a)h + \frac{f''(a)}{2} h^2 + \frac{f'''(\xi_1)}{6} h^3 \qquad a - h < \xi_1 < a + h$$

and

$$f(a - h) = f(a) - f'(a)h + \frac{f''(a)}{2} h^2 - \frac{f'''(\xi_2)}{6} h^3 \qquad a - h < \xi_2 < a + h$$

So

$$f(a + h) - f(a - h) = 2f'(a)h + \frac{h^3}{6} [f'''(\xi_1) + f'''(\xi_2)]$$

and

$$\frac{f(a + h) - f(a - h)}{2h} = f'(a) + \frac{h^2}{6} \left[\frac{f'''(\xi_1) + f'''(\xi_2)}{2} \right]$$

But if f''' is continuous as assumed on $(a - c, a + c)$, then by the intermediate value theorem for continuous functions, we see that

$$\frac{f'''(\xi_1) + f'''(\xi_2)}{2} = f'''(\xi)$$

for some ξ between ξ_1 and ξ_2. It follows that

$$f'(a) = \frac{f(a + h) - f(a - h)}{2h} - \tfrac{1}{6}h^2 f'''(\xi) \qquad a - h < \xi < a + h$$

and we have rederived Equation D2 using an alternate method. Taylor formulas also provide a convenient method for the derivation of formulas for higher-order derivatives. We will illustrate by deriving a formula for the second derivative of f. We use Taylor formulas

$$f(a + h) = f(a) + f'(a)h + \frac{f''(a)}{2} h^2 + \frac{f'''(a)}{6} h^3 + \frac{f^{iv}(\xi_1)}{24} h^4$$

and

$$f(a - h) = f(a) - f'(a)h + \frac{f''(a)}{2} h^2 - \frac{f'''(a)}{6} h^3 + \frac{f^{iv}(\xi_2)}{24} h^4$$

From these formulas we find that

$$\frac{f(a + h) - 2f(a) + f(a - h)}{h^2} = f''(a) + \frac{f^{iv}(\xi_1) + f^{iv}(\xi_2)}{24} h^2$$

Again, if $f^{iv}(x)$ is continuous on an interval $(a - c, a + c)$ and $|h| < c$, then

$$\frac{f^{iv}(\xi_1) + f^{iv}(\xi_2)}{2} = f^{iv}(\xi) \qquad \xi_1 < \xi < \xi_2$$

so we find that

$$f''(a) = \frac{f(a + h) - 2f(a) + f(a - h)}{h^2} - \tfrac{1}{12}h^2 f^{iv}(\xi) \qquad a - h < \xi < a + h \quad \textbf{(D4)}$$

Equation D4 is a very useful formula for making estimates of second derivatives, complete with error estimates.

EXAMPLE 9.3   Estimate $f''(1)$ from the following table of values:

x	$f(x)$
0.8	2.225541
0.9	2.459603
1.0	2.718283
1.1	3.004166
1.2	3.320117

Solution   Using $h = 0.2$ with Equation D4, we obtain

$$\frac{3.320117 + 2.225541 - 2(2.718283)}{0.04} = 2.727300$$

Using $h = 0.1$ with the same equation, we obtain

$$\frac{3.004166 + 2.459603 - 2(2.718283)}{0.01} = 2.720300$$

Function f in this example is exp x, so the true value of the derivative is 2.718283. Both estimates are "in the neighborhood," with the value obtained using $h = 0.1$ being a slightly better estimate, as expected.   ■

■ *Exercise Set 9.1*

1. Use Equation D1 to estimate the derivative $f'(a)$ of $f(x) = 1/x$ at the point a indicated in each case using the specified values of h. Find error bounds for each estimate. Compare estimates with values of $f'(x) = -1/x^2$.
   a. $a = 1.1$; $h = 0.1, 0.2$, and 0.4   **b.** $a = 1.3$; $h = 0.1, 0.2$, and -0.1

2. Sketch the graph of $f(x) = 1/x$ on $[1, 2]$. By interpreting the estimates in exercise 1 as slopes of line segments, decide whether each estimate is too high or too low. Do your numerical results agree with these conclusions?

3. Use Equation D2 with the table of values of $f(x)$ shown to estimate values of $f'(a)$ at the points a indicated with the specified values of h.

   a. $a = 0.6$; $h = 0.1$
   b. $a = 0.7$; $h = 0.1$ and 0.2
   c. $a = 0.8$; $h = 0.1$
   d. $a = 0.75$; $h = 0.05$

x	$f(x)$
0.5	0.877583
0.6	0.825336
0.7	0.764842
0.8	0.696707
0.9	0.621610

4. Given that the function $f(x)$ in exercise 3 is cos x, find error bounds for each of the derivative estimates, and compare with values of the true error computed from the difference between your estimate and the value of $-\sin a$ at the point a in each case.

5. Assuming that f is continuous on $[a, a + h]$ and differentiable on $(a, a + h)$, the mean value theorem states that $[f(a + h) - f(a)]/h$ is the value of $f'(c)$ for some c in $(a, a + h)$. If we estimate c to be $a + h/2$, we have the result that the difference quotient is an estimate of $f'(a + h/2)$.

   a. Use this fact to estimate $f'(1.25)$ for $f(x) = 1/x$ using values of the function at $x = 1.2$ and $x = 1.3$.

   b. Use this fact to estimate $f'(1.45)$ for $f(x) = 1/x$ using values of the function at $x = 1.4$ and $x = 1.5$.

   c. Show that the estimate of $f'(1.25)$ obtained in part (a) is in fact the value of $f'(x)$ at $x = 1.2490$.

   d. Show that for $f(x) = 1/x$, if $a > 0$ and $h > 0$, then

$$\frac{f(a + h) - f(a)}{h} = f'(c)$$

   where $c = \sqrt{a(a + h)}$, so c is a geometric mean of a and $a + h$.

   e. Show that for $f(x) = x^2$, if $a > 0$ and $h > 0$, then in fact

$$\frac{f(a + h) - f(a)}{h} = f'\left(a + \frac{h}{2}\right).$$

6. Show that the derivative estimate given by Equation D2 is in fact the exact value of $f'(a)$ if f is a polynomial of degree 2 or less.

7. Show that the derivative estimate given by Equation D3 is in fact the exact value of $f'(a)$ if f is a polynomial of degree 2 or less.

8. Given $f(x) = (x - 1)(x - 1.7)(x - 1.9)$, estimate $f'(1.5)$ by doing the following.

   a. Compute $\dfrac{f(1.6) - f(1.4)}{0.2}$.

   b. Use the fact that $f'''(\xi) = 3! = 6$ in the error term for Equation D2 to compute the error.

   c. Compute $f'(1.5)$ by combining results in parts (a) and (b).

9. Use Equation D4 to estimate $f''(0.6)$ and $f''(0.8)$ using the table of values in exercise 3. Find error bounds using the fact that $f(x) = \cos x$.

10. Use Taylor formula expansions to derive the formula

$$f'(a) = \frac{-25f(a) + 48f(a + h) - 36f(a + 2h) + 16f(a + 3h) - 3f(a + 4h)}{12h}$$

   with error term $\frac{1}{5}f^{(5)}(\xi)h^4$.

9.2   Some Error Analysis

Suppose we wish to estimate $f'(a)$ using Equation D2. We obtain

$$f'(a) \cong \frac{f(a + h) - f(a - h)}{2h}$$

with truncation error $-\frac{1}{6}h^2 f'''(\xi)$. So if $|f'''(x)|$ is bounded on an interval $(a - c, a + c)$ and $|h| < c$, then the truncation error goes to zero as $|h|$ decreases to zero. However, the total error in estimation of the derivative by Equation D2 consists of truncation error plus other errors that do not go to zero as $|h|$ is decreased.

EXAMPLE 9.4   Estimate $f'(0.1)$ for $f(x) = \exp x$ using Equation D2 with $h = 0.02$, 0.01, and 0.001 and six-decimal arithmetic.

Solution   The results are

$$f'(0.1) = \frac{f(0.12) - f(0.08)}{0.04} = \frac{1.127497 - 1.083287}{0.04} = 1.105250$$

$$f'(0.1) = \frac{f(0.11) - f(0.09)}{0.02} = \frac{1.116278 - 1.094174}{0.02} = 1.105200$$

$$f'(0.1) = \frac{f(0.101) - f(0.099)}{0.01} = \frac{1.106277 - 1.104066}{0.01} = 1.105500$$

The correct value is $\exp 0.1 = 1.1051709$. So the true error in the first estimate, obtained using $h = 0.02$, is 7.91×10^{-5}. The truncation error formula gives an error bound of 7.52×10^{-5} in this case, which is consistent with the actual error but indicates that some other error is present.

The true error using $h = 0.01$ is 2.91×10^{-5}, but the error bound obtained from the theoretical error term is 1.87×10^{-5}. The actual error exceeds this bound, which indicates that besides truncation error, we have a significant amount of other error involved.

The true error using $h = 0.001$ is 3.29×10^{-4}, which is actually larger than the error obtained using either $h = 0.01$ or $h = 0.02$. The theoretical error bound in this case is 1.1844×10^{-7}, which is much too small. It follows that most of the error in this case is something other than truncation error. ■

It is important to remember that the theoretical error terms represent only truncation error and give the true error only in cases where all function values are exact and all arithmetic is performed without error. This is, of course, a very rare situation. In estimating derivatives by use of Equation D2 it is clear that error is introduced when we replace function values $f(a + h)$ and $f(a - h)$ by approximate values. Further error enters in when the subtraction in the numerator is performed and the division is carried out. The smaller the value of h, the more serious the error introduced by the arithmetic operations will be.

A simple theoretical and informative analysis of this situation can be accomplished as follows. First let

$$f(a + h) + E_+ \quad \text{and} \quad f(a - h) + E_-$$

denote the actual values used in the formula as values of the function, while $f(a + h)$ and $f(a - h)$ are the true values. Terms E_+ and E_- are the errors in the function values used to perform the computations. Our computed value of the derivative estimate is

$$\frac{f(a+h) + E_+ - (f(a-h) + E_-)}{2h} = \frac{f(a+h) - f(a-h)}{2h} + \frac{E_+ - E_-}{2h}$$

$$= f'(a) + \tfrac{1}{6}h^2 f'''(\xi) + \frac{E_+ - E_-}{2h}$$

So the total error is a sum of truncation error and another error that is a form of propagated error, since its primary source is the error introduced when we replaced true function values by approximate values. Terms E_+ and E_- could represent other kinds of errors in the function values, but at this point, we are regarding these terms as error introduced by the fact that we are representing the function values by numbers with a limited number of decimals (e.g., four or six decimal places). This limitation may come about because we do not have the information required to produce more accurate values, or it may be the result of limitations on the number of decimals our machine can use. It is clear that the propagated error grows in magnitude as h is decreased, so eventually, most of the error is propagated error. At this point, the truncation error terms in our formulas tell us very little about the accuracy of our derivative estimate.

Typically, then, as h is decreased, the total error in a derivative estimate decreases at first and then begins to increase as the propagated error becomes predominant. At some point, an optimum h is reached that produces the most accurate estimate of the desired derivative.

Table 9.1   DERIVATIVE OF SIN a
FOR a = 0.2

TRUE VALUE OF THE
DERIVATIVE IS 0.9800666

H	DERIVATIVE	ERROR
0.100000	0.9784339	$-1.63275E - 03$
0.010000	0.9800516	$-1.496077E - 05$
0.001000	0.9800718	$5.185604E - 06$
0.000100	0.9800495	$-1.710653E - 05$
0.000010	0.9790064	$-1.060188E - 03$
0.000001	0.9834768	$3.410161E - 03$

In Table 9.1, we see the numerical results of estimating the derivative of the sine function using a sequence of decreasing values of h. There is clearly an optimum value around $h = 0.001$.

■ *Exercise Set 9.2*

1. Use Equation D2 with h values 0.01, 0.005, 0.001, 0.0005, and 0.0001 to estimate $f'(0.5)$ for each of the listed functions. Use a calculator capable of evaluating the functions involved.

Compare results with values of $f'(0.5)$ known to be correct to at least six decimal places.

a. $f(x) = \sin x$ **b.** $f(x) = e^{-x}$ **c.** $f(x) = \tan x$

d. $f(x) = xe^{-x}$ **e.** $f(x) = \arctan x$

2. Repeat exercise 1 using Equation D1.

3. Repeat exercise 1 using Equation D3.

4. Use the following table of values of $f(x) = \sin x$ to do the indicated computations.

x	$\sin x$
0.95000	0.81342
0.96000	0.81919
0.97000	0.82489
0.98000	0.8305
0.99000	0.83603
1.00000	0.84147
1.01000	0.84683
1.02000	0.85211
1.03000	0.8573
1.04000	0.8624
1.05000	0.86742

a. Use Equation D2 with h values 0.05, 0.04, 0.03, 0.02, and 0.01 to estimate $f'(1)$. Which h gives the best estimate of the true value 0.540302?

b. According to the reasoning in this section, the total error in the estimates in part (a) is given by the formula

$$\tfrac{1}{6}h^2 f'''(\xi) + \frac{E_+ - E_-}{2h}$$

In this case, $|E_+| < \delta$ and $|E_-| < \delta$, where $\delta = 5 \times 10^{-6}$. Show that the total error in part (a) is bounded by $E = h^2/6 + \delta/h$.

c. Find the value of h that minimizes E from part (b). Does this value agree with the results in part (a)?

d. Find the minimum value of E. Compare this value with the minimum value of the error in part (a). Are these results consistent?

5. Use the following values of $\tan x$ to do the indicated computations.

x	$\tan x$
0.95000	1.39838
0.96000	1.42836
0.97000	1.4592
0.98000	1.49096
0.99000	1.52368
1.00000	1.55741
1.01000	1.59221
1.02000	1.62813
1.03000	1.66524
1.04000	1.70361
1.05000	1.74332

a. Use Equation D2 with h values 0.05, 0.04, 0.03, 0.02, and 0.01 to estimate $f'(1)$. Compare with the true value 3.425519. Which value of h yields the best estimate of the derivative?

b. Use tan $1.1 < 2$ to show that $f'''(x) < 130$ on $[0.9, 1.1]$, so an error bound for the total error in the estimates in part (a) is

$$E = \tfrac{65}{3}h^2 + \frac{\delta}{h}$$

if δ is a bound on the size of errors $|E_+|$ and $|E_-|$.

c. Use $\delta = 5 \times 10^{-6}$ to find the value of h that minimizes E. Does this value agree with the results in part (a)?

6. Suppose for a function f we have $|f'''(x)| \le M$ on $[a-h, a+h]$ and errors E_+ and E_- do not exceed δ in absolute value.

a. Show that a bound on the total error in the estimation of $f'(a)$ by

$$D_h = \frac{f(a+h) - f(a-h)}{2h}$$

is

$$E = \frac{Mh^2}{6} + \frac{\delta}{h}$$

b. Show that E is minimized if $h^3 = 3\delta/M$ and the minimum value of E is $(3/2)\,(M\delta^2/3)^{1/3}$.

c. Does the bound in part (b) agree with the size of the minimum error found in exercise 4a?

d. Does the bound in part (b) agree with the size of the minimum error found in exercise 5a?

7. Find the results corresponding to the results in exercise 6a and b if $f'(a)$ is estimated by $(f(a+h) - f(a))/h$.

9.3   Richardson Extrapolation

Theoretical error terms can sometimes be used to estimate errors or at least put bounds on the errors in our numerical estimates, as we have seen in several types of problems. These error terms also have another use. From knowledge of the functional form of the error, we can sometimes develop procedures for improving our numerical estimates. The extrapolation method discussed in this section is based on that idea.

Equation D2 in Section 9.1 states that

$$f'(a) = D_h f - \tfrac{1}{6}h^2 f'''(\xi_1) \qquad a - h < \xi_1 < a + h \tag{1}$$

where

$$D_h f = \frac{f(a+h) - f(a-h)}{2h}$$

Then it also follows that

$$f'(a) = D_{h/2}\, f - \frac{1}{6}\left(\frac{h}{2}\right)^2 f'''(\xi_2) \qquad a - h < \xi_2 < a + h$$

so

$$4f'(a) = 4D_{h/2}\, f - \tfrac{1}{6}h^2 f'''(\xi_2) \tag{2}$$

If we assume that $f'''(\xi_1) \cong f'''(\xi_2)$, then it follows by subtraction of Equation 1 from Equation 2 that

$$3f'(a) \cong 4D_{h/2}\, f - D_h\, f$$

and

$$f'(a) \cong \frac{4D_{h/2}\, f - D_h\, f}{3} = D^{(1)}_{h/2}\, f$$

In general, because of this manipulation of the error, we find that $D^{(1)}_{h/2}\, f$ usually gives a more accurate estimate of $f'(a)$ than does $D_{h/2}\, f$. We will give a theoretical proof of this soon, but first let us examine some numerical results.

EXAMPLE 9.5   Given $f(x) = \sin x$, use both $D_h\, f$ and $D^{(1)}_{h/2}\, f$ to estimate $f'(1)$ using $h = 0.4$. Compare results with the correct result $\cos 1 = 0.540302$.

Solution

$$f'(1) \cong D_h\, f = \frac{\sin 1.4 - \sin 0.6}{0.8}$$

$$= \frac{0.985450 - 0.564642}{0.8} = 0.526010$$

Also,

$$D_{h/2}\, f = \frac{\sin 1.2 - \sin 0.8}{0.4} = 0.536707$$

So

$$f'(1) \cong \frac{4D_{h/2}\, f - D_h\, f}{3} = D^{(1)}_{h/2}\, f = 0.540273$$

We find that $D_{h/2}\, f$ is more accurate than $D_h\, f$, but $D^{(1)}_{h/2}\, f$ is much more accurate than either of these. The improvement attained by extrapolation is quite noticeable.

■

Results similar to those obtained in Example 9.5 would be obtained using most functions. Rather than do more arithmetic at this point, however, we will move on to the theoretical proof, which is based on a comparison of error terms for $D_h f$ and $D_{h/2}^{(1)} f$.

The Taylor formulas for $f(a + h)$ and $f(a - h)$ are, respectively,

$$f(a) + hf'(a) + \frac{h^2}{2!} f''(a) + \cdots + \frac{h^n}{n!} f^{(n)}(a) + \frac{h^{n+1}}{(n+1)!} f^{n+1}(\xi_1)$$

and

$$f(a) - hf'(a) + \frac{h^2}{2!} f''(a) - \cdots + \frac{(-h)^n}{n!} f^{(n)}(a) + \frac{(-h)^{n+1}}{(n+1)!} f^{n+1}(\xi_2)$$

So

$$f(a + h) - f(a - h) = 2\left[hf'(a) + \frac{h^3}{3!} f'''(a) + \frac{h^5}{5!} f^5(a) + \cdots \right]$$

Then

$$D_h f = f'(a) + \frac{h^2}{3!} f'''(a) + \cdots + \frac{h^{2n}}{(2n+1)!} f^{2n+1}(a) + 0(h^{2n+2})$$

That is, the error in estimation of $f'(a)$ by $D_h f$ is a sum of the form

$$A_1 h^2 + A_2 h^4 + \cdots + A_n h^{2n} + 0(h^{2n+2})$$

Briefly, in this case we write

$$D_h f = f'(a) + 0(h^2)$$

to indicate that fact. Now we see that

$$D_{h/2} f = f'(a) + \tfrac{1}{4} A_1 h^2 + 0(h^4)$$

so

$$4D_{h/2} f - D_h f = 3f'(a) + 0(h^4)$$

and

$$D_{h/2}^{(1)} f = \frac{4D_{h/2} f - D_h f}{3} = f'(a) + 0(h^4)$$

Since the error in estimate $D_h f$ is $0(h^2)$, while the error in estimate $D_{h/2}^{(1)} f$ is $0(h^4)$, we expect estimate $D_{h/2}^{(1)} f$ to be more accurate for a given h.

Continuing this argument, we find that

$$D_h f = f'(a) + A_1 h^2 + A_2 h^4 + 0(h^6)$$

and

$$D_{h/2} f = f'(a) + \tfrac{1}{4} A_1 h^2 + \tfrac{1}{16} A_2 h^4 + 0(h^6)$$

So

$$4 D_{h/2} f - D_h f = 3 f'(a) - \tfrac{3}{4} A_2 h^4 + 0(h^6)$$

Then

$$D_{h/2}^{(1)} f = f'(a) - \tfrac{1}{4} A_2 h^4 + 0(h^6)$$

Now

$$D_{h/4}^{(1)} f = f'(a) - \tfrac{1}{4} A_2 \frac{h^4}{16} + 0(h^6)$$

$$16 D_{h/4}^{(1)} f - D_{h/2}^{(1)} f = 15 f'(a) + 0(h^6)$$

and

$$D_{h/4}^{(2)} f = \frac{16 D_{h/4}^{(1)} f - D_{h/2}^{(1)} f}{15} = f'(a) + 0(h^6)$$

So $D_{h/4}^{(2)} f$ should be a better estimate of $f'(a)$ than is $D_{h/2}^{(1)} f$. Numerical calculations confirm this fact.

EXAMPLE 9.6   Estimate $f'(1.25)$ for $f(x) = \ln x$ using $h = 0.2$ and estimates $D_h f$, $D_{h/2}^{(1)} f$, and $D_{h/4}^{(2)} f$.

Solution   We find that $D_h f = 0.806933$ and $D_{h/2} f = 0.801713$, so $D_{h/2}^{(1)} f = 0.799973$. Then $D_{h/4} f = 0.800427$, so $D_{h/4}^{(1)} f = 0.799998$. Finally we find that

$$D_{h/4}^{(2)} f = \frac{16(0.799998) - 0.799973}{15} = 0.800000$$

The final estimate is exactly correct through six decimal places, so the improvement obtained by the added extrapolation is clear.   ■

Generalizing the extrapolation results, we obtain the **Richardson extrapolation** method. We generate the following array row by row from top to bottom:

$$D_h^{(0)} f$$

$$D_{h/2}^{(0)} f \qquad D_{h/2}^{(1)} f$$

$$D_{h/4}^{(0)} f \qquad D_{h/4}^{(1)} f \qquad D_{h/4}^{(2)} f$$

$$D_{h/8}^{(0)} f \qquad D_{h/8}^{(1)} f \qquad D_{h/8}^{(3)} f \qquad D_{h/8}^{(3)} f$$

and so on, where

$$D_{h/2^k}^{(0)} f = D_{h/2^k} f$$

$$D_{h/2^k}^{(m)} f = \frac{2^{2m} D_{h/2^k}^{(m-1)} f - D_{h/2^{k-1}}^{(m-1)} f}{2^{2m} - 1} \qquad m \geq 1$$

For the estimate in the kth row and mth column, we have

$$D_{h/2^k}^{(m)} f = f'(a) + 0(h^{2m+2})$$

For example,

$$D_{h/8}^{(3)} f = \frac{64 D_{h/8}^{(2)} f - D_{h/4}^{(2)} f}{63} = f'(a) + 0(h^8)$$

The array of estimates can easily be generated by a machine algorithm. In that case, it is convenient to view the array as having the form

$$D(1, 0)$$
$$D(2, 0) \qquad D(2, 1)$$
$$D(3, 0) \qquad D(3, 1) \qquad D(3, 2)$$

and so on, where

$$D(k, 0) = D_{h/2^k} f \qquad k = 1, 2, 3, \ldots$$

and

$$D(k, m) = \frac{4^m D(k, m-1) - D(k-1, m-1)}{4^m - 1}$$

$$k = 2, 3, 4, \ldots \quad \text{and} \quad m = 1, \ldots, k-1$$

Some typical machine output from such an algorithm is shown in the following table.

Initial h is 0.2

DHF EST.	EXTRAPOLATED VALUES		
0.079603			
0.077428	0.076703		
0.076891	0.076713	0.076713	
0.076757	0.076713	0.076713	0.076713

The function is $f(x) = \ln x / x$, and the object was to compute $f'(2)$.

The values of h used to generate the table shown were 0.2, 0.1, 0.05, and 0.025, of course. The value of $D_{h/2}^{(1)} f$ is 0.076703. The value of $D_{h/8}^{(3)} f$ is the last value printed, which is 0.076713. Since

$$f'(x) = \frac{1 - \ln x}{x^2}$$

in this case, then $f'(2) = 0.0767132$. We see that even estimate $D_{h/4}^{(1)} f$ is correct to six decimal places. The accuracy of the interpolated values greatly exceeds the accuracy of the estimates obtained directly from $D_h f$.

■ *Exercise Set 9.3*

1. Verify all arithmetic in Example 9.6.

2. Use a pocket calculator to verify the extrapolated values given at the end of this section for the derivative of $\ln x/x$ at $x = 2$.

3. Beginning with $h = 0.2$, compute estimates $D_h f$, $D_{h/2}^{(1)} f$, and $D_{h/4}^{(2)} f$ at point $a = 1$ for each of the following functions. Compare estimates with the true value of $f'(1)$ in each case.
   a. $f(x) = \ln x$ **b.** $f(x) = \arctan x$
   c. $f(x) = \sqrt{x}$ **d.** $f(x) = \ln (\sec x)$

4. Beginning with the values of $D_{0.04} f$, $D_{0.02} f$, and $D_{0.01} f$, computed in exercise 4a of Exercise Set 9.2, compute extrapolated estimates $D_{0.02}^{(1)} f$ and $D_{0.04}^{(2)} f$. Compare with the true value, which is 0.540302 in this case. Which estimate is most accurate? How do you explain this?

5. Repeat exercise 4 using values of $D_{0.04} f$, $D_{0.02} f$, and $D_{0.01} f$ from exercise 5a of Exercise Set 9.2.

6. Show that if $E_h f$ is *any* estimator of $f'(a)$ with the property that

$$E_h f = f'(a) + C_1 h^2 + C_2 h^4 + \cdots + C_k h^{2k} + 0(h^{2k+2})$$

then

$$E_{h/2}^{(1)} f = \frac{4E_{h/2} f - E_h f}{3}$$

is an estimate of $f'(a)$ for which

$$E_{h/2}^{(1)} f = f'(a) + 0(h^4)$$

7. Let

$$L_h f = \frac{f(a + h) - f(a)}{h}$$

   a. Show that $L_h f = f'(a) + 0(h)$.
   b. Show that

$$L_{h/2}^{(1)} f = 2L_{h/2} f - L_h f = f'(a) + 0(h^2)$$

   c. Show that

$$L_{h/2}^{(1)} f = \frac{4f\left(a + \dfrac{h}{2}\right) - 3f(a) - f(a + h)}{h}$$

d. How do the estimates obtained from $L_{h/2}^{(1)} f$ compare with the estimates obtained from Equation D3 in Section 9.1?

8. If we used the estimates from Equation D3 (Section 9.1) with step sizes h, $h/2$, $h/4$, and so on, and then extrapolated on these values using the same computational procedure that we used with the estimates obtained from $D_h f$ in this section, would we obtain improved estimates of $f'(a)$? Support your answer with some theory.

9. Use the procedure described in exercise 8 to estimate $f'(1)$ for each of the functions in exercise 3. Begin with $h = 0.2$. Compare results with those obtained in exercise 3, and explain any differences in values of $D_{h/4}^{(2)} f$.

10. Let

$$S_h f = \frac{f(a + h) + f(a - h) - 2f(a)}{h^2}$$

a. Show that

$$S_h f = f''(a) + \tfrac{1}{12}h^2 f^{(4)}(a) + \tfrac{1}{360}h^4 f^{(6)}(a) + 0(h^6)$$

b. Show that

$$S_{h/2}^{(1)} f = \frac{4S_{h/2} f - S_h f}{3} = f''(a) + 0(h^4)$$

c. Continue the extrapolation process begun in part (b) to define estimates of $f''(a)$ with error terms of order $0(h^6)$ and $0(h^8)$.

11. Use the extrapolation procedure of exercise 10 to estimate the derivative $f''(0.5)$ for each of the following functions. Begin with $h = 0.2$ in each case. Compare with true values of $f''(0.5)$ for each function.
a. $f(x) = \ln x$ **b.** $f(x) = e^{-x}$
c. $f(x) = \sin (x/2)$ **d.** $f(x) = x^2 + 1/x$

9.4   Numerical Integration Formulas

In beginning calculus courses, you are taught to evaluate definite integrals

$$\int_a^b f(x)\, dx$$

by use of the fundamental theorem of calculus. It is not always possible, however, to find an antiderivative of f as required for application of the fundamental theorem. Try to find an antiderivative of e^{-x^2} or $\sqrt{1 + x^3}$, for example. Most students have been introduced to the Simpson or trapezoid rules as methods for approximating such integrals.

In some cases, all the information available about a function f is a set of values

$$y_k = f(x_k) \qquad k = 0, 1, \ldots, n$$

where $a = x_0$ and $b = x_n$. In most such cases, the Simpson or trapezoid rule can be used to estimate the definite integral $\int_a^b f(x)\, dx$. The Simpson and trapezoid rules are examples of numerical integration formulas. In this section, we will examine methods for producing such formulas, complete with error terms. Later we will return to the Simpson and trapezoid rules as special cases and examine these elementary rules in some depth.

In general, we assume that a function f is given on interval $[a, b]$ with values $y_k = f(x_k)$ known for $k = 0, 1, \ldots, n$, where $a = x_0$ and $b = x_n$. Let $P_n(x)$ denote the interpolation polynomial for f on the given set of points. Then

$$f(x) = P_n(x) + R_n(x)$$

and

$$\int_a^b f(x)\, dx = \int_a^b P_n(x)\, dx + \int_a^b R_n(x)\, dx$$

So $\int_a^b f(x)\, dx$ is approximated by $\int_a^b P_n(x)\, dx$ with error

$$E_n = \int_a^b R_n(x)\, dx = \int_a^b \frac{f^{n+1}(\xi)}{(n+1)!}\, \psi_n(x)\, dx$$

The following **mean value theorem** for **integrals**, from calculus, is useful in the analysis of error term E_n.

Theorem 9.1

Suppose that functions h and g are continuous on interval $[a, b]$ and g does not change sign on $[a, b]$. Then

$$\int_a^b h(x)g(x)\, dx = h(c) \int_a^b g(x)\, dx$$

for some number c in (a, b).

Case 1   Let $n = 1$, $x_0 = a$, and $x_1 = b$. In this case,

$$P_1(x) = f(x_0) + f[x_0, x_1](x - x_0)$$

so

$$\int_a^b P_1(x)\, dx = (b - a)f(a) + f[a, b]\, \frac{(x - x_0)^2}{2}\, \Big|_a^b$$

$$= (b - a)\, \frac{f(a) + f(b)}{2}$$

Then

$$\int_a^b f(x)\, dx \cong (b - a)\, \frac{f(a) + f(b)}{2}$$

with error

$$E_1 = \int_a^b \frac{f''(\hat{\xi})}{2}(x-a)(x-b)\,dx$$

In this integral, we see that $g(x) = (x-a)(x-b)$ does not change sign on $[a, b]$, so we may apply Theorem 9.1 to obtain

$$E_1 = \frac{f''(\xi)}{2}\int_a^b (x-a)(x-b)\,dx$$

Now we let $h = b - a$ and define s by $x = a + sh$. Then the integral becomes

$$\int_0^1 sh(s-1)(h)h\,ds = -\tfrac{1}{6}h^3$$

Hence,

$$E_1 = -\tfrac{1}{12}f''(\xi)h^3$$

where $h = b - a$ and $a < \xi < b$.

This is a special case of the trapezoid rule from calculus.

Case 2 Let $n = 2$, $x_0 = a$, $x_1 = a + h$, and $x_2 = a + 2h = b$, where $h = (b-a)/2$. Using the Newton forward difference polynomial form for interpolation polynomial $P_2(x)$, we find that

$$P_2(x) = y_0 + s\,\Delta y_0 + \frac{s(s-1)}{2}\Delta^2 y_0$$

where $s = (x - x_0)/h$ and $x = x_0 + sh$. So

$$\int_a^b P_2(x)\,dx = \int_0^2 \left\{ y_0 + s\,\Delta y_0 + \frac{s(s-1)}{2}\Delta^2 y_0 \right\}h\,ds$$

Carrying out the integrations and simplifying results gives

$$\int_a^b P_2(x)\,dx = \frac{h}{3}(y_0 + 4y_1 + y_2)$$

The formula

$$\int_a^b f(x)\,dx \cong \frac{h}{3}(y_0 + 4y_1 + y_2)$$

is called **Simpson's formula** for estimating the definite integral. The error in estimating $\int_a^b f(x)\,dx$ by $\int_a^b P_2(x)\,dx$ is

$$\int_0^2 \left\{ \binom{s}{3} \Delta^3 y_0 + \frac{f^{iv}(\xi)}{24} h^4 s(s-1)(s-2)(s-3) \right\} h\, ds$$

Now

$$\int_0^2 \binom{s}{3} ds = 0$$

So the error is

$$h^5 \int_0^2 \frac{f^{iv}(\xi)}{4!} s(s-1)(s-2)(s-3)\, ds$$

Let

$$\phi(s) = s(s-1)(s-2)(s-3)$$

Then $\phi(s) < 0$ on $(0, 1)$ and $\phi(s) > 0$ on $(1, 2)$. That is, $\phi(s)$ does not change sign on $(0, 1)$, and $\phi(s)$ does not change sign on $(1, 2)$. Now

$$E_2 = \frac{h^5}{24} \left\{ \int_0^1 f^{iv}(\xi)\phi(s)\, ds + \int_1^2 f^{iv}(\xi)\phi(s)\, ds \right\}$$

By Theorem 9.1, we see that

$$E_2 = \frac{h^5}{24} \left\{ f^{iv}(\eta_1) \int_0^1 \phi(s)\, ds + f^{iv}(\eta_2) \int_1^2 \phi(s)\, ds \right\}$$

Now

$$\int_0^1 \phi(s)\, ds = -\tfrac{19}{30} \quad \text{and} \quad \int_1^2 \phi(s)\, ds = \tfrac{11}{30}$$

So

$$E_2 = \frac{h^5}{24} \left\{ -\tfrac{19}{30}\, f^{iv}(\eta_1) + \tfrac{11}{30}\, f^{iv}(\eta_2) \right\}$$

For small h,

$$f^{iv}(\eta_1) \cong f^{iv}(\eta_2) \cong f^{iv}(a)$$

Hence, if h is sufficiently small, then the error is approximately

$$\frac{h^5}{24}\left(-\tfrac{8}{30}\right)f^{iv}(a) = -\tfrac{1}{90}f^{iv}(a)h^5$$

By using a more rigorous argument and some real analysis, it can be shown that in general, the error is

$$E = -\tfrac{1}{90}f^{iv}(\eta)h^5$$

for some η in $[a, b]$.

We can now continue to derive integration formulas for other values of n, just as we did for $n = 1$ and $n = 2$. Formulas derived in this way are called **Newton-Cotes integration formulas**. In general, we let

$$x_k = a + kh$$

where

$$h = \frac{b-a}{n}$$

and define

$$y_k = f(x_k) \qquad k = 0, 1, \ldots, n$$

Then

$$P_n(x) = \sum_{k=0}^{n} L_k(x)y_k$$

if we use the Lagrange form of the interpolation polynomial. So

$$\int_a^b f(x)\, dx \cong \int_a^b \sum_{k=0}^{n} L_k(x)y_k\, dx = \sum_{k=0}^{n} A_k y_k$$

where

$$A_k = \int_a^b L_k(x)\, dx$$

The error is

$$\int_a^b R_n(x)\, dx$$

where $R_n(x)$ is the Lagrange error term. Using $n = 3$, we obtain

$$\int_a^b f(x)\,dx \cong \frac{3h}{8}\,(y_0 + 3y_1 + 3y_2 + y_3)$$

with error

$$-\tfrac{3}{80}f^{iv}(\xi)h^5$$

for example. With $n = 4$, we obtain

$$\int_a^b f(x)\,dx \cong \frac{2h}{45}\,(7y_0 + 32y_1 + 12y_2 + 32y_3 + 7y_4)$$

with error

$$-\tfrac{8}{945}f^{vi}(\xi)h^7$$

The higher-order Newton-Cotes formulas ($n \geq 3$) are rarely used in practice, so they are primarily of academic interest. It is important for the student of numerical analysis to know that such formulas exist, however. It is even more important to know that using higher-order Newton-Cotes formulas is not the way to obtain more accurate results when estimating integrals. In the next section, we turn to composite integration formulas to obtain accurate estimates of integrals.

■ *Exercise Set 9.4*

1. Derive the trapezoid formula with error term by integrating the Lagrange formula

$$f(x) = \sum_{k=0}^{1} L_k(x)y_k + R_1(x)$$

using $x_0 = a$ and $x_1 = b$, of course.

2. Derive the Simpson numerical integration formula by integration of the Lagrange formula for interpolation polynomial $P_2(x)$.

3. Show that integration of the Lagrange formula for $P_n(x)$ always produces a numerical integration formula of the form

$$\int_a^b f(x)\,dx \cong \sum_{k=0}^{n} A_k y_k$$

for some constants $A_0, A_1, \ldots, A_n$.

4. Suppose that $x_k = a + kh$ for $k = 0, 1, \ldots, n$ and $x_n = b$. Let $s = (x - x_0)/h$. Show that

$$\int_a^b f(x)\,dx \cong h \sum_{k=0}^{n} C_k^n y_k$$

where

$$C_k^n = \frac{(-1)^{n-k}}{k!(n-k)!} \int_0^n s(s-1)(s-2)\ldots(s-n)\,ds$$

5. Compute coefficients C_k^n, as defined in exercise 4, for $n = 2$, and write down the resulting integration formula.

6. Assume that $|f^{(n+1)}(x)| \le M$ on $[a, b]$. Show that the error in the numerical integration formula of exercise 4 is order $0(h^{n+2})$.

7. Explain why the error for Simpson's formula is $0(h^5)$ rather than $0(h^4)$ as indicated by the result in exercise 6.

8. Derive the formula (three-eighths rule)

$$\int_{x_0}^{x_3} f(x)\,dx = \frac{3h}{8}(y_0 + 3y_1 + 3y_2 + y_3) + 0(h^5)$$

9. Show that the error in the three-eighths rule estimate of $\int_a^b f(x)\,dx$ is zero if $f(x)$ is a polynomial of degree 3 or less.

10. Given that $f(x) = (\sin x)/x$ if $0 < x \le 1$ and $f(0) = 1$, do the following.
    a. Show that $f(x) = 1 - x^2/6 + 0(x^4)$ on $[0, 1]$.
    b. Estimate $\int_0^1 f(x)\,dx$ by integrating $1 - x^2/6$.
    c. Estimate $\int_0^1 f(x)\,dx$ by Simpson's formula with $h = 0.5$.
    d. The value of the integral in parts (b) and (c) is 0.946083 to six decimal places. Both estimates employ a quadratic approximation of $f(x)$ on $[0, 1]$. Which estimate is most accurate? Explain why.

11. A theoretical comparison of the two quadratic estimates in exercise 10 can be obtained as follows.
    a. Show that the Simpson approximation of

$$\int_0^1 \frac{\sin x}{x}\,dx \text{ is } S_h = h + (4 + \cos h)\sin h$$

    b. Use the Maclaurin approximations

$$\cos h = 1 - \frac{h^2}{2} + \frac{h^4}{24} + 0(h^6)$$

    and

$$\sin h = h - \frac{h^3}{6} + \frac{h^5}{120} + 0(h^7)$$

    to show that S_h is

$$2h - \tfrac{4}{9}h^3 + \tfrac{1}{18}h^5 + 0(h^7)$$

    c. Show that

$$I_h = \int_0^1 \left(1 - \frac{x^2}{6}\right)dx = 2h - \tfrac{4}{9}h^3$$

so

$$S_h = I_h + \tfrac{1}{18}h^5 + 0(h^7)$$

d. Verify that the difference between the estimates in exercise 10b and c is approximately $h^5/18$, as claimed in part (c) of this exercise.

9.5   Simpson and Trapezoid Composite Formulas

It is clear from inspection of the error terms in Section 9.4 that in order to obtain small error in estimation of an integral, we need to have a small value of h. Now h can be made small by increasing n and using higher-order Newton-Cotes formulas. In this case, the derivatives in the error formulas can become large and cancel out the benefit obtained from decreasing h. Some numerical experimentation shows that this does indeed happen, since as n is increased, the higher-order interpolation polynomials oscillate rapidly.

This fact motivates the use of composite integration formulas. Instead of trying to fit an interpolation polynomial to a large number of equally spaced points throughout the integration interval $[a, b]$, we first subdivide the interval into N subintervals. Then we apply a low-order integration formula to each subinterval and add up the results.

Given interval $[a, b]$ let

$$h = \frac{(b - a)}{N} \quad \text{and} \quad x_k = a + kh \qquad k = 0, 1, \ldots, N$$

Then

$$\int_a^b f(x)\, dx = \sum_{k=1}^{N} \int_{x_{k-1}}^{x_k} f(x)\, dx$$

If we use the Newton-Cotes formula for $n = 1$ from Section 9.4 to estimate the integrals over the subintervals, we find that

$$\int_{x_{k-1}}^{x_k} f(x)\, dx = \frac{h}{2}(y_{k-1} + y_k) - \tfrac{1}{12}f''(\xi_k)h^3$$

So

$$\int_a^b f(x)\, dx = \sum_{k=1}^{N} \frac{h}{2}(y_{k-1} + y_k) - \frac{h^3}{12}\sum_{k=1}^{N} f''(\xi_k)$$

and

$$\int_a^b f(x)\, dx = \frac{h}{2}(y_0 + 2y_1 + 2y_2 + \cdots + 2y_{N-1} + y_N) - \frac{h^3}{12}\sum_{k=1}^{N} f''(\xi_k)$$

Assuming that f'' is continuous on $[a, b]$, we find that

$$\frac{f''(\xi_1) + f''(\xi_2) + \cdots + f''(\xi_N)}{N} = f''(\xi)$$

for some ξ in (a, b). It follows that $\int_a^b f(x)\, dx$ is equal to

$$\frac{h}{2}(y_0 + 2y_1 + 2y_2 + \cdots + 2y_{N-1} + y_N) - \frac{N}{12} f''(\xi)h^3 \qquad (1)$$

We will call Equation 1 the **composite trapezoid rule**. Note that the error term in Equation 1 can be written

$$\frac{-N}{12} \cdot \frac{(b-a)^3}{N^3} f''(\xi) = -\frac{(b-a)^3}{12N^2} f''(\xi)$$

This error is usually called the **discretization error** for the integral estimate. We see that for the composite trapezoid rule, the discretization error goes to zero like $1/N^2$.

EXAMPLE 9.7   Estimate $\int_0^1 e^{-x}\, dx$.

Solution   We will use $N = 4$, so $h = 0.25$. We construct the following table of values of $f(x)$ (column 2) and weighted values (column 3):

x_k	y_k	$A_k y_k$
0.0	1.000000	1.000000
0.25	0.778801	1.557602
0.50	0.606531	1.213062
0.75	0.472367	0.944734
1.00	0.367879	0.367879

The sum of the weighted values in column 3 is 5.083277. So the estimate of the integral is

$$\frac{0.25}{2}(5.083277) = 0.635410$$

The absolute error is $\frac{4}{12}f''(\xi)(0.25)^3$, where $0 < \xi < 1$. Since $f''(x) = e^{-x}$, then $|f''(\xi)| < 1$. So the absolute error is less than $(0.25)^3/3 < 5.21 \times 10^{-3}$. The true value of the integral is 0.632121, and the true absolute error is 3.29×10^{-3}. These values agree well with the estimate and the theoretical error bound.   ∎

Now suppose that we form N subintervals, where N is even. Then

$$\int_a^b f(x)\, dx = \int_{x_0}^{x_2} f(x)\, dx + \int_{x_2}^{x_4} f(x)\, dx + \cdots + \int_{x_{N-2}}^{x_N} f(x)\, dx$$

If we apply the Simpson formula from Section 9.4 to each of the integrations over subintervals, we obtain

$$\int_a^b f(x)\, dx \cong \sum_{k=0}^{N-2} \frac{h}{3} (y_k + 4y_{k+1} + y_{k+2})$$

Collecting terms in the sum, we find

$$\int_a^b f(x)\, dx \cong \frac{h}{3} (y_0 + 4y_1 + 2y_2 + 4y_3 + \cdots + 2y_{N-2} + 4y_{N-1} + y_N)$$

The error in this estimate is

$$-\tfrac{1}{90} \sum_{k=0}^{N-2} f^{iv}(\xi_k) h^5$$

But if $f^{iv}(x)$ is continuous on $[a, b]$, then

$$\sum_{k=0}^{N-2} f^{iv}(\xi_k) = \frac{N}{2} f^{iv}(\xi)$$

for some ξ in (a, b). So the error is

$$\frac{-N}{180} f^{iv}(\xi) h^5 = \frac{-(b-a)}{180} f^{iv}(\xi) h^4$$

Collecting results, we have

$$\int_a^b f(x)\, dx = \frac{h}{3} (y_0 + 4y_1 + 2y_2 + 4y_3 + \cdots + 2y_{N-2} + 4y_{N-1} + y_N)$$

$$- \frac{b-a}{180} f^{iv}(\xi) h^4 \tag{2}$$

Equation 2 will be called the **composite Simpson rule** in the following.

EXAMPLE 9.8   Estimate $\int_1^2 \ln\, dx$.

Solution   We will use $N = 4$, so $h = 0.25$. We construct the following table of values, including the weighted values in column 3:

x_k	y_k	$A_k y_k$
1.0	0.000000	0.000000
1.25	0.223144	0.892576
1.50	0.405465	0.810930
1.75	0.559616	2.238464
2.00	0.603147	0.693147

The sum of the values in column 3 is 4.635117. The integral estimate is

$$\left(\frac{0.25}{3}\right)(4.635117) = 0.386260$$

The absolute error is

$$\left(\frac{4}{180}\right)|f^{iv}(\xi)|(0.25)^5$$

Since $f^{iv}(x) = -6/x^4$, then $|f^{iv}(\xi)| < 6$. So

$$|\text{error}| < (\tfrac{4}{180})(6)(0.25)^5 < 1.302 \times 10^{-4}$$

The true value of the integral is 0.386294, and the true error is 3.4×10^{-5}. Our theoretical results are agreeable with these values. ■

If we use either the trapezoid rule or the Simpson rule to estimate an integral, then the total error in such estimates consists of a sum of discretization error and round-off error. The discretization error can be estimated by use of the theoretical error terms in Equations 1 and 2. The discretization error decreases as N is increased, but the round-off error increases as N is increased. Under these conditions, we expect the total error to decrease for a while as N is increased, and then, because of round-off error accumulation, the total error should begin to increase.

In Table 9.2, we see the results from the estimation of the integral $\int_0^3 x^2 \, dx$ by use of the trapezoid rule. The exact value of the integral is 9, so we can compute the total error in the estimates as N is increased. The behavior is exactly as expected. In this case, the discretization error is

$$-\frac{(b-a)^3}{12N^2} f''(\xi) = -\frac{9}{2N^2}$$

At $N = 50$, the discretization error is -1.8×10^{-3} and the total error is -1.79×10^{-3}, so the total error is primarily discretization error. At $N = 200$, the discretization error is -1.125×10^{-4} and the total error is -1.173×10^{-4}, so most of the total error is discretization error. However, when we reach $N = 3200$, we find

Table 9.2

N	TRAPEZOID EST.	ERROR
50	9.001791	-0.001791
100	9.000437	$-4.367829E - 04$
200	9.000118	$-1.173019E - 04$
400	9.000043	$-4.291535E - 05$
800	9.000033	$-3.33789E - 05$
1600	9.000022	$-2.193451E - 05$
3200	8.999711	$2.889633E - 04$
6400	8.999498	$5.025864E - 04$

Table 9.3

N	SIMPSON EST.	ERROR
10	0.2000133	−1.332164E − 05
20	0.2000009	−8.791685E − 07
40	0.2000002	−1.639128E − 07
80	0.1999998	2.384186E − 07
160	0.1999995	4.917383E − 07
320	0.2000008	−8.046627E − 07

a total error of 2.89×10^{-4}, while the discretization error is only -8.79×10^{-7}. At this point, almost all the error is round-off error. Inspection of the results in Table 9.2 shows that the total absolute error clearly passes through a minimum value.

Similar results are obtained with the Simpson rule. In Table 9.3, we see the results from machine estimation of $\int_0^1 x^4 \, dx$ using the Simpson rule with various values of N. The exact value of the integral is 0.2, so we can compute the total error in each estimate as N is varied. The discretization error in this case is given by

$$-\frac{N}{180} f^{iv}(\xi) h^5 = -\frac{(b-a)^5}{180N^4} f^{iv}(\xi) = \frac{-2}{15N^4}$$

At $N = 10$, the discretization error is -1.33×10^{-5}, which accounts for most of the total error. At $N = 40$, the discretization error is -5.208×10^{-8}, while the total error is -16.39×10^{-8}. So a sizable amount of round-off error is already present. At $N = 320$, the discretization error is -1.27×10^{-11}, while the total error is -8.047×10^{-7}, so the error is almost entirely round-off error at this point. In the table of estimates, the total error is smallest for the relatively low value $N = 40$. In the case of the Simpson rule, the discretization error decreases rapidly while the round-off error grows quite rapidly with increasing N. Hence, the best estimate of the integral is frequently obtained with values of N around 50 using machines with a six- or seven-digit mantissa length.

When operating with a binary machine (base 2 machine number system), we get some decrease in round-off error by using a power of 2 as our choice of N. In Table 9.4, we see some results using $N = 2, 4, 8, \ldots$. The total error decreases steadily in absolute value as N is increased up to $N = 128$. The discretization error at

Table 9.4

N	SIMPSON EST.	ERROR
2	0.2083333	−8.333326E − 03
4	0.2005208	−5.208254E − 04
8	0.2000326	−3.254414E − 05
16	0.200002	−2.026558E − 06
32	0.2000001	−1.192093E − 07
64	0.2	2.980232E − 08
128	0.2	−2.980232E − 08
256	0.2	0

$N = 64$ should be -7.95×10^{-9}. The total error is observed to be 2.98×10^{-8} at this point. Most of the error is round-off error, but both round-off error and discretization error are very small, so a very accurate estimate of the integral is obtained. At $N = 256$, the total error is virtually zero because at this point the discretization error is negligible while the round-off error is still quite small.

Let S_N denote the estimate of an integral using the Simpson rule with N subintervals. Assume that N is sufficiently small, so that the error is mostly discretization error. We will denote this error by E_N. Then we may assume that

$$\int_a^b f(x)\, dx = S_N + E_N$$

where

$$E_N = \frac{-(b-a)^5}{180N^4} f^{\mathrm{iv}}(\xi)$$

Also,

$$\int_a^b f(x)\, dx = S_{2N} + E_{2N}$$

where

$$E_{2N} \cong \frac{E_N}{16}$$

Then it is approximately true that

$$S_N + E_N = S_{2N} + \frac{E_N}{16}$$

It follows that

$$E_N \cong \tfrac{16}{15}(S_{2N} - S_N) \quad \text{and} \quad E_{2N} \cong \frac{S_{2N} - S_N}{15}$$

So we can approximate the errors E_N and E_{2N} from the difference of S_{2N} and S_N.

EXAMPLE 9.9 In estimation of the integral

$$\int_{-4}^4 \frac{dx}{1 + x^2}$$

a machine program gave estimates

$$S_{40} = 2.651634711 \quad \text{and} \quad S_{80} = 2.651635310$$

Estimate the errors in estimates S_{40} and S_{80}.

Solution   Assuming negligible round-off error, the errors should be approximately

$$E_N = \tfrac{16}{15}(2.651635310 - 2.651634711) = 6.389 \times 10^{-7}$$

and

$$E_{2N} = \frac{E_N}{16} = 3.993 \times 10^{-8}$$

As a check on this calculation, we note that the true value of the integral is 2 arctan 4 = 2.651635328, so

$$E_N = 6.17 \times 10^{-7} \quad \text{and} \quad E_{2N} = 1.8 \times 10^{-8}$$

Our numerical estimates are very close to the correct errors. ∎

Similarly, let T_N denote the estimate of an integral using the trapezoid rule with N subintervals. Assume that round-off error is negligible. Then

$$\int_a^b f(x)\, dx = T_N + E_N$$

where

$$E_N = \frac{(b-a)^3}{12N^2}\, f''(\xi)$$

Then

$$\int_a^b f(x)\, dx = T_{2N} + E_{2N}$$

also where

$$E_{2N} \cong \frac{E_N}{4}$$

It follows that

$$T_N + E_N \cong T_{2N} + \frac{E_N}{4}$$

So

$$E_N \cong \tfrac{4}{3}(T_{2N} - T_N) \quad \text{and} \quad E_{2N} \cong \frac{E_N}{4}$$

EXAMPLE 9.10 Using the trapezoid rule to estimate the integral $\int_0^1 e^{-x^2}\, dx$, a machine produced estimates

$$T_{50} = 0.74679961 \quad \text{and} \quad T_{100} = 0.74681800$$

Estimate the errors in T_{50} and T_{100}.

Solution We find that the error E_{50} is approximately

$$\tfrac{4}{3}(0.74681800 - 0.74679961) = 2.45 \times 10^{-5}$$

and the error E_{100} is approximately

$$\frac{E_{50}}{4} = 6.125 \times 10^{-6}$$

It is known that the true value of the integral in this example is 0.74682413 to eight decimal places. So the true value of error E_{50} is

$$0.74682413 - 0.74679961 = 2.452 \times 10^{-5}$$

while the value of E_{100} is 6.1×10^{-6}. The numerical estimates of the error are very accurate in this case. ∎

Other composite integration formulas can be derived by applying higher-order Newton-Cotes formulas to the subintervals of $[a, b]$. Equations 1 and 2 are sufficient for many integral estimation problems. Accurate estimates of a large number of definite integrals, complete with error estimates, can be obtained by use of these two formulas.

■ *Exercise Set 9.5*

1. Use the composite trapezoid rule (Equation 1) with the indicated value of N to estimate each of the following integrals.

   a. $\displaystyle\int_1^2 \frac{1}{x}\, dx; \quad N = 10$ **b.** $\displaystyle\int_0^4 \sqrt{x}\, dx; \quad N = 8$ **c.** $\displaystyle\int_0^1 \frac{1}{1 + x^2}\, dx; \quad N = 10$

   d. $\displaystyle\int_0^1 f(x)\, dx; \quad N = 8$ $f(x) = \dfrac{\sin x}{x}$ on $(0, 1]$ and $f(0) = 1$.

2. Repeat exercise 1 using the composite Simpson rule (Equation 2).

3. Use a machine program to evaluate each of the following integrals with Equation 1 and values of $N = 20$, 50, and 100.

   a. $\int_0^1 e^{-x^2}\, dx$ **b.** $\int_0^1 f(x)\, dx$, where $f(x)$ is the function in exercise 1d

c. $\int_0^1 g(x)\, dx$, where $g(x) = x \ln x$ on $(0, 1]$ and $g(0) = 0$

d. $\int_0^{1.5} \sqrt{1 - e \cos^2 \theta}\, d\theta$, where $e = 0.01$ **e.** $\int_1^2 \sqrt{x + 1/x}\, dx$

4. Repeat exercise 3 using the Simpson formula (Equation 2)

5. Show that the arc length from point $(0, 0)$ to $(1, 1)$ on the graph of $y = x^2$ is given by integral $\int_0^1 \sqrt{1 + 4x^2}\, dx$. Estimate the arc length using the trapezoid formula (Equation 1) with $N = 10$. Find a bound for the error in this estimate.

6. By imitating the procedure followed in this section to derive the composite rules (Equations 1 and 2) from the trapezoid and Simpson rules of Section 9.4, derive a composite rule from the three-eighths rule given in exercise 8 of Exercise Set 9.4.

7. Use $N = 9$ with the composite rule derived in exercise 6 to estimate each of the integrals in exercise 1. Compare with your results in exercises 1 and 2.

8. Estimate $\int_0^{0.5} f(x)\, dx$ for the function with values listed in the following table
   a. using the trapezoid rule (Equation 1)
   b. using the Simpson rule (Equation 2)

x	$f(x)$
0.00	0.000000
0.05	0.070389
0.10	0.137503
0.15	0.201634
0.20	0.263034
0.25	0.321928
0.30	0.378512
0.35	0.432960
0.40	0.485427
0.45	0.536053
0.50	0.584963

9. Use the Simpson formula (Equation 2) with $N = 20$ to estimate the area under the graph of $y = e^{-x^2}$ between $x = -1$ and $x = 1$. Find a bound for the error in your estimate.

10. Find the value of N required in Equation 1 to estimate $\int_0^1 \sin \pi x\, dx$ with a discretization error less than 10^{-6} in absolute value.

11. Repeat exercise 10 for Equation 2.

12. We wish to estimate $\ln 2 = \int_1^2 (1/x)\, dx$ with an absolute error less than 10^{-6}. Determine the minimum value of N needed in Equation 2 to accomplish this. Compute the estimate using Equation 2 with your minimum value of N. Compare your result with the accurate value $\ln 2 = 0.69314718$, which is correct through eight decimal places. Does your estimate have the desired accuracy? If not, why not?

13. Use Equation 1 with a machine program and single-precision arithmetic to estimate $\ln 2$ by the integral in exercise 12. Use $N = 50, 100, 200, 500, 1000$, and 2000. Which estimate is most accurate? Explain why increasing N does not necessarily reduce the error in estimates from Equation 1 even though the theoretical error goes to zero as $1/N^2$.

14. The integral $\int_1^2 (1/x)\, dx$ was estimated by the trapezoid rule. Using $N = 25$, the estimate was $T_{25} = 0.6932472$, while $N = 50$ yielded estimate $T_{50} = 0.6931722$.
   a. Estimate the error in T_{25}.
   b. Given that $T_{100} = 0.6931534$, estimate the error in T_{50}.

15. Use the numerical data in Table 9.2 (estimation of $\int_0^3 x^2 \, dx$) to do the following.
    a. Estimate the error in the trapezoid estimates obtained for $N = 100$ and $N = 200$. Compare with the total error given in Table 9.2.
    b. Estimate the error in the estimate for $N = 3200$. Compare with the total error. Would you expect the error estimate to agree very well with the actual total error in this case?

16. Use the numerical data in Table 9.3 (estimation of $\int_0^1 x^4 \, dx$) to do the following.
    a. Estimate the error in the Simpson estimates obtained for $N = 20$ and $N = 40$. Compare with the stated total error.
    b. Estimate the error in the Simpson estimate for $N = 160$, and compare with the total error.
    c. Estimate the percent of the total error that is round-off error in the cases where $N = 40$ and $N = 160$.

17. The following table gives trapezoid estimates of $\int_0^1 e^{-x^2} \, dx$ using single- versus double-precision arithmetic:

N	T_N (double)	T_N (single)
50	0.746799607	0.7467997
100	0.746818001	0.7468180
200	0.746822604	0.7468228
400	0.746823754	0.7468245

    a. Use the procedure illustrated in this section to estimate the error in the single-precision values of T_{100} and T_{200}.
    b. Repeat part (a) using the double-precision values.
    c. The accepted value of the integral correct to eight decimals is 0.74682413. Use this value to estimate the total error in double- and single-precision values of T_{100} and T_{200}.

18. To obtain a very precise value of integral $\int_0^1 e^{-x^2} \, dx$, the Simpson rule with double-precision arithmetic was used. With $N = 200$, the estimate was 0.746824132817536, while with $N = 400$, the estimate was 0.7468241328127463. How many digits in the S_{200} estimate are correct in your opinion? Make some error analysis to substantiate your claim.

19. Using another machine with $N = 160$ and $N = 320$, the double-precision estimates of the integral in exercise 18 were 0.7468241328249012 and 0.7468241328132066. How many digits in estimate S_{160} should be correct? Compare with your conclusion in exercise 18.

20. It is desired to estimate $\int_0^1 (\sin x/x) \, dx$ with absolute error less than 10^{-11}. Use the double-precision estimates given in the following table to make such an estimate; that is, state the value correctly to ten decimal places. Justify your answer.

N	Simpson Est.
40	0.9460830707515
80	0.9460830703912
160	0.9460830703686
320	0.9460830703672

21. We wish to estimate $\int_0^\pi (\sin x/x)\, dx$ as accurately as possible. Using double-precision arithmetic and appropriate values of N, try to determine the value correctly to at least ten decimal places (absolute error less than 5×10^{-11}). More correct digits are desirable if possible. (*Caution:* You will need a precise value of π.)

22. We wish to evaluate $\int_0^{\pi/4} \sec^5 \theta\, d\theta$. We can use integration by parts, but that is time-consuming. A very accurate value can be found by use of the Simpson rule with double-precision arithmetic and $N < 200$. This can be done very quickly if the integration software is already available. It is assumed that this is true.

    a. Compute the Simpson estimates of the integral for $N = 160$ and $N = 320$. Use these values to determine the value of the integral correct to eight decimal places.

    b. Compute the integral by using integration by parts, and compare the Simpson estimate with this theoretical value.

    c. Discuss pros and cons of these two methods of evaluating the integral.

9.6 Adaptive Quadrature Methods (Optional)

Suppose that we want to use the Simpson composite rule (Equation 2 in Section 9.5) to estimate the value of a definite integral. We select a value of N and compute step size h. Then the same step size is used at all points of the interval $[a, b]$.

For some integration problems, there are reasons to believe that allowing the step size to vary over the interval might be wiser. Consider the integral

$$\int_0^1 100xe^{-20x^2}\, dx$$

The graph of the integrand is shown in Figure 9.1. We see that the values of this function change rapidly over the interval $[0.1, 0.3]$ but vary slowly over the interval $[0.6, 0.8]$. To obtain an accurate estimate of this integral using the Simpson composite rule, a small value of h (large N) is required because of the behavior of the integrand around the point $x = 0.158$, where the function attains its maximum value. The small step size required there is smaller than necessary for integration

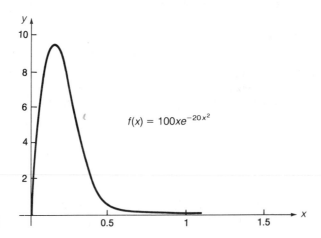

Figure 9.1

over the interval [0.6, 1]. In this case, we need an algorithm that allows us to use different step sizes in these two different parts of the interval [0, 1]. That is the motivation behind the design of **adaptive quadrature methods.**

Suppose that we want to estimate the value of the integral

$$\int_{0.5}^{3} \frac{1}{x^2} \, dx$$

using the Simpson composite formula with a fixed step size h. We would like to have an error smaller than 0.001. We could use the discretization error term for the Simpson formula to determine the step size needed. We need

$$\frac{(b - a)}{180} |f^{iv}(\xi)| h^4 < 0.001$$

Since $f^{iv}(x) = 120/x^6$ in this case, then

$$|f^{iv}(x)| < \frac{120}{(0.5)^6} = 7680$$

We need $(2.5/180)7680h^4 < 0.001$. Solving this for the permissible values of h, we find that we need an $h < 0.05533$. Since $h = 2.5/N$, we would need a value of $N > 2.5/0.05533 = 45.183$. As a minimum, $N = 46$ is suggested. Actually, this value of N is larger than needed because the large values of the derivative $f^{iv}(x)$ near 0.5 in the interval [0.5, 3] lead to the large upper bound used in the estimate. The values of the fourth derivative are much smaller on the subinterval [1, 3]. If we compute this integral using the Simpson formula for values of N from 6 to 46, we obtain the values shown in Table 9.5. We see from these machine results that a value of $N \geq 20$ will suffice. This corresponds to a value of $h \leq 0.125$.

By removing the restriction to subintervals of fixed length, we can achieve the desired accuracy with fewer subintervals. Suppose that we form a partition of the interval [0.5, 3] using node points

$$\{0.6, 0.7, 0.9, 1.1, 1.7, 2.2\}$$

and then apply the Simpson formula with $N = 2$ to each subinterval. We will add the results to obtain the estimate of the integral over [0.5, 3], of course. A summary of the results is given in machine-generated Table 9.6. Columns 1 and 2 are the subinterval end points. In column 3, we have the Simpson estimate over that subinterval. As indicated, the total of these subestimates is 1.667297, and this is our "Simpson" variable-step estimate. Since the exact value is $\frac{5}{3}$, we see that the error in this estimate is about 6.306172×10^{-4}. This is less than 0.001, as required, but only seven subintervals were required. To achieve the desired accuracy, the fixed-step

Table 9.5

N	SIMPSON ESTIMATES	ERROR
6	1.712219	$-4.555226E-02$
8	1.686252	$-1.958501E-02$
10	1.676357	$-9.690046E-03$
12	1.671962	$-5.295515E-03$
14	1.669785	$-3.118038E-03$
16	1.668613	$-1.94633E-03$
18	1.667941	$-1.273871E-03$
20	1.667533	$-8.659363E-04$
22	1.667275	$-6.079674E-04$
24	1.667106	$-4.389286E-04$
26	1.666991	$-3.244877E-04$
28	1.666912	$-2.448559E-04$
30	1.666855	$-1.882315E-04$
32	1.666813	$-1.466274E-04$
34	1.666783	$-1.161099E-04$
36	1.66676	$-9.334088E-05$
38	1.666742	$-7.557869E-05$
40	1.666728	$-6.175041E-05$

Simpson formula requires at least 20 subintervals. Moreover, the fixed-step Simpson formula requires at least 22 subintervals to provide a *better* estimate than that obtained with the 7 subintervals of variable length.

After seeing an example such as this one, there should be no doubt that a variable step size has some advantages. A smaller number of subintervals means fewer function evaluations and faster, more efficient computation. However, we need an automatic procedure for selecting the subintervals and associated step sizes to be used. The following procedure is widely used and will be chosen in this discussion.

Before we begin, let us agree that in the following discussion, $S_1[x_1, x_2]$ will denote the result of applying the Simpson formula with $h = (x_2 - x_1)/2$ over the interval $[x_1, x_2]$.

Table 9.6

INTERVAL		ESTIMATE
0.5	0.6	0.3333488
0.6	0.7	0.2381008
0.7	0.9	0.3175128
0.9	1.1	0.2020339
1.1	1.7	0.3213283
1.7	2.2	0.1337143
2.2	3	0.1212585

INTEGRAL = 1.667297
ERROR IS 6.306172E $-$ 04

Select a value ε for the permissible error in the final estimate. We begin by applying the Simpson formula with $h = (b - a)/2$ to the interval $[a, b]$. Let us call this result $S_1[a, b]$. Let $c = (a + b)/2$. Now we apply the Simpson formula to the subintervals $[a, c]$ and $[c, b]$. We call these results $S_1[a, c]$ and $S_1[c, b]$. We compute

$$S_2[a, b] = S_1[a, c] + S_1[c, b]$$

In Section 9.5, we found that the error in estimating the integral over $[a, b]$ by the number $S_2[a, b]$ is approximately

$$\frac{S_2[a, b] - S_1[a, b]}{15}$$

If $(S_2[a, b] - S_1[a, b])/15$ is less than ε, we have the desired result and we can stop. If this number is larger than ε, we put the right half of $[a, b]$ aside and concentrate on estimating the integral over $[a, c]$. We apply the same procedure to the interval $[a, c]$ with the error bound decreased to $\varepsilon/2$.

We have already computed $S_1[a, c]$. We proceed to the computation of $S_1[a, c']$, $S_1[c', c]$, and $S_2[a, c]$, where $c' = (c + a)/2$. If the error estimate

$$\frac{S_2[a, c] - S_1[a, c]}{15}$$

is greater than ε, then we put interval $[c', c]$ aside and apply the outlined procedure again. We repeat this process of halving the interval and putting aside right half subintervals until our error bound is satisfied.

If

$$\frac{S_2[a, c] - S_1[a, c]}{15}$$

is less than ε, then we choose $S_2[a, c]$ as an acceptable estimate of the integral over that subinterval and add $S_2[a, c]$ to the sum, which will eventually become our estimate of the integral.

Note that if at any point our error requirement is met, then we add in that result to our total and then move back one step to the right to the last subinterval that we set aside. At this point, we already have computed the first Simpson estimate S_1 for that interval, so we proceed to compute the Simpson estimates over the two half intervals. We only need two new function evaluations to do that (values at midpoints of the half intervals). Each time that we move to the left to an interval of half the present length because the error is too large, we should store the function evaluations and other computed information, so that it can be retrieved when we move back to the right. We do not want to repeat function evaluations at the same points or any other computations already performed. Every machine algorithm contains some provision for efficiently storing and retrieving previous computations as they are needed. The following algorithm implements the described procedure.

Algorithm 9.1

Assumed input is A, B, and δ.

Step 1 Set $x_0 = A$, $x_2 = B$, $I = 0$, and $k = 0$.
Step 2 Compute $h = (x_2 - x_0)/2$, $x_1 = (x_0 + x_2)/2$, $y_0 = f(x_0)$, $y_1 = f(x_1)$, $y_2 = f(x_2)$, and $S_1 = h(y_0 + 4y_1 + y_2)/3$.
Step 3 Compute $x_3 = x_0 + h/2$, $x_4 = x_0 + 3h/2$, $y_3 = f(x_3)$, $y_4 = f(x_4)$, $S_{21} = h(y_0 + 4y_3 + y_1)$, $S_{22} = h(y_1 + 4y_4 + y_2)$, $S_2 = S_{21} + S_{22}$, and $E = (S_2 - S_1)$.
Step 4 If $E < 10\delta$, then go to step 9.
Step 5 Set $k = k + 1$. If $k \geq 10$, then output message "TOO MANY ITERATIONS" and go to step 9.
Step 6 (Store data) Set $A_k = x_1$, $B_k = x_2$, $f_k = y_1$, $g = y_2$, $h = y_4$, $e_k = \delta$, and $S_k = S_{22}$.
Step 7 Set $x_2 = x_1$, $\delta = \delta/2$, $h = (x_2 - x_0)/2$, $x_1 = (x_0 + x_2)/2$, $y_2 = y_1$, and $y_1 = y_3$.
Step 8 Return to step 3.
Step 9 Set $I = I + S_2$. If $x_2 \geq B$, then stop. Output I.
Step 10 (Retrieve data) Set $x_0 = A_k$, $x_2 = B_k$, $\delta = e_k$, $y_0 = f_k$, $y_2 = g_k$, $y_1 = h_k$, $S_1 = S_k$, and $k = k - 1$.
Step 11 Return to step 3.

Now let us consider some machine results from the implementation of Algorithm 9.1, the adaptive quadrature algorithm.

EXAMPLE 9.11   Compute the integral

$$\int_{0.5}^{3} \frac{1}{x^2}\, dx$$

using Algorithm 9.1.

Solution   The machine output is as follows.

```
EXACT VALUE 1.666666667
ENTER INTEGRATION LIMITS A, B 0.5, 3
ERROR TOLERANCE IS 0.000001
```

INTERVAL	S2	H	ERROR EST.
[0.5, 0.5195313]	0.0751880	0.0097656	4.967054E − 10
[0.5195313, 0.5390625]	0.0697396	0.0097656	4.967054E − 10
[0.5390625, 0.578125]	0.1253427	0.0195313	7.947286E − 09
0.578125, 0.6171875]	0.1094766	0.0195313	5.960465E − 09
[0.6171875, 0.65625]	0.0964436	0.0195313	4.470349E − 09
[0.65625, 0.6953125]	0.0856073	0.0195313	2.980232E − 09
[0.6953125, 0.734375]	0.0765001	0.0195313	2.483527E − 09
[0.734375, 0.8125]	0.1309329	0.0390625	3.576279E − 08
[0.8125, 0.890625]	0.1079622	0.0390625	2.086163E − 08
0.890625, 0.96875]	0.0905490	0.0390625	1.142422E − 08
[0.96875, 1.046875]	0.0770342	0.0390625	6.953876E − 09
[1.046875, 1.125]	0.0663350	0.0390625	4.470349E − 09
[1.125, 1.28125]	0.1084012	0.0781250	8.046627E − 08
[1.28125, 1.4375]	0.0848357	0.0781250	3.824631E − 08
[1.4375, 1.59375]	0.0682012	0.0781250	2.036492E − 08
[1.59375, 1.75]	0.0560224	0.0781250	1.092752E − 08
[1.75, 2.0625]	0.0865803	0.1562500	1.639128E − 07
[2.0625, 2.375]	0.0637959	0.1562500	6.556511E − 08
[2.375, 2.6875]	0.0489596	0.1562500	2.930562E − 08
2.6875, 3]	0.0387597	0.1562500	1.514951E − 08

```
INTEGRAL = 1.666667

ERROR 5.960465E − 07    ERROR EST. = 5.282462E − 07

FUNCTION EVALUATIONS = 81
```

In the first column, the subintervals that were used are displayed. In the next column, the corresponding values of the final Simpson estimates for those intervals are given. The H column is the step size used, and the final column on the right is the estimated error for each Simpson estimate listed. The value of the integral is estimated to be 1.666667 (rounded to six decimal places). This estimate is the sum of the numbers in the S2 column. The ERROR EST. given is the sum of the errors in the last column. This is only an estimate of the error, of course. In this case, we know that the exact value of the integral is $\frac{5}{3}$, which allows us to compute the true error, which is listed as ERROR on the output for comparison. The error estimate is quite close in this case. An error tolerance $\delta = 0.000001$ was selected at the input. The resulting error is less than δ, as required. Note that 81 function evaluations were required to obtain the results shown. The Simpson composite formula applied to this integral gave an estimate of 1.666668 for $N = 128$ (129 function evaluations) and gave 1.666667 for $N = 256$ (257 function evaluations). The improvement in speed and efficiency given by adaptive quadrature is clear in this case. ∎

EXAMPLE 9.12   Compute

$$\int_0^1 100xe^{-20x^2}\, dx$$

Solution   Applying Algorithm 9.1, the machine results are as follows.

```
EXACT VALUE  ?  2.500000
ENTER INTEGRATION LIMITS A, B  ?  0, 1
ERROR TOLERANCE IS  ?  0.00001
```

INTERVAL	S2	H	ERROR EST.
[0, 0.03125]	0.0483544	0.015625	2.384186E − 08
[0.03125, 0.0625]	0.1395237	0.015625	6.655852E − 08
[0.0625, 0.09375]	0.2151185	0.015625	8.940697E − 08
[0.09375, 0.125]	0.2679647	0.015625	9.139379E − 08
[0.125, 0.15625]	0.2948385	0.015625	7.748604E − 08
[0.15625, 0.1875]	0.296611	0.015625	4.172325E − 08
[0.1875, 0.25]	0.5213277	0.03125	1.231829E − 07
[0.25, 0.28125]	0.2023691	0.015625	3.476938E − 08
[0.28125, 0.3125]	0.1593174	0.015625	4.172325E − 08
[0.3125, 0.375]	0.2044377	0.03125	1.057983E − 06
[0.375, 0.4375]	9.576039E − 02	0.03125	3.705422E − 07
[0.4375, 0.5]	3.753111E − 02	0.03125	7.996957E − 08
[0.5, 0.5625]	0.0123816	0.03125	1.822288E − 07
[0.5625, 0.625]	3.451957E − 03	0.03125	1.240987E − 07
[0.625, 0.75]	9.802789E − 04	0.0625	1.08971E − 06
[0.75, 1]	3.381661E − 05	0.125	6.735742E − 07

```
INTEGRAL = 2.500002

ERROR 1.907349E − 06    ERROR EST. = 4.168192E − 06

FUNCTION EVALUATIONS = 65
```

The estimated value of the integral is 2.500002. The true value is 2.50000000 correct to eight decimal places. The true error is about 2×10^{-6}, and the estimated error is a little higher, but close. Note that in step 4 of the algorithm, we use the test E less than 10 times δ, rather than 15 times δ, as the preceding analysis suggested we should do. It is common practice to use the smaller bound to allow for round-off error also. This often leads to estimates even more accurate than required by the choice of δ. This choice slows down the computation some, so some experimentation with the multiplier might be advisable for a given integration problem. The machine makes choices at all times to keep the estimated error below δ. We are only working with an estimate, so the choice of 10 rather than 15 is a "safety factor." ∎

EXAMPLE 9.13   Compute

$$\int_1^3 \frac{100}{x^2} \sin \frac{10}{x}\, dx$$

with an error less than 0.00001.

Solution   Applying Algorithm 9.1, we obtain the results shown.

```
EXACT VALUE  ?  −1.42602491
ENTER INTEGRATION LIMITS A, B ? 1, 3
ERROR TOLERANCE IS ? 0.00001
```

INTERVAL	S2	H	ERROR EST.
[1, 1.015625]	−0.7345548	0.0078125	1.986822E − 08
[1.015625, 1.03125]	−0.506569	0.0078125	1.589457E − 08
[1.03125, 1.046875]	−0.2870386	0.0078125	4.172325E − 08
[1.046875, 1.0625]	−8.027433E − 02	0.0078125	7.450581E − 09
[1.0625, 1.078125]	0.1105716	0.0078125	1.88748E − 08
[1.078125, 1.09375]	0.2833541	0.0078125	0
[1.09375, 1.125]	1.007079	0.015625	3.417333E − 07
[1.125, 1.140625]	0.6838698	0.0078125	3.973643E − 09
[1.140625, 1.15625]	0.7779762	0.0078125	7.947286E − 09
[1.15625, 1.1875]	1.764666	0.015625	1.11262E − 07
[1.1875, 1.25]	3.916638	0.03125	0
[1.25, 1.28125]	1.945839	0.015625	4.768372E − 08
[1.28125, 1.3125]	1.83695	0.015625	7.152558E − 08
[1.3125, 1.34375]	1.67775	0.015625	6.357828E − 08
[1.34375, 1.375]	1.485189	0.015625	7.152558E − 08
[1.375, 1.40625]	1.273322	0.015625	4.768372E − 08
[1.40625, 1.4375]	1.053405	0.015625	3.973643E − 08
[1.4375, 1.5]	1.456222	0.03125	1.176198E − 06
[1.5, 1.53125]	0.421781	0.015625	3.774961E − 08
[1.53125, 1.5625]	0.2363911	0.015625	2.284845E − 08
[1.5625, 1.625]	−0.0153762	0.03125	4.120792E − 07
[1.625, 1.6875]	−0.547886	0.03125	2.22524E − 07
[1.6875, 1.75]	−0.9436449	0.03125	6.755193E − 08
[1.75, 1.875]	−2.606907	0.0625	6.039938E − 07
[1.875, 2]	−2.981412	0.0625	1.859665E − 06
[2, 2.0625]	−1.479861	0.03125	5.5631E − 08
[2.0625, 2.125]	−1.421827	0.03125	3.973643E − 08
[2.125, 2.25]	−2.582432	0.0625	1.255671E − 06
[2.25, 2.375]	−2.163094	0.0625	7.78834E − 07
[2.375, 2.5]	−1.725842	0.0625	5.165736E − 07
[2.5, 2.625]	−1.314611	0.0625	2.78155E − 07
[2.625, 2.75]	−0.9497275	0.0625	1.589457E − 07
[2.75, 3]	−1.015964	0.125	1.549721E − 06

```
INTEGRAL = −1.426016

ERROR 8.583069E − 06     ERROR EST. = 9.946339E − 06

FUNCTION EVALUATIONS = 133
```

This integral is easily evaluated by the fundamental theorem of calculus and found to have the value

$$10(\cos \tfrac{10}{3} - \cos 10) = -1.4260249078$$

correct to ten decimal places. The machine estimate is −1.426016, with an estimated error of about 9.95×10^{-6}, which is below the bound set. The true error is about

8.58×10^{-6}, as indicated. This is also below the required bound. The machine required 133 function evaluations to accomplish this. An inspection of the graph of this function shows it to be oscillatory over the interval owing to the sine factor. The behavior is quite complicated. Such functions are difficult to handle with the trapezoid or Simpson composite formulas. The results from the adaptive quadrature algorithm are quite satisfactory, however. ■

In this section, we have considered only one form of an adaptive quadrature method. The idea, in principle, could be applied to other Newton-Cotes formulas, particularly the trapezoid rule. The final results can also be improved by applying extrapolation to the estimates on each subinterval. This is often done. There are a variety of possible variations and extensions of the method presented here. You are encouraged to explore the methods presented by other authors in the current literature.

■ *Exercise Set 9.6*

1. We want to compute the integral $\int_0^4 \sqrt{x}\, dx$ by the trapezoid rule.
   a. Use the composite trapezoid rule with $N = 10$
   b. Use the trapezoid rule with $N = 1$ applied to each of the subintervals for the partition

   $$\{0, 0.02, 0.03, 0.06, 0.125, 0.25, 0.5, 1, 2, 4\}$$

   adding the results
   c. Find the exact value of the integral, and use that value to compute the error in each of the estimates in parts (a) and (b).

2. Verify the Simpson estimates for each subinterval given in Table 9.7. Add your results to obtain the estimate of the integral.

3. Estimate the error in the Simpson estimate of the integral over the interval $[0.5, 0.6]$ given in Table 9.6.

4. If you use Algorithm 9.1 for the integration of $f(x) = x^2$ over the interval $[0, 1]$ with $\delta = 0.001$, what is the first subinterval for which the error test in step 4 is satisfied?

5. Repeat exercise 4 for $f(x) = x^3$ on the interval $[0, 1]$.

6. Estimate the integral of $f(x) = x^3$ over the interval $[0, 1]$ using the adaptive quadrature method given in Algorithm 9.1. Record all computations carefully.

7. Write a machine program (your choice of language) to implement Algorithm 9.1. Use your program with $\delta = 0.001$ and 0.00001 to estimate the value of each of the following integrals.

   a. $\displaystyle\int_{1/2}^1 \frac{2}{x^2} \cos\left(\frac{10}{x}\right) dx$ b. $\displaystyle\int_0^1 50 x e^{-10x^2}\, dx$ c. $\displaystyle\int_0^4 x e^{-x^2}\, dx$

   d. $\displaystyle\int_0^3 \frac{\sin x}{1 + x^2}\, dx$ e. $\displaystyle\int_0^{0.5} e^{-x^2} \sin 20x\, dx$

8. Compute each of the integrals in exercise 7 using the Simpson rule. Determine the value of N needed in each case to obtain estimates with errors less than 0.00001. Compare results with the number of subintervals needed in exercise 7 to achieve the same accuracy.

9.7   The Romberg Algorithm

In this section, we will see that the Richardson extrapolation technique used in Section 9.3 to obtain better estimates of derivatives can also be applied to the problem of estimating integrals. Richardson extrapolation was applicable in the derivative problem because the error in estimation of $f'(a)$ by D_h had the form of a sum of even powers of h. We were able to exploit our knowledge of the algebraic form of the error to produce a sequence of improved estimates. Curiously enough, we have a similar situation for the trapezoid estimate of a definite integral.

We know that

$$\int_a^b f(x)\,dx = I = T_N + E_N$$

where

$$E_N = \frac{-(b-a)h^2}{12}\,f''(\xi)$$

Briefly,

$$I = T_N + Ch^2$$

where

$$C = \frac{-(b-a)}{12}\,f''(\xi)$$

Also,

$$I = T_{2N} + \frac{\hat{C}h^2}{4}$$

So

$$4I = 4T_{2N} + \hat{C}h^2$$

where $\hat{C} \cong C$ is assumed. Subtracting the first equation for I from the equation for $4I$ gives

$$3I \cong 4T_{2N} - T_N$$

and

$$I \cong \frac{4T_{2N} - T_N}{3} = T_{2N}^{(1)}$$

These algebraic manipulations are of course analogous to those used to derive the estimate $D_{h/2}^{(1)}$ for $f'(a)$ in Section 9.3. We also find in this case that $T_{2N}^{(1)}$ is a better estimate of the integral than T_{2N}.

For example, given estimates $T_{50} = 0.74679961$ and $T_{100} = 0.74681800$ for the integral $\int_0^1 e^{-x^2}\, dx$, we find that

$$T_{100}^{(1)} = \frac{4(0.74681800) - 0.74679961}{3} = 0.74682413$$

which is the correct value of the integral to eight decimal places—a tremendous improvement!

More generally, it can be shown that if f is sufficiently smooth on $[a, b]$, then the error in estimation of the integral $\int_a^b f(x)\, dx$ by trapezoid estimate T_N has the form

$$A_2 h^2 + A_4 h^4 + \cdots + A_{2k} h^{2k} + 0(h^{2k+2})$$

For a proof of this fact, you may consult Ralston (1965). It follows, just as in Section 9.3, that starting with any positive integer N, we can construct an array of integral estimates

$$
\begin{array}{lllll}
T_N & & & & \\
T_{2N} & T_{2N}^{(1)} & & & \\
T_{4N} & T_{4N}^{(1)} & T_{4N}^{(2)} & & \\
T_{8N} & T_{8N}^{(1)} & T_{8N}^{(2)} & T_{8N}^{(3)} & \cdots
\end{array}
$$

where the numbers in the first column are estimates of the integral using the trapezoid rule. The numbers in the second column are computed by the formula

$$T_{2k}^{(1)} = \frac{4T_{2k} - T_k}{3} \qquad k = N, 2N, 4N, \ldots$$

The numbers in the third column are computed by the formula

$$T_{4k}^{(2)} = \frac{16T_{4k}^{(1)} - T_{2k}^{(1)}}{15} \qquad k = N, 2N, 4N, \ldots$$

The numbers in the mth column are computed from the numbers in the $(m-1)$th column by the general rule

$$T_{2k}^{(m)} = \frac{4^m T_{2k}^{(m-1)} - T_k^{(m-1)}}{4^m - 1} \qquad k = 2^{m-1}N, 2^m N, \ldots \qquad (m \geq 1)$$

Actually, the array of estimates should be computed row by row from top down rather than column by column from left to right. As we proceed from left to right in a row or from the top downward in the array, the estimates should improve. All of this depends on the basic assumption of negligible round-off error in the trapezoid estimates down the leftmost column, however. Usually, only a few rows are needed before we reach the point where the estimates agree to several decimal places. If we continue to compute more rows in the array, eventually the accuracy of the estimates will deteriorate because of round-off error.

It is relatively easy to write a machine program to compute the elements of this **Romberg array**. We will now examine some sample output from such a program.

EXAMPLE 9.14   Estimate $\int_0^1 e^{-x^2}\, dx$.

Solution   Machine output is as follows.

INITIAL VALUE OF N?   2

TRAP. EST.   EXTRAPOLATED VALUES

0.73137
0.742984   0.746855
0.745865   0.746826   0.746824
0.746584   0.746824   0.746824   0.746824

The numbers in the left column are T_2, T_4, T_8, and T_{16}. Then

$$T_4^{(1)} = \frac{4(0.742984) - 0.73137}{3} = 0.746855$$

Also,

$$T_8^{(2)} = \frac{16(0.746826) - 0.746855}{15} = 0.746824$$

Finally,

$$T_{16}^{(3)} = \frac{64(0.746824) - 0.746824}{63} = 0.746824$$

The estimates quickly converge on 0.746824, which is the correct value of the integral to six decimal places. Note that this value is obtained in row 3 by extrapolation from the trapezoid estimate using only $N = 8$. Very little round-off error has been introduced in the trapezoid estimates because of the small values of N used.   ∎

EXAMPLE 9.15   Estimate $\int_0^1 \cos x\, dx$.

Solution   Machine output is as follows.

INITIAL VALUE OF N?   1

TRAP. EST.   EXTRAPOLATED VALUES

0.7701511
0.8238668   0.841772
0.8370838   0.841490   0.841471
0.8403751   0.841472   0.841471   0.841471

The values in the left column are T_1, T_2, T_4, and T_8. The value of the integral is $\sin 1 = 0.84147098$, so we see that the estimates have converged on the correct value to six decimal places. We first reach this value in row 3, extrapolating from T_4. Only $N = 4$ is needed. Round-off error is negligible at this point.   ∎

As Example 9.15 shows, the Romberg estimates can be initiated with extremely small values of N. That is important because the theory of the method requires that round-off error in the trapezoid estimates must be negligible.

EXAMPLE 9.16   Estimate $\int_0^1 e^{-x^2}\, dx$ again.

Solution   Machine output is as follows.

INITIAL VALUE OF N? 50

TRAP. EST.   EXTRAPOLATED VALUES

0.7467997			
0.746818	0.846824		
0.7468227	0.746824	0.746824	
0.7468245	0.746825	0.746825	0.746825

In this example, larger values of N are used to make the trapezoid estimates. We have computed T_{50}, T_{100}, T_{200}, and T_{400}. $T_{200}^{(1)}$ and $T_{200}^{(2)}$ have produced the correct value of the integral. If we proceed to the next row in the table, corresponding to using $N = 400$, we find that the extrapolated values do not improve and in fact are less accurate estimates of the integral. At this point, round-off error in the estimate T_{400} has had its effect.   ■

The lesson in Example 9.16 is to keep N small. Large values of N are not needed and in fact can cause less accurate estimates to be obtained.

In the implementation of the Romberg algorithm, we need to compute a sequence of trapezoid estimates T_N, T_{2N}, T_{4N}, and so on. It is worth noting that some computing time can be saved by using the following shortcut procedure in computing T_{2N}. We have

$$T_N = \frac{h}{2}\,(y_0 + 2y_1 + \cdots + 2y_{N-1} + y_N) \qquad h = \frac{b-a}{N}$$

$$T_{2N} = \frac{h}{4}\,(Y_0 + 2Y_1 + \cdots + 2Y_{2N-1} + Y_{2N})$$

where

$$y_k = f(a + kh) \qquad k = 0, 1, 2, \ldots, N$$

and

$$Y_k = f\!\left(a + \frac{kh}{2}\right) \qquad k = 0, 1, 2, \ldots, 2N$$

But

$$Y_{2k} = y_k \qquad k = 0, 1, 2, \ldots, N$$

and

$$Y_{2k+1} = f\!\left(x_k + \frac{h}{2}\right) = f(a + (k + 0.5)h)$$

Then

$$T_{2N} = \frac{h}{4}(Y_0 + 2Y_2 + 2Y_4 + \cdots + 2Y_{2N-2} + Y_{2N}) + \frac{h}{4}(2Y_1 + 2Y_3 + \cdots + 2Y_{2N-1})$$

$$= \tfrac{1}{2}T_N + \frac{h}{2}(Y_1 + Y_3 + \cdots + Y_{2N-1})$$

Briefly,

$$T_{2N} = \tfrac{1}{2}T_N + \frac{h}{2}\sum_{k=0}^{N-1} f(a + (k + 0.5)h)$$

where $h = (b - a)/N$. We have already computed T_N. To proceed to T_{2N}, we only need to compute the sum of the N values $Y_1, Y_3, \ldots, Y_{2N-1}$ and then compute T_{2N} from the formula.

EXAMPLE 9.17 Given $T_2 = 0.8238668$ for $f(x) = \cos x$ on interval $[0, 1]$, compute T_4.

Solution From the formula, we see that

$$T_4 = \frac{T_2}{2} + \frac{h}{2}(f(0.25) + f(0.75))$$

where $h = 0.5$. This is

$$T_4 = \frac{0.8238668}{2} + 0.25(0.9689124 + 0.7316889) = 0.8370837$$

You may wish to compare this calculation with the direct calculation of T_4 from the trapezoid formula to fully appreciate the difference. ■

In summary, we have the following algorithm for estimating the integral $\int_a^b f(x)\,dx$.

Algorithm 9.2

Step 1 Input a, b, δ, and positive even integer N.

Step 2 Set $k = 1$.

Step 3 Compute $T_{N/2}^{(0)} = T_{N/2}$, $T_N^{(0)} = T_N$, and $T_N^{(1)} = (4T_N^{(0)} - T_{N/2}^{(0)})/3$.

Step 4 Set $T = T_N^{(k)}$.

Step 5 Set $N = 2N$ and $k = k + 1$.

Step 6 Compute $T_N^{(0)} = T_N$.

Step 7 Compute $\quad T_N^{(m)} = \dfrac{4^m T_N^{(m-1)} - T_{N/2}^{(m-1)}}{4^m - 1} \qquad m = 1, \ldots, k$

Step 8 If $|T_N^{(k)} - T| < \delta$, then go to step 10.

Step 9 Go to step 4.

Step 10 Output $T_N^{(k)}$ and stop.

■ *Exercise Set 9.7*

1. Do the following by hand with calculator assistance.
   a. Compute T_1, T_2, T_4, and T_8 for $\int_1^2 (1/x)\, dx$
   b. Use the results of part (a) to compute extrapolated estimates $T_2^{(1)}$, $T_4^{(2)}$, and $T_8^{(3)}$.

2. Write a machine program to implement Algorithm 9.2, and use the machine to estimate each of the following integrals.

   a. $\displaystyle\int_0^1 \sin \pi x\, dx$ b. $\displaystyle\int_0^1 \sec x\, dx$ c. $\displaystyle\int_0^1 \frac{\sin x}{x}\, dx$

   d. $\displaystyle\int_{-2}^2 \frac{1}{1 + x^2}\, dx$ e. $\displaystyle\int_0^2 \frac{1}{1 + x}\, dx$

3. The following exercises illustrate a relation between Simpson estimates and Romberg estimates.
   a. Compute T_2, T_4, and $T_4^{(1)}$ for

   $$\int_0^1 \frac{1}{1 + x^2}\, dx$$

   Now compute the Simpson estimate S_4 and compare with $T_4^{(1)}$.
   b. Compute T_2, T_4, and $T_4^{(1)}$ for integral $\int_{-1}^1 e^x\, dx$. Now compute S_4 and compare with $T_4^{(1)}$.
   c. Show that theoretically, $S_N = T_N^{(1)}$.

4. If function f does not have enough continuous derivatives on $[a, b]$, then Romberg estimates of $\int_a^b f(x)\, dx$ may not be satisfactory because the theoretical development is no longer valid. To see this, use the Romberg algorithm to estimate each of the following integrals, and compare with the theoretical values.
   a. $\int_0^1 \sqrt{x}\, dx = \frac{2}{3}$ b. $\int_0^1 \sqrt{1 - x^2}\, dx = \pi/4 = 0.785398164$

5. Estimate the integrals in exercise 4 using the Simpson rule with $N = 40$ and $N = 80$. Are these results any more accurate?

6. Let $I = \int_a^b f(x)\, dx$. Then $I = S_N + E_N$ with

   $$E_N = -\frac{(b - a)^5 f^{iv}(\xi)}{180N^4}$$

   Extrapolate the error to show that $\hat{S}_{2N} = (16S_{2N} - S_N)/15$ should be a better estimate of I than S_N or S_{2N}.

7. Compute $\hat{S}_{80}$ as defined in exercise 6 for the integrals in exercise 4. Compare with values of $T_{80}^{(2)}$. What is your conclusion?

8. In Section 8.3, we found that if $P(x) = \sum_{k=0}^2 c_k T_k(x)$ is a Chebyshev least squares approximation of $f(x)$, then

   $$c_k = \frac{2}{\pi} \int_0^\pi f(\cos \theta) \cos k\theta\, d\theta$$

Use the Romberg algorithm to compute c_0, c_1, and c_2 for each of the following functions, and construct the quadratic least squares Chebyshev approximation.

a. $f(x) = e^x$ **b.** $f(x) = \ln(x + 2)$
c. $f(x) = \cosh x$ **d.** $f(x) = e^{-x^2/2}$

9.8 Gaussian Integration Formulas

In the preceding discussion of numerical integration methods, we assumed that numerical values $y_k = f(x_k)$ of a function f are given at $n + 1$ specified points $x_0, x_1, \ldots, x_n$. In most cases, we assumed that the points x_k were equally spaced with spacing h. We then proceeded to construct the interpolation polynomial for f on the given set of points and by integration of that polynomial determined an integration formula. All of our formulas have the general form

$$\int_a^b f(x)\, dx = \sum_{k=0}^n A_k f(x_k) + \text{(error term)}$$

where the error is zero if f is a polynomial of degree n or less. Coefficients A_k are determined by integration of the interpolation polynomial.

The basic idea of Gaussian integration methods is to try to produce more accurate integration estimates by proper selection of points x_k when the function f and interval $[a, b]$ are given. By a change of variables, it is always possible to transform any integral over $[a, b]$ to an integral over $[-1, 1]$, so in the following, we will only consider integrals over $[-1, 1]$. In general, if $x_0, x_1, \ldots, x_n$ is any set of $n + 1$ distinct points in $[-1, 1]$ and $P_n(x)$ interpolates f on this set of points, then

$$f(x) = P_n(x) + g(x)(x - x_0)(x - x_1) \ldots (x - x_n)$$

where

$$g(x) = \frac{f^{(n+1)}(\xi)}{(n + 1)!}$$

So

$$\int_{-1}^1 f(x)\, dx = \int_{-1}^1 P_n(x)\, dx + \int_{-1}^1 g(x)(x - x_0) \ldots (x - x_n)\, dx$$

Now suppose that f is a polynomial of degree $2n + 1$ or less. Then it follows that the product

$$g(x)(x - x_0)(x - x_1) \ldots (x - x_n)$$

is a polynomial of degree $2n + 1$ or less. We conclude that $g(x)$ is a polynomial of degree n or less in this case. Now the error in estimation of $\int_{-1}^1 f(x)\, dx$ by $\int_{-1}^1 P_n(x)\, dx$ is

$$\int_{-1}^1 g(x)(x - x_0)(x - x_1) \ldots (x - x_n)\, dx$$

This error will be zero if polynomial $g(x)$ is orthogonal to polynomial $(x - x_0)$ $(x - x_1)\ldots(x - x_n)$ on $[-1, 1]$ as defined in Section 8.3. The weight function in this case is $w(x) = 1$ on $[-1, 1]$. Can we select points $x_0, x_1, \ldots, x_n$ so that the polynomial $(x - x_0)(x - x_1)\ldots(x - x_n)$ of degree $n + 1$ is orthogonal to every polynomial $g(x)$ of degree n or less on $[-1, 1]$? We know from Section 8.3 that the Legendre polynomial of degree $n + 1$ has precisely that desired property. Hence, we need to choose points x_k to be the zeros of the Legendre polynomial of degree $n + 1$.

Then using the Lagrange form of the interpolating polynomial, we find that

$$\int_{-1}^{1} f(x)\, dx = \sum_{k=0}^{n} \int_{-1}^{1} L_k(x)f(x_k)\, dx = \sum_{k=0}^{n} A_k\, f(x_k)$$

where

$$A_k = \int_{-1}^{1} L_k(x)\, dx$$

The error in the estimate of the integral is zero if function f is a polynomial of degree $2n + 1$ or less.

Case 1   Two-Point Formula

The Legendre polynomial of degree 2 is

$$\tfrac{3}{2}(x^2 - \tfrac{1}{3})$$

with zeros $\pm 1/\sqrt{3} = \pm 0.5773502692$. So we choose $x_0 = -1/\sqrt{3}$ and $x_1 = 1/\sqrt{3}$. Then

$$A_0 = \int_{-1}^{1} \frac{x - x_1}{x_0 - x_1}\, dx = 1$$

and

$$A_1 = \int_{-1}^{1} \frac{x - x_0}{x_1 - x_0}\, dx = 1$$

Our integration formula is

$$\int_{-1}^{1} f(x)\, dx \cong f\left(-\frac{1}{\sqrt{3}}\right) + f\left(\frac{1}{\sqrt{3}}\right)$$

EXAMPLE 9.18   Estimate $\int_{-1}^{1} e^{-x^2}\, dx$.

Solution   We have

$$\int_{-1}^{1} e^{-x^2}\, dx \cong \exp\left(-\tfrac{1}{3}\right) + \exp\left(-\tfrac{1}{3}\right) = 1.4330626$$

The true value, correct to six decimals, is 1.493648, so the Gaussian estimate using two points is low and does not seem very close. In comparison, however, the trapezoid rule with $N = 1$ (actually two points) gives 0.735758 and with $N = 2$ (three points) gives 1.367879. Viewed in this way, the Gaussian estimate is surprisingly accurate. ∎

Case 2 Three-Point Formula
The Legendre polynomial of degree 3 is

$$\tfrac{1}{2}(5x^3 - 3x)$$

with zeros $0, \pm\sqrt{3/5} = \pm 0.7745966692$. We use $x_0 = -\sqrt{3/5}$, $x_1 = 0$, and $x_2 = \sqrt{3/5}$. The coefficients are

$$A_0 = \int_{-1}^{1} \frac{(x - x_1)(x - x_2)}{(x_0 - x_1)(x_0 - x_2)}\, dx = \tfrac{5}{9}$$

$$A_1 = \int_{-1}^{1} \frac{(x - x_0)(x - x_2)}{(x_1 - x_0)(x_1 - x_2)}\, dx = \tfrac{8}{9}$$

$$A_2 = \int_{-1}^{1} \frac{(x - x_0)(x - x_1)}{(x_2 - x_0)(x_2 - x_1)}\, dx = \tfrac{5}{9}$$

So we have the Gaussian three-point integration formula

$$\int_{-1}^{1} f(x)\, dx \cong \tfrac{5}{9} f(-\sqrt{\tfrac{3}{5}}) + \tfrac{8}{9} f(0) + \tfrac{5}{9} f(\sqrt{\tfrac{3}{5}})$$

Similarly, higher-order formulas can be produced by using zeros of higher-degree Legendre polynomials. In the case of the five-point formula, we find, for example, that $x_0 = -0.90617985$, $x_1 = -0.53846931$, $x_2 = 0$, $x_3 = -x_1$, and $x_4 = -x_0$, while $A_0 = A_4 = 0.23692689$, $A_1 = A_3 = 0.47862867$, and $A_2 = 0.56888889$.

Values of points x_k and weights A_k for various other numbers of points can be found in most mathematical handbooks. An excellent source is Stroud and Secrist (1966).

If we wish to integrate over an interval other than $[-1, 1]$, we must first make the change of variables

$$x = \frac{(b - a)t + (a + b)}{2}$$

Then

$$f(x) = f\left(\frac{(b - a)t + (b + a)}{2}\right) = F(t)$$

so

$$\int_a^b f(x)\,dx = \frac{b-a}{2}\int_{-1}^1 F(t)\,dt$$

We apply one of the Gaussian integration formulas to estimate $\int_{-1}^1 F(t)\,dt$, of course.

EXAMPLE 9.19   Estimate $\int_0^1 e^{-x^2}\,dx$.

Solution   We use

$$x = \frac{t+1}{2}$$

so

$$F(t) = \exp\left(-\frac{(t+1)^2}{4}\right)$$

Then

$$\int_0^1 e^{-x^2}\,dx = \frac{1}{2}\int_{-1}^1 F(t)\,dt$$

Using the five-point Gauss formula, we find that

$$\sum_{k=0}^4 A_k F(t_k) = 1.49364826$$

Then

$$\int_0^1 e^{-x^2}\,dx \cong 1.49364826/2 = 0.74682413 \qquad \blacksquare$$

Error analysis for Gaussian integration is quite difficult and will not be attempted here. There are also other Gaussian integration formulas based on using zeros of orthogonal polynomials other than the Legendre polynomials. If you are interested in pursuing error analysis or alternate Gaussian integration formulas, you may want to consult Chapter 8 of Hildebrand (1956) or Chapter 5 of Conte and DeBoor (1972).

In this chapter, you have been introduced to a number of algorithms for approximation of a definite integral. All of these algorithms are easy to convert to machine programs using BASIC, FORTRAN, or Pascal. For FORTRAN users, it is important to know that the IMSL library contains several FORTRAN subroutines for computing integrals. These routines are accurate and efficient in their operation. Some routines are included for handling the integration of functions that may have singularities. There are also routines for the approximation of infinite integrals.

More sophisticated algorithms and programming techniques are needed for such problems. The routines QDNG or DQDNG (double-precision version) can be used in the case of a smooth function on a finite interval. The routine QDAGI can be used to integrate a function over an infinite interval. If the function to be integrated has singularities, the routine QDAGP should be used. The subroutine GQRUL can be used if Gaussian quadrature is preferred. There are several other routines for other integration problems. Some experimentation with these IMSL routines is recommended.

■ *Exercise Set 9.8*

1. Carry out the integrations $\int_{-1}^{1} L_0(x)\, dx$ and $\int_{-1}^{1} L_1(x)\, dx$ to show that $A_0 = A_1 = 1$ in the Gauss two-point formula.

2. Verify the details of the computation of A_0, A_1, and A_2 for the three-point Gauss formula discussed in case 2 of this section.

3. Verify all details of the arithmetic in Example 9.19.

4. Compute $\int_{-1}^{1} e^{-x^2}\, dx$ using the four-point Gauss formula. Compare results with the results of Example 9.19.

5. Estimate each of the following integrals using the three-point Gauss formula.
   a. $\int_{-1}^{1} e^{-x}\, dx$ b. $\int_{-1}^{1} \sec x\, dx$

   c. $\displaystyle\int_{-1}^{1} \frac{1}{1+x^2}\, dx$ d. $\int_{-1}^{1} (1+x^2)^{3/2}\, dx$

6. Repeat exercise 5 using the six-point formula.

7. Repeat exercise 5 using the eight-point formula.

8. Estimate each of the following integrals by first making a change of variables and then applying the three-point Gauss formula.
   a. $\int_0^1 \sqrt{x}\, dx$ b. $\int_0^1 x \ln x\, dx$ c. $\int_0^\pi \sqrt{\sin x}\, dx$
   d. $\int_0^1 \sqrt{1+4x^2}\, dx$ e. $\int_{0.5}^1 \sin \sqrt{x}\, dx$

9. Repeat exercise 8 using the four-point Gauss formula.

10. Estimate $\int_0^1 e^{-x^2}\, dx$ by the following procedure.
    a. Compute $\int_0^{1/2} e^{-x^2}\, dx$ by using a change of variables followed by an application of the Gauss two-point formula.
    b. Compute $\int_{1/2}^1 e^{-x^2}\, dx$ by using a change of variables followed by an application of the Gauss two-point formula.
    c. Add the results in parts (a) and (b). Compare the results with the estimate obtained in Example 9.19.

11. Apply the procedure of exercise 10 to the integral $\int_a^b f(x)\, dx$ and show that

$$\int_a^b f(x)\, dx \cong \frac{h}{2} F\left(\frac{-1}{\sqrt{3}}\right) + F\left(\frac{1}{\sqrt{3}}\right) + G\left(\frac{-1}{\sqrt{3}}\right) + G\left(\frac{1}{\sqrt{3}}\right)$$

where $h = (b - a)/2$ and

$$F(t) = f\left(\frac{h}{2} t + \left(a + \frac{h}{2}\right)\right)$$

$$G(t) = f\left(\frac{h}{2} t + \left(a + \frac{3h}{4}\right)\right)$$

12. Use the result in exercise 10 to estimate $\int_1^2 (1/x)\, dx$. Compare with the true value 0.693147181.

9.9 Applied Problems

Applications of the numerical methods discussed in this chapter are widespread throughout other areas of mathematics and the sciences. Derivatives occur wherever variables are changing, as in population growth, chemical reactions, and motion problems. In a large number of practical problems, we have functional relationships for which analytic formulas and equations for differentiation are unknown or unavailable. If numerical values of the variables are available, then estimation of derivatives is at least possible. It is equally true that definite integrals often appear that cannot be evaluated by standard textbook methods, so we turn to numerical estimates. This is a very familiar situation for people who work in the physical sciences or engineering. We will begin this section with some uses for numerical estimates of derivatives and then consider several problems leading to integrals that must be evaluated by numerical methods.

Suppose that $N(t)$ is the size of a population at time t. If the population is growing exponentially, then

$$N(t) = N(0)e^{rt}$$

where $N(0)$ is the initial size of the population and r is the *growth rate*. Since $N'(t) = rN(t)$ in this case, the growth rate can be computed from the formula

$$r = \frac{1}{N} \frac{dN}{dt}$$

Furthermore, it is clear that if $t = (\ln 2)/r$, then $N(t) = 2N(0)$. So $T = (\ln 2)/r$ is called the *doubling time* for the population. In the case of human populations, demographers frequently treat the population growth as exponential over short periods of time (a few decades perhaps). In such studies, the number r computed from the preceding formula is called the growth rate for the population, and the corresponding T is called the doubling time. Suppose now that we are given the values in Table 9.7 for the population of the United States from 1900 to 1970. We wish to estimate growth rates for the years 1920 and 1940.

Using the symmetric difference formula, we can estimate $N'(1920)$ to be

$$\frac{131669275 - 75994575}{40} = 1391867.5$$

Table 9.7

Year	Population
1900	75,994,575
1920	105,710,620
1940	131,669,275
1960	179,323,175
1970	203,211,926

The growth rate is

$$r = \frac{1391867.5}{105710620} = 0.0132$$

In popular terms, this corresponds to a 1.32% rate of growth.

Similarly, for 1940 we find that

$$N'(1940) = \frac{179323175 - 105710620}{40} = 1840314$$

and in 1940,

$$r = \frac{1840314}{131669275} = 0.0140$$

If these estimates are reliable, then the rate of growth had increased to about 1.4% by 1940.

As our next application, let us consider the problem of estimating velocity and acceleration of a rocket in the first few seconds after lift-off while the rocket is still in the vertical ascent stage. Table 9.8 gives some typical values for a Saturn V lift-off. We want to use this data to estimate the velocity 1 second after lift-off. The velocity and acceleration are also desired at the 3-second mark.

Table 9.8

t (sec.)	Altitude (ft)	Velocity (ft/sec)
1	6.801	6.801
2	20.575	13.773
3	41.504	20.929
4	69.771	28.267
5	105.560	35.789

Using our three-point formula for estimating derivatives, we see that the velocity at time $t = 1$ is approximately

$$\frac{4(20.575) - 3(6.801) - 41.504}{2} = 10.20 \text{ ft/sec}$$

The measured value of the velocity was 6.801 ft/sec, so our numerical estimate is too high. There is a physical reason for this, as we shall see.

At 3 seconds, the velocity can best be estimated by the symmetric difference formula, using altitudes at $t = 2$ and $t = 4$ as data. The velocity is

$$\frac{69.771 - 20.575}{2} = 24.60 \text{ ft/sec}$$

The measured value at this time was only 20.93 ft/sec. Again, our estimated velocity is too high. Note that in the case of both numerical differentiation formulas, the error depends on the third derivative of the altitude as a function of time. To obtain accurate estimates, the third derivative must be small. If the third derivative were zero, we would obtain the desired velocities exactly. Physically, the third derivative of the altitude is the rate of change of acceleration. If the acceleration were constant, the derivative estimation formulas employed would give us exact results. However, in the time period selected, the thrust is high and the rate of increase of acceleration is large because the mass of the rocket is decreasing. In general, when estimating velocities in motion problems, we will obtain accurate estimates from position data only if the rate of change of acceleration is small.

Using Equation 4 from Section 9.1, we find that the acceleration at time $t = 3$ is approximately

$$69.771 + 20.575 - 2(41.504) = 7.34 \text{ ft/sec}^2$$

Similar calculations show that the acceleration at $t = 2$ is 7.15 ft/sec², and at $t = 4$ it is 7.52 ft/sec². We have a steady increase in the acceleration, which is required for proper lift-off of a rocket such as the Saturn V.

Now let us turn to the estimation of integrals.

All students of probability are familiar with standard normal probability tables. These tables are used to find probabilities for so-called standard normal random variables. Their construction is based on the fact that if X is a standard normal random variable, then the probability that X assumes a value between 0 and a is

$$\frac{1}{\sqrt{2\pi}} \int_0^a e^{-x^2/2} \, dx$$

So to compute such probabilities, we must compute the value of a definite integral. If we want the probability that X assumes a value between 0 and 1, we need to compute

$$\frac{1}{\sqrt{2\pi}} \int_0^1 e^{-x^2/2} \, dx$$

We have a number of ways to do this. We will use the Simpson rule with $N = 20$. The computation should be performed by a machine program, of course. Using such a program, we find that the value of the integral is 0.8556244. To check the accuracy, we make a new estimate using $N = 40$. The estimate is again 0.8556244. From the discussion in Section 9.6, we know that the error in the first estimate is proportional to the difference of the two estimates. We can safely conclude that the integral is 0.855624 to six decimal places. The desired probability is achieved by dividing this result by the square root of 2π. The result is 0.341345 correct to six decimal places.

In general, if X is a standard normal random variable, the probability that X assumes a value between a and b is given by

$$\frac{1}{\sqrt{2\pi}} \int_a^b e^{-x^2/2} \, dx$$

All such probabilities are easily estimated using the Simpson or trapezoid rule or even the Romberg algorithm if desired. Error estimates can also be made.

Numerous problems in physics lead to definite integrals that cannot be evaluated analytically by textbook formulas. An interesting but not too well known example occurs in the theory of blackbody radiation. Theoretically, a blackbody radiates electromagnetic energy at all frequencies from zero to infinity. Some of this is in the form of visible light, some is infrared, some is in the form of microwaves, and so forth. In Arfken (1966), the author states that the fraction of blackbody radiation in the range of frequencies from 0 to v_0 is given by

$$\frac{15}{\pi^4} \int_0^{x_0} \frac{x^3 \, dx}{e^x - 1}$$

where $x_0 = hv_0/kT$. Number h is Planck's constant, k is Boltzmann's constant, and T is the temperature of the blackbody in degrees Kelvin. From standard tables, we find that

$$h = 6.625 \times 10^{-27} \text{ erg-sec}$$

and

$$k = 1.38 \times 10^{-16} \text{ erg/deg}$$

Astrophysicists treat stars as blackbody radiators, so the preceding statement by Arfken applies to radiation from our sun. The surface of the sun is believed to have a temperature of about $6000°$ K, so we use $T = 6000$. We wish to compute the fraction of the radiant energy emitted from the sun's surface with frequencies between 0 and 5×10^{14} Hz. Then

$$x_0 = \frac{(6.625 \times 10^{-27})(5 \times 10^{14})}{(1.38 \times 10^{-16})(6000)} = 4.0006 \cong 4.0$$

We must compute the integral

$$\int_0^4 \frac{x^3 \, dx}{e^x - 1}$$

Using the Romberg algorithm beginning with $N = 20$, we obtain the array

3.875977
3.876786   3.877056
3.876987   3.877054   3.977053
3.877037   3.877054   3.877054   3.877054

The close agreement of the extrapolated values in the last row indicates that the value is 3.877054 correct to at least five places and possibly six. The fraction of the energy in the specified frequency range is

$$\frac{15}{\pi^4} (3.87705) = (0.1539897)(3.87705) = 0.5970264$$

So about 59.70% of the energy is radiated at frequencies lower than 5×10^{14}. This is energy in the infrared and radio range (nonvisible). So most of the energy radiated by our sun is not in the visible part of the spectrum.

A similar calculation for a blackbody with temperature $T = 1000$ (hot iron bar?) and the same frequency range indicates that practically 99.99% of the radiation is in the specified range. There would be virtually no visible radiation in this case (possibly some red light).

We will conclude this section with an application to a calculus problem that must be left unsolved in a beginning calculus course. Suppose we wish to compute the circumference of an ellipse with semimajor axis a and eccentricity e. In that case, the semiminor axis is $b = a\sqrt{1 - e^2}$, and we may represent the ellipse with parametric equations

$$x = a \cos t \quad \text{and} \quad y = \sin t$$

The circumference is

$$\int_0^{2\pi} \sqrt{\left(\frac{dx}{dt}\right)^2 + \left(\frac{dy}{dt}\right)^2} \, dt = 4a \int_0^{\pi/2} (1 - e^2 \cos^2 t)^{1/2} \, dt$$

The integral cannot be evaluated by the fundamental theorem, so we turn to numerical methods. We will evaluate the integral using the Simpson rule for various values of the eccentricity e.

For $e = 0.1$, we find with $N = 30$, that

$$\int_0^{\pi/2} (1 - 0.01 \cos^2 t)^{1/2} \, dt = 1.566862$$

Repeating with $N = 60$ gives the same result, so apparently, the true value correct to five decimals is 1.566862. The circumference of the ellipse in this case is

$$4a(1.566862) = 6.26745a$$

With eccentricity $e = 0.1$, the ellipse is not far from circular, so the circumference should be slightly less than $6.28a$. Our result is reasonable.

Selecting $e = 0.95$ (as in the case of an elongated comet orbit, such as the orbit of Halley's comet), we find that the integral to be evaluated is

$$\int_0^{\pi/2} (1 - 0.9025 \cos^2 t)^{1/2} \, dt$$

Using $N = 20$ with the Simpson rule, we obtain 1.102722, while with $N = 40$, we obtain exactly the same result. The value of the integral is then 1.102722, and the circumference of the ellipse is

$$4a(1.102722) = 4.41089a$$

In this case, the circumference is much less than the circumference of a circle of radius a.

■ *Exercise Set 9.9*

1. Given the population figures shown for a certain population, estimate growth rates and doubling times for the years 1960 and 1970.

Year	Population
1950	11,231,671
1960	12,502,731
1970	13,341,842
1980	14,578,925

2. Using the population figures given in exercise 1, estimate the growth rate in 1950 using the three-point formula (Equation D3) from Section 9.1. Now assuming exponential growth, estimate the population figures for the years 1960, 1970, and 1980. Compare with the actual population figures. Is population projection by use of the exponential formula always reliable?

3. Using the rocket launch data in Table 9.8, verify that the accelerations at times $t = 2$ and $t = 4$ are 7.15 ft/sec^2 and 7.52 ft/sec^2, respectively, as claimed.

4. Compute accelerations at times $t = 2$ and $t = 4$ for the rocket launch data of Table 9.8, using Equation D2 (from Section 9.1) with the velocity values given. Compare with the values in exercise 3.

5. The heat loss through a glass windowpane is estimated by the formula

$$Q = kA\left(\frac{T_1 - T_0}{L}\right)$$

   where k is the coefficient of heat conductivity for the type of glass used, A is the area of the pane, L is the thickness of the glass, T_1 is the indoor temperature, and T_0 is the outdoor temperature.

   a. Show that factor $(T_1 - T_0)/L$ is an estimate of a derivative of a certain function. Identify the function and the numerical differentiation formula being used.

b. Suppose that $k = 0.002$ cal/sec $\cdot$ cm $\cdot$ deg, $L = 0.5$ cm, and the window size is 150 cm $\times$ 90 cm. Given that 1 Btu $= 252$ cal, find the heat loss in Btu/hr for this window-pane.

6. Use numerical integration to estimate the probability that a standard normal random variable assumes a value between 0.5 and 0.6. Make an error estimate.

7. Given that a certain star has a surface temperature of $20000°$ K, use numerical integration to estimate the percent of the total radiation with frequencies from 10^{18} to 10^{21} Hz.

8. Estimate the circumference of the ellipse with the given semimajor axis and eccentricity.
    a. $a = 5, e = 0.02$
    b. $a = 1.5, e = 0.6$
    c. $a = 3, e = 0.9$

9. A student read in her astronomy text that Halley's comet travels in an elliptical orbit with the sun at one focal point. The orbit has eccentricity $e = 0.97$, and the minimum distance from the sun is 0.59 AU (astronomical units). She knows that 1 AU is about 92,900,000 miles. Use this information to help her find the comet's maximum distance from the sun and estimate the total distance (in miles) traveled by the comet during the 77-year circuit of its orbit.

10. Values of the Bessel function $J_0(x)$ can be computed from the integral form

$$J_0(x) = \frac{1}{\pi} \int_0^\pi \cos (x \sin \theta) \, d\theta$$

Use numerical integration to construct a table of values of $J_0(x)$ for $x = 0, 0.1, 0.2, 0.3, 0.4,$ and 0.5. Choose your method so as to yield values correct to at least five decimal places.

11. Using $t = \sin \theta$, we find that

$$J_0(x) = \frac{2}{\pi} \int_0^1 \frac{\cos xt}{\sqrt{1 - t^2}} \, dt$$

Use this formula with numerical integration to construct a table of values for $x = 0, 0.1, 0.2, 0.3, 0.4,$ and 0.5. Compare results with the values found in exercise 10. What is wrong?

12. The growth rate of a population is a function $r(t)$ of time for real populations. If $x_0 = x(0)$ is the population at some initial time $t = 0$, then the population at time $t > 0$ is

$$x(t) = x_0 \exp\left(\int_0^t r(u) \, du \right)$$

Use this fact, along with the table of values for the growth rate of the population, to estimate the population in 1980, given that the 1970 population was 210,000,000.

Year	Growth Rate
1970	0.0087
1972	0.0085
1974	0.0089
1976	0.0091
1978	0.0095
1980	0.0110

CHAPTER ■

10 NUMERICAL SOLUTION OF ORDINARY DIFFERENTIAL EQUATIONS

In this chapter, we will study the problem of finding a function $y(x)$ on interval (a, b) such that

$$y'(x) = f(x, y(x)) \quad \text{and} \quad y(x_0) = y_0$$

The equation $y'(x) = f(x, y(x))$ is called a first-order differential equation, and $y(x_0) = y_0$ is an initial condition. The combination is called an **initial-value problem**. Most numerical methods for differential equations are designed to approximate the solution of such initial-value problems. There are many well-known numerical methods for doing this. In this chapter, we will study a few of the best-known and most often used methods. We will also consider some error analysis and some of the difficulties encountered in the attempt to construct an accurate solution function $y(x)$.

10.1   Basic Principles and Theory

The solution of an initial-value problem is a function. Functions can be represented by formulas, graphs, numerical tables, or even integrals of other functions. In a conventional introductory course in differential equations, the emphasis is placed on finding solutions in some analytical form. This usually means a closed formula, expressing $y(x)$ as a combination of so-called elementary functions, or a power series expansion of $y(x)$ about some convenient point in the domain of the solution. When that can be accomplished, and when the form selected is usable to study and evaluate the function, then such an analytical solution is best. However, there are numerous differential equations for which the solution cannot be expressed in any such simple, usable manner. The following differential equations are examples:

$$y' = x^2 + y^2$$
$$y' = xy + \sin y$$
$$y' = \sqrt{x} + \sqrt{y}$$

401

In many such cases, solutions can be expressed in the form of a power series that partially represents the solution, but all too often in applications, these series solutions do not provide a usable form for $y(x)$ because of slow convergence or other problems.

Fortunately, in most cases, whether the solution of the differential equation can be expressed in a usable analytical form or not, we can find a representation of the solution of the initial-value problem as a table of numerical values on an interval containing the initial point x_0.

Apart from the question of what form we can use to represent the solution, we must consider the question of whether a unique solution to the initial-value problem exists. Let us consider some examples.

EXAMPLE 10.1   Solve the initial-value problem

$$y' = \sqrt{1 - y^2} \qquad y(0) = 2$$

Solution   The differential equation is equivalent to

$$\frac{dy}{\sqrt{1 - y^2}} = dx$$

Integration gives

$$\arcsin(y) = x + c$$

so

$$y = \sin(x + c)$$

Then $y(0) = 2$ implies

$$y(0) = \sin c = 2$$

Clearly, no c exists such that $\sin c = 2$. Hence, this initial-value problem has no solution. ∎

EXAMPLE 10.2   Solve the initial-value problem

$$y' = 3y^{2/3} \qquad y(0) = 0$$

Solution   Rearranging the differential equation gives

$$\tfrac{1}{3}y^{-2/3} \, dy = dx$$

so

$$y^{1/3} = x + c$$

Then

$$y = (x + c)^3$$

Now $y(0) = 0$ implies $0 = c$, so $y = x^3$ is the solution. It is true that $y = x^3$ is a solution of the problem. But $y = 0$ is also a solution. Hence, the solution to this problem is not unique. ∎

EXAMPLE 10.3 Solve the initial-value problem

$$y' = -y^2 \qquad y(0) = 0$$

Solution The differential equation can be rearranged to

$$y^{-2}y' = -1$$

Integration gives

$$-y^{-1} = x + \hat{c}$$

so

$$y^{-1} = c - x$$

or

$$y = \frac{1}{c - x}$$

Now $y(0) = 0$ implies $0 = 1/c$. No such number c exists, so no solution is determined. Actually, $y = 0$ is easily seen to be a solution. The analytical approach does not produce the solution in this case, even though a solution does exist. ∎

A number of uniqueness and existence theorems are found in the literature. The following theorem concerning the initial-value problem

$$y' = f(x, y) \qquad y(x_0) = y_0$$

will suffice for our present purposes.

Theorem 10.1 If $f(x, y)$ and first partial derivative $\partial f/\partial y$ are continuous on an open rectangular region R containing initial point (x_0, y_0), then a unique solution $y(x)$ exists for the initial-value problem. Furthermore the function $y(x)$ is defined on some interval centered on point x_0.

This theorem with a proof is found in most introductory differential equations texts. There are numerous generalizations and extensions, some of which you may wish to explore by consulting such texts. All the numerical methods discussed in this chapter assume that the initial-value problem in question has a unique solution, so the question of existence and uniqueness of solutions is important in numerical analysis. The user of numerical methods should be aware of the fact that some initial-value problems do not have a unique solution, so numerical methods applied without consideration of that question could produce puzzling results. As Theorem 10.1 shows, however, most initial-value problems do have unique solutions. There is some danger here, but a reasonable amount of caution is sufficient to guard against it.

Note that $y(x)$ is a solution of the initial-value problem

$$y' = f(x, y) \qquad y(x_0) = y_0$$

if and only if $y(x)$ satisfies the integral equation

$$y = y_0 + \int_{x_0}^{x} f(t, y(t))\, dt$$

This last equation also can be considered as an integral formula for the solution of the initial-value problem. This relation between solutions of initial-value problems and integral equations is important and quite useful. Solution function $y(x)$ can be approximated if we can approximate the integral in the integral equation. The effect on the solution $y(x)$ produced by small changes in initial value y_0 or the function f can be analyzed using the integral formula. The integral formula is often used in constructing uniqueness and existence proofs for the initial-value problem.

■ Exercise Set 10.1

1. Apply Theorem 10.1 to each of the following initial-value problems to show that a unique solution exists on some interval containing the initial value of x.

a. $y' = 1 + y^2; \quad y(0) = 0$ **b.** $y' = \sin xy; \quad y(1) = 1$

c. $y' = \sqrt{x + y}; \quad y(1) = 0$ **d.** $y' = 2y - x + 1; \quad y(0) = 2$

2. Given the initial-value problem

$$y' + 2xy = x \qquad y(0) = 1$$

find the solution as follows.
a. Compute

$$g(x) = \int_{0}^{x} s e^{s^2}\, ds$$

b. Show that $y = e^{-x^2}(g(x) + C)$ is a solution of the differential equation for every choice of constant C.

c. Find the value of C for which $y(0) = 1$, and write down a solution for the initial-value problem.

d. Show that the solution is unique and that no other solution to the problem is possible.

3. Given initial-value problem

$$y' = x^2 + y^2 \qquad y(0) = 1$$

do the following.

a. By differentiation of the differential equation, find $y'(0)$, $y''(0)$, and $y'''(0)$.

b. Write down the first four nonzero terms of the Maclaurin series for $y(x)$.

4. Let $y(x)$ be the unique solution of initial-value problem

$$y' + x^2 y = 0 \qquad y(0) = 1$$

a. Show that $y'(0) = 0$ and $y'(x) < 0$ if $x \neq 0$ and $y > 0$.

b. Compute $y''(x)$ from the differential equation, and show that $y''(x) < 0$ if $0 < x < \sqrt[3]{2}$ with $y > 0$ while $y''(x) > 0$ if $x > \sqrt[3]{2}$ with $y > 0$.

c. Show that $y > 0$ for all x.

d. Sketch the graph of $y(x)$ by using the preceding information.

5. Imitate the procedure of exercise 4 to sketch the graph of the solution of initial-value problem

$$y' + e^x y = 0 \qquad y(0) = 1$$

6. Let $y(x)$ be the unique solution of initial-value problem

$$y' = 1 + y^2 \qquad y(0) = 0$$

a. Show that y is an increasing function on every interval contained in the domain of y.

b. Show that the graph of $y(x)$ is concave upward over all intervals for which $y > 0$ and concave downward over all intervals for which $y < 0$.

c. Show that $y(c) = 0$ iff $y'(c) = 1$.

d. Given that $y(x)$ is a trigonometric function, guess which function. Verify that your guess is in fact the solution of the initial-value problem.

7. Suppose that $y(x)$ is the unique solution of initial-value problem

$$y' = \sqrt{1 - y^2} \qquad y(0) = 0$$

a. Show that $y'' = -y$.

b. Show that $y'(0) = 1$ and $y''(0) = 0$.

c. Show that $(y')^2 + y^2 = 1$.

d. What is the function $y(x)$? Verify that your guess is the solution of the initial-value problem.

8. We wish to solve the functional equation

$$f(x + y) = f(x)f(y)$$

We are given that f is a nonconstant function that is differentiable on the whole real line and that $f(0) = 1$.

a. Show that $f'(x) = f(x)f'(0)$.

b. Show that $f'(0) = 1$, so f is a solution of the initial-value problem

$$f'(x) = f(x) \qquad f(0) = 1$$

c. Write down (by inspection) a function that is a solution of the initial-value problem.

d. Show that the initial-value problem has a unique solution so your solution is the only solution and $f(x)$ has been uniquely determined.

10.2   The Euler Algorithm

Given the initial-value problem

$$y' = f(x, y) \qquad y(x_0) = y_0$$

let $a = x_0$ and suppose that $b > a$. All of our numerical methods will begin by selecting a step size h and then proceeding to estimate values of the solution $y(x_k)$ at points $x_k = a + kh$ for $k = 1, 2, \ldots, n$, where $b = nh$. We will agree to use y_k to denote the estimate and $y(x_k)$ to denote the exact value of the solution at point x_k. Then the error at the kth step is $y(x_k) - y_k$.

The simplest of our numerical methods employs the fact that

$$y(x + h) = y(x) + hy'(x) + \tfrac{1}{2}h^2 y''(\xi)$$

where ξ lies between x and $x + h$. Since $y'(x) = f(x, y)$ in this case, a valid approximation of $y(x + h)$ is

$$y(x + h) \cong y(x) + hf(x, y(x))$$

The following simple algorithm can be constructed from the preceding.

Algorithm 10.1

Step 1 Input x_0, y_0, h, and b.
Step 2 Compute $y_0 = y_0 + hf(x_0, y_0)$, and set $x_0 = x_0 + h$.
Step 3 Output x_0, y_0.
Step 4 If $x_0 < b$, go to step 2.
Step 5 Stop.

This algorithm is known as the **Euler method**.

EXAMPLE 10.4   Use the Euler method to approximate the solution of the problem

$$y' = x + y - 3 \qquad y(0) = 3$$

on the interval $[0, 0.5]$.

Solution   We choose step size $h = 0.1$, so $x_1 = 0.1$, $x_2 = 0.2, \ldots, x_5 = 0.5$. Then

$$y_1 = 3 + (0.1)(0) = 3.0000$$

$$y_2 = 3 + (0.1)(0.1) = 3.0100$$

$$y_3 = 3.01 + (0.1)(0.21) = 3.0310$$

$$y_4 = 3.031 + (0.1)(0.331) = 3.0641$$

$$y_5 = 3.0641 + (0.1)(0.4641) = 3.110510 \qquad \blacksquare$$

It is easy to verify that the exact solution to the initial-value problem in Example 10.4 is

$$y(x) = \exp(x) - x + 2$$

Hence, $y(0.1) = 3.00517$, and the error in y_1 is 0.00517. Also, $y(0.4) = 3.09182$, so the error in y_4 is 0.2772.

The Euler method does not produce a very accurate approximation to the solution $y(x)$, and the error grows noticeably as we move away from the initial point. Nevertheless, the Euler method is usable when great accuracy is not required, and it is extremely simple to use. The Euler method can also be used to illustrate some basic ideas underlying all so-called *one-step numerical methods* for approximating the solution of an initial-value problem.

A method for solving an initial-value problem is a one-step method if it has the form

$$y_{i+1} = y_i + h\phi(x_i, y_i, h)$$

In all of these methods, we replace the differential equation $y' = f(x, y)$ by a difference equation

$$y_{i+1} = y_i + h\phi(x_i, y_i, h) \qquad y_0 = y(x_0)$$

The **local truncation error** at x_{i+1}, usually denoted by τ_{i+1}, is defined by the equation

$$y(x_{i+1}) = y(x_i) + h\phi(x_i, y(x_i), h) + h\tau_{i+1}$$

Then

$$\tau_{i+1} = \frac{y(x_{i+1}) - y(x_i)}{h} - \phi(x_i, y(x_i), h)$$

For the Euler method,

$$\phi(x, y, h) = f(x, y) = y'(x)$$

so we see that τ_{i+1} is a measure of the error in approximation of $y'(x_i)$ by the quotient

$$\frac{y(x_{i+1}) - y(x_i)}{h}$$

The local truncation error is always a function of h. In the Euler method, we see that

$$\tau_{i+1} = \tfrac{1}{2}hy''(\xi_i)$$

where

$$x_i < \xi_i < x_i + h$$

If τ_i goes to zero as h goes to zero, we say that the method is **consistent**. Clearly, the Euler method is a consistent method.

We will use the term **global error** at x_i to denote the difference $y(x_i) - y_i$. This is the error we would like to keep small when approximating $y(x)$ at the points $x = x_i$. We need to distinguish this error from the **local error** at x_i, which is the new error introduced by the approximation process as we move from point x_{i-1} to x_i. The local error is a combination of truncation error and round-off error. Decreasing the truncation error decreases the local error at each point and thereby decreases the global errors. We will use this fact to try to maintain some control over the size of the errors in our approximations. An accurate approximation is, of course, one in which the global error is small. Theoretical estimation of the global error is extremely difficult, so we will not attempt to do that in this introductory treatment. We will content ourselves with estimates of the size of the truncation error at each point and simply reason that the smaller the error, the better our approximation should be.

Suppose that $x_n = a + nh = b$ is fixed. Since $h = (b - a)/n$, h is decreased if we increase n. We expect the global error at x_n to become smaller if h is decreased in size. We say that a one-step method for approximating the solution of an initial-value problem, which has a unique solution defined at x_n, is **convergent** if

$$\lim_{n \to \infty} | y(x_n) - y_n | = 0$$

In the case of a convergent algorithm, we can theoretically make the error in our estimate of $y(x_n)$ as small as we wish by choosing a value of n sufficiently large. In practice we cannot do this because of the limitations of machine arithmetic. There is a practical limit to how small h can be, and the problem of round-off error becomes serious if h is extremely small. Nevertheless, we have reason to expect a convergent algorithm to give better results when h is decreased within these practical limits. Theorems such as the following are important for that reason.

Theorem 10.2   Suppose that f and $\partial f/\partial y$ are continuous on the rectangle described by inequalities

$$a < x < b \quad \text{and} \quad c < y < d$$

Assume that

$$a < x_0 < b \quad \text{and} \quad c < y(x) < d$$

for solution $y(x)$. Then the Euler algorithm is convergent for the initial-value problem

$$y' = f(x, y) \qquad y(x_0) = y_0$$

Proof   Under the stated hypotheses, we know that a unique solution exists on some interval about x_0. We choose $z < b$ in that interval. We define $h = (z - x_0)/n$ for some positive integer n. Then, as usual,

$$x_k = x_0 + kh \qquad k = 1, 2, \ldots, n$$

Also, we have $z = x_n$. Now,

$$y(x_{k+1}) = y(x_k) + hy'(x_k) + \tfrac{1}{2}y''(\xi_k)h^2$$

and

$$y_{k+1} = y_k + hf(x_k, y_k)$$

It follows that

$$y(x_{k+1}) - y_{k+1} = y(x_k) - y_k + hf(x_k, y(x_k)) - hf(x_k, y_k) + \tfrac{1}{2}y''(\xi_k)h^2$$

So

$$|y(x_{k+1}) - y_{k+1}| < |y(x_k) - y_k| + h|f(x_k, y(x_k)) - f(x_k, y_k)| + \tfrac{1}{2}y''(\xi_k)h^2$$

Now,

$$f(x_k, y(x_k)) - f(x_k, y_k) = f_y(x_k, \eta_k)(y(x_k) - y_k)$$

for some η_k between y_k and $y(x_k)$. Since f_y is continuous on $[x_0, x_n]$, $|f_y(x, y)| \leq L$ on $[x_0, x_n]$ for some constant L. Also, the assumptions on function f imply that $y''(x)$ is continuous on its domain, so $|y''(x)| \leq M$ on $[x_0, x_n]$ for some constant M. It follows that

$$|y(x_{k+1}) - y_{k+1}| \leq |y(x_k) - y_k| + hL|y(x_k) - y_k| + \tfrac{1}{2}Mh^2$$

Letting

$$E_k = |y(x_k) - y_k|$$

we have

$$E_{k+1} \leq E_k(1 + hL) + \tfrac{1}{2}Mh^2$$

So,

$$E_1 \leq E_0(1 + hL) + \tfrac{1}{2}Mh^2$$

$$E_2 \leq E_1(1 + hL) + \tfrac{1}{2}Mh^2$$

$$\leq E_0(1 + hL)^2 + (1 + hL)\tfrac{1}{2}Mh^2 + \tfrac{1}{2}Mh^2$$

Continuing this argument, we arrive at

$$E_{k+1} \leq E_0(1 + hL)^{k+1} + \{(1 + hL)^k + (1 + hL)^{k-1} + \cdots + 1\}\tfrac{1}{2}Mh^2$$

Since $E_0 = 0$, we have, in simpler form,

$$E_{k+1} \leq \frac{Mh^2}{2}\left(\frac{(1 + hL)^{k+1} - 1}{hL}\right)$$

Then

$$E_n \leq \frac{Mh}{2L}[(1 + hL)^n - 1]$$

But

$$(1 + hL)^n = \left[1 + \frac{zL}{n}\right]^n = (1 + t)^{zL/t}$$

if we set $t = zL/n$. Then

$$(1 + hL)^n = [(1 + t)^{1/t}]^{zL}$$

Since

$$\lim_{t \to 0}(1 + t)^{1/t} = e$$

we have

$$\lim_{n \to \infty}(1 + hL)^n = e^{zL}$$

It follows that the limit of $(1 + hL)^n - 1$ is $e^{zL} - 1$, and the expression

$$\frac{Mh}{2L}[(1 + hL)^n - 1]$$

has a limit of zero as n goes to infinity. Then $\lim_{n \to \infty} E_n = 0$ and the theorem is proved. $\square$

In theory, then, the solution of most initial-value problems could be determined with any accuracy you wish by using the Euler algorithm with a large enough value of n. In practice, this is not feasible, since with large values of n, the time consumed becomes excessive and round-off error becomes large. In the preceding arguments, we have only considered truncation error at each step. The global error also includes round-off errors and propagated error.

EXAMPLE 10.5   Solve the initial-value problem

$$y' = \tfrac{1}{2}y \qquad y(1) = 1$$

Solution   An approximate solution using a step size $h = 0.05$ and Algorithm 10.1 gave the results shown in the following table.

```
ENTER X0 1
ENTER Y0 1
ENTER STEP SIZE H .05
ENTER FINAL X VALUE 2
```

X	Y	ERROR EST	TRUE ERROR
1.050	1.0250000	−0.0003011	−0.0003049
1.100	1.0493900	−0.0005811	−0.0005814
1.150	1.0732140	−0.0008426	−0.0008330
1.200	1.0965080	−0.0010873	−0.0010631
1.250	1.1193080	−0.0013171	−0.0012739
1.300	1.1416430	−0.0015334	−0.0014677
1.350	1.1635410	−0.0017375	−0.0016465
1.400	1.1850280	−0.0019305	−0.0018116
1.450	1.2061240	−0.0021133	−0.0019648
1.500	1.2268520	−0.0022869	−0.0021069
1.550	1.2472290	−0.0024520	−0.0022392
1.600	1.2672730	−0.0026093	−0.0023626
1.650	1.2870010	−0.0027593	−0.0024778
1.700	1.3064260	−0.0029026	−0.0025856
1.750	1.3255620	−0.0030397	−0.0026866
1.800	1.3444220	−0.0031711	−0.0027815
1.850	1.3630170	−0.0032970	−0.0028706
1.900	1.3813590	−0.0034180	−0.0029544
1.950	1.3994570	−0.0035342	−0.0030334
2.000	1.4173210	−0.0036461	−0.0031080
2.050	1.4349600	−0.0037538	−0.0031784

It is easy to verify that $y(x) = \sqrt{x}$ is the exact solution for this initial-value problem. This fact was used to compute the TRUE ERROR column in the table. To compute the numbers in the ERROR EST column, the value of $\tfrac{1}{2}h^2 y''(x_i - h/2)$ was used as an estimate of the local error in the computation of y_i. These values were accumulated as the computations proceeded. For example, the error estimate given for y_4 is the sum of the estimated local errors for y_1, y_2, y_3, and y_4. From the differential equation, we see that

$$y'' = -\frac{1}{2y^2} \, y' = -\frac{1}{4y^3}$$

For $y''(x_i - h/2)$, an estimate was obtained by using this formula for y'' with y replaced by

$$y_{i-1} + 0.5f(x_{i-1}, y_{i-1})$$

This is only an estimate, but the estimates are quite close in view of the fact that only an approximate value of y'' could be obtained and the TRUE ERROR also includes round-off error, which was not included in the error estimate.   ∎

EXAMPLE 10.6   Approximate the solution of the problem

$$y' = 1 + y^2 \qquad y(0) = 0$$

Solution   Using Algorithm 10.1 with $h = 0.01$ over the interval from 0 to 0.2, we obtain the results shown in the following table.

```
ENTER X0 0
ENTER Y0 0
ENTER STEP SIZE H .01
ENTER FINAL X VALUE .2
```

X	Y	ERROR EST	TRUE ERROR
0.010	0.0100000	0.0000005	0.0000003
0.020	0.0200010	0.0000020	0.0000017
0.030	0.0300050	0.0000045	0.0000040
0.040	0.0400140	0.0000080	0.0000073
0.050	0.0500300	0.0000125	0.0000117
0.060	0.0600550	0.0000180	0.0000171
0.070	0.0700911	0.0000246	0.0000234
0.080	0.0801402	0.0000321	0.0000309
0.090	0.0902045	0.0000407	0.0000393
0.100	0.1002858	0.0000503	0.0000488
0.110	0.1103864	0.0000610	0.0000594
0.120	0.1205083	0.0000727	0.0000711
0.130	0.1306535	0.0000855	0.0000838
0.140	0.1408242	0.0000993	0.0000977
0.150	0.1510225	0.0001142	0.0001127
0.160	0.1612506	0.0001302	0.0001289
0.170	0.1715106	0.0001473	0.0001462
0.180	0.1818047	0.0001656	0.0001648
0.190	0.1921353	0.0001849	0.0001846
0.200	0.2025044	0.0002054	0.0002056

The error was estimated as described in Example 10.5, using

$$y'' = 2yy' = 2y(1 + y^2)$$

to estimate the local errors. The exact solution is $y(x) = \tan x$. This fact was used to compute the TRUE ERROR at each step. Again, the estimates agree well with the true errors and illustrate the fact that the error at each step is an accumulation of local error, most of which originated in the truncation error that resulted from approximating the derivative by a finite difference quotient.   ■

■ *Exercise Set 10.2*

1. Approximate the solution of each initial-value problem on the interval $[0, 0.5]$ using the Euler algorithm with $h = 0.1$ (as in Example 10.4).

    a. $y' = 2y$; $y(0) = 0.5$ **b.** $y' = 1 + y^2$; $y(0) = 0$

    c. $y' = x - y$; $y(0) = 0.2$ **d.** $y' = x + \sin y$; $y(0) = 0$

2. Given initial-value problem

$$y' = -y^2 \quad y(1) = 1$$

do the following.

    a. Estimate $y(1.2)$ using the Euler algorithm with $h = 0.2$. Repeat with $h = 0.1$ and then again with $h = 0.05$.

    b. Compute $y(1.2)$ from the exact solution $y(x) = 1/x$ and compare with estimates from part (a). Did reducing h decrease the error as expected?

3. Repeat exercise 2 using a machine program that implements the Euler algorithm using step sizes $h = 0.05$, 0.01, 0.005, 0.001, 0.0005, 0.0001, 0.00005, 0.00001, and 0.000001. Compare machine estimates with $y(1.2)$. Do the estimates "converge" on $y(1.2)$ as expected?

4. Compute a numerical solution for each initial-value problem on the interval $[0, 1]$. Use the Euler algorithm with $h = 0.05$. Compare the machine estimates with values of the exact solution $y(x)$ given in each case.

    a. $y' = 1 + y^2$; $y(0) = 0$ **b.** $y' = -y$; $y(0) = 2$

        $y(x) = \tan x$ $y(x) = 2e^{-x}$

    c. $y' = \cos^2 y$; $y(0) = 0$ **d.** $y' = 0.5y^{-1}$; $y(0) = 1$

        $y(x) = \arctan x$ $y(x) = \sqrt{1 + x}$

5. Use the Euler algorithm with $h = 0.05$ to generate a machine solution of initial-value problem

$$y' = \frac{2}{3\sqrt{y}} \qquad y(0) = 1$$

on interval $[0, 2]$. Compare results with the exact solution $y(x) = (x - 1)^{2/3}$. Is the numerical solution an acceptable approximation of the exact solution? What is the difficulty?

10.3   Taylor Algorithms

If the solution $y(x)$ of an initial-value problem can be expanded about point x_0 in a Taylor series, then

$$y(x) = y(x_0) + y'(x_0)(x - x_0) + \frac{y''(x_0)(x - x_0)^2}{2!} + \cdots$$

Since

$$y'(x) = f(x, y(x))$$

we have

$$y'(x_0) = f(x_0, y_0)$$

Also,

$$y''(x) = f_x + f_y f$$

So

$$y''(x_0) = f_x(x_0, y_0) + f_y(x_0, y_0) f(x_0, y_0)$$

All derivatives of $y(x)$ can be computed in terms of f and partial derivatives of f. We find that

$$y'(x) = f(x, y) = f^{(0)}(x, y)$$
$$y''(x) = f_x + ff_y = f^{(1)}(x, y)$$
$$y'''(x) = f_{xx} + 2ff_{xy} + f^2 f_{yy} + f_x f_y + f(f_y)^2 = f^{(2)}(x, y)$$

and in general

$$y^{(k+1)}(x) = f^{(k)}(x, y)$$

where $f^{(k)}(x, y)$ is some combination of f and partials of f. We define

$$T_p(x, y, h) = f(x, y) + \frac{h^2}{2!} f^{(1)}(x, y) + \cdots + \frac{h^{p-1}}{p!} f^{(p-1)}(x, y)$$

Then

$$y(x_0 + h) = y_0 + hy'(x_0) + \frac{h^2}{2!} y''(x_0) + \frac{h^3}{3!} y'''(x_0) + \cdots$$

$$= y_0 + hf^{(0)}(x_0, y_0) + \frac{h^2}{2!} f^{(1)}(x_0, y_0) + \frac{h^3}{3!} f^{(2)}(x_0, y_0) + \cdots$$

So

$$y(x_1) = y_0 + hT_p(x_0, y_0, h) + \frac{h^{p+1}}{(p+1)!} y^{(p+1)}(\xi)$$

if $y^{(p+1)}(x)$ is continuous at x_0. These calculations suggest an algorithm of the following form.

Let

$$x_k = x_0 + kh \qquad k = 0, 1, 2, \ldots$$

Then compute

$$y_{k+1} = y_k + hT_p(x_k, y_k, h) \quad \text{with} \quad y_0 = y(x_0)$$

Such an algorithm is called a **Taylor algorithm of order** p. This is a one-step method for approximating the solution of an initial-value problem. In fact, $p = 1$ yields the Euler method. Theoretically, such an algorithm with large p is capable of producing very accurate estimates of the solution values. The local truncation error τ_{i+1} at x_{i+1} is defined by

$$y(x_{i+1}) = y(x_i) + hT_p(x_i, y(x_i), h) + h\tau_{i+1}$$

That is, the error in approximating $y(x_{i+1})$ by

$$y(x_i) + hT_p(x_i, y(x_i), h)$$

is h times the truncation error. A simple computation shows that the truncation error in this case is

$$\frac{h^p y^{(p+1)}(\xi_i)}{(p + 1)!}$$

If p is selected to be 4 or larger and h is of the order of magnitude of 0.05 or smaller, it is clear that the local errors resulting from such a truncation error will be very small.

The Taylor algorithms corresponding to values of p larger than 2 are very difficult to use directly because of the need to compute high-order partial derivatives of $f(x, y)$. By modifying this algorithm, we can produce usable algorithms and will do so in subsequent sections.

EXAMPLE 10.7   Use a Taylor algorithm of order 5 to estimate $y(0.1)$ and $y(0.2)$ for the problem

$$y' = 1 + y^2 \qquad y(0) = 1$$

Solution   We have

$$f^{(0)}(x, y) = 1 + y^2$$

Then

$$f^{(1)}(x, y) = y''(x) = 2yy' = 2y(1 + y^2)$$
$$f^{(2)}(x, y) = y'''(x) = 2(1 + y^2)(1 + 3y^2)$$
$$f^{(3)}(x, y) = y^{iv}(x) = 8y(1 + y^2)(2 + 3y^2)$$
$$f^{(4)}(x, y) = y^v(x) = 8(1 + y^2)(2 + 15y^2 + 15y^4)$$

So

$$T_5(x, y, h) = (1 + y^2) + \frac{h}{2} 2y(1 + y^2) + \frac{h^2}{6} 2(1 + y^2)(1 + 3y^2)$$

$$+ \frac{h^3}{24} 8y(1 + y^2)(2 + 3y^2) + \frac{h^4}{120} 8(1 + y^2)(2 + 15y^2 + 15y^4)$$

We will use $h = 0.1$ and compute y_1 and y_2. Now

$$y_1 = y_0 + hT_5(x_0, y_0, h)$$

where

$$T_5(x_0, y_0, h) = 2 + 0.05(2)(2) + \frac{0.01(16)}{6}$$

$$+ \frac{0.001(80)}{24} + \frac{0.0001(16)(32)}{120} = 2.2304267$$

Hence,

$$y_1 = 1 + (0.1)(2.2304267) = 1.2230427$$

$$y_2 = y_1 + hT_5(x_1, y_1, h)$$

$$= 1.2230427 + (0.1)T_5(0.1, 1.2230427, 0.1)$$

$$T_5(0.1, 1.2230427, 0.1) = 2.4958334 + 0.05(6.1050218)$$

$$+ \frac{0.01(27.391774)}{6} + \frac{0.001(158.42532)}{24}$$

$$+ \frac{0.0001(1158.0721)}{120} = 2.854304$$

So

$$y_2 = 1.2230427 + (0.1)(2.854304) = 1.508473$$

Then, approximately, we have

$$y(0.1) = 1.2230427 \quad \text{and} \quad y(0.2) = 1.508473$$ ∎

The exact solution of the initial-value problem in Example 10.7 is

$$y(x) = \tan\left(x + \frac{\pi}{4}\right)$$

So exact values are

$$y(0.1) = 1.2230489 \quad \text{and} \quad y(0.2) = 1.5084976$$

The accuracy of the Taylor algorithm used in Example 10.7 is quite good despite the large values of the derivatives. However, this is a tedious computational procedure. In the next section, we will investigate some algorithms that can produce the same accuracy employing a more usable computational procedure.

The Taylor algorithm of order 2 is relatively simple to use, however, and it does give results more accurate than those obtained with the Euler method. The following algorithm implements the computational procedure required in this case. It is assumed that the partial derivatives $f_x(x, y)$ and $f_y(x, y)$ have been computed in advance.

Algorithm 10.2

Step 1 Input x_0, y_0, h, x_n.
Step 2 Compute $f_0 = f(x_0, y_0)$ and

$$f_1 = f_x(x_0, y_0) + f_0 f_y(x_0, y_0)$$

Step 3 Compute $T_2 = f_0 + 0.5hf_1$.
Step 4 Compute $y_0 = y_0 + hT_2$ and set $x_0 = x_0 + h$.
Step 5 Output x_0, y_0.
Step 6 If $x_0 < x_n$, return to step 2.
Step 7 Stop.

EXAMPLE 10.8   Estimate the solution of the initial-value problem

$$y' = x + y - 3 \qquad y(0) = 3$$

on the interval $[0, 1]$ using $h = 0.1$.

Solution   Using Algorithm 10.2 with $h = 0.1$, we obtain the following results.

```
ENTER X0                0
ENTER Y0                3
ENTER STEP SIZE H      .1
ENTER FINAL X VALUE  1
```

X	Y	Y(X)	ERROR
0.100	3.0050000	3.0051710	0.0001707
0.200	3.0210250	3.0214030	0.0003777
0.300	3.0492330	3.0498590	0.0006261
0.400	3.0909020	3.0918250	0.0009224
0.500	3.1474470	3.1487210	0.0012743
0.600	3.2204290	3.2221190	0.0016901
0.700	3.3115740	3.3137530	0.0021791
0.800	3.4227890	3.4255410	0.0027518
0.900	3.5561820	3.5596030	0.0034213
1.000	3.7140810	3.7182820	0.0042007

The exact solution is

$$y(x) = \exp(x) - x + 2$$

This fact was used to compute the ERROR column. The errors over the interval [0, 1] are seen to be less than 5×10^{-3}. If we compare these results with the numerical estimates obtained in Example 10.4 for the same initial-value problem using the Euler method, the improvement is quite clear. Moving from the Taylor algorithm of order 1 to the Taylor algorithm of order 2 produces enough improvement in accuracy to justify the added computations.   ∎

We now have an algorithm with enough accuracy for many applications.

EXAMPLE 10.9   Approximate the solution of the problem

$$y' = 1 + y^2 \qquad y(0) = 0$$

over [0, 0.2].

Solution   Again, we use Algorithm 10.2 with $h = 0.01$. The results are seen in the following table.

```
ENTER X0              0
ENTER Y0              0
ENTER STEP SIZE H    .01
ENTER FINAL X VALUE .2
```

X	Y	Y(X)	ERROR
0.010	0.0100000	0.0100003	0.0000003
0.020	0.0200020	0.0200027	0.0000007
0.030	0.0300080	0.0300090	0.0000010
0.040	0.0400200	0.0400213	0.0000013
0.050	0.0500400	0.0500417	0.0000017
0.060	0.0600701	0.0600721	0.0000020
0.070	0.0701122	0.0701146	0.0000023
0.080	0.0801684	0.0801711	0.0000027
0.090	0.0902407	0.0902438	0.0000030
0.100	0.1003313	0.1003347	0.0000034
0.110	0.1104421	0.1104458	0.0000037
0.120	0.1205752	0.1205793	0.0000041
0.130	0.1307329	0.1307373	0.0000044
0.140	0.1409171	0.1409219	0.0000048
0.150	0.1511300	0.1511352	0.0000052
0.160	0.1613739	0.1613795	0.0000056
0.170	0.1716508	0.1716568	0.0000060
0.180	0.1819632	0.1819695	0.0000064
0.190	0.1923131	0.1923199	0.0000068
0.200	0.2027028	0.2027101	0.0000072

The exact solution is $y(x) = \tan x$. The errors are all less than 10^{-5}. Using a smaller step size decreases the size of the errors, as expected. The step size $h = 0.01$ is not small enough to cause any computational problems and provides very accurate estimates of the solution values. The error grows as the computations proceed owing to the accumulation of local error, but the growth is not rapid. ∎

For a Taylor algorithm of order p, the local truncation error is $0(h^p)$. In general, we will say that an algorithm for approximating the solution of an initial-value problem is of order p if the local truncation error is $0(h^p)$. The Euler method is of order 1. In the next section, we will examine some higher-order methods.

Higher-order Taylor algorithms can be implemented directly, as was done with the algorithm of order 2 in Algorithm 10.2. But there is a better way. That is the subject of the next section.

∎ *Exercise Set 10.3*

1. For each initial-value problem, find the first three nonzero terms of the Taylor series expansion about point x_0.
    a. $y' = x + y$; $y(0) = 1$ **b.** $y' = -y^2$; $y(1) = 1$
    c. $y' = 1 + y^2$; $y(\pi/4) = 1$ **d.** $y' = -xy + \sin x$; $y(0) = 1$

2. Find $T_p(x_k, y_k, h)$ for the stated p in each case.
    a. $y' = x^2 + y^2$; $y(1) = 1$ **b.** $y' = x + y^2$; $y(0) = 1$
    $p = 3$ $p = 4$
    c. $y' = \sqrt{1 - y^2}$; $y(0) = 1$ **d.** $y' = (y + x)/x$; $y(1) = 1$
    $p = 3$ $p = 2$

3. For each initial-value problem in exercise 2, use the corresponding Taylor algorithm with $h = 0.1$ to compute estimates y_1 and y_2.

4. For each initial-value problem in exercise 1, use Algorithm 10.2 with $h = 0.05$ and a hand calculator to find estimates y_1 and y_2.

5. Repeat exercise 4 for the initial-value problems in exercise 2.

6. Use Algorithm 10.2 with $h = 0.05$ to estimate the solution of each of the following initial-value problems on $[0, 1]$. You will need a machine program to do this.
    a. $y' = x + y$; $y(0) = 1$ **b.** $y' = -xy + \sin x$; $y(0) = 1$
    c. $y' = y^2 + \sqrt{x}$; $y(1) = 2$ **d.** $y' = \sqrt{1 - y^2}$; $y(0) = 0$

7. Verify the formula

$$y'''(x) = f_{xx} + 2ff_{xy} + f^2 f_{yy} + f_x f_y + f(f_y)^2$$

given in this section.

8. Assume that

$$y_1 = y_0 + hT_3(x_0, y_0, h)$$

is a Taylor algorithm (order 3) estimate of $y(x_1)$ for an initial-value problem. Explain why

$$y_1 = y(x_1) + 0(h^4)$$

9. Let y_1^h denote the estimate of $y(x_0 + h) = y(x_1)$ obtained using Algorithm 10.2 with step size h. Let the estimate of $y(x_0 + h)$ obtained using step size $h/2$ be denoted by $y_1^{h/2}$. Show that the error $|y_1^h - y(x_1)|$ is approximately $\frac{4}{3}(y_1^h - y_1^{h/2})$.

10. Given initial-value problem

$$y' = x + y \qquad y(0) = 1$$

do the following.
a. Use Algorithm 10.2 with $h = 0.1$ to estimate $y(0.1)$.
b. Recompute (estimate) $y(0.1)$ using step size 0.05.
c. Apply the result in exercise 9 to approximate the error in the estimate obtained in part (a).
d. Use exact solution $y(x) = 2e^x - x - 1$ to find the actual error in the estimate of part (a). Compare with your result in part (c).

10.4 Runge-Kutta Methods

The Taylor algorithms with $p > 2$ are very accurate but computationally tedious to use. The so-called Runge-Kutta algorithms were developed to obtain the accuracy of a high-order Taylor algorithm using a computational procedure that does not require computation of the partial derivatives of $f(x, y)$. A **Runge-Kutta algorithm of order p** produces estimates equivalent to those obtained with a Taylor algorithm of order p. We will begin by considering Runge-Kutta algorithms of order 2.

The general form of the iteration equation is assumed to be

$$y_1 = y_0 + h\{a_1 f(x_0, y_0) + a_2 f(x_0 + b_1 h, y_0 + b_2 h y_0')\}$$

for constants a_1, a_2, b_1, and b_2 with $a_2 > 0$. We need to choose these constants so that

$$y_1 = y_0 + hT_2(x_0, y_0, h) + 0(h^3)$$

We have

$$y_1 = y_0 + ha_1 f(x_0, y_0)$$
$$+ ha_2\{f(x_0, y_0) + b_1 hf_x(x_0, y_0) + b_2 hy_0' f_y(x_0, y_0) + 0(h^2)\}$$

Then

$$y_1 = y_0 + h(a_1 + a_2)f(x_0, y_0)$$
$$+ \{a_2 b_1 f_x(x_0, y_0) + a_2 b_2 y_0' f_y(x_0, y_0)\}h^2 + 0(h^3)$$

Now

$$y_0 + hT_2(x, y, h) = y_0 + hf(x_0, y_0) + \frac{h^2}{2}[f_x(x_0, y_0) + y_0' f_y(x_0, y_0)]$$

So we need to choose the constants so that $a_1 + a_2 = 1$, $a_2 b_1 = \frac{1}{2}$, and $a_2 b_2 = \frac{1}{2}$ with $a_2 > 0$. Let $a_2 = w > 0$ be selected arbitrarily. Then we need $a_1 = 1 - w$, $b_1 = b_2 = \frac{1}{2}w$. Any such choice of w will yield a Runge-Kutta algorithm of order 2. There are clearly infinitely many Runge-Kutta algorithms of order 2.

Case 1 Given $w = 1 \Rightarrow a_2 = 1$, $a_1 = 0$, $b_1 = b_2 = \frac{1}{2}$. Then

$$y_{i+1} = y_i + hf\left(x_i + \frac{h}{2}, y_i + \frac{k_i}{2}\right)$$

where $k_i = hf(x_i, y_i)$. This algorithm is sometimes called the **modified Euler method**.

Case 2 Given $w = \frac{1}{2} \Rightarrow a_2 = a_1 = \frac{1}{2}$ and $b_1 = b_2 = 1$. Then

$$y_{i+1} = y_i + h\{\tfrac{1}{2}f(x_i, y_i) + \tfrac{1}{2}f(x_i + h, y_i + k_i)\}$$

where $k_i = hf(x_i, y_i)$. Some authors call this algorithm the **improved Euler method**.

There are a number of useful algorithms included in the collection of order 2 Runge-Kutta algorithms. For all order 2 Runge-Kutta alogrithms, we have

$$y(x_i) = y_i + 0(h^3)$$

In programming these algorithms, or in computing with a calculator, it is common practice to compute

$$k_1 = hf(x_i, y_i)$$
$$k_2 = hf(x_i + b_1 h, y_i + b_2 k_1)$$

and then

$$y_{i+1} = y_i + a_1 k_1 + a_2 k_2$$

for $i = 0, 1, 2, 3, \ldots$.

EXAMPLE 10.10 Use the improved Euler method to estimate $y(0.1)$, $y(0.2)$, and $y(0.3)$ for initial-value problem

$$y' = 2x + y \qquad y(0) = 1$$

Solution The algorithm is described by

$$k_1 = hf(x_0, y_0)$$
$$k_2 = hf(x_0 + h, y_0 + k_1)$$
$$y_1 = y_0 + 0.5k_1 + 0.5k_2$$

We will use $h = 0.1$ in the following. We find that

$$k_1 = hf(0, 1) = (0.1)(0 + 1) = 0.1$$
$$k_2 = hf(0.1, 1.1) = (0.1)(1.3) = 0.13$$

So

$$y_1 = 1 + (0.5)(0.1) + (0.5)(0.13) = 1.115000$$
$$k_1 = hf(0.1, 1.115) = 0.1315$$
$$k_2 = hf(0.2, 1.2465) = 0.16465$$

So

$$y_2 = 1.115 + (0.5)(0.1315 + 0.16465) = 1.263075$$
$$k_1 = (0.1)f(0.2, 1.263075) = 0.1663075$$
$$k_2 = (0.1)f(0.3, 1.429383) = 0.2029383$$

So

$$y_3 = 1.263075 + (0.5)(0.1663075 + 0.2029383) = 1.447698$$

The exact solution is

$$y(x) = 3 \exp(x) - 2x - 2$$

So $y(0.1) = 1.1155128$, $y(0.2) = 1.2642083$, and $y(0.3) = 1.4495764$. ■

Algorithm 10.3

Step 1 Input x_0, y_0, h, and XN.
Step 2 Set K1 $= hf(x_0, y_0)$.
Step 3 Compute K2 $= hf(x_0 + h, y_0 + \text{K1})$.
Step 4 Set $y_0 = y_0 + h(\text{K1} + \text{K2})/2$ and $x_0 = x_0 + h$.
Step 5 Output the pair x_0, y_0.
Step 6 If $x_0 <$ XN, then go to step 2.
Step 7 Stop.

EXAMPLE 10.11   Approximate the solution of the problem

$$y' = x + y - 3 \qquad y(0) = 3$$

Solution   Using Algorithm 10.3 with step size $h = 0.1$, the machine results are as follows.

```
ENTER X0 ? 0
ENTER Y0 ? 3
ENTER STEP SIZE H ? .1
ENTER FINAL X VALUE ? 1
```

X	Y	Y(X)	ERROR
0.10	3.0050000	3.0051710	0.0001707
0.20	3.0210250	3.0214030	0.0003777
0.30	3.0492330	3.0498590	0.0006261
0.40	3.0909020	3.0918250	0.0009224
0.50	3.1474470	3.1487210	0.0012743
0.60	3.2204290	3.2221190	0.0016901
0.70	3.3115740	3.3137530	0.0021791
0.80	3.4227890	3.4255410	0.0027518
0.90	3.5561820	3.5596030	0.0034213
1.00	3.7140810	3.7182820	0.0042007

These results should be compared with the numerical results in Example 10.8, which were obtained by applying the Taylor algorithm of order 2 to the same initial-value problem. The improved Euler method, Algorithm 10.3, produces identical results. There is no need to compute partial derivatives in this case, however.   ∎

There are, of course Runge-Kutta methods of orders 3, 4, 5, and so on. The most popular higher-order Runge-Kutta method is a fourth-order method described by the iteration formulas

$$k_1 = hf(x_i, y_i)$$

$$k_2 = hf\left(x_i + \frac{h}{2}, y_i + \frac{k_1}{2}\right)$$

$$k_3 = hf\left(x_i + \frac{h}{2}, y_i + \frac{k_2}{2}\right)$$

$$k_4 = hf(x_i + h, y_i + k_3)$$

$$y_{i+1} = y_i + \frac{k_1 + 2k_2 + 2k_3 + k_4}{6}$$

In this case,

$$y(x_i) = y_i + 0(h^5)$$

so the estimates are quite accurate, although the computational procedure is not difficult, even with pocket calculators. A machine program would normally be used, of course, but it is educational to try a few steps of the computations with the aid of only a calculator.

EXAMPLE 10.12   Use the fourth-order Runge-Kutta algorithm to estimate $y(0.1)$ and $y(0.2)$ for initial-value problem

$$y = 2x + y \qquad y(0) = 1$$

Solution   We will use $h = 0.1$ in the following. We find that

$$k_1 = (0.1)(1) = 0.100000$$

$$k_2 = (0.1)(0.1 + 1.05) = 0.115000$$

$$k_3 = (0.1)(0.1 + 1.0575) = 0.115750$$

$$k_4 = (0.1)(0.2 + 1.11575) = 0.131575$$

So

$$y_1 = y_0 + \frac{0.1 + 0.23 + 0.2315 + 0.131575}{6} = 1.1155125$$

$$k_1 = (0.1)(0.2 + 1.1155125) = 0.13155120$$

$$k_2 = (0.1)(0.3 + 1.1812876) = 0.14812876$$

$$k_3 = (0.1)(0.3 + 1.18957638) = 0.14895763$$

$$k_4 = (0.1)(0.4 + 1.26446937) = 0.16644696$$

So

$$y_2 = 1.1155125 + \frac{0.89217094}{6} = 1.26420715$$

In this case, $y(0.2) = 1.264208274$, so the error in y_2 is 1.124×10^{-6}. This is more than enough accuracy for most applications. If more accuracy is required, we can use a smaller h. Using a machine program with $h = 0.01$, for example, we obtain $y_1 = 1.115512$ and $y_2 = 1.264208$. These estimates are accurate to six decimals.   ∎

For machine purposes, the following algorithm is a summary of the necessary computational steps.

Algorithm 10.4   Runge-Kutta Order 4

Step 1 Input x_0, y_0, h, and XN.

Step 2 Compute

$$K1 = hf(x_0, y_0)$$

$$K2 = hf\left(x_0 + \frac{h}{2}, y_0 + \frac{K1}{2}\right)$$

$$K3 = hf\left(x_0 + \frac{h}{2}, y_0 + \frac{K2}{2}\right)$$

$$K4 = hf(x_0 + h, y_0 + K)$$

Step 3 Set

$$y_0 = y_0 + \frac{K1 + 2K2 + 2K3 + K4}{6}$$

$$x_0 = x_0 + h$$

Step 4 Output the pair x_0, y_0.

Step 5 If $x_0 <$ XN, then go to step 2.

Step 6 Stop.

EXAMPLE 10.13   Solve the problem

$$y' = 2x + y \qquad y(0) = 1$$

over the interval $[0, 1]$.

Solution   Using Algorithm 10.4 with $h = 0.1$ and a machine program, we obtain the following results.

```
ENTER INITIAL VALUES OF X0, Y0 ? 0, 1
ENTER STEP SIZE H ? .1
ENTER LAST VALUE OF X ? 1
```

X	Y	Y(X)	ERROR
0.10	1.115513	1.115513	0.000000
0.20	1.264208	1.264208	0.000001
0.30	1.449576	1.449576	0.000001
0.40	1.675473	1.675474	0.000001
0.50	1.946162	1.946164	0.000002
0.60	2.266354	2.266357	0.000003
0.70	2.641255	2.641259	0.000004
0.80	3.076619	3.076623	0.000004
0.90	3.578804	3.578810	0.000005
1.00	4.154840	4.154846	0.000007

The error is very small throughout the interval $[0, 1]$. When the computations are repeated using $h = 0.05$, the following results are obtained.

```
ENTER INITIAL VALUES OF X0, Y0 ? 0, 1
ENTER STEP SIZE H ? .05
ENTER LAST VALUE OF X ? 1
```

X	Y	Y(X)	ERROR
0.05	1.053813	1.053814	0.000000
0.10	1.115513	1.115513	0.000000
0.15	1.185503	1.185503	0.000000
0.20	1.264208	1.264208	0.000000
0.25	1.352076	1.352076	0.000000
0.30	1.449577	1.449576	−0.000000
0.35	1.557203	1.557203	0.000000
0.40	1.675474	1.675474	0.000000
0.45	1.804937	1.804936	−0.000000
0.50	1.946164	1.946164	0.000000
0.55	2.099759	2.099759	0.000000
0.60	2.266357	2.266357	0.000000
0.65	2.446623	2.446623	0.000000
0.70	2.641258	2.641259	0.000000
0.75	2.851000	2.851001	0.000001
0.80	3.076623	3.076623	0.000000
0.85	3.318941	3.318941	0.000000
0.90	3.578809	3.578810	0.000000
0.95	3.857129	3.857130	0.000001
1.00	4.154845	4.154846	0.000001

The estimates are accurate to six decimal places.   ■

A detailed discussion of Runge-Kutta methods, including a derivation of the iteration formulas used for the fourth-order Runge-Kutta algorithm in this section, can be found in Ralston (1965, Chap. 5). Another well-written article that may be of interest to the serious student is in Rosser (1967).

■ *Exercise Set 10.4*

1. Show that the iteration formula for second-order Runge-Kutta methods has the general form

$$y_{i+1} = y_i + h\left\{(1-w)f(x_i, y_i) + wf\left(x_i + \frac{h}{2w}, y_i + \frac{h}{2w} y_i'\right)\right\}$$

with $w > 0$.

2. Clearly, $w = a_2$ must be nonzero in the formula in exercise 1. Explain why $a_2 \neq 0$ is an essential assumption in the initial formula given in this section for all Runge-Kutta methods.

3. Write out the Taylor series expansion of the second term in the formula

$$y_1 = y_0 + hf\left(x_0 + \frac{h}{2}, y_0 + \frac{h}{2} f(x_0, y_0)\right)$$

and verify directly that $y_1 = y(x_1) + 0(h^3)$.

4. Verify that $y_1 = y(x_1) + 0(h^3)$ for

$$y_1 = y_0 + \frac{h}{2} \{f(x_0, y_0) + f(x_0 + h, y_0 + hf(x_0, y_0))\}$$

5. Use the modified Euler method with $h = 0.01$ to compute estimates y_1 and y_2 for each initial-value problem.
   a. $y' = 2y;$ $y(0) = 1$ **b.** $y' = 1 + \sqrt{y};$ $y(0) = 0$
   c. $y' = 1 + y + y^2;$ $y(0) = 0.5$ **d.** $y' = y + e^{-x};$ $y(0) = 2$

6. Use the improved Euler method with $h = 0.01$ to compute estimates y_1 and y_2 for each initial-value problem in exercise 5 and compare results with those obtained in exercise 5.

7. Repeat exercise 5 using the fourth-order Runge-Kutta algorithm.

8. Given initial-value problem

$$y' = -y \qquad y(0) = 1$$

do the following.
   a. Use the Taylor algorithm of order 4 with $h = 0.1$ to compute solution estimates y_1 and y_2.
   b. Use the fourth-order Runge-Kutta method with $h = 0.1$ to compute solution estimates y_1 and y_2.
   c. Compare results of parts (a) and (b). How well do they agree? Would you expect them to be exactly the same? Discuss.
   d. Compare results of parts (a) and (b) with values $y(0.1)$ and $y(0.2)$ computed from the solution $y(x) = e^{-x}$.

9. Use a machine program with $h = 0.05$ to compute the modified Euler method estimates of the solution on $[0, 1]$ for each initial-value problem in exercise 5.

10. Repeat exercise 9 using the fourth-order Runge-Kutta method.

10.5   Multistep Methods

Suppose that the solution of an initial-value problem has been estimated at points $x_0, x_1, x_2, \ldots, x_n$ for some $n \geq 1$. It would seem reasonable to try to use this added information to speed up the computation of subsequent estimates $y_{n+1}, y_{n+2}, \ldots$, and so on. One simple way to do this is as follows. Recall that

$$y'(a) = \frac{y(a+h) - y(a-h)}{2h} - \tfrac{1}{6}h^2 y'''(\xi)$$

It follows that

$$y'(x_n) = \frac{y(x_{n+1}) - y(x_{n-1})}{2h} - \tfrac{1}{6}h^2 y'''(\xi_n)$$

and

$$y(x_{n+1}) = y(x_{n-1}) + 2hy'(x_n) + \tfrac{1}{3}h^3 y'''(\xi_n)$$

If $y(x_{n-1}) = y_{n-1}$ were known exactly, we could compute

$$y_{n+1} = y_{n-1} + 2hf(x_n, y_n)$$

and conclude that

$$y(x_{n+1}) = y_{n+1} + O(h^3)$$

This suggests using the **two-step formula**

$$y_{n+1} = y_{n-1} + hf(x_n, y_n) \tag{1}$$

to estimate solution values after two estimates y_n and y_{n-1} of $y(x_n)$ and $y(x_{n-1})$, respectively, are available. The local trunction error $\tau_{n+1}(h)$ in this case is defined by

$$y(x_{n+1}) - y(x_{n-1}) - 2hf(x_n, y(x_n)) = \tau_{n+1}(h)h$$

It follows that

$$\tau_{n+1}(h) = \tfrac{1}{3}y'''(\xi_n)h^2 = O(h^2)$$

so this is a second-order method. The numerical results should have about the same accuracy as a second-order Runge-Kutta algorithm but are much faster to compute because there are fewer arithmetic computations and only one function evaluation per step rather than two.

**EXAMPLE
10.14**

For the initial-value problem

$$y' = 2x + y \qquad y(0) = 1$$

estimate the solution at points $x_1 = 0.1$, $x_2 = 0.2$, $x_3 = 0.3$, and $x_4 = 0.4$. Suppose that estimate $y_1 = 1.115513$ has been determined.

Solution We want to use the two-step formula (Equation 1) to find y_2, y_3, and y_4. The computation is as follows:

$$y_2 = y_0 + 2hf(x_1, y_1) = 1 + (0.2)(1.315513) = 1.263103$$

$$y_3 = y_1 + 2hf(x_2, y_2) = 1.115513 + (0.2)(1.663103) = 1.448133$$

$$y_4 = y_2 + 2hf(x_3, y_3) = 1.263103 + (0.2)(2.048133) = 1.672730$$

In fact, $y(x_2) = 1.26421$, $y(x_3) = 1.44958$, and $y(x_4) = 1.67547$, so the estimates are all slightly low. The improved Euler algorithm gives estimates $y_2 = 1.263080$, $y_3 = 1.447700$, and $y_4 = 1.672710$. So we see that the two-step formula actually performs better than the improved Euler algorithm in this case. ∎

The two-step formula used in Example 10.14 is an example of a **two-step continuation formula**. This leads us to consider the possible construction of multistep continuation formulas with the hope of attaining greater accuracy. In general, these could have the form

$$y_{n+1} = F(y_n, y_{n-1}, y_{n-2}, \ldots, y_{n-p})$$

assuming that values $y_n, y_{n-1}, \ldots, y_{n-p}$ have been determined. It is usually assumed that $F(y_n, \ldots, y_{n-p})$ is a sum of a linear combination of $y_n, y_{n-1}, \ldots, y_{n-p}$ and a linear combination of the derivatives $y'_n, y'_{n-1}, \ldots, y'_{n-p}$. We say that the method is a linear $(p + 1)$-step method in that case. In the following discussion, we will consider the special case of formulas of the form

$$y_{n+1} = \sum_{k=0}^{p} (A_k y_{n-k} + hB_k y'_{n-k}) + hB_{-1} y'_{n+1}$$

for constants $A_0, \ldots, A_p$ and $B_{-1}, B_0, \ldots, B_p$. A large number of multistep continuation formulas have been derived from this assumed general form. The local truncation error $\tau_{n+1}(h)$ is defined by

$$y(x_{n+1}) - \sum_{k=0}^{p} [A_k y(x_{n-k}) + hB_k f(x_{n-k}, y(x_{n-k}))]$$

$$- hB_{-1} f(x_{n+1}, y(x_{n+1})) = \tau_{n+1}(h)h$$

If we choose $A_0 = 1$ and $A_k = 0$ for $k > 0$, we obtain

$$y_{n+1} = y_n + h\left\{ \sum_{k=0}^{p} B_k\, y'(x_{n-k}) + B_{-1}y'(x_{n+1}) \right\} \tag{2}$$

If $B_{-1} = 0$, we get *explicit* formulas involving only the known values $y_n, y_{n-1}, \ldots, y_{n-p}$ and applicable after $n > p$. Formulas of this type are called **Adams-Bashforth $(p + 1)$-step formulas**. If $B_{-1} \neq 0$, we get *implicit* formulas known as **Adams-Moulton formulas**. In this case, a preliminary estimate of y_{n+1} must be made to use in the right side of the formula. The formula is then used to compute an improved estimate (y_{n+1} on the left).

The traditional Adams-Bashforth formula is an explicit formula with $p = 3$ (a four-step continuation formula). The necessary constants are derived as follows.

We have the general form

$$y_{n+1} = y_n + h[B_0\, y'_n + B_1 y'_{n-1} + B_2\, y'_{n-2} + B_3\, y'_{n-3}]$$

We assume that the formula gives exact results if $y(x)$ is a polynomial of degree 4 or less in $(x - x_n)$. Hence, the formula is exact if $y(x)$ is either $(x - x_n)$, $(x - x_n)^2$, $(x - x_n)^3$, or $(x - x_n)^4$. Setting $y(x) = x - x_n$, we find that

$$h = 0 + h(B_0 + B_1 + B_2 + B_3)$$

So

$$B_0 + B_1 + B_2 + B_3 = 1$$

Setting $y(x) = (x - x_n)^2$, we find that

$$h^2 = 0 + h(0 - 2hB_1 - 4hB_2 - 6hB_3)$$

So

$$B_1 + 2B_2 + 3B_3 = -\tfrac{1}{2}$$

Setting $y(x) = (x - x_n)^3$, we find that

$$h^3 = 0 + h(0 + 3h^2 B_1 + 12h^2 B_2 + 27h^2 B_3)$$

So

$$B_1 + 4B_2 + 9B_3 = \tfrac{1}{3}$$

Setting $y(x) = (x - x_n)^4$, we find that

$$h^4 = 0 + h(0 - 4h^3 B_1 - 32h^3 B_2 - 108h^3 B_3)$$

So

$$B_1 + 8B_2 + 27B_3 = -\tfrac{1}{4}$$

Solving the linear system, we obtain $B_0 = \tfrac{55}{24}$, $B_1 = -\tfrac{59}{24}$, $B_2 = \tfrac{37}{24}$, and $B_3 = -\tfrac{9}{24}$. The resulting continuation formula is

$$y_{n+1} = y_n + \frac{h}{24}(55y'_n - 59y'_{n-1} + 37y'_{n-2} - 9y'_{n-3}) \tag{3}$$

It can be shown that the truncation error for the resulting four-step method is

$$\tau_{n+1}(h) = \tfrac{251}{720}h^4 y^{(5)}(\xi_n) \qquad x_{n-3} < \xi_n < x_{n+1}$$

This means that if the values used in computing y_{n+1} were known exactly and there was no round-off error, we would have

$$y(x_{n+1}) = y_{n+1} + \tfrac{251}{720}h^5 y^{(5)}(\xi_n) = y_{n+1} + 0(h^5)$$

When $B_{-1} \neq 0$ in Equation 2 and $p = 3$ as before, we obtain the implicit four-step formula

$$y_{n+1} = y_n + \frac{h}{720}(251y'_{n+1} + 646y'_n - 264y'_{n-1} + 106y'_{n-2} - 19y'_{n-3}) \tag{4}$$

The truncation error in this case is

$$\tau_{n+1}(h) = -\tfrac{3}{160}h^5 y^{(6)}(\xi_n)$$

The implicit formula (Equation 4) has a smaller truncation error than does the explicit formula (Equation 3). Equation 4 is traditionally called the Adams-Moulton formula, while Equation 3 is known as the Adams-Bashforth formula.

EXAMPLE 10.15   For the initial-value problem

$$y' = 2x + y \qquad y(0) = 1$$

the fourth-order Runge-Kutta algorithm has been used to find the values

$$
\begin{aligned}
y_1 &= 1.11551250 & y'_1 &= 1.31551250 \\
y_2 &= 1.26420771 & y'_2 &= 1.66420771 \\
y_3 &= 1.44957549 & y'_3 &= 2.04957549 \\
y_4 &= 1.67547272 & y'_4 &= 2.47547272
\end{aligned}
$$

for $x_1 = 0.1$, $x_2 = 0.2$, $x_3 = 0.3$, and $x_4 = 0.4$. Use the Adams-Bashforth continuation formula to estimate $y(0.5)$ and $y(0.6)$.

Solution   Computation is as follows:

$$y_5 = y_4 + \left(\frac{0.1}{24}\right)(55y_4' - 59y_3' + 37y_2' - 9y_1')$$

$$= 1.67547272 + \left(\frac{0.1}{24}\right)(69.9621185) = 1.946148214$$

Then

$$y_5' = 2(0.5) + 1.946148214 = 2.946148214$$

$$y_6 = y_5 + \left(\frac{0.1}{24}\right)(55y_5' - 59y_4' + 37y_3' - 9y_2')$$

$$= 1.946148214 + \left(\frac{0.1}{24}\right)(76.84168504) = 2.266321902 \qquad \blacksquare$$

Computations such as those in Example 10.15 are easily continued to more points, but a machine program is desirable if that is to be done. We will now look at some machine output from such a program.

EXAMPLE 10.16   Solve the initial-value problem

$$y' = 2x + y \qquad y(0) = 1$$

Solution   We select $h = 0.1$. Our machine program uses the fourth-order Runge-Kutta algorithm to estimate four starting values and then continues the solution using the four-step Adams-Bashforth method. The results are as follows:

x	y	y'
0.10	1.115512	1.315512
0.20	1.264207	1.664207
0.30	1.449575	2.048575
0.40	1.675472	2.475472
0.50	1.946148	2.946148
0.60	2.266322	3.466322
0.70	2.641202	4.041202
0.80	3.076541	4.676541
0.90	3.578697	5.378697
1.00	4.154697	6.154697

The values above the line are starting values from the fourth-order Runge-Kutta method. Beginning at $x = 0.5$, the solution values are estimated using the four-step Adams-Bashforth algorithm. The starting values are accurate to six decimal places.

Since $y(1) = 4.15484548\ldots$, the error in the estimate of $y(1)$ is 1.485×10^{-4}. Much better accuracy can be attained by using a smaller value of h if that is desired. ∎

EXAMPLE 10.17

Solve the initial-value problem

$$y' = \frac{1}{1 + \tan^2 x} \qquad y(0) = 0$$

Solution   Again we use $h = 0.1$ and estimate the solution on the interval $[0, 1]$. The machine results are as follows:

x	y	y'
0.10	0.099668	0.990099
0.20	0.197395	0.961538
0.30	0.291456	0.917431
0.40	0.380506	0.862069
0.50	0.463636	0.800008
0.60	0.540426	0.735288
0.70	0.619759	0.671109
0.80	0.674804	0.609694
0.90	0.732901	0.552400
1.00	0.785499	0.499898

The exact solution to the initial-value problem is $y(x) = \tan^{-1}(x)$. So we should have $y(0.4) = 0.3805064$, $y(0.5) = 0.4636476$, $y(0.8) = 0.6747409$, and $y(1) = 0.7853982$. The estimate of $y(0.4)$ from the Runge-Kutta algorithm is correct to six decimals. The error in the estimate of $y(0.5)$ from the Adams-Bashforth algorithm is -1.16×10^{-5}. The error in the estimate of $y(0.8)$ is -6.31×10^{-5}, and the error in the estimate of $y(1)$ is -1.01×10^{-4}. These estimates from the Adams-Bashforth algorithm are quite satisfactory considering the choice of h. A smaller h would give better results. In fact, repeating the procedure with $h = 0.05$, we obtain $y(1) = 0.785403$ with an error of only -4.8×10^{-6}. ∎

■ *Exercise Set 10.5*

1. Two starting values are given for each of the following initial-value problems. Use these values with the two-step formula (Equation 1) to estimate $y(x_2)$ in each case.

   a. $y' = \sqrt{1 - y^2}$ **b.** $y' = -y^2$
       $y(0) = 0$ $y(1) = 1$
       $y(0.1) = 0.0998334$ $y(1.1) = 0.0909091$

2. Use Equation 1 with a machine program to find a numerical solution on intervals $[0, 1]$ and $[1, 2]$, respectively, for the initial-value problems in exercise 1.

3. The exact solution of the initial-value problem in exercise 1a is $y(x) = \sin x$, while the solution of the initial-value problem in exercise 1b is $y(x) = 1/x$. After examining the graphs of these two functions, give a geometric reason for the fact that both applications of the two-step formula in exercise 2 produced estimates that were too high.

4. Compute y''' by differentiation of the differential equation in exercise 1a. Apply this result to the error term for the two-step formula to show analytically that the estimate y_2 should be higher than $y(0.2)$.

5. Compute y''' by differentiation of the differential equation in exercise 1b. Apply this result to the error term for the two-step formula to show analytically that the estimate y_2 should be higher than $y(1.1)$.

6. Verify the numerical results given in Example 10.16.

7. Verify the numerical results given in Example 10.17.

8. Use the fourth-order Runge-Kutta method with $h = 0.1$ to obtain starting values for each initial-value problem. Continue the numerical solution over the indicated interval by using the Adams-Bashforth formula (Equation 3). You will need to use a machine program for this exercise.

   a. $y' = x - 2xy$; $y(0) = 1$; $[0, 2]$ b. $y' = \sqrt{x} + \sqrt{y}$; $y(0) = 0$; $[0, 1]$

   c. $y' = \cos^2 y$; $y(0) = 0$; $[0, 1]$ d. $y' = \sqrt{1 + y^2}$; $y(0) = 1$; $[0, 2]$

9. Derive the explicit Adams-Bashforth two-step formula using Equation 2 with $B_{-1} = 0$. Assume exact results for $y(x) = x - x_n$ and $y(x) = (x - x_n)^2$.

10. Derive the explicit Adams-Bashforth three-step formula from Equation 2 with $B_{-1} = 0$.

10.6   Predictor-Corrector Methods

Suppose once more that accurate estimates of solution values $y_1, y_2, \ldots, y_n$ have been determined for some $n \geq 1$. Now

$$\int_{x_{n-p}}^{x_{n+1}} y'(x)\, dx = y(x_{n+1}) - y(x_{n-p})$$

so

$$y(x_{n+1}) = y(x_{n-p}) + \int_{x_{n-p}}^{x_{n+1}} y'(x)\, dx \tag{1}$$

Since $y'(x_k) = f(x_k, y_k)$, approximate values of $y'(x)$ are known for points $x_0, x_1, \ldots, x_n$. It follows that the integral in Equation 1 can be estimated by use of numerical integration formulas. This leads to an estimate of $y(x_{n+1})$. Suppose, for example, we use $p = 0$ and the trapezoid rule. We obtain

$$y(x_{n+1}) = y(x_n) + \frac{h}{2}\left(y'(x_{n+1}) + y'(x_n)\right) - \tfrac{1}{12} y'''(\eta) h^3$$

So $y(x_{n+1})$ can be estimated by

$$y(x_n) + \frac{h}{2}\left(f(x_{n+1}, y_{n+1}) + f(x_n, y_n)\right) \tag{2}$$

with error

$$-\tfrac{1}{12}y'''(\eta)h^3$$

Unfortunately, the value of $y(x_{n+1})$ appears on both sides of Equation 2. So we need a preliminary estimate of $y(x_{n+1})$ before we can use Equation 2. Equation 2 is an implicit formula for $y(x_{n+1})$. We need to use an explicit formula to make a preliminary estimate. Even the Euler formula could be used to do this, but a more accurate estimate can be made using the midpoint formula

$$y_{n+1} = y_{n-1} + 2hf(x_n, y_n)$$

from Section 10.5.

EXAMPLE 10.18

Estimate the solution of the initial-value problem

$$y' = 2x + y \qquad y(0) = 1$$

Solution We will choose step size $h = 0.1$ and estimate the solution on interval $[0, 1]$. Using a machine program, we first estimate $y(0.1)$ using the fourth-order Runge-Kutta method. Then we continue the solution by using the midpoint formula to make a preliminary estimate of $y(x_{n+1})$ and correcting that estimate with an application of Equation 2. The results are as follows:

x	y (midpoint)	Corrected
0.10	1.115513	(No Correction)
0.20	1.263103	1.264443
0.30	1.448133	1.450005
0.40	1.672729	1.676048
0.50	1.942679	1.946819
0.60	2.261265	2.267016
0.70	2.634932	2.641826
0.80	3.068251	3.076985
0.90	3.568582	3.578827
1.00	4.141968	4.154354

The true value of $y(1)$ is 4.1548455, so the corrected estimate is clearly more accurate than the preliminary estimate. A second application of the correction formula would produce some further improvement. ∎

The procedure used in Example 10.18 is called a **predictor-corrector algorithm**. There are many such algorithms. To form such an algorithm, we select an explicit formula (the predictor) to make a preliminary estimate of $y(x_{n+1})$. Then we use an implicit formula with smaller error term (the corrector) to make a corrected estimate. For this procedure to be effective, the corrector formula must have a smaller error, of course. There is an advantage to choosing a predictor-corrector pair in which both formulas have error terms involving the same order of derivative of $y(x)$ and the same power of h. When that is done, we can estimate the error in the corrected value by use of the difference between the predicted value and the corrected value. To see how this works, let us consider the predictor-corrector pair used in Example 10.18.

Predictor:　$y_{n+1}^{(0)} = y_{n-1} + 2hf(x_n, y_n)$

Corrector:　$y_{n+1} = y_n + \dfrac{h}{2}(f(x_{n+1}, y_{n+1}^{(0)}) + f(x_n, y_n))$

Predictor error:　$\frac{1}{3}y'''(\xi)h^3$

Corrector error:　$-\frac{1}{12}y'''(\eta)h^3$

We have

$$y(x_{n+1}) = y_{n+1}^{(0)} + \tfrac{1}{3}y'''(\xi)h^3$$

and

$$y(x_{n+1}) = y_{n+1} - \tfrac{1}{12}y'''(\eta)h^3$$

Subtracting, we find that

$$0 = y_{n+1} - y_{n+1}^{(0)} - \tfrac{1}{12}y'''(\eta)h^3 - \tfrac{1}{3}y'''(\xi)h^3$$

So

$$y_{n+1} - y_{n+1}^{(0)} \cong \tfrac{5}{12}y'''(\eta)h^3$$

and

$$-\tfrac{1}{12}y'''(\eta)h^3 \cong \tfrac{1}{5}(y_{n+1}^{(0)} - y_{n+1})$$

This is an estimate of the error in the corrected value. Applying this result to the estimate of $y(0.2)$ in Example 10.18, we see that the error in the corrected estimate is

$$\frac{1.263103 - 1.264443}{5} = -2.68 \times 10^{-4}$$

The true error is -2.35×10^{-4}, so the approximation is quite close.

By using higher-order formulas, we can construct predictor-corrector algorithms that provide more accurate results than those obtained with this simple initial illustration. One of the most widely used such algorithms is the **Adams-**

Moulton algorithm, which uses the four-step Adams-Bashforth formula as predictor and the three-step Adams-Moulton implicit formula as corrector:

Predictor: $\quad y_{n+1}^{(0)} = y_n + \dfrac{h}{24}(55y_n' - 59y_{n-1}' + 37y_{n-2}' - 9y_{n-3}')$

Error term: $\quad \frac{251}{720}h^5 y^{(5)}(\xi)$

Corrector: $\quad y_{n+1} = y_n + \dfrac{h}{24}(9y_{n+1}' + 19y_n' - 5y_{n-1}' + y_{n-2}')$

Error term: $\quad -\frac{19}{720}h^5 y^{(5)}(\eta)$

For this predictor-corrector combination, we have

$$y(x_{n+1}) = y_{n+1}^{(0)} + \tfrac{251}{720}h^5 y^{(5)}(\xi)$$

and

$$y(x_{n+1}) = y_{n+1} - \tfrac{19}{720}h^5 y^{(5)}(\eta)$$

Subtraction yields

$$0 \cong y_{n+1}^{(0)} - y_{n+1} + \tfrac{270}{720}h^5 y^{(5)}(\eta)$$

so

$$-\tfrac{19}{720}h^5 y^{(5)}(\eta) \cong \tfrac{19}{270}(y_{n+1}^{(0)} - y_{n+1})$$

This is an estimate of the error in the corrected estimate y_{n+1}. It is, of course, important to remember that this estimate is computed from the truncation error terms, so only the truncation error being introduced at each step is estimated. Round-off error is also present and accumulates as the iterations proceed.

The Adams-Moulton predictor-corrector algorithm is easily programmed. We will now look at some machine output that includes computation of the preceding error estimate at each step.

EXAMPLE 10.19   Estimate the solution of initial-value problem

$$y' = \sqrt{1 - y^2} \qquad y(0) = 0$$

on interval $[0, 1]$.

Solution   We will use $h = 0.05$, but only estimates for $x = 0.1, 0.2, 0.3$, and so on, will be printed. The machine output is as follows.

RUNGE-KUTTA STARTING VALUES

X	Y
0.05	0.049979
0.10	0.099833
0.15	0.149438
0.20	0.198669

ADAMS-MOULTON ESTIMATES

X	Y0	Y	ERROR
0.30	0.295520	0.295520	0.0000000
0.40	0.389418	0.389418	0.0000000
0.50	0.479426	0.479425	0.0000000
0.60	0.564643	0.564642	0.0000000
0.70	0.644218	0.644218	0.0000000
0.80	0.717356	0.717356	0.0000000
0.90	0.783327	0.783327	0.0000000
1.00	0.841471	0.841471	0.0000000

The solution of this initial-value problem is $y(x) = \sin x$. Hence, $y(0.4) = 0.38941834$ and $y(1) = 0.84147098$. The true errors in the estimates of these values are 3.4×10^{-7} and 2×10^{-8}. The errors are small but of course not zero as the last column seems to indicate. The column of error estimates does indicate that all corrected estimates should be accurate to at least six decimal places, and that is true. ■

We will consider one more well-known predictor-corrector algorithm. First note that by using the Simpson formula to estimate the integral of $y'(x)$, we obtain

$$\int_{x_{n-1}}^{x_{n+1}} y'(x)\, dx = \frac{h}{3}\,(y'_{n-1} + 4y'_n + y'_{n+1}) - \frac{h^5}{90}\,y^{(5)}(\eta)$$

so

$$y(x_{n+1}) = y(x_{n-1}) + \frac{h}{3}\,(y'_{n+1} + 4y'_n + y'_{n-1})$$

with error

$$-\frac{h^5}{90}\,y^{(5)}(\eta)$$

We will call this implicit formula the **Simpson correction formula**. Before this formula can be applied, we need an estimate of y_{n+1} from a predictor formula. It is desirable to have a formula with an error term of the form of a constant times $y^{(5)}(\xi)h^5$. Such a formula can be produced as follows. Approximate $y'(x)$ on the set of points x_0, x_1, x_2, and x_3 using the Newton forward difference polynomial $P_3(x)$.

Then

$$\int_{x_0}^{x_4} y'(x)\, dx \cong \int_{x_0}^{x_4} P_3(x)\, dx$$

You can verify that the integral on the right is

$$\frac{4h}{3}(2y_3' - y_2' + 2y_1')$$

It follows that

$$y(x_4) \cong y(x_0) + \frac{4h}{3}(2y_3' - y_2' + 2y_1')$$

The error term is found to be

$$\tfrac{14}{45}h^5 y^{(5)}(\xi)$$

so it is of the desired form. The explicit formula

$$y_{n+1} = y_{n-3} + \frac{4h}{3}(2y_n' - y_{n-1}' + 2y_{n-2}')$$

is known as the **Milne formula** and can be used as a predictor with the Simpson corrector to form a predictor-corrector algorithm that we will call the Milne predictor-corrector algorithm. In summary, we have the following:

Predictor: $\quad y_{n+1}^{(0)} = y_{n-3} + \dfrac{4h}{3}(2y_n' - y_{n-1}' + 2y_{n-2}')$

Corrector: $\quad y_{n+1} = y_{n-1} + \dfrac{h}{3}(y_{n+1}' + 4y_n' + y_{n-1}')$

We have

$$y(x_{n+1}) = y_{n+1}^{(0)} + \tfrac{28}{90}h^5 y^{(5)}(\xi)$$

and

$$y(x_{n+1}) = y_{n+1} - \tfrac{1}{90}h^5 y^{(5)}(\eta)$$

Subtracting and rearranging, we find that

$$-\tfrac{1}{90}h^5 y^{(5)}(\eta) \cong \frac{y_{n+1}^{(0)} - y_{n+1}}{29}$$

is our error estimation formula for the corrected estimate.

EXAMPLE   Use the Milne predictor-corrector algorithm to estimate the solution of initial-value
10.20   problem

$$y' = \sqrt{1 - y^2} \qquad y(0) = 0$$

Solution   We will use a machine program with $h = 0.1$. The results are as follows.

RUNGE-KUTTA STARTING VALUES

X	Y
0.10	0.099833
0.20	0.198669
0.30	0.295520

MILNE ESTIMATES

X	Y0	Y	ERROR
0.40	0.389415	0.389418	−0.0000001
0.50	0.479423	0.479426	−0.0000001
0.60	0.564640	0.564643	−0.0000001
0.70	0.644215	0.644218	−0.0000001
0.80	0.717354	0.717357	−0.0000001
0.90	0.783325	0.783327	−0.0000001
1.00	0.841469	0.841472	−0.0000001

The results obtained in this example are quite satisfactory. As we will see in the next
section, however, sometimes there are difficulties with the Milne algorithm that we
do not encounter with the Adams-Moulton algorithm.   ■

■ *Exercise Set 10.6*

1. Use a machine program that computes values of the midpoint predictor with trapezoid
   correction to verify the numerical results given in Example 10.18.

2. Use a machine program that implements the Adams-Moulton predictor-corrector algorithm to verify the numerical results given in Example 10.19.

3. Use a machine program that implements the Milne predictor-corrector algorithm to
   verify the numerical results given in Example 10.20.

4. Let $P_3(x)$ be the Newton forward difference polynomial $P_3(x)$ that interpolates $y'(x)$ on
   points x_0, x_1, x_2, and x_3. Perform the integration and verify that

   $$\int_{x_3}^{x_4} P_3(x)\, dx = \frac{4h}{3}(2y_3' - y_2' + 2y_1')$$

5. Use the approximation $(y_{n+1}^{(0)} - y_{n+1})/5$ to estimate the error in the corrected midpoint
   estimate of $y(0.7)$ given in Example 10.18.

6. The fourth-order Runge-Kutta algorithm gave the values shown for the solution of
   initial-value problem

   $$y' = 1 + y^2 \qquad y(0) = 0$$

   a. Estimate $y(0.2)$ and $y(0.25)$ using the Adams-Moulton algorithm.
   b. Estimate $y(0.2)$ and $y(0.25)$ using the Milne predictor-corrector algorithm.

c. Compare these results with values of the solution $y(x) = \tan x$.

x	y	y'
0	0	1
0.05	0.050041	1.002504
0.10	0.100334	1.010067
0.15	0.151135	1.022841

7. Make error estimates for each of your corrected values in exercise 6a and b.

8. Suppose that for a certain predictor-corrector algorithm, the error terms were as follows:

Predictor: $\frac{3}{8}y^{iv}(\xi)h^4$
Corrector: $-\frac{1}{32}y^{iv}(\xi)h^4$

Find a formula for estimating the error in the corrector values from values of y_{n+1} and $y_{n+1}^{(0)}$.

9. A predictor-corrector method known as *Hamming's method* is described by the following formulas:

Predictor: $y_{n+1}^{(0)} = y_{n-3} + \dfrac{4h}{3}(2y_n' - y_{n-1}' + 2y_{n-2}')$

Corrector: $y_{n+1} = \frac{1}{8}(9y_n - y_{n-2}) + \dfrac{3h}{8}(y_{n+1}' + 2y_n' - y_{n-1}')$

Apply Hamming's method to the initial-value problem in exercise 6, and compare results with the results obtained in exercise 6.

10. Consider the initial-value problem

$$y' + y = g(x) \qquad y(0) = 0$$

where $g(x) = 0$ if $0 \le x < 1$ and $g(x) = 1$ if $x \ge 1$.
a. Using step size $h = 0.1$, find a numerical solution on $[0, 2]$ using the fourth-order Runge-Kutta method.
b. Again using step size $h = 0.1$, find a numerical solution on $[0, 2]$ using the Adams-Moulton predictor-corrector method.

11. Find the analytic solution of the initial-value problem in exercise 10, and compare values with the estimates in exercise 10.

12. Compare the fourth-order Runge-Kutta method with the Adams-Moulton method in regard to strengths and weaknesses. Consider accuracy, speed, number of function evaluations required, and any other factors you consider important when selecting an algorithm for solving a specified initial-value problem.

10.7 Stability (Optional)

Multistep algorithms for estimating the solution of an initial-value problem sometimes suffer from a peculiar difficulty known as instability. When an algorithm is implemented on a machine, the machine results may be much more inaccurate than can be accounted for by analysis of local truncation error and round-off error. When that happens, we say that the algorithm is *unstable* for that particular problem. What we have in these cases is a bad case of propagated error. Small errors in the first few estimates have propagated and grown to the point that they dominate the solution and destroy the accuracy. The two-step algorithm employing the midpoint formula given by Equation 1 in Section 10.5 is notorious for this difficulty, as the following example illustrates.

EXAMPLE 10.21 Solve the initial-value problem

$$x' = -2x + 1 \qquad x(0) = 1$$

on the interval [0, 3].

Solution A machine program using the two-step midpoint formula is used with a step size of $h = 0.1$. The machine output is as shown here.

H = .1

T	X(t)	X'(t)
0.10	0.909367	−0.818733
0.20	0.836253	−0.672507
0.30	0.774865	−0.549731
0.40	0.726307	−0.452614
0.50	0.684342	−0.368685
0.60	0.652570	−0.305141
0.70	0.623314	−0.246629
0.80	0.603245	−0.206489
0.90	0.582017	−0.164033
1.00	0.570438	−0.140876
1.10	0.553841	−0.107683
1.20	0.548901	−0.097803
1.30	0.534281	−0.068562
1.40	0.535189	−0.070378
1.50	0.520205	−0.040410
1.60	0.527107	−0.054214
1.70	0.509362	−0.018725
1.80	0.523362	−0.046724
1.90	0.500018	−0.000035
2.00	0.523355	−0.046710
2.10	0.490676	0.018649
2.20	0.527085	−0.054170
2.30	0.479842	0.040317
2.40	0.535148	−0.070296
2.50	0.465782	0.068435
2.60	0.548835	−0.097671
2.70	0.446248	0.107504
2.80	0.570336	−0.140672
2.90	0.418114	0.163772
3.00	0.603090	−0.206181
3.10	0.376878	0.246245

The values of $x(t)$ decrease for a while toward a value around 0.5, but then some kind of wild oscillation seems to develop. From the form of the differential equation, it is clear that the solution should be essentially a decaying exponential function of some type, and certainly not oscillatory. Something is wrong with the machine output.   ■

The exact solution for the initial-value problem in Example 10.21 is easily seen to be

$$x(t) = \frac{1 + e^{-2t}}{2}$$

So the solution should decay steadily toward a value of 0.5. At $t = 2$, the value should be 0.5091578; at $t = 3$, we should have 0.5012394; and so on.

In Example 10.21, only the two-step formula was used with no corrector. Can better results be obtained if we use the predictor-corrector method that employs the two-step formula only as a predictor and then corrects by using the trapezoid corrector (the algorithm employed in Example 10.18)? The results of doing this are shown in Table 10.1. Not only is there no improvement, but the "corrected" estimates are in fact more inaccurate than the predicted values were. Whatever the difficulty may be, using the corrector formula certainly does not eliminate the problem or even help.

The problem in this case originates in the fact that all of our numerical methods replace the differential equation with a *difference equation*. The solution of the difference equation is a sequence of numbers y_n, where y_n approximates the true solution value $y(x_n)$ for the initial-value problem. For some algorithms, the solution of the difference equation converges to the solution of the initial-value problem as the step size h is decreased, and there is no stability problem if h is sufficiently small. Theoretically, the difference between y_n and $y(x_n)$ can be made arbitrarily small by choosing h sufficiently small. This is the case for the Euler and Runge-Kutta algorithms. Such algorithms employ a first-order difference equation in which we compute y_{n+1} using only values at the preceding step. In multistep algorithms, we are employing a higher-order difference equation to approximate the solution of the initial-value problem. In Example 10.21, we have replaced the differential equation by the difference equation

$$x_{n+1} = x_{n-1} + 2(-2x_n + 1)h$$

or

$$x_{n+1} + 4hx_n - x_{n-1} = 2h \tag{1}$$

Equation 1 is a second-order difference equation, since we use both x_n and x_{n-1} to compute x_{n+1}.

The theory for difference equations is very similar to the theory for differential equations. In particular, the general solution of Equation 1 is the sum of a particular solution of the nonhomogeneous equation and the general solution of the homogeneous difference equation

$$x_{n+1} + 4hx_n - x_{n-1} = 0 \tag{2}$$

Table 10.1   H = .1

T	X(t)	X'(t)	Y(t)
0.10	0.909367	−0.818733	
0.20	0.836253	−0.672507	0.834805
0.30	0.774865	−0.549731	0.773693
0.40	0.726307	−0.452614	0.723576
0.50	0.684342	−0.368685	0.682511
0.60	0.652570	−0.305141	0.648819
0.70	0.623314	−0.246629	0.621231
0.80	0.603245	−0.206489	0.598575
0.90	0.582017	−0.164033	0.580049
1.00	0.570438	−0.140876	0.564803
1.10	0.553841	−0.107683	0.552375
1.20	0.548901	−0.097803	0.542101
1.30	0.534281	−0.068562	0.533783
1.40	0.535189	−0.070378	0.526836
1.50	0.520205	−0.040410	0.521297
1.60	0.527107	−0.054214	0.516565
1.70	0.509362	−0.018725	0.512918
1.80	0.523362	−0.046724	0.509646
1.90	0.500018	−0.000035	0.507308
2.00	0.523355	−0.046710	0.504971
2.10	0.490676	0.018649	0.503568
2.20	0.527085	−0.054170	0.501792
2.30	0.479842	0.040317	0.501099
2.40	0.535148	−0.070296	0.499600
2.50	0.465782	0.068435	0.499507
2.60	0.548835	−0.097671	0.498045
2.70	0.446248	0.107504	0.498537
2.80	0.570336	−0.140672	0.496878
2.90	0.418114	0.163772	0.498033
3.00	0.603090	−0.206181	0.495913
3.10	0.376878	0.246245	0.497916

A particular solution is easily seen to be $x_n = \frac{1}{2}$ in this case. The general solution of Equation 2 is a linear combination of solutions of the form $x_n = r^n$ for some constant r. Substituting $x_n = r^n$ in the difference equation (Equation 2), we find that r must be a solution of the quadratic equation

$$r^2 + 4hr - 1 = 0 \qquad\qquad (3)$$

Equation 3 is called the **characteristic equation** for the difference equation. Let us assume that Equation 3 has two distinct real solutions r_1 and r_2. Then the general solution of Equation 2 will be of the form

$$Ar_1^n + Br_2^n$$

for some constants A and B determined by the initial values. For a second-order equation such as Equation 2, the constants A and B are determined by specifying values of x_0 and x_1.

The roots of Equation 3 are

$$r_1 = -2h + \sqrt{1 + 4h^2} \quad \text{and} \quad r_2 = -2h - \sqrt{1 + 4h^2}$$

If h is small, then

$$r_1 \cong 1 - 2h \quad \text{and} \quad r_2 \cong -1 - 2h$$

Then

$$r_1^n \cong (1 - 2h)^n = \left(1 + \frac{-2t_n}{n}\right)^n \cong e^{-2t_n}$$

Similarly,

$$r_2^n \cong (-1 - 2h)^n \cong (-1)^n e^{2t_n}$$

if h is small. So, for small h, an approximation to the solution of the difference equation (Equation 1) is

$$x_n = \tfrac{1}{2} + C_1 e^{-2t_n} + C_2 e^{2t_n} \tag{4}$$

Constants C_1 and C_2 are determined from the conditions that $x_0 = 1$ and $x_1 = x(h)$. By comparison with the exact solution, we can see that we should have $C_2 = 0$ and $C_1 = \tfrac{1}{2}$. Because there is some error in our value of x_1 and because round-off errors are introduced as a machine performs the computations, the value of C_2 is not zero for our computed values of x_n. So we have picked up an extraneous solution in addition to the desired solution represented by the first two terms in Equation 4.

The stability problem occurs because the homogeneous difference equation (Equation 2) has a general solution consisting of the sum of two solutions. One of these solutions is a part of the numerical solution to the initial-value problem, while the second solution should be nonexistent or possibly small with values that decay as the solution progresses. In cases where instability occurs, we have a second "extraneous" solution that eventually dominates the solution. If all arithmetic were exact, the extraneous solution would not be present, but owing to round-off errors in the first few computed values of the sequence $\{x_n\}$, the constants A and B will both be nonzero in the machine arithmetic. If the nonzero extraneous term does not decrease in value as n is increased, then instability results.

A slightly more general analysis of the two-step continuation method can be carried out by considering the difference equation corresponding to the initial-value problem

$$y' = \lambda y \qquad y(0) = y_0$$

The associated difference equation is

$$y_{n+1} = y_{n-1} + 2h\lambda y_n$$

The first term of the y_n sequence $y_0 = y(0)$ is known. We also assume that the second term $y_1 = y(h)$ is known. Substitution of $y_n = r^n$ in the difference equation yields

$$r^{n+1} = r^{n-1} + 2h\lambda r^n$$

which simplifies to

$$r^2 - 2\lambda hr - 1 = 0$$

This quadratic equation has solutions

$$r_1 = \lambda h + \sqrt{1 + \lambda^2 h^2} \quad \text{and} \quad r_2 = \lambda h - \sqrt{1 + \lambda^2 h^2}$$

The general solution of the difference equation is

$$y_n = Ar_1^n + Br_2^n$$

for constants A and B determined by the values of y_0 and y_1. That is, A and B satisfy equations

$$A + B = y_0 \quad \text{and} \quad Ar_1 + Br_2 = y_1$$

The solution is found to be

$$A = \frac{y_1 - r_2 y_0}{r_1 - r_2} \quad \text{and} \quad B = \frac{y_1 - r_1 y_0}{r_2 - r_1}$$

If h is very small, then

$$r_1 \cong \lambda h + 1 + \frac{h^2 \lambda^2}{2} \cong e^{\lambda h}$$

since

$$\sqrt{1 + \lambda^2 h^2} \cong 1 + \frac{h^2 \lambda^2}{2}$$

Similarly,

$$r_2 \cong \lambda h - 1 - \frac{h^2 \lambda^2}{2} \cong -e^{-\lambda h}$$

So the general solution of the difference equation is approximately

$$y = Ae^{\lambda nh} + B(-1)^n e^{-\lambda nh}$$

when h is small. We can also approximate A and B when h is small by using the preceding approximations for r_1 and r_2 with the fact that $y_1 = y_0 e^{\lambda h}$. The result is

$$A \cong y_0\left(\frac{-e^{-\lambda h} - e^{\lambda h}}{-e^{-\lambda h} - e^{\lambda h}}\right) = y_0 \quad \text{and} \quad B \cong y_0\left(\frac{e^{\lambda h} - e^{\lambda h}}{e^{\lambda h} - e^{-\lambda h}}\right) = 0$$

So for small h, we should have

$$y_n \cong y_0 e^{\lambda nh}$$

Now,

$$y(x_n) = y_0 e^{\lambda nh}$$

so it is clear that in theory, the first term in the general solution of the difference equation provides the numerical solution. In fact, in the numerical computations, the second term is not zero, and even in the theory, B is not quite zero. If $\lambda > 0$, then a nonzero B is not serious because the exponential factor multiplying B decreases to zero and the effect of the nonzero B decreases as n is increased. If $\lambda < 0$, we have a different situation. In that case, B is multiplied by an exponential term that grows as n increases.

The conclusion is that the two-step algorithm is unstable when applied to the given initial-value problem if $\lambda < 0$, but the algorithm is not unstable if $\lambda > 0$. In Example 10.21, we find that λ is negative, so instability is expected. If we use the algorithm to solve the problem

$$x' = 0.1x + 1 \qquad x(0) = 1$$

the algorithm should be stable. In Table 10.2, we see the machine output for this problem. There is no indication of instability. In this case, the exact solution is

$$x(t) = 11e^{0.1t} - 10$$

so we find that

$$x(0.5) = 1.563982$$
$$x(1) = 2.156880$$
$$x(2) = 3.435430$$

The agreement with the machine results is very good.

The reason for instability has been illustrated by use of the simplest of our multistep algorithms for the sake of clarity. The basic idea is the same for other

Table 10.2   H = .05

T	X(t)	X'(t)
0.05	1.055138	1.105514
0.10	1.110551	1.111055
0.15	1.166243	1.116624
0.20	1.222214	1.122221
0.25	1.278466	1.127847
0.30	1.334998	1.133500
0.35	1.391816	1.139182
0.40	1.448917	1.144892
0.45	1.506305	1.150631
0.50	1.563980	1.156398
0.55	1.621945	1.162195
0.60	1.680199	1.168020
0.65	1.738747	1.173875
0.70	1.797587	1.179759
0.75	1.856722	1.185672
0.80	1.916154	1.191615
0.85	1.975884	1.197589
0.90	2.035913	1.203591
0.95	2.096243	1.209624
1.00	2.156875	1.215688
1.05	2.217812	1.221781
1.10	2.279053	1.227905
1.15	2.340602	1.234060
1.20	2.402459	1.240246
1.25	2.464627	1.246463
1.30	2.527105	1.252711
1.35	2.589898	1.258990
1.40	2.653004	1.265300
1.45	2.716428	1.271643
1.50	2.780169	1.278017
1.55	2.844230	1.284423
1.60	2.908611	1.290861
1.65	2.973316	1.297332
1.70	3.038344	1.303835
1.75	3.103699	1.310370
1.80	3.169381	1.316938
1.85	3.235393	1.323539
1.90	3.301735	1.330174
1.95	3.368410	1.336841
2.00	3.435419	1.343542
2.05	3.502764	1.350277

algorithms, such as the Adams-Bashforth algorithms. Suppose we use a linear multi-step method defined by

$$y_{n+1} = \sum_{k=0}^{p} (A_k y_{n-k} + hB_k y'_{n-k})$$

Stability properties of this method are frequently analyzed by applying the method to the initial-value problem

$$y' = 0 \qquad y(0) = 1$$

The solution of this problem is $y(x) = 1$. Any large deviation from this solution in an approximation of the solution indicates instability. The difference equation for this problem is

$$y_{n+1} = \sum_{k=0}^{p} A_k y_{n-k}$$

The associated characteristic equation is found by replacing y_n by r^n. The result is the polynomial equation

$$r^{p+1} = \sum_{k=0}^{p} A_k r^{p-k}$$

This equation has $(p + 1)$ roots. Assume, for simplicity, that all of the roots are simple. The method should give the exact value of y_{p+1} for the given initial-value problem when there is no error in the preceding estimates. Theoretically, then,

$$y_{p+1} = \sum_{k=0}^{p} A_k = 1$$

It follows that $r_0 = 1$ is a root of the preceding polynomial equation. This root provides the solution to the initial-value problem. The general solution of the difference equation in this case is

$$1 + \sum_{k=1}^{p} C_k r_k^n$$

where the numbers $r_1, r_2, \ldots, r_p$ are the other p roots of the characteristic equation and the numbers $C_1, C_2, \ldots, C_p$ are constants determined by the values $y_0, y_1, \ldots,$ y_p. Theoretically, all these constants should be zero. Actually, in the numerical solution, they are not zero because of errors in the starting values and round-off error in the machine computations. If $|r_j| \leq 1$ for $j = 1, \ldots, p$, the effect of these extraneous terms will not be serious. But if any one of these numbers has an absolute value greater than 1, that term can contribute an extraneous solution that dominates the total solution. These facts motivate the following definitions.

Given any linear multistep method, we have a characteristic equation for the difference equation corresponding to the initial-value problem

$$y' = 0 \qquad y(0) = 1$$

Denote this equation by $P(r) = 0$ in the following.

Definition 10.1

If all the roots of $P(r) = 0$ with absolute value 1 are simple and $|r| \leq 1$ for every root r, then we say that the method satisfies the **root condition**.

Definition 10.2

If a linear multistep method satisfies the root condition, then we say that the method is stable. If $r = 1$ is the only root with absolute value 1, then we say that the method is *strongly stable*. Otherwise, the method is *weakly stable*.

For the Adams-Bashforth $(p + 1)$-step method, we have $P(r) = r^{p+1} - r^p$. The roots of $P(r) = 0$ are 1 and 0, where 0 has multiplicity p. This method is strongly stable. For the two-step method used at the beginning of this section, we find that $P(r) = r^2 - 1$. The roots of $P(r) = 0$ are -1 and $+1$. This method is evidently weakly stable. For the Milne method, we find that $P(r) = r^4 - 1$, so the roots of $P(r) = 0$ are 1, -1, i and $-i$. This method is weakly stable. We also find that the Adams-Moulton method is strongly stable. So the Adams-Moulton predictor-corrector method is strongly stable. The Milne predictor-corrector method is weakly stable. Clearly, the Milne method should be used with care. This is one of the reasons why the Adams-Moulton method is usually preferred for general-purpose machine programs.

▪ Exercise Set 10.7

1. Verify the numerical results given in Example 10.21.

2. Verify the numerical results given in Table 10.2.

3. Use the two-step method derived from the midpoint formula to find a numerical solution for each initial-value problem on the interval $[0, 2]$. Use $h = 0.1$. Decide in each case whether the algorithm is stable for that problem.
   a. $x' + x = 3; \quad x(0) = 3$ **b.** $x' + x = 3; \quad x(0) = 1$
   c. $x' + 3x = 2; \quad x(0) = 1$ **d.** $2x' + x = 3; \quad x(0) = 1$

4. Use the Adams-Moulton method to find a numerical solution on $[0, 3]$ for

$$x' = -2x + 1 \qquad x(0) = 1$$

Compare with the results in Example 10.21 and values of the solution

$$x(t) = \frac{1 + e^{-2t}}{2}$$

5. Find the general solution of each difference equation given that the general solution is of the form $Ar_1^n + Br_2^n$ for some constants A and B. Determine possible values for r by substituting $x_n = r^n$ in the difference equation.

   a. $x_{n+2} = x_{n+1} + x_n$ **b.** $8x_n - 6x_{n-1} + x_{n-2} = 0$ **c.** $x_n - 5x_{n-1} + 6x_{n-2} = 0$

6. We wish to use the improved Euler method to find a numerical solution of initial-value problem

$$y' = -10y \qquad y(0) = 1$$

   on interval $[0, 3]$. Show that if we use step size $h = 0.1$, we obtain a satisfactory approximation of solution $y(x) = e^{-10x}$, while step sizes $h = 0.2$ and $h = 0.3$ produce unusable results.

7. Show that the iteration formula employed in the improved Euler method reduces to

$$y_{n+1} = y_n(1 - \lambda h + 0.5\lambda^2 h^2)$$

   when the differential equation is $y' = -\lambda y (\lambda > 0)$. Use this result to show that the estimates are merely values of

$$y_n = (1 - \lambda h + 0.5\lambda^2 h^2)^n y_0$$

8. Apply the result in exercise 7 to the initial-value problem in exercise 6. Find the values of h for which

$$1 - \lambda h + 0.5\lambda^2 h^2 > 1$$

   in this case. Does this explain, to your satisfaction, the peculiar numerical output obtained in exercise 6?

10.8 Systems of Differential Equations

By a system of differential equations, we will mean a set of n first-order differential equations

$$y_i' = f_i(x, y_1, y_2, \ldots, y_n) \qquad i = 1, 2, \ldots, n \tag{1}$$

where $y_i = y_i(x)$. We will assume that initial values

$$y_i = y_i(x_0) \qquad i = 1, 2, \ldots, n$$

are given. We want to find n functions $y_i(x)$ that satisfy the differential equations and the initial conditions. A vector formulation of the problem is convenient. Let $Y(x)$ denote the column vector whose elements are the solution functions $y_i(x)$. Let Y^0 denote the column matrix whose elements are the constants y_i. Let $F(x, Y(x))$

denote the column matrix whose elements are the functions

$$f_i(x, y_1(x), y_2(x), \ldots, y_n(x))$$

Then the statement of our problem in vector form is

$$\mathbf{Y}'(x) = \mathbf{F}(x, \mathbf{Y}(x)) \qquad \mathbf{Y}(x_0) = \mathbf{Y}^0 \tag{2}$$

If

$$\mathbf{F}(x, Y(x)) = \mathbf{A}\mathbf{Y}(x)$$

for some $n \times n$ matrix of constants $\mathbf{A}$, then the system is a linear system. Most introductory differential equations texts are restricted to this special case. Some useful theory and methods can be developed in this special case, however, and if you are not familiar with this body of material, you may wish to consult the introductory text by D. Zill (1986). The current theory for the general case described by Equation 1 or 2 is primarily qualitative with the emphasis on theorems that describe general behavior of solutions rather than methods of solution. There are, in fact, no general methods for finding explicit analytic solutions of the system represented by Equation 2. Only very specialized cases can be solved. In the case of most such systems, we must be content to know that under certain conditions, a unique solution exists and the solution functions have certain general characteristics relating to global behavior or questions concerning stability of solutions near equilibrium points. An excellent introduction to this body of theory is found in Borrelli (1987).

In most cases where a unique solution exists, a numerical solution is the only method for obtaining quantitative information. This is a common situation in problems in the physical and biological sciences, where mathematical models based on systems of differential equations are quite widely used. Very few such systems can be solved explicitly by analytic methods. If the problem involves motion of particles, we would like to know locations (of planets, satellites, lunar probes, etc.). If the problem involves interacting populations, such as in the predator-prey problem, we would like to have population numbers or graphs (usually plotted by computers using numerical values of solution functions).

Fortunately, most such systems can be solved (approximately, of course) by numerical methods. In fact, the numerical methods most often used are merely vector forms of the Euler method or the Runge-Kutta methods or possibly one of our predictor-corrector methods. We will illustrate these methods by using systems with only two to four unknown functions.

Let us begin with a system of two equations and solve by use of a vector form of the Euler method. Given the system

$$y_1' = f(x, y_1, y_2)$$
$$y_2' = g(x, y_1, y_2)$$

with initial values $y_i(x_0) = y_i$, $i = 1, 2$, we use the iterative procedure described by

$$\mathbf{Y}^{n+1} = \mathbf{Y}^n + h\mathbf{F}(x_n, \mathbf{Y}^n) \qquad n = 0, 1, 2, \ldots$$

with $Y^n = Y(x_n)$. Vector function $F(x, Y(x))$ is described by the column matrix

$$\begin{bmatrix} f(x, y_1, y_2) \\ g(x, y_1, y_2) \end{bmatrix}$$

Explicitly, we have

$$y_1^{n+1} = y_1^n + hf(x_n, y_1^n, y_2^n)$$
$$y_2^{n+1} = y_2^n + hg(x_n, y_1^n, y_2^n)$$

with $y_1^0 = y_1$ and $y_2^0 = y_2$.

EXAMPLE 10.22   Solve the system

$$x'(t) = -0.5x \qquad\qquad x(0) = 10$$
$$y'(t) = 0.5x - 0.25y \qquad y(0) = 0$$

Solution   We will use the Euler method for systems with $h = 0.1$. Some of the machine output is as follows (only selected values are listed):

t	x	y
0.00	10	0
0.10	9.5	0.5
0.50	7.7378	2.1462
1.00	5.9873	3.5518
1.50	4.6329	4.4145
1.90	3.7735	4.8157

The qualitative behavior of $x(t)$ and $y(t)$ is clearly shown, but the numerical estimates are not very accurate. In this case, it is possible to show that the exact solution to the problem is

$$x(t) = 10e^{-t/2} \quad\text{and}\quad y(t) = 20(e^{-t/4} - e^{-t/2})$$

So $x(0.5) = 7.788$, $y(0.5) = 2.073922$, $x(1) = 6.065307$, and $y(1) = 3.445402$. The estimates are in the neighborhood, but not close enough for a serious problem in physics.   ∎

It is very easy to modify the preceding algorithm to produce a very simple easy-to-use algorithm that will produce results accurate enough for a large number of problems.

After y_1^{n+1} and y_2^{n+1} are estimated, we can use these values as predicted values and recompute using the formulas

$$y_1^{n+1} = y_1^n + h(f(x_n, y_1^n, y_2^n) + f(x_{n+1}, y_1^{n+1}, y_2^{n+1}))/2$$
$$y_2^{n+1} = y_2^n + h(g(x_n, y_1^n, y_2^n) + g(x_{n+1}, y_1^{n+1}, y_2^{n+1}))/2$$

This is, of course, a predictor-corrector method for systems and a vector form of the corrector in Equation 2 of Section 10.6. The improvement in numerical results is quite noticeable.

EXAMPLE 10.23   Solve the system in Example 10.22 using the predictor-corrector method just described.

Solution   A sample of the machine output is as follows:

t	x	y
0.10	9.5125	0.4812
0.20	9.048701	0.9271
0.50	7.7888	2.0724
1.00	6.0666	3.4431
2.00	3.6803	4.7704

The agreement with the exact values at $t = 0.5$ and $t = 1$ is much better than before. Also, we note that $x(2) = 3.6788$ and $y(2) = 4.7730$, so even at $t = 2$, we have very good estimates.   ∎

If we really want to obtain accurate estimates, we should use a vector form of the fourth-order Runge-Kutta method or possibly the Adams-Moulton predictor-corrector method. For the Runge-Kutta method, the formulas are

$$k_1 = hf(x_n, y_1^n, y_2^n)$$

$$l_1 = hg(x_n, y_1^n, y_2^n)$$

$$k_2 = hf\left(x_n + \frac{h}{2}, y_1^n + \frac{k_1}{2}, y_2^n + \frac{l_1}{2}\right)$$

$$l_2 = hg\left(x_n + \frac{h}{2}, y_1^n + \frac{k_1}{2}, y_2^n + \frac{l_1}{2}\right)$$

$$k_3 = hf\left(x_n + \frac{h}{2}, y_1 + \frac{k_2}{2}, y_2 + \frac{l_2}{2}\right)$$

$$l_3 = hg\left(x_n + \frac{h}{2}, y_1 + \frac{k_2}{2}, y_2 + \frac{l_2}{2}\right)$$

$$k_4 = hf(x_n + h, y_1 + k_3, y_2 + l_3)$$

$$l_4 = hg(x_n + h, y_1 + k_3, y_2 + l_3)$$

$$y_1^{n+1} = y_1^n + \frac{k_1 + 2k_2 + 2k_3 + k_4}{6}$$

$$y_2^{n+1} = y_2^n + \frac{l_1 + 2l_2 + 2l_3 + l_4}{6}$$

where

$$x_0, \; y_1^0 = y_1(x_0) \quad \text{and} \quad y_2^0 = y_2(x_0)$$

EXAMPLE   Solve the system in Example 10.22 using the Runge-Kutta algorithm for systems.
10.24   Here is some sample machine output:

t	x	y
0.20	9.048374	0.92784
0.30	8.607081	1.34071
0.50	7.788007	2.073922
1.00	6.065306	3.445402
1.50	4.723665	4.162055
2.00	3.678794	4.773024

From the analytic solution (see Example 10.22), we know that $x(0.5) = 7.788008$ and $y(0.5) = 2.073922$, so the estimated values agree almost perfectly to six decimal places. Also, we find that $x(1) = 6.065307$, $y(1) = 3.445402$, $x(2) = 3.678794$, and $y(2) = 4.773024$, so all of the estimates seem to be correct through six decimal places. That is enough accuracy for most purposes.   ∎

Similarly, the Adams-Moulton predictor-corrector method is easily adapted for use with systems of differential equations. The details are left to the reader.

Now suppose that we wish to solve the second-order differential equation

$$y'' = f(x, y, y')$$

with initial values $y(x_0) = c_1$ and $y'(x_0) = c_2$. If we let $y_1(x) = y(x)$ and $y_2(x) = y'(x)$, then the single second-order differential equation is converted to a system of two first-order differential equations:

$$y_1' = y_2 \qquad\qquad y_1(x_0) = c_1$$
$$y_2' = f(x, y_1, y_2) \qquad y_2(x_0) = c_2$$

So we can solve the second-order equation by solving a system for the two functions $y_1 = y$ and $y_2 = y'$. Similarly, if we wish to solve a third-order differential equation, we define $y_1 = y$, $y_2 = y'$, and $y_3 = y''$. Then the third-order equation corresponds to a system of three equations. This is a standard method for solving higher-order equations.

In this chapter, you have been introduced to several standard algorithms for numerically solving a differential equation or possibly a system of differential equations. The algorithms are easy to program. Still, it is helpful to know that professionally written software is available that is very efficient and accurate. A great deal of effort has been put into the problem of developing reliable software that gives accurate results for a wide range of problems. The IMSL MATH/LIBRARY contains several high-quality subroutines for solving initial-value problems. The routine IVPAG solves an initial-value problem using an Adams-Moulton method. The routine IVPRK uses a Runge-Kutta type of algorithm known as the Runge-Kutta-Verner method, which provides more accuracy than is available from the Runge-Kutta method of order 4 studied in this chapter. The more ambitious students of numerical analysis may wish to do some research to find out more about this method and experiment with its operation using the IMSL subroutine.

■ *Exercise Set 10.8*

1. Use the Euler method for two-variable systems to verify the numerical output presented in Example 10.22.

2. Use the predictor-corrector method for two-variable systems that employs the two-step midpoint formula as predictor with the trapezoid formula as corrector to find a numerical solution for the system discussed in Example 10.23. Find the solution for $0 \le t \le 2$ using $h = 0.1$.

3. Use the Runge-Kutta method for two-variable systems as described in this section to verify the numerical output presented in Example 10.24.

4. Construct an algorithm for solving a two-variable system of differential equations using the Adams-Moulton predictor-corrector method. Apply the algorithm to the system described in Example 10.22 to find a numerical solution on [0, 2]. Use step size $h = 0.1$.

5. Given the second-order differential equation $y'' + y = 0$ with initial values $y(0) = 0$ and $y'(0) = 1$, do the following.
   a. Convert to a two-variable system for $y_1(x) = y(x)$ and $y_2(x) = y'(x)$.
   b. Solve the system in part (a) on interval [0, 1] using the fourth-order Runge-Kutta method with $h = 0.05$.
   c. Solve the system in part (a) on interval [0, 1] using the Adams-Bashforth four-step method with $h = 0.05$.
   d. Solve the system in part (a) on interval [0, 1] using the Adams-Moulton method with $h = 0.05$.
   e. The solution of the initial-value problem is $y(x) = \sin x$. Compare numerical values of this solution with the numerical solutions obtained in parts (b)–(d).

6. Construct an algorithm for solving a two-variable system of differential equations using the improved Euler method. Use this algorithm with $h = 0.05$ to find a numerical solution on [0, 1] for each of the following initial-value problems.
   a. $y'' + 2y' - 3y = 0;$ $y(0) = 1,$ $y'(0) = 0$
   b. $y'' + xy = \sin x;$ $y(0) = 1,$ $y'(0) = 0.5$
   c. $y'' - y' - 6y = x + 1;$ $y(0) = 2,$ $y'(0) = -1$

7. Construct an algorithm using the Euler method to find a numerical solution on [0, 100] for the system

$$
\begin{aligned}
x'(t) &= -0.0001xy & x(0) &= 998 \\
y'(t) &= 0.0001xy - 0.05y & y(0) &= 2 \\
z'(t) &= 0.05y & z(0) &= 0
\end{aligned}
$$

Use step size $h = 1$.

8. Repeat exercise 7 using the fourth-order Runge-Kutta method for systems. Is there a significant difference between this numerical solution and the solution of exercise 7? Given that we only need values of $x(t)$, $y(t)$, and $z(t)$ correct to the nearest whole number, do you think it is worth the extra effort to use the Runge-Kutta method?

10.9 Applications

In this section, we will examine a variety of situations in which differential equations are used to describe the relationship between the variables involved. This happens, for example, in the study of population growth, rate of chemical reactions, launch of a space shuttle, and growth of a tumor. We begin with a population growth model.

If $N(t)$ denotes the size of a population at time t, then it is common practice to assume that $N(t)$ is the solution of an initial-value problem

$$N'(t) = f(N, t) \qquad N(0) = N_0$$

for some function f. The simplest such model is the exponential growth model in which $f(N, t) = rN$ for some constant r that we call the growth rate. All calculus students have seen this model and know that the solution is

$$N(t) = N_0 \, e^{rt}$$

A more interesting model is the logistic model in which

$$f(N, t) = aN(b - N)$$

for some positive constants a and b. Students of differential equations usually encounter this model and know that an analytic solution can be found. But in this case, the formula for $N(t)$ is complicated enough that it is not clear from the formula how the population is behaving. Often, in such cases, a numerical solution is preferred. People who deal with population problems prefer to see numbers rather than obscure formulas.

Suppose, for a certain population problem, a researcher proposes the model

$$N'(t) = 0.1N(10 - N) \qquad N(0) = 2$$

where the population numbers have been appropriately scaled. We would like to see numerical values for the resulting population. The modified Euler method is very easy to use, fast, and sufficiently accurate for this purpose. The researcher in this case is well aware that the differential equation model is only an approximation to reality and that numerical values of $N(t)$ are only approximations to real population values. So time-consuming, highly accurate algorithms are unnecessary.

The time scale is in years, so we will use $h = 0.5$, which represents semiannual estimates of the population size. The machine output for a five-year projection is shown in Table 10.3. Theoretically, graphs of solutions of the logistic equation

Table 10.3

T	N	NPRIME
0.500	1.53494	1.23975
1.000	2.28661	1.70735
1.500	3.26869	2.16456
2.000	4.43379	2.46016
2.500	5.66463	2.45541
3.000	6.81404	2.14183
3.500	7.77160	1.65929
4.000	8.49877	1.17685
4.500	9.01493	0.78877
5.000	9.36488	0.51178

follow an S-shaped pattern, since the rate of increase of N increases for a while and then decreases as the population size N approaches the limiting size b, which is 10 in this case. It is not easy to determine from the analytic solution the time at which $N'(t)$ reaches its maximum. The numerical output in this case shows clearly that for this population, this occurred around $t = 2$. A more precise value of the time could easily be determined by decreasing the step size. Using $h = 0.01$ and recomputing, we find that N' reaches a maximum value of 2.5 when $t = 2.2$. At this time, the population size is 5.000, which agrees with the theoretical fact that the value of N should be $b/2 = 5$ at this time.

Most students have at least heard of the so-called Lotka-Volterra predator-prey model used in the study of predator-prey relationships in the biological sciences. It has been used in the study of wolf (predator) and moose (prey) populations and many others. The model is a system of first-order equations based on assumptions about how the individual populations would develop separately and how their interaction affects their growth rates. The equations are

$$x'(t) = a_1 x - b_1 xy \qquad x(0) = x_0$$

$$y'(t) = -a_2 y + b_2 xy \qquad y(0) = y_0$$

Function $x(t)$ represents the prey population size and $y(t)$ represents the predator population size. Positive constants a_1 and a_2 represent the growth rates for the populations, and positive constants b_1 and b_2 represent the effect of their interaction. The rate of mixing is assumed to be proportional to the product xy. From the viewpoint of a biologist, there are some questions about the reasonableness of the assumptions in the model, but it is an interesting model and one that has been widely discussed, even in the biological literature. From the viewpoint of a mathematician, it is at least reasonable to consider what kind of solutions the system might have and what kinds of qualitative behavior they exhibit.

In this general case, no analytic solution for $x(t)$ and $y(t)$ can be found. In the literature, it is common practice to apply qualitative techniques to study the behavior of the populations without being concerned about actual numerical values of the population sizes. Given assumed values for the constants for a given predator-prey combination, it is fairly easy to find numerical values, so we shall do this. Choosing some typical values of the constants, we will solve the system

$$x' = x - 0.1xy \qquad x(0) = 25$$

$$y' = -0.5y + 0.02xy \qquad y(0) = 2$$

Using a step size of 0.5 and the vector form of the modified Euler method, we obtain the output shown in Table 10.4.

Table 10.4

T	PREY	PREDATOR
0.00	25	2
0.50	37	2.1
1.00	54.1762	2.5389
1.50	76.7789	3.7192
2.00	99.8188	6.8242
2.50	102.1629	14.7855
3.00	58.4754	27.3945
3.50	27.9874	28.8018
4.00	14.0078	25.7729
4.50	7.5264	21.8285
5.00	4.6846	17.9466
5.50	3.4504	14.5379
6.00	2.9652	11.698
6.50	2.899	9.392801
7.00	3.1434	7.5495
7.50	3.6977	6.0911
8.00	4.6344	4.9491
8.50	6.0988	4.0669
9.00	8.3284	3.4007
9.50	11.6869	2.9199
10.00	16.7101	2.6094
10.50	24.1497	2.4759
11.00	34.9612	2.5665
11.50	50.0636	3.0183
12.00	69.31301	4.1994
12.50	88.171	7.0818
13.00	90.6744	13.7118
13.50	58.7827	23.7626
14.00	28.5708	26.7166
14.50	14.5585	24.3837
15.00	8.1135	20.8274
15.50	5.2372	17.2271
16.00	3.9713	14.0282
16.50	3.4859	11.3441
17.00	3.4579	9.155599
17.50	3.7846	7.401
18.00	4.4763	6.0117

It is a well-known fact that the population values x and y should rise and fall periodically. That type of behavior is exhibited by the numerical values in the output. From the numerical values, we can see in detail how this occurs in the time relationship. It is commonly believed that after a period of time, the population sizes will return to their initial values. If the solution points $(x(t), y(t))$ for the system were plotted in the xy plane, the graph should be a closed curve. It can be shown that the analytic solution of the system of differential equations exhibits this property. The numbers in Table 10.4 do not seem to agree with this, but that is because of the error resulting from the step size of $h = 0.5$. As shown in Table 10.5, if the computations are performed using a step size $h = 0.05$, it is found that x and y return to their initial values when t is approximately 10.75. The time required to complete the population cycle is most conveniently found by use of a numerical solution such as this one.

Table 10.5

T	PREY	PREDATOR
10.45	19.5548	2.0355
10.50	20.3496	2.0252
10.55	21.1777	2.0166
10.60	22.0403	2.0098
10.65	22.9388	2.0047
10.70	23.8743	2.0015
→ 10.75	24.8484	2.0002 ←
10.80	25.8622	2.0009
10.85	26.9171	2.0037
10.90	28.0145	2.0086
10.95	29.1558	2.0158

We will now consider a problem from physics requiring us to solve a second-order differential equation. In the physical sciences, second-order equations are very common, and the following procedure is an illustration of how to solve such problems numerically.

Let us consider the pendulum motion problem. A mass m is suspended by a cord of length l from a rigid support, and we want to investigate the resulting motion if the mass (pendulum bob) is pulled to one side and released. Let θ denote the angle between the support cord and the vertical line through the support point A as shown in Figure 10.1. Some elementary physics can be used to show that we

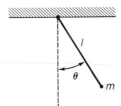

Figure 10.1

must have

$$\theta'' + \frac{g}{l} \sin \theta = 0 \quad \text{or} \quad \theta'' = -\frac{g}{l} \sin \theta$$

If we assume that the initial value of θ is θ_0 and $\theta'(0) = 0$, then we have an initial-value problem to solve. We convert to a system by setting

$$y_1 = \theta \quad \text{and} \quad y_2 = \theta'$$

We then need to solve the following system:

$$y_1' = y_2 \qquad\qquad y_1(0) = \theta_0$$

$$y_2' = -\frac{g}{l} \sin y_1 \qquad y_2(0) = 0$$

Let us suppose that $l = 60$ cm. We use textbook value $g = 980$ cm/sec^2 in this case. We will assume that the initial value of θ is 0.2 radian. The system to solve is

$$y_1' = y_2 \qquad\qquad y_1(0) = 0.2$$

$$y_2' = -\frac{49}{3} \sin y_1 \qquad y_2(0) = 0$$

In this case, we will use the Runge-Kutta method for systems because the accuracy of that method is desirable. We choose $h = 0.01$ second. The machine results at times $t = 0.1, 0.2, 0.3$, and so on, are shown in Table 10.6. No interpretation is

Table 10.6

TIME	THETA	THETAPRIME
0.10	.1839	− .3159
0.20	.1384	− .5817
0.30	.0707	− .7546
0.40	− .0084	− .8063
0.50	− .0862	− .7281
0.60	− .1501	− .5331
0.70	− .1899	− .253
0.80	− .1994	6.710001E − 02
0.90	− .1769	.3766
1.00	− .126	.6263
1.10	− .055	.7758
1.20	.025	.8005
1.30	.1008	.6964
1.40	.1605	.4806
1.50	.1944	.1883
1.60	.1972	− .1338
1.70	.1684	− .4348
1.80	.1125	− .6666
1.90	.0386	− .7917
2.00	− .0415	− .7894

Table 10.7

TIME	THETA	THETAPRIME
1.51	.1961	.1566
1.52	.1975	.1246
1.53	.1986	.0925
1.54	.1994	.0602
1.55	.1998	.0278
1.56	.1999	−.0047
1.57	.1997	−.0371
1.58	.1992	−.0695
1.59	.1984	−.1017
1.60	.1972	−.1338
1.61	.1957	−.1657

necessary, since the meaning of the numbers is clear. It is interesting to note that the period for the motion seems to be about 1.6 seconds from inspection of Table 10.6. It would be of interest to know the period more precisely. In Table 10.7, we see a close-up look at the numerical results for t in the range from 1.5 to 1.6 seconds. The data show that the period is very close to 1.56 seconds, since θ returns to the original value of 0.2 radian at that time.

In this section, we have sampled only a few representative applied problems requiring the solution of an initial-value problem. Differential equation systems and single equations have been considered. You now possess the knowledge required to produce usable solutions for a large number of such problems. In the exercises that follow, you will have an opportunity to examine a number of initial-value problems from applied areas to test your skill at doing this.

■ *Exercise Set 10.9*

1. Find a numerical solution for the predator-prey problem using $a_1 = 0.9$, $b_1 = 0.075$, $a_2 = -0.7$, and $b_2 = 0.03$. Use $x(0) = 50$ and $y(0) = 5$. Assume that time is measured in weeks. Find a solution covering a 52-week period of time.

2. If $N(t)$ is the size of a population at time t, the logistic law of population growth tells us that $N'(t) = aN(b - N)$ for some constants a and b. We wish to apply this to the population of the United States. In 1910, the population was 92 million. A demographer estimated the constants to be $a = 1.5 \times 10^{-4}$ and $b = 250$ million. Use this information to construct a numerical solution for the years from 1910 to 1990. Choose an appropriate algorithm. Use step size $h = 1$ year. Round off computer estimates to the nearest 100,000. Compare your results with U.S. Census Bureau figures.

3. Newton's law of cooling states that

$$\frac{dT}{dt} = -k(T - T_0)$$

where $T(t)$ is the temperature of an object at time t and T_0 is the ambient air temperature. A freshly poured cup of coffee has a temperature of 200° F. The room temperature is 70° F.

Given $k = 0.080$ in this case, make a table of values of the temperature of the coffee over the next 10 minutes at 30-second intervals. If you like your coffee at a temperature of $150°$ F, how long should you wait before taking a drink?

4. A room with volume 120 m^3 contains cigar smoke with a concentration of 6×10^{-5} g per cubic centimetres of air. Clear air enters the room at the rate of 10 m^3/min, and room air is removed at the same rate.
  a. Find a differential equation for cigar smoke concentration $C(t)$.
  b. Solve the resulting initial-value problem to construct a table of values over the next 5 minutes. Use the fourth-order Runge-Kutta method with a step size $h = 0.25$ minutes.
  c. Find the rate of air flow needed to produce a value of $C(5)$ less than 10^{-5}.

5. If $x(t)$ is the distance from the moon's center to an object high above the moon that is falling toward the surface of the moon, then

$$\frac{d^2x}{dt^2} = \frac{-gR^2}{x^2}$$

where $g = 5.36$ ft/sec^2 and R is about 1080 miles. Assume that $x(0) = 1100$ miles (object is 20 miles above the moon's surface) and $x'(0) = 0$. Solve the resulting initial-value problem numerically to determine the following.
  a. The number of seconds required to reach the surface of the moon ($x = 1080$ miles)
  b. The velocity (feet per second) with which the object will strike the moon.

6. In a certain deterministic model of the spread of infectious diseases, the following system of differential equations appears:

$$x'(t) = -axy$$
$$y'(t) = axy - by$$
$$z'(t) = by$$

Time t is measured in days. Function $x(t)$ represents the number of individuals who are susceptible to the disease but have not yet been infected. The number of individuals who are infected at time t is given by $y(t)$. Function $z(t)$ represents those individuals who have died or have recovered from the disease and are now immune. Suppose in a population of 500 people 2 are infected. Then $x(0) = 498$, $y(0) = 2$, and $z(0) = 0$. Solve this initial-value problem numerically for each of the possible values of a and b given. Use the Euler algorithm with step size $h = 1$ day.
  a. $a = 0.00001$, $b = 0.25$
  b. $a = 0.0002$, $b = 0.10$
  c. $a = 0.00015$, $b = 0.05$
Make a "community health" interpretation of the results in each of these cases; that is, describe the health situation represented by the epidemic figures.

7. In the study of diffraction of light, we encounter the so-called Fresnel integral

$$C(x) = \int_0^x \cos(t^2)dt$$

Function $C(t)$ is the solution of the initial-value problem

$$C'(t) = \cos(t^2) \qquad C(0) = 0$$

Use the Runge-Kutta method with $h = 0.05$ to construct a table of values of $C(t)$ on interval $[0, 1]$.

8. Use the procedure of exercise 7 to construct a table of values for Fresnel integral

$$C(x) = \int_0^x \sin(t^2)dt$$

11 BOUNDARY VALUE PROBLEMS FOR ORDINARY DIFFERENTIAL EQUATIONS

In Chapter 10, we examined several methods that can be used to approximate the solution of the differential equation $y'' = f(x, y, y')$ that satisfies the initial conditions $y(a) = A$, $y'(a) = C$. In this chapter, we will study the problem of approximating a solution of the differential equation $y'' = f(x, y, y')$ that satisfies end point conditions $y(a) = A$ and $y(b) = B$ or $y'(a) = C$ and $y(b) = B$ for some given constants A, B, or C. Such a problem is called a **two-point boundary value problem**. More general boundary value problems sometimes occur containing other kinds of information about the values of $y(x)$ and $y'(x)$ at points a and b. In this text, we will restrict ourselves to methods for these two special cases.

Some basic theoretical facts about solutions of the problem

$$y'' = f(x, y, y') \qquad y(a) = A, \quad y(b) = B$$

are important in connection with the study of numerical methods. First in importance is the fact that a solution may or may not exist.

EXAMPLE 11.1   Solve the boundary value problem

$$y'' + y = 0 \quad y(0) = 0, \quad y\left(\frac{\pi}{2}\right) = 1$$

Solution   Clearly, $y = \sin x$ is a solution of this problem. We can easily determine this by inspection. In this case, a solution exists.   ■

EXAMPLE 11.2   Solve the boundary value problem

$$y'' + y = 0 \qquad y(0) = 0, \quad y(\pi) = -1$$

Solution   The general solution of the differential equation is

$$y(x) = A \sin x + B \cos x$$

Since $y(0) = 0$, we have $0 = B$, so $B = 0$. Since $y(\pi) = -1$, we have $-1 = -B$, so $B = 1$. These contradictory results show that no solution exists satisfying the desired end point conditions.   ∎

Next we note that when a solution exists, it may not be unique.

EXAMPLE 11.3   Solve the boundary value problem

$$y'' + y = 0 \qquad y(0) = 0, \quad y(\pi) = 0$$

Solution   The general solution is the same as in Example 11.2. The equations for A and B are $0 = B$ and $0 = -B$. We conclude that B must be 0, but A is arbitrary. That means that $y(x) = A \sin x$ is a solution for every real number A. There are infinitely many solutions.   ∎

The theory of existence and uniqueness of solutions is much more complicated for boundary value problems than for initial-value problems. For more information, you may wish to consult Roberts and Shipman (1972) or Keller (1968). In the following discussion, we will assume that the boundary value problem under consideration has a unique solution. Our efforts will be concentrated on numerical procedures that can be used to produce a numerical approximation for the solution of such a problem.

11.1   Shooting Methods

Suppose that the problem

$$y'' = f(x, y, y') \qquad y(a) = A, \quad y(b) = B$$

has the unique solution $y(x)$. Let $y'(a) = \alpha$. If we know the value of α, we can construct a numerical approximation of $y(x)$ by solving the initial-value problem

$$y'' = f(x, y, y') \qquad y(a) = A, \quad y'(a) = \alpha$$

The procedures given in Section 10.8 can be used to accomplish this.

We do not know the value of α, however. In the so-called **shooting methods**, we begin with one or more estimates of α. We then proceed to compute better estimates of α using some algorithm, until a satisfactory estimate is found. Various algorithms have been developed for producing a sequence of improved estimates of α. This leads to a variety of different shooting methods. Most of these algorithms employ the following basic idea. Let the solution of the initial-value problem using $y'(a) = \alpha$ be $y(x, \alpha)$ to denote the dependence on α. We need to find that α for which $y(b, \alpha) = B$. This means that α can be viewed as a solution of the equation $y(b, \alpha) = B$. This suggests that we try the algorithms given in Chapter 2 for solving an equation to determine α.

If we wish to employ the regula falsi method to do this, we could proceed as follows. We choose two values α_1 and α_2 so that $\alpha_1 < \alpha < \alpha_2$. As shown in Section 10.8, to solve our initial-value problem, we first replace the second-order differential equation with a system of two first-order differential equations by setting $y_1(x) = y(x)$ and $y_2(x) = y'(x)$. Then we have

$$
\begin{aligned}
y_1' &= y_2 & y_1(a) &= A \\
y_2' &= f(x, y_1, y_2) & y_2(a) &= \alpha
\end{aligned}
$$

Using the Runge-Kutta method of order 4 for systems, we can solve this system using $\alpha = \alpha_1$ to obtain $y(b, \alpha_1)$. Now we repeat with $\alpha = \alpha_2$ to obtain $y(b, \alpha_2)$. Using the regula falsi method to obtain a better estimate of α, we compute

$$
\alpha_3 = \frac{\alpha_1(y(b, \alpha_2) - B) - \alpha_2(y(b, \alpha_1) - B)}{y(b, \alpha_2) - y(b, \alpha_1)}
$$

Now, using $\alpha = \alpha_3$, we solve the initial-value problem to obtain $y(b, \alpha_3)$. If $y(b, \alpha_3)$ is sufficiently close to B, then we use $\alpha = \alpha_3$ and output the solution. If not, we proceed as follows. If

$$
(y(b, \alpha_1) - B)(y(b, \alpha_3) - B) < 0
$$

then we set $\alpha_2 = \alpha_3$. Otherwise, we set $\alpha_1 = \alpha_3$. We now proceed to compute a new α_3, as above. This process is continued until $|y(b, \alpha_3) - B|$ is as small as desired.

EXAMPLE 11.4   Solve the boundary value problem

$$
y'' = -y \qquad y(0) = 0, \quad y\!\left(\frac{\pi}{2}\right) = 1
$$

Solution   The unique solution of this problem is $y(x) = \sin x$. A numerical solution using the procedure just described produces the following results.

```
ENTER VALUES OF a and b 0, 1.570796
ENTER VALUES OF X(a) and X(b) 0, 1
ENTER STEP SIZE H ? .05
ENTER INITIAL ESTIMATES OF Y'(A) ? .8, .95

USING y'(a) = 1.000217
```

X	Y	YPRIME
0.05	0.04999	0.99897
0.10	0.09986	0.99522
0.15	0.14947	0.98899
0.20	0.19871	0.98028
0.25	0.24746	0.96912
0.30	0.29558	0.95554
0.35	0.34297	0.93958
0.40	0.38950	0.92126
0.45	0.43506	0.90064
0.50	0.47953	0.87777
0.55	0.52280	0.85271
0.60	0.56476	0.82551
0.65	0.60532	0.79626
0.70	0.64436	0.76501
0.75	0.68179	0.73185
0.80	0.71751	0.69686
0.85	0.75144	0.66013
0.90	0.78350	0.62174
0.95	0.81359	0.58181
1.00	0.84165	0.54042
1.05	0.86761	0.49768
1.10	0.89140	0.45369
1.15	0.91296	0.40858
1.20	0.93224	0.36244
1.25	0.94919	0.31539
1.30	0.96377	0.26756
1.35	0.97593	0.21905
1.40	0.98566	0.17000
1.45	0.99293	0.12053
1.50	0.99771	0.07075
1.55	1.00000	0.02080

Comparing estimates with accurate values of sin x, we see that the agreement is very good throughout the interval $[0, \pi/2]$. ∎

Other algorithms can be used in place of the method of regula falsi. Also, the initial-value problem can be solved by various algorithms from Chapter 10. In the following algorithm, the secant method is used to find estimates for $\alpha = y'(a)$, and the fourth-order Runge-Kutta algorithm is used to solve the initial-value problem for each α.

Algorithm 11.1

Step 1 Input a, b, A, B, h, TOL, α_1 and α_2.
Step 2 Compute $y(b, \alpha_1)$ by solving the initial-value problem

$$y'' = f(x, y, y') \qquad y(a) = A, \quad y'(a) = \alpha_1$$

using the fourth-order Runge-Kutta method. Set $y_1(b) = y(b, \alpha_1)$.
Step 3 Compute $y(b, \alpha_2)$ by solving the initial-value problem

$$y'' = f(x, y, y') \qquad y(a) = A, \quad y'(a) = \alpha_2$$

using the fourth-order Runge-Kutta method. Set $y_2(b) = y(b, \alpha_2)$.
Step 4 Compute

$$\alpha_3 = \alpha_1 + \frac{B - y_1(b)}{y_2(b) - y_1(b)} (\alpha_2 - \alpha_1)$$

Step 5 Compute $y(b, \alpha_3)$ by solving the initial-value problem

$$y'' = f(x, y, y') \qquad y(a) = A, \quad y'(a) = \alpha_3$$

using the fourth-order Runge-Kutta method. Set $y_3(b) = y(b, \alpha_3)$.
Step 6 If $|y_3(b) - B| <$ TOL, then go to step 8.
Step 7 Set $\alpha_1 = \alpha_2$, $\alpha_2 = \alpha_3$, $y_1(b) = y_2(b)$, and $y_2(b) = y_3(b)$. Go to step 4.
Step 8 Output solution $y(x, \alpha_3)$.

EXAMPLE 11.5 Solve the problem

$$y'' + y = 0 \qquad y(0) = 0, \quad y(1) = 0.84147$$

Solution The results from applying Algorithm 11.1 to this boundary value problem with $h = 0.05$ are displayed in the following table.

```
ENTER VALUES OF a and b 0, 1
ENTER VALUES OF X(a) and X(b) 0, .841471
ENTER STEP SIZE H ? .05
ENTER INITIAL ESTIMATES OF Y'(A) ? .8, .95

USING y'(a) = 1
```

X	Y	YPRIME
0.00	0.00000	1.00000
0.05	0.04998	0.99875
0.10	0.09983	0.99500

X	Y	YPRIME
0.15	0.14944	0.98877
0.20	0.19867	0.98007
0.25	0.24740	0.96891
0.30	0.29552	0.95534
0.35	0.34290	0.93937
0.40	0.38942	0.92106
0.45	0.43497	0.90045
0.50	0.47943	0.87758
0.55	0.52269	0.85252
0.60	0.56464	0.82534
0.65	0.60519	0.79608
0.70	0.64422	0.76484
0.75	0.68164	0.73169
0.80	0.71736	0.69671
0.85	0.75128	0.65998
0.90	0.78333	0.62161
0.95	0.81342	0.58168
1.00	0.84147	0.54030

The solution of the problem is $y = \sin x$. The estimate of $\sin 0.1$ is 0.099833, while the true value is 0.0998334. The estimate of $\sin 0.5$ is 0.479425, while the true value is 0.4794255. The estimate of $\sin 0.8$ is 0.717356, while the true value is 0.7173561. The error is extremely small throughout the interval $[0, 1]$.   ∎

We might expect to obtain more rapid convergence in a shooting method if we employed the Newton algorithm to find estimates of $y'(a)$. Let

$$g(\alpha) = y(b, \alpha) - B$$

We need to find a root of the equation $g(\alpha) = 0$. If α_0 is an initial estimate of a root, then we need to compute

$$\alpha_1 = \alpha_0 - \frac{g(\alpha_0)}{g'(\alpha_0)}$$

Note that

$$g'(\alpha) = \frac{\partial y(b, \alpha)}{\partial \alpha}$$

Now from the differential equation

$$y''(x, \alpha) = f(x, y(x, \alpha), y'(x, \alpha))$$

we find that

$$\frac{\partial y''(x, \alpha)}{\partial \alpha} = \frac{\partial f}{\partial y}\frac{\partial y}{\partial \alpha} + \frac{\partial f}{\partial y'}\frac{\partial y'}{\partial \alpha}$$

Let

$$u(x, \alpha) = \frac{\partial y(x, \alpha)}{\partial \alpha}$$

Then

$$g'(\alpha_0) = u(b, \alpha_0)$$

This is the number we need to compute. If we assume that the interchange of differentiations is legitimate, then we find that

$$\frac{\partial y''(x, \alpha)}{\partial \alpha} = \frac{d^2}{dx^2} \frac{\partial y}{\partial \alpha} = u''(x, \alpha)$$

So

$$u''(x, \alpha) = \frac{f}{y} u + \frac{f}{y'} u'$$

Also, $u(a, \alpha) = 0$ and differentiation of $y'(a, \alpha) = \alpha$ gives

$$\frac{\partial y'(a, \alpha)}{\partial \alpha} = 1$$

So

$$u'(a, \alpha) = 1$$

Then $u(x, \alpha)$ is the solution of the initial-value problem

$$u''(x, \alpha) = \frac{\partial f}{\partial y} u(x, \alpha) + \frac{\partial f}{\partial y'} u'(x, \alpha)$$

$$u(a, \alpha) = 0, \quad u'(a, \alpha) = 1$$

Given an estimate α_0, we can solve this initial-value problem for $u(b, \alpha_0) = g'(\alpha_0)$. Then the new estimate of α is α_1, where

$$\alpha_1 = \alpha_0 - \frac{y(b, \alpha_0) - B}{u(b, \alpha_0)}$$

Using this value for $y'(a)$, we solve the initial-value problem

$$y'' = f(x, y, y') \qquad y(a) = A, \quad y'(a) = \alpha_1$$

to determine $y(b, \alpha_1)$. If $| y(b, \alpha_1) - B |$ is sufficiently small, we print out the values of $y(x, \alpha)$ as the numerical solution of the boundary value problem. If not, we replace α_0 by α_1 and repeat the process. That is the basic idea behind the following algorithm for solving the boundary value problem.

Algorithm 11.2

Step 1 Input a, b, A, B, TOL, α_0, and h.
Step 2 Use the fourth-order Runge-Kutta method to solve the initial-value problem

$$y''(x, \alpha_0) = f(x, y, y') \qquad y(a) = A, \quad y'(a) = \alpha_0$$

on the interval $[a, b]$ to obtain $y(b, \alpha_0)$.
Step 3 Use the fourth-order Runge-Kutta method to solve the initial-value problem

$$u'' = uf + u'f \qquad u(a) = 0, \quad u'(a) = 1$$

on the interval $[a, b]$ to obtain $u(b)$.

Step 4 Set $\alpha_1 = \alpha_0 - \dfrac{y(b, \alpha_0) - B}{u(b)}$

Step 5 Use the fourth-order Runge-Kutta method to solve the initial-value problem

$$y''(x, \alpha_1) = f(x, y, y') \qquad y(a) = A, \quad y'(a) = \alpha_1$$

on the interval $[a, b]$ to obtain $y(b, \alpha_1)$.
Step 6 If $| y(b, \alpha_1) - B | <$ TOL, then go to step 8.
Step 7 Set $\alpha_0 = \alpha_1$ and return to step 2.
Step 8 Output computed values of $y(x, \alpha_1)$ and stop.

EXAMPLE 11.6   Solve the boundary value problem

$$yy'' + (y')^2 + 1 = 0 \qquad y(0) = 1, \quad y(2) = 2$$

Solution   Using Algorithm 11.2 with $h = 0.1$ and TOL $= 0.000015$, we obtain the results shown in the following table.

```
ENTER VALUES OF a and b 0, 1
ENTER VALUES OF X(a) and X(b) 1, 2
ENTER STEP SIZE H ? .05
ENTER INITIAL ESTIMATE OF Y'(A) ? 1.5
```

USING $y'(a) = 2.000021$

X	Y	YPRIME
0.00	1.000000	2.000021
0.05	1.094302	1.781969
0.10	1.178981	1.611569
0.15	1.255984	1.472955
0.20	1.326649	1.356808
0.25	1.391940	1.257243
0.30	1.452583	1.170333
0.35	1.509138	1.093343
0.40	1.562049	1.024299
0.45	1.611676	0.961736
0.50	1.658312	0.904538
0.55	1.702204	0.851840
0.60	1.743560	0.802958
0.65	1.782554	0.757343
0.70	1.819341	0.714548
0.75	1.854050	0.674203
0.80	1.886797	0.636002
0.85	1.917682	0.599685
0.90	1.946793	0.565035
0.95	1.974210	0.531861
1.00	2.000001	0.500003

The exact value of $y'(0)$ is 2, so there is a noticeable amount of error in the estimate of $y'(0)$. Using a smaller h would provide a better estimate, with an increase in the amount of computing time required. ∎

In Algorithms 11.1 and 11.2, other algorithms can be substituted for the fourth-order Runge-Kutta method to solve the initial-value problems. To gain some speed, we might want to use a Runge-Kutta algorithm of order 2, such as the improved Euler method. There is some loss of accuracy in doing this, but for some purposes, the numerical solution produced might be accurate enough with a properly chosen h.

■ *Exercise Set 11.1*

1. Use Algorithm 11.1 to approximate the solution of each of the following boundary value problems. Use $h = 0.05$.
   a. $x^2 y'' - 5xy' + 8y = 24$; $y(1) = 3$, $y(2) = 15$
   b. $y'' - 2y' + y = x^2 - 1$; $y(0) = 5$, $y(1) = 10$
   c. $y'' + xy' + y = 2x$; $y(0) = 1$, $y(1) = 0$
   d. $y'' + 2y(y')^3 = 0$; $y(0) = 0$, $y(1) = 0.5$
   e. $yy'' + (y')^2 + 1 = 0$; $y(0) = 1$, $y(1) = 2$
   f. $y'' = \dfrac{2y}{x^2}$; $y(1) = 2$, $y(2) = 4.5$

2. Repeat exercise 1 using Algorithm 11.2.

3. Use Algorithm 11.2 to approximate the solution of the boundary value problem

$$y'' - 2y' + y = x^2 - 1 \qquad y(0) = 5, \quad y(1) = 10$$

   a. Use $h = 0.05$ with initial estimate 2 for $y'(0)$.
   b. Use $h = 0.05$ with initial estimate 3 for $y'(0)$.
   c. Use $h = 0.05$ with initial estimate 5 for $y'(0)$.
   d. Compare the estimates obtained in parts (a)–(c) with values of the exact solution $y(x) = x^2 + 4x + 5$. Is there any significant difference in the approximate solutions? If so, which one is most accurate?

4. Approximate the solution of the boundary value problem

$$x^2 y'' - 4xy' + 6y = 0 \qquad y(1) = 0, \quad y(2) = -4$$

   using Algorithm 11.1 first with $h = 0.1$ and then with $h = 0.05$. Use TOL = 0.000005. In both cases, use as initial estimates for $y'(1)$ the values -0.5 and -0.8. Compare results with values of the exact solution $y(x) = x^2 - x^3$. Verify that this is the solution to the given boundary value problem.

5. Verify that the differential equation

$$y'' + 6y(y')^3 = 0$$

   is satisfied by

$$x = y^3 + c_1 y + c_2$$

   for all choices of the constants c_1 and c_2. Then do the following.
   a. Show that the boundary value problem consisting of the differential equation

$$y'' + 6y(y')^3 = 0$$

   with boundary conditions $y(1) = 0$ and $y(3) = 0$ is satsified by $x = y^3 + y + 1$.
   b. Use Algorithm 11.1 with $h = 0.05$ to approximate the solution to the boundary value problem in part (a). Substitute the approximations for $y(1.5)$, $y(2)$, and $y(2.5)$ into the formula $x = y^3 + y + 1$, and compare results with the true values 1.5, 2, and 2.5 for x.
   c. Verify that the relation $x = y^3 + y + 1$ does define y as an implicit function of x. Sketch the graph.
   d. Show that the boundary value problem consisting of the differential equation

$$y'' + 6y(y')^3 = 0$$

   with the boundary conditions $y(0) = 0$ and $y(1) = 2$ has the formal "solution" $x = y^3 - 3.5y$, in that all the conditions appear to be satisfied. Then show that the given "solution" does not define y as a function of x.
   e. Apply Algorithm 11.1 to the problem in part (d), and attempt to find an approximate solution to this problem. Use various initial estimates for $y'(0)$. Discuss the numerical results.

11.2   Linear Shooting Methods

If the differential equation is linear, then our boundary value problem assumes the form

$$y'' + p(x)y' + q(x)y = f(x) \qquad y(a) = A, \quad y(b) = B$$

In this special case, it is possible to develop simpler shooting methods for solving our boundary value problem. Iteration methods such as the secant method or the Newton method are not necessary. We need only solve two initial-value problems. The solutions of these two problems can be used to determine the solution of the boundary value problem. A variety of algorithms are possible. In this section, we will explore two of these.

Let $y_1(x)$ and $y_2(x)$ be the solutions of the initial-value problem

$$y'' + p(x)y' + q(x)y = f(x) \qquad y(a) = A, \quad y'(a) = \alpha$$

corresponding to choosing $\alpha = \alpha_1$ and α_2, respectively. The values α_1 and α_2 may be chosen arbitrarily, so long as they are not equal. Next,

$$\text{let } y(x) = c_1 y_1(x) + c_2 y_2(x)$$

We want to choose constants c_1 and c_2 so that $y(x)$ is the solution to our boundary value problem.

We find that

$$y'' + py' + qy = c_1(y_1'' + py_1' + qy_1) + c_2(y_2'' + py_2' + qy_2)$$
$$= c_1 f(x) + c_2 f(x) = f(x)$$

so long as $c_1 + c_2 = 1$. Also,

$$y(a) = c_1 y_1(a) + c_2 y_2(a) = c_1 A + c_2 A = A$$

as desired in this case. Finally, we need

$$y(b) = c_1 y_1(b) + c_1 y_2(b) = B$$

The constants c_1 and c_2 are determined by the system of equations

$$c_1 + c_2 = 1$$
$$c_1 y_1(b) + c_2 y_2(b) = B$$

We find that

$$c_1 = \frac{B - y_2(b)}{y_1(b) - y_2(b)}$$

and

$$c_2 = 1 - c_1$$

Now we have a possible algorithm. We can solve the initial-value problem

$$y'' + py' + qy = f(x) \qquad y(a) = A, \quad y'(a) = 0$$

over the interval $[a, b]$ and call the solution $y_1(x)$. Then we solve the initial-value problem using $y'(a) = 1$ and call this solution $y_2(x)$. Next we compute

$$c_1 = \frac{B - y_2(b)}{y_1(b) - y_2(b)} \quad \text{and} \quad c_2 = 1 - c_1$$

The solution of our boundary value problem is

$$y(x) = c_1 y_1(x) + c_2 y_2(x)$$

The computation of solutions $y_1(x)$ and $y_2(x)$ on $[a, b]$ can be accomplished using any of the algorithms from Chapter 10 for solving initial-value problems. We need to choose a step size h so that $(b - a)/h = N$ is an integer, and we need to define mesh points $x_i = a + ih$ for $i = 0, 1, \ldots, N$. Numerical estimates for $y_1(x)$ and $y_2(x)$ at points x_i have to be stored in an array. Then we compute values of $y(x)$ at these same points using the linear relation

$$y(x_i) = c_1 y_1(x_i) + c_2 y_2(x_i) \quad i = 0, 1, \ldots, N$$

That is the idea behind the following algorithm.

Algorithm 11.3

Step 1 Input a, b, A, B, and h.

Step 2 Set $N = (b - a)/h$.

Step 3 Using the fourth-order Runge-Kutta method with step size h, compute the solution of the initial-value problem

$$y'' + py' + qy = f(x) \qquad y(a) = A, \quad y'(a) = 0$$

on $[a, b]$. Set $y_1(i) = y(a + ih)$ at each step.

Step 4 Using the fourth-order Runge-Kutta method with step size h, compute the solution of the initial-value problem

$$y'' + py' + qy = f(x) \qquad y(a) = A, \quad y'(a) = 1$$

on $[a, b]$. Set $y_2(i) = y(a + ih)$ at each step.

Step 5 Compute

$$c_1 = \frac{B - y_2(N)}{y_1(N) - y_2(N)}$$

and $c_2 = 1 - c_1$.

Step 6 For $i = 0$ to N, set $y(i) = c_1 y_1(i) + c_2 y_2(i)$, and output values of $y(i)$.

EXAMPLE 11.7   Find the solution of the problem

$$x^2 y'' - xy' + y = x^2 \qquad y(1) = 1, \quad y(2) = 4 - 2 \ln 2$$

Solution   We use Algorithm 11.3 in this case with $h = 0.1$. We approximate $4 - 2 \ln 2$ by 2.6137056. The resulting output appears in the following table.

```
ENTER VALUES OF a and b 1, 2
ENTER VALUES OF X(a) and X(b) 1, 2.6137056
ENTER STEP SIZE H ? .1
```

X	Y ESTIMATE	TRUE Y
1.00	1.000000	1.000000
1.10	1.105159	1.105159
1.20	1.221214	1.221214
1.30	1.348926	1.348927
1.40	1.488939	1.488939
1.50	1.641802	1.641803
1.60	1.807994	1.807994
1.70	1.987932	1.987932
1.80	2.181984	2.181985
1.90	2.390478	2.390478
2.00	2.613706	2.613706

You can easily verify that the exact solution of this problem is

$$y(x) = x^2 - x \ln x$$

Values computed from this formula agree very well with the output from the algorithm.   ∎

EXAMPLE 11.8   Find the solution of the problem

$$y'' + y' + xy = 0 \qquad y(0) = 1, \quad y(1) = 0$$

Solution   Using Algorithm 11.3 with a step size $h = 0.1$, we get the following results.

```
ENTER VALUES OF a and b 0, 1
ENTER VALUES OF X(a) and X(b) 1, 0
ENTER STEP SIZE H ? .1
```

X	Y ESTIMATE
0.00	1.000000
0.10	0.857164
0.20	0.727120
0.30	0.608081
0.40	0.498648
0.50	0.397744
0.60	0.304563
0.70	0.218523
0.80	0.139227
0.90	0.066428
1.00	0.000000

In this case, the exact solution can only be represented by a power series. So we have no closed-form analytic solution with which to compare our numerical estimates. Some idea of the accuracy of our estimates can be obtained by solving the problem again using $h = 0.05$. (See exercise 9.)   ■

Differentiation of

$$y(x) = c_1 y_1(x) + c_2 y_2(x)$$

gives

$$y'(x) = c_1 y_1'(x) + c_2 y_2'(x)$$

It follows that

$$y'(a) = c_1 y_1'(a) + c_2 y_2'(a) = c_2$$

As an alternative to the earlier algorithm, we could compute $y_1(x)$, $y_2(x)$, and c_2 as before. Then, using $y'(a) = c_2$, we could solve the resulting initial-value problem for $y(x)$. In this case, we do not need to store values of $y_1(x_i)$ and $y_2(x_i)$. We only need to compute $y_1(b)$ and $y_2(b)$ to compute c_2. We do have to solve three initial-value problems rather than two. We have a trade-off on time and storage requirements. If $(b - a)/h$ is large, then the size of the arrays required in Algorithm 11.3 for values of $y_1(x)$ and $y_2(x)$ becomes large. If this is a problem, our alternate approach might be preferred. We sum up this alternate procedure in the following algorithm.

Algorithm 11.4

Step 1 Input a, b, A, B, and h.

Step 2 Using the fourth-order Runge-Kutta method with step size h, compute the solution of the initial-value problem

$$y'' + py' + qy = f(x) \qquad y(a) = A, \quad y'(a) = 0$$

on the interval $[a, b]$. Set $y_1(b) = y(b)$.

Step 3 Using the fourth-order Runge-Kutta method with step size h, compute the solution of the initial-value problem

$$y'' + py' + qy = f(x) \qquad y(a) = A, \quad y'(a) = 1$$

on the interval $[a, b]$. Set $y_2(b) = y(b)$.

Step 4 Compute

$$c_2 = \frac{B - y_1(b)}{y_2(b) - y_1(b)}$$

Algorithm 11.4, continued

Step 5 Using the fourth-order Runge-Kutta method with step size h, compute the solution of the initial-value problem

$$y'' + py' + qy = f(x) \qquad y(a) = A, \quad y'(a) = c_2$$

on the interval $[a, b]$. Output values of $y(x)$.

EXAMPLE 11.9   Find the solution of the problem

$$x^2 y'' - xy' + y = x^2 \qquad y(1) = 1, \quad y(2) = 4 - 2 \ln 2$$

Solution   We use Algorithm 11.4 in this case, with $h = 0.05$. As in Example 11.7, we will approximate $4 - 2 \ln 2$ by 2.6137056. The resulting numerical output is as follows.

```
ENTER VALUES OF a and b 1, 2
ENTER VALUES OF x(a) and x(b) 1, 2.6137056
ENTER STEP SIZE H ? .1

USING y'(a) = 1
```

X	Y	YPRIME
1.10	1.105159	1.104690
1.20	1.221214	1.217678
1.30	1.348926	1.337636
1.40	1.488939	1.463528
1.50	1.641802	1.594535
1.60	1.807994	1.729997
1.70	1.987932	1.869372
1.80	2.181984	2.012214
1.90	2.390478	2.158147
2.00	2.613706	2.306853

These values agree exactly with the results listed in Example 11.7 for the output from Algorithm 11.3 when applied to this same boundary value problem.   ∎

EXAMPLE 11.10   Find the solution of the problem

$$y'' + y' + xy = 0 \qquad y(0) = 1, \quad y(1) = 0$$

Solution   Using Algorithm 11.4 with $h = 0.05$, we obtain the following results.

```
ENTER VALUES OF a and b 0, 1
ENTER VALUES OF x(a) and x(b) 1, 0
ENTER STEP SIZE H ? .05
```

USING $y'(a) = -1.499388$

X	Y	YPRIME
0.05	0.926854	−1.427431
0.10	0.857164	−1.361070
0.15	0.790664	−1.299704
0.20	0.727119	−1.242784
0.25	0.666320	−1.189808
0.30	0.608080	−1.140315
0.35	0.552237	−1.093884
0.40	0.498647	−1.050132
0.45	0.447185	−1.008705
0.50	0.397744	−0.969282
0.55	0.350229	−0.931571
0.60	0.304562	−0.895303
0.65	0.260678	−0.860237
0.70	0.218522	−0.826154
0.75	0.178050	−0.792856
0.80	0.139227	−0.760167
0.85	0.102026	−0.727928
0.90	0.066429	−0.696003
0.95	0.032422	−0.664269
1.00	0.000000	−0.632624

It is interesting to note that with $h = 0.1$, we obtain almost exactly the same estimates at the points where $x = 0.1, 0.2, 0.3, \ldots, 0.9$. The numerical results are shown in the following table.

ENTER VALUES OF a and b 0, 1
ENTER VALUES OF x(a) and x(b) 1, 0
ENTER STEP SIZE H ? .1

USING $y'(a) = -1.499389$

X	Y	YPRIME
0.10	0.857164	−1.361070
0.20	0.727119	−1.242785
0.30	0.608081	−1.140315
0.40	0.498648	−1.050132
0.50	0.397744	−0.969283
0.60	0.304562	−0.895304
0.70	0.218522	−0.826154
0.80	0.139226	−0.760167
0.90	0.066428	−0.696003
1.00	−0.000000	−0.632624

This indicates that the error in these estimates is probably of the order of 10^{-6}.   ∎

Numerous variations on Algorithms 11.3 and 11.4 are possible. You will be invited to investigate some of these in the exercises.

■ *Exercise Set 11.2*

1. Use Algorithm 11.3 to approximate the solution of each of the following boundary value problems. Use $h = 0.05$.

   a. $x^2 y'' - 5y' + 8y = 24$; $\quad y(1) = 3$, $\quad y(2) = 15$
   b. $y'' - 2y' + y = x^2 - 1$; $\quad y(0) = 5$, $\quad y(1) = 10$
   c. $y'' + xy' + y = 2x$; $\quad y(0) = 1$, $\quad y(1) = 0$
   d. $y'' + 2y' - 8y = 0$; $\quad y(0) = 0$, $\quad y(1) = 0.5$
   e. $y'' + 2y' + 1 = 0$; $\quad y(0) = 1$, $\quad y(1) = 2$
   f. $y'' = 2y/x^2$; $\quad y(1) = 2$, $\quad y(2) = 4.5$

2. Repeat exercise 1 using Algorithm 11.4.

3. Use Algorithm 11.4 to approximate the solution of the boundary value problem

$$y'' - 2y' + y = x^2 - 1 \qquad y(0) = 5, \quad y(1) = 10$$

   a. Use $h = 0.1$.
   b. Use $h = 0.05$.
   c. Use $h = 0.02$.
   d. Compare the estimates obtained in parts (a)–(c) with values of the exact solution $y(x) = x^2 + 4x + 5$. Is there any significant difference in the approximate solutions? If so, which one is most accurate?
   e. Compare the approximate solutions obtained in parts (a)–(c) with the approximate solution found in exercise 3 of Exercise Set 11.1. Is there any noticeable difference with regard to accuracy?

4. Approximate the solution of the boundary value problem

$$x^2 y'' - 4xy' + 6y = 0 \qquad y(1) = 0, \quad y(2) = -4$$

   using Algorithm 11.3 with $h = 0.05$. Repeat with Algorithm 11.4. Compare results with values of the exact solution $y(x) = x^2 - x$. Compare these approximate solutions with those found in exercise 4 of Exercise Set 11.1.

5. Verify that the general solution of the differential equation

$$(1 + 2x)y'' + 4xy' - 4y = 0$$

   is $y = c_1 x + c_2 e^{-2x}$. Find values of c_1 and c_2 so that $y(0) = -1$ and $y(2) = 2 - e^{-4}$. Now find an approximate solution using Algorithm 11.3 with $h = 0.1$. Compute the error in the numerical estimates of $y(0.2)$, $y(0.5)$, $y(1)$, and $y(1.5)$. Repeat with $h = 0.05$.

6. Repeat exercise 5 using Algorithm 11.4.

7. Verify that $y(x) = 3x + 2e^{-x}$ is the solution of the boundary value problem

$$(1 + x)y'' + xy' - y = 0 \qquad y(0) = 2, \quad y(1) = 3 + \frac{2}{e}$$

   Now find an approximate solution using Algorithm 11.3 with $h = 0.05$. Compute the error at the points $x = 0.2$, 0.4, 0.6, and 0.8 by comparing estimates with values of the exact solution.

8. Repeat exercise 7 using Algorithm 11.4.

9. The exact solution for the boundary value problem

$$y'' + 4y' + 5y = 8 \sin x \qquad y(0) = -1, \quad y(\pi/2) = 1$$

   is $y(x) = \sin x - \cos x$.

a. Use Algorithm 11.3 with $h = 0.1$ to approximate the solution. Use the exact solution to compute errors at $x = 0.2, 0.6, 1.0, 1.2,$ and 1.5.

b. Repeat part (a) using $h = 0.05$.

c. Repeat part (a) using $h = 0.01$.

d. Change the boundary conditions to $y(0) = -1$ and $y(1.5) = 0.9267578$. Use Algorithm 11.3 with $h = 0.05$ to approximate the solution. Compare estimates with values of the exact solution $y(x) = \sin x - \cos x$.

e. How do you explain the fact that the errors obtained in part (d) are so much smaller than those obtained in part (b)?

11.3   Finite Difference Methods

It is sometimes possible to solve a boundary value problem by replacing the differential equation with a difference equation approximation. This can be done by using the numerical differentiation formulas in Section 9.1 to approximate the derivatives in the equation. We define mesh points $x_i + a + ih$ and apply the approximation formulas at the point x_i. The result is a system of equations to be solved for values y_i of the solution function. We will confine ourselves to second-order linear differential equations in this section. With this restriction, the system of equations for the solution values is a linear system that can be easily solved. In theory, the method could be used on higher-order linear or nonlinear equations. If the differential equation is nonlinear, then the difference approximation leads to a nonlinear system. This increases the difficulty of finding the solution.

We will assume that the differential equation has the form

$$y'' + p(x)y' + q(x)y = f(x)$$

in the following. We will use the numerical formulas

$$y'(x_i) = \frac{y(x_{i+1}) - y(x_{i-1})}{2h} - \frac{h^2}{6} y'''(\xi_i)$$

$$y''(x_i) = \frac{y(x_{i+1}) - 2y(x_i) + y(x_{i-1})}{h^2} - \frac{h^2}{12} y^{(4)}(\eta_i)$$

to approximate the derivatives in the differential equation. We substitute the approximations resulting from dropping the error terms. After simplifying the results, we obtain the system of equations

$$\left[1 - \frac{h}{2} P(x_i)\right] y_{i-1} + [h^2 q(x_i) - 2] y_i$$

$$+ \left[1 + \frac{h}{2} P(x_i)\right] y_{i+1} = h^2 f(x_i) \qquad i = 1, 2, \ldots, n-1$$

We use $y_0 = A$ and $y_n = B$. Then we have a linear system of $n - 1$ equations for the unknowns $y_1, y_2, \ldots, y_{n-1}$. This is a tridiagonal system and can be easily solved.

The algorithm that follows employs this method to solve linear boundary value problems.

Algorithm 11.5

Step 1 Input a, b, A, B, h.

Step 2 Let $N = (b - a)/h$.

Step 3 Set $x_0 = a$, $x_N = b$, $x_1 = a + h$, $x_{N-1} = b - h$. For $i = 1, 2, \ldots,$ $N - 1$ and $j = 1, 2, \ldots, N$, set $A_{ij} = 0$ (initialize array).

Step 4 Set

$$A_{11} = h^2 q(x_1) - 2$$

$$A_{12} = 1 + \left(\frac{h}{2}\right)p(x_1)$$

and

$$A_{1N} = h^2 f(x_1) - \left(1 - \left(\frac{h}{2}\right)p(x_1)\right)A$$

Step 5 For $i = 2, 3, \ldots, N - 2$, set

$$x_i = x_0 + ih$$

$$A_{i, i-1} = 1 - \left(\frac{h}{2}\right)p(x_i)$$

$$A_{ii} = h^2 q(x_i) - 2$$

$$A_{i, i+1} = 1 + \left(\frac{h}{2}\right)p(x_i)$$

$$A_{iN} = h^2 f(x_i)$$

Step 6 Set

$$A_{N-1, N-2} = 1 - \left(\frac{h}{2}\right)f(x_{N-1})$$

$$A_{N-1, N-1} = h^2 q(x_{N-1}) - 2$$

$$A_{N-1, N} = h^2 f(x_{N-1}) - \left(1 + \left(\frac{h}{2}\right)p(x_{N-1})\right)B$$

Step 7 Set $j = 1$.

Step 8 Set $m_{j+1, j} = A_{j+1}/A_{jj}$.

Step 9 For $i = j, \ldots, N$, set

$$A_{j+1, i} = A_{j+1, i} - m_{j+1, j}A_{ji}$$

(*continued*)

Algorithm 11.5, continued

Step 10 Set $j = j + 1$. If $j > N - 2$, then go to step 12.
Step 11 Go to step 8.
Step 12 Set $y_{N-1} = A_{N-1,N}/A_{N-1,N-1}$.
Step 13 For $k = N - 2, N - 3, \ldots, 1$, set

$$y_k = \frac{A_{k,N} - \sum_{j=k+1}^{N-1} A_{kj} y_j}{A_{kk}}$$

Step 14 Output pairs x_k, y_k for $i = 0, 1, 2, \ldots, N$.
Step 15 Stop.

EXAMPLE 11.11

Solve the boundary value problem

$$x^2 y'' - xy' + y = x^2 \qquad y(1) = 1, \quad y(2) = 4 - 2 \ln 2$$

Solution

Using Algorithm 11.5 with $h = 0.05$, we obtain the following results.

X	Y(X)	TRUE VALUE	ERROR
1.050	1.0512620	1.0512700	0.0000083
1.100	1.1051430	1.1051590	0.0000155
1.150	1.1617530	1.1617740	0.0000211
1.200	1.2211890	1.2212140	0.0000257
1.250	1.2835420	1.2835710	0.0000292
1.300	1.3488950	1.3489260	0.0000316
1.350	1.4173260	1.4173590	0.0000333
1.400	1.4889050	1.4889390	0.0000341
1.450	1.5636980	1.5637330	0.0000346
1.500	1.6417690	1.6418020	0.0000339
1.550	1.7231720	1.7232050	0.0000327
1.600	1.8079630	1.8079940	0.0000308
1.650	1.8961920	1.8962210	0.0000284
1.700	1.9879060	1.9879320	0.0000257
1.750	2.0831500	2.0831720	0.0000224
1.800	2.1819650	2.1819840	0.0000188
1.850	2.2843920	2.2844070	0.0000148
1.900	2.3904680	2.3904780	0.0000105
1.950	2.5002280	2.5002330	0.0000057

The exact solution in this case is $y = x^2 - x \ln x$. Values computed from this solution are given in the TRUE VALUE column with corresponding absolute errors in the ERROR column. We see that the absolute error is less than 5×10^{-5} throughout the interval $[1, 2]$. ■

EXAMPLE 11.12

Solve the boundary value problem

$$x^2 y'' + xy' + x^2 y = 0 \qquad y(0) = 1, \quad y(1) = 0.765198$$

Solution Using a step size $h = 0.05$ with Algorithm 11.5, we obtain the following results.

```
ENTER a, b AND STEP SIZE h ? 0, 1, .05
ENTER f(a) AND f(b) ? 1, .765198
```

X	Y
0.05	0.999403
0.10	0.997538
0.15	0.994424
0.20	0.990069
0.25	0.984481
0.30	0.977672
0.35	0.969654
0.40	0.960443
0.45	0.950055
0.50	0.938511
0.55	0.925831
0.60	0.912040
0.65	0.897164
0.70	0.881230
0.75	0.864267
0.80	0.846308
0.85	0.827385
0.90	0.807535
0.95	0.786793

In this case, the differential equation is the Bessel equation, and the end point values are values of $J_0(x)$ correct to six decimal places. Some precise values of the function $J_0(x)$ in the interval $[0, 1]$ are $J_0(0.3) = 0.9776263$, $J_0(0.5) = 0.9384698$, and $J_0(0.8) = 0.8462874$. If we compare the corresponding values in the table with these values, we see that the absolute errors are all less than 6×10^{-5}. Note that $p(x) = 1/x$ in this example, so the function $p(x)$ is not defined at the boundary point $x = 0$. This causes no difficulty when we use Algorithm 11.5, since the value of $p(0)$ is never needed in the computations. If we attempted to solve this problem using a shooting method, the singularity at $x = 0$ would cause difficulties. ∎

■ *Exercise Set 11.3*

1. Use Algorithm 11.5 to approximate the solution of each of the following boundary value problems. Use $h = 0.05$.
   a. $y'' - 2y' + y = x^2 - 1$; $y(0) = 5$, $y(1) = 10$
   b. $y'' + xy' + y = 2x$; $y(0) = 1$, $y(1) = 0$
   c. $y'' + 2y' - 8y = 0$; $y(0) = 0$, $y(1) = 0.5$
   d. $y'' + 2y' + 1 = 0$; $y(0) = 1$, $y(1) = 2$
   e. $y'' = 2y/x^2$; $y(1) = 2$, $y(2) = 4.5$

2. Repeat exercise 1 using Algorithm 11.5 with $h = 0.1$.

3. Use Algorithm 11.5 to approximate the solution of the boundary value problem

$$y'' - 2y' + y = x^2 - 1 \qquad y(0) = 5, \quad y(1) = 10$$

a. Use $h = 0.1$.

b. Use $h = 0.05$.

c. Compute the error in the estimates obtained in parts (a) and (b) for values of $y(0.2)$, $y(0.5)$ and $y(0.8)$ by comparing with values of the exact solution $y(x) = x^2 + 4x + 5$ at these points.

d. Compare the approximate solutions obtained in parts (a) and (b) with the corresponding solutions in exercise 3 of Exercise Set 11.2. Is there any noticeable difference with regard to accuracy?

4. Verify that $y(x) = 3x + 2e^{-x}$ is the solution of the boundary value problem

$$(1 + x)y'' + xy' - y = 0 \qquad y(0) = 2, \quad y(1) = 3 + \frac{2}{e}$$

Now find an approximate solution using Algorithm 11.5 with $h = 0.05$. Compute the error at the points $x = 0.2, 0.4, 0.6$, and 0.8 by comparing estimates with values of the exact solution.

5. Approximate the solution of the boundary value problem

$$x^2 y'' - 4xy' + 6y = 0 \qquad y(1) = 0, \quad y(2) = -4$$

using Algorithm 11.5 with $h = 0.05$. Repeat with $h = 0.1$. Compare these numerical results with values of the exact solution $y(x) = x^2 - x^3$ by finding the maximum and minimum errors for each value of h.

6. Verify that the general solution of the differential equation

$$(1 + 2x)y'' + 4xy' - 4y = 0$$

is $y = c_1 x + c_2 e^{-2x}$. Find values of c_1 and c_2 so that $y(0) = -1$ and $y(2) = 2 - e^{-4}$. Now find an approximate solution using Algorithm 11.5 with $h = 0.1$. Compute the error in the numerical estimates of $y(0.2)$, $y(0.5)$, $y(1)$, and $y(1.5)$. Repeat with $h = 0.05$.

7. It can be shown that the error in the estimates obtained with Algorithm 11.5 is of the order of h^2. It follows that if the value of the solution of a boundary value problem at a point x is estimated using a certain value of h and is then recomputed using $h/2$, we can extrapolate on these values $y_h(x)$ and $y_{h/2}(x)$ to obtain a more accurate estimate

$$\hat{y}_{h/2}(x) = \frac{4y_{h/2}(x) - y_h(x)}{3}$$

a. Apply this extrapolation procedure to the estimates of $y(1.3)$ and $y(1.7)$ in exercise 5 to obtain more accurate estimates. Compare with exact values of $y(1.3)$ and $y(1.7)$. Did extrapolation improve the estimates?

b. Repeat part (a) using the numerical results in exercise 6. Apply extrapolation to estimates at 0.5, 1, and 1.5. Find the errors in the extrapolated values.

8. Assume that the differential equation in a boundary value problem has the form $y'' + q(x)y = f(x)$. Substitute the same difference approximation for $y''(x_i)$ as was used in this section to find the system of difference equations to be solved. Show that the result is a linear system $\mathbf{Ay} = \mathbf{b}$, where matrix $\mathbf{A}$ is tridiagonal with 1s above and below the diagonal.

9. Given the boundary value problem

$$y'' = \frac{2y}{x^2} \qquad y(1) = 2, \quad y(2) = 4.5$$

write the system of difference equations described in exercise 8 for the case where $h = 0.2$. Solve this system for y_1, y_2, y_3, and y_4. The exact solution is $y(x) = x^2 + 1/x$. Calculate the error in each of your estimates.

11.4   Derivative Boundary Conditions (Optional)

In the preceding sections, we considered boundary value problems in which the boundary conditions were values of the solution function at the end points of an interval $[a, b]$. The boundary conditions for a boundary value problem may require that the end point derivatives of the solution function assume certain specified values. In this section, we will consider some algorithms for solving some boundary value problems of this type. We will begin with a shooting method.

Assume that we wish to solve the boundary value problem

$$y'' = f(x, y, y') \qquad y'(a) = C, \quad y(b) = B$$

Let $y(x, \alpha)$ be the solution of the initial-value problem

$$y'' = f(x, y, y') \qquad y'(a) = C, \quad y(a) = \alpha$$

We need to determine that α for which $y(b, \alpha) = B$. We are again faced with the problem of solving an equation. As in Section 11.1, we can use either the secant method or the Newton method to do this. The following algorithm employs the secant method to do the job.

Algorithm 11.6

Step 1 Input a, b, C, B, h, TOL, α_1, and α_2.

Step 2 Compute $y(b, \alpha_1)$ by solving the initial-value problem

$$y'' = f(x, y, y') \qquad y(a) = \alpha_1, \quad y'(a) = C$$

on $[a, b]$ using the fourth-order Runge-Kutta method. Set $y_1(b) = y(b, \alpha_1)$.

Step 3 Compute $y(b, \alpha_2)$ by solving the initial-value problem

$$y'' = f(x, y, y') \qquad y(a) = \alpha_2, \quad y'(a) = C$$

on $[a, b]$ using the fourth-order Runge-Kutta method. Set $y_2(b) = y(b, \alpha_2)$.

Step 4 Compute

$$\alpha_3 = \alpha_1 + \frac{B - y_1(b)}{y_2(b) - y_1(b)} (\alpha_2 - \alpha_1)$$

(*continued*)

Algorithm 11.6, continued

Step 5 Compute $y(b, \alpha_3)$ by solving the initial-value problem

$$y'' = f(x, y, y') \qquad y(a) = \alpha_3, \quad y'(a) = C$$

using the fourth-order Runge-Kutta method. Set $y_3(b) = y(b, \alpha_3)$.

Step 6 If $|y_3(b) - B| < $ TOL, then go to step 8.

Step 7 Set $\alpha_1 = \alpha_2$, $\alpha_2 = \alpha_3$, $y_1(b) = y_2(b)$, and $y_2(b) = y_3(b)$, and return to step 4.

Step 8 Output solution $y(x, \alpha_3)$.

EXAMPLE 11.13

Solve the problem

$$x^2 y'' - xy' + y = x^2 \qquad y'(1) = 1, \quad y(3) = 6$$

Solution   The numerical output from application of Algorithm 11.6 to this problem is shown in the following table.

```
ENTER VALUES OF a and b 1, 3
ENTER VALUES OF y'(a) and y(b) 1, 6
ENTER STEP SIZE H ? .1
ENTER TWO INITIAL ESTIMATES OF y(A) ? .5, 1

USING X(A) = 6.437302E − 06
```

X	Y	YPRIME
1.00	0.00001	1.00000
1.10	0.11001	1.20000
1.20	0.24001	1.40000
1.30	0.39001	1.60000
1.40	0.56001	1.80000
1.50	0.75001	2.00000
1.60	0.96001	2.20000
1.70	1.19001	2.40000
1.80	1.44001	2.60000
1.90	1.71001	2.80000
2.00	2.00001	3.00000
2.10	2.31000	3.20000
2.20	2.64000	3.39999
2.30	2.99000	3.59999
2.40	3.36000	3.79999
2.50	3.75000	3.99999
2.60	4.16000	4.19999
2.70	4.59000	4.39999
2.80	5.04000	4.59999
2.90	5.51000	4.79999
3.00	6.00000	4.99999

The exact solution to the problem is $y(x) = x^2 - x$. The value of $y(1.5)$ is 0.75, while the estimate is 0.75001. The value of $y(2.5)$ is 3.75, and the estimate is also 3.75000. The errors are of the order of 10^{-5}, as these comparison values show. ∎

EXAMPLE 11.14

Solve the problem

$$t^2 x'' - tx' + x = t^2 \qquad x'(1) = 1, \quad x(2) = 4 - 2 \ln 2$$

Solution Use the approximation $4 - 2 \ln 2 = 2.6137056$. The numerical results are as follows.

```
ENTER VALUES OF a and b 1, 2
ENTER VALUES OF X'(a) and X(b) 1, 2.6137056
ENTER STEP SIZE H ? .05
ENTER TWO INITIAL ESTIMATES OF X(A) ? 1, 1.5

USING X(A) = 1
```

t	X	XPRIME
1.00	1.00000	1.00000
1.05	1.05127	1.05121
1.10	1.10516	1.10469
1.15	1.16177	1.16024
1.20	1.22121	1.21768
1.25	1.28357	1.27686
1.30	1.34893	1.33764
1.35	1.41736	1.39990
1.40	1.48894	1.46353
1.45	1.56373	1.52844
1.50	1.64180	1.59453
1.55	1.72321	1.66174
1.60	1.80799	1.73000
1.65	1.89622	1.79922
1.70	1.98793	1.86937
1.75	2.08317	1.94038
1.80	2.18198	2.01221
1.85	2.28441	2.08481
1.90	2.39048	2.15815
1.95	2.50023	2.23217
2.00	2.61371	2.30685

The true solution is $x(t) = t^2 - t \ln t$. Values calculated from this formula may be compared with the numerical estimates. The errors are found to be quite small. ∎

It is also possible to use difference methods for boundary value problems with derivative boundary conditions. We will illustrate how this can be done with the linear problem

$$y'' + p(x)y' + q(x)y = f(x) \qquad y'(a) = C, \quad y(b) = B$$

We choose a step size h and let $n = (b - a)/h$. We assume that h has been selected so that n is a positive integer. We define mesh points by $x_i = a + ih$, for $i = 0, 1, \ldots, n$. We want to find numerical estimates of $y_i = y(x_i)$ for $i = 0, \ldots, n - 1$. We introduce an additional point $x_{-1} = a - h$. We assume that x_{-1} is in the domain of the solution $y(x)$. We can then approximate $y'(x_0)$ by

$$\frac{y(x_1) - y(x_{-1})}{2h}$$

This leads to the difference equation

$$y_1 - y_{-1} = 2Ch$$

since we know that $y'(x_0) = C$. Now using the same approximations for derivatives $y''(x_i)$ and $y'(x_i)$ as in Section 11.3, we again arrive at a system of difference equations

$$\left(1 - \frac{h}{2} p_i\right) y_{i-1} + (h^2 q_i - 2) y_i + \left(1 + \frac{h}{2} p_i\right) y_{i+1} = h^2 f_i$$

where $p_i = p(x_i)$, $q_i = q(x_i)$, and $f_i = f(x_i)$. In this case, however, the subscript i assumes values from 0 to $n - 1$, rather than from 1 to $n - 1$, because we have introduced a new point x_{-1}. We have n equations for the $n + 1$ unknowns $y_{-1}, y_0, y_1, \ldots, y_{n-1}$. The additional equation needed to complete the set of $n + 1$ linear equations for $n + 1$ unknowns is the equation

$$y_{-1} - y_1 = -2Ch$$

We can approximate the solution of the boundary value problem by solving this system of linear equations. The system in this case is not tridiagonal because of the equation that arises from approximation of $y'(x_0)$. That presents no great difficulty, however. The following algorithm can be used to implement the procedure described to approximate the solution of a linear boundary value problem with the boundary conditions $y'(a) = C$ and $y(b) = B$.

Algorithm 11.7

Step 1 Input a, b, h, C, and B. Set $n = (b - a)/h$.
Step 2 Set $A_{11} = 1$, $A_{12} = 0$, and $A_{13} = -1$. Set $A_{1, n+2} = -2Ch$.
Step 3 For $i = 1, \ldots, n - 1$, set

$$x_i = a + (i - 1)h$$

$$A_{i+1, i} = 1 - 0.5hp(x_i)$$

$$A_{i+1, i+1} = h^2 q(x_i) - 2$$

$$A_{i+1, i+2} = 1 + 0.5hp(x_i)$$

$$A_{i+1, n+2} = h^2 f(x_i)$$

Algorithm 11.7, continued

Step 4 Set

$$x_{n-1} = b - h$$
$$A_{n+1, n} = 1 - 0.5hp(x_{n-1})$$
$$A_{n+1, n+1} = h^2 q(x_{n-1}) - 2$$
$$A_{n+1, n+2} = h^2 f(x_{n-1}) - (1 + 0.5hp(x_{n-1})B$$

Step 5 Solve the linear system whose augmented matrix is the $(n + 1) \times (n + 2)$ matrix $\mathbf{A} = (A_{ij})$ for unknowns $y_{-1}, y_0, y_1, \ldots, y_{n-1}$.
Step 6 Output solution pairs x_i, y_i for $i = 1, \ldots, n$.

The following examples illustrate some applications of this algorithm.

EXAMPLE 11.15   Solve the boundary value problem

$$x^2 y'' - xy' + y = x^2 \qquad y'(1) = 1, \quad y(3) = 6$$

Solution   We use Algorithm 11.7 with $h = 0.1$. The numerical results are as follows.

```
ENTER a, b AND STEP SIZE h ? 1, 3, .1
ENTER y'(a) AND y(b) ? 1, 6

X        Y

1.00    −0.000032
1.10     0.109968
1.20     0.239968
1.30     0.389969
1.40     0.559970
1.50     0.749971
1.60     0.959973
1.70     1.189974
1.80     1.439976
1.90     1.709978
2.00     1.999981
2.10     2.309983
2.20     2.639985
2.30     2.989988
2.40     3.359991
2.50     3.749993
2.60     4.159996
2.70     4.589998
2.80     5.039999
2.90     5.510000
3.00     6.000000
```

These results should be compared with the results in Example 11.13. The errors are larger for most of the estimates here, but they are still only of the order of 10^{-5}. ∎

EXAMPLE   Solve the boundary value problem
11.16

$$t^2x'' - tx' + x = t^2 \qquad x'(1) = 1, \quad x(2) = 4 - 2\ln 2$$

Use the approximation $4 - 2\ln 2 = 2.6137056$.

Solution   We use Algorithm 11.7 with $h = 0.05$ to generate the approximate solution shown in the following table.

```
ENTER a, b AND STEP SIZE h ? 1, 2, .05
ENTER x'(a) AND x(b) ? 1, 2.6137056
```

t	X EST.	TRUE X	ERROR
1.00	1.000539	1.000000	−0.000539
1.05	1.051788	1.051270	−0.000518
1.10	1.105655	1.105159	−0.000496
1.15	1.162248	1.161774	−0.000474
1.20	1.221666	1.221214	−0.000452
1.25	1.283999	1.283571	−0.000429
1.30	1.349332	1.348926	−0.000405
1.35	1.417739	1.417359	−0.000381
1.40	1.489294	1.488939	−0.000355
1.45	1.564063	1.563733	−0.000330
1.50	1.642106	1.641802	−0.000303
1.55	1.723481	1.723205	−0.000276
1.60	1.808242	1.807994	−0.000248
1.65	1.896440	1.896221	−0.000220
1.70	1.988122	1.987932	−0.000190
1.75	2.083333	2.083172	−0.000160
1.80	2.182114	2.181984	−0.000129
1.85	2.284505	2.284407	−0.000098
1.90	2.390544	2.390478	−0.000066
1.95	2.500266	2.500233	−0.000033
2.00	2.613706	2.613706	0.000000

As noted in our discussion of Example 11.14, we are approximating values of $x(t) = t^2 - t\ln t$. This formula was used to calculate the values displayed in the TRUE X column of the table. The last column lists the error in each estimate. These seem to range from slightly over 5×10^{-4} to 3×10^{-5}. ∎

One might think that the errors in Example 11.16 are larger than they were in Example 11.15 because we had to use an estimate of $x(2)$ to begin the calculations. This should produce some propagated error in the output. We used the same estimate in preparing the table for Example 11.14, however, with noticeably better results. This suggests that the additional error may be due to the choice of algorithm itself. Comparison of the tables in Examples 11.13 and 11.15 also leads us to suspect that the finite difference method in Algorithm 11.7 does not yield results quite as accurate as those obtained from the shooting method in Algorithm 11.6. The difference in accuracy in the preceding examples was not large, however, and the difference method in Algorithm 11.7 does produce results more rapidly. It is also possible to have convergence problems with a shooting method such as Algorithm 11.6.

This concludes our study of methods for solving two-point boundary value problems. Some of the most interesting boundary value problems involve systems of three or more differential equations. The algorithms in this section can be modified to handle such problems. It would be better, however, to investigate the IMSL subroutines for solving such problems. The subroutine BVPFD uses a finite difference method, and routine BVPMS uses a shooting method to solve two-point boundary value problems. An investigation of two-point boundary value problems for systems, combined with use of these IMSL subroutines, would make an interesting and worthwhile project for any ambitious student of numerical analysis.

■ *Exercise Set 11.4*

1. Use Algorithm 11.6 to approximate the solution of each of the following boundary value problems. Use $h = 0.05$.
   a. $y'' - xy' + y = -x + 1/x$; $y'(1) = 1$, $y(2) = 2 \ln 2$
   b. $xy'' + y = x^2$; $y'(\pi/2) = 1 + \pi$, $y(\pi) = \pi^2$
   c. $x^2 y'' + xy' + y = 8 - 10x^3$; $y'(1) = -3$, $y(2) = 0$
   d. $2(1 + x)y'' + y' + y = \sqrt{1 + x}$; $y'(0) = 0.5$, $y(3) = 0.25$
   e. $y'' + 2y' + 1 = 0$; $y(0) = 1$, $y(1) = 2$
   f. $y'' = 2y/x^2$; $y'(1) = 1$, $y(2) = 4.5$
   g. $xy'' - (1 + x)y' + y = 0$; $y'(1) = e$, $y(2) = e^2$
   h. $xy'' - (1 + x)y' + y = 0$; $y'(1) = 1$, $y(2) = 3$

2. Repeat exercise 1 using Algorithm 11.7.

3. We can approximate the solution of the boundary value problem

   $$x^2 y'' + xy' + y = 8 - 10x^3 \qquad y(1) = 7, \quad y'(2) = -12$$

   by use of Algorithm 11.6 if we set $a = 2$ and $b = 1$ with $h = -0.05$. Then $y'(a) = -12$ and $y(b) = 7$ is used, of course. Run the computations from right to left rather than from left to right. Make any other necessary modifications, and compute an approximate solution to this boundary value problem. Compare the numerical results with the solution to exercise 1c.

4. Use Algorithm 11.6 to approximate the solution of the boundary value problem

   $$y'' - 2y' + y = x^2 - 1 \qquad y'(0) = 4, \quad y(1) = 10$$

   a. Use $h = 0.1$.
   b. Use $h = 0.05$.
   c. Compute the error in the estimates obtained in parts (a) and (b) for values of $y(0.2)$, $y(0.5)$, and $y(0.8)$ by comparing with values of the exact solution $y(x) = x^2 + 4x + 5$ at these points.
   d. Compare the approximate solutions obtained in parts (a) and (b) with the corresponding solutions in exercise 3 of Exercise Set 11.2. Is there any noticeable difference with regard to accuracy?

5. Repeat exercise 4 using Algorithm 11.7.

6. Verify that $y(x) = 3x + 2e^{-x}$ is the solution of the boundary value problem

$$(1 + x)y'' + xy' - y = 0 \qquad y'(0) = 1, \quad y(1) = 3 + \frac{2}{e}$$

Now find an approximate solution using Algorithm 11.6 with $h = 0.05$. Compute the error at the points $x = 0.2, 0.4, 0.6,$ and 0.8 by comparing estimates with values of the exact solution.

7. Approximate the solution of the boundary value problem

$$x^2 y'' - 4xy' + 6y = 0 \qquad y'(1) = -1, \quad y(2) = -4$$

Using Algorithm 11.7 with $h = 0.05$. Repeat with $h = 0.1$. Compare these numerical results with values of the exact solution $y(x) = x^2 - x^3$ by finding the maximum and minimum errors for each value of h.

8. Verify that the general solution of the differential equation

$$(1 + 2x)y'' + 4xy' - 4y = 0$$

is

$$y = c_1 x + c_2 e^{-2x}$$

Find values of c_1 and c_2 so that $y'(0) = 3$ and $y(2) = 2 - e^{-4}$. Now find an approximate solution using Algorithm 11.6 with $h = 0.1$. Compute the error in the numerical estimates of $y(0.2), y(0.5), y(1),$ and $y(1.5)$. Repeat with $h = 0.05$.

11.5 Applications

In this section, we will examine some applied problems that require us to solve a boundary value problem. Boundary value problems occur frequently in the physical sciences. We have examined only a few of the many known methods for solving boundary value problems, and we have studied only some of the simpler types of boundary value problems; but even with this small amount of material, we can solve some useful problems.

Let us begin with a problem from engineering mechanics. Suppose that a flexible cable is hanging between two supports, as shown in Figure 11.1. Assume that

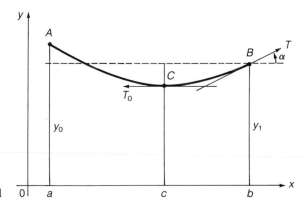

Figure 11.1

the weight per unit length is ρ and the tension at the low point on the cable is T_0. Let α denote the angle of inclination of the tangent line at the end point B, as shown. If we view the cable as the graph of a function $y(x)$, then by using some elementary mechanics, it can be shown that

$$y''(x) = \frac{\rho}{T_0} \sqrt{1 + (y')^2}$$

It is, of course, also true that $y(a) = y_0$ and $y(b) = y_1$. So $y(x)$ is the solution of a boundary value problem. If we are given values of ρ, T_0, a, b, y_0, and y_1 we should be able to compute values of $y(x)$ by use of the algorithms given in Sections 11.1 and 11.2. Suppose that $a = 0$, $b = 48$, and $y_0 = y_1 = 20$ ft. Let the cable have a weight of 1 lb/ft, and assume that the tension T_0 is 100 lb. When Algorithm 11.1 is applied to this problem, the results are as shown in Table 11.1. To use this algorithm, we need

Table 11.1   ENTER VALUES OF a and b 0, 48
ENTER VALUES OF X(a) and X(b) 20, 20
ENTER STEP SIZE H ? .5
ENTER INITIAL ESTIMATES OF Y'(A) ? $-.2$, $-.3$

USING Y'(A) = $-.2423105$

X	Y	YPRIME
0.00	20.00000	-0.24231
2.00	19.53593	-0.22178
4.00	19.11282	-0.20134
6.00	18.73053	-0.18097
8.00	18.38888	-0.16068
10.00	18.08775	-0.14046
12.00	17.82701	-0.12029
14.00	17.60656	-0.10017
16.00	17.42631	-0.08009
18.00	17.28620	-0.06004
20.00	17.18615	-0.04001
22.00	17.12614	-0.02000
24.00	17.10614	0.00000
26.00	17.12614	0.02000
28.00	17.18615	0.04001
30.00	17.28620	0.06004
32.00	17.42632	0.08009
34.00	17.60656	0.10017
36.00	17.82701	0.12029
38.00	18.08775	0.14046
40.00	18.38888	0.16068
42.00	18.73053	0.18097
44.00	19.11283	0.20134
46.00	19.53593	0.22178
48.00	20.00000	0.24231

to supply estimates of $y'(a)$. In this case,

$$y'(a) = y'(0) = -\tan \alpha$$

We use the two facts that (1) $T \sin \alpha =$ the weight of the cable from C to B and (2) $T \cos \alpha = T_0$. In this example, the support heights are of equal length, so the point C is at the midpoint of the cable. It follows that $T \sin \alpha$ is approximately 25 lb. The two equations

$$T \sin \alpha = 25 \quad \text{and} \quad T \cos \alpha = 100$$

imply that $\tan \alpha = 0.25$. This suggests that we use as our initial estimates of $y'(0)$ values near -0.25. We chose -0.2 and -0.3 to generate Table 11.1. We see from the numerical results that the correct value is -0.24231.

Several interesting facts can be inferred from the numerical output. Apparently, the tangent of α is 0.24231, so we conclude that angle $\alpha = 13.62°$. Since the tension T at the cable support is T_0 divided by $\cos \alpha$ we see that $T = 100/\cos 13.62° = 102.89$ lb. The sag in the cable is 20 minus the low value of $y(x)$. We find that the sag is about 2.89 ft. According to one engineering mechanics text (Poorman 1949), if the sag is less than 5% of the span (48 ft in this case), we may assume that T and T_0 are the same in practice. In our case, the sag is about 6% of the span, and the difference between T and T_0 is less than 3%. This is fair confirmation of the engineering approximation. The tangent of α is clearly the weight of the cable divided by twice the tension T_0, so we conclude that the cable weight is 200 times 0.24231. This gives a weight of 48.46 lb. Since the cable weighs 1 lb/ft, we conclude that the length of the cable is 48.46 ft. All of this is useful engineering information that becomes available from the numerical solution of the boundary value problem.

Boundary value problems are common in heat flow problems. For example, consider the temperature distribution in a spherical shell with inner radius a and outer radius b, as shown in Figure 11.2. Let $T(r, t)$ denote the temperature at time t at distance r from the center. We assume that the temperature depends only on this distance and is independent of direction from the center. Then $T(r, t)$ satisfies the

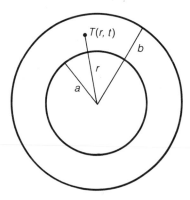

Figure 11.2

partial differential equation

$$\frac{\partial T}{\partial t} = k\left(\frac{\partial^2 T}{\partial r^2} + \frac{2}{r}\frac{\partial T}{\partial r}\right)$$

where k is a constant. Assume now that a steady-state flow has been reached, so the time derivative of T is zero and T is a function $T(r)$ of r only. Then $T(r)$ satisfies the ordinary differential equation

$$\frac{d^2 T}{dr^2} + \frac{2}{r}\frac{dT}{dr} = 0$$

If we assume that the interior and the exterior temperatures, $T(a)$ and $T(b)$, are known, then we can determine $T(r)$ on the interval $[a, b]$ as the solution of a boundary value problem.

For example, suppose that $a = 5$ cm, $b = 8$ cm, $T(a) = 90°$ C, and $T(b) = 20°$ C. Using Algorithm 11.2, we obtain the numerical results shown in Table 11.2 for $T(r)$ and $T'(r)$. In this case, values of $T'(r)$ are also included in the output because of their physical interpretation. It is possible in this case to solve the boundary value problem analytically. The solution is found to be

$$T(r) = -96.66667 + \frac{2800}{3r}$$

The numerical solution agrees very well with the values computed from this formula, as you can easily verify with selected values of r.

As our final application, let us consider the problem of a basketball player shooting for a basket. Assume that the horizontal distance from his feet to the point on the floor directly below the basket is 40 ft. He releases the ball at a point 7 ft above the floor, and the basket opening is 10 ft above the floor. The velocity of the ball at the moment of release is 50 ft/sec. What should the release angle θ in Figure 11.3 be in order for the player to make the basket?

Let $y(x)$ denote the height of the ball above the floor, where $x = 0$ at the location of the player and $x = 40$ at the point on the floor below the basket. We have $y(0) = 7$ and we want $y(40) = 10$. From elementary mechanics, we find that

$$y''(x) = \frac{-32}{V^2 \cos^2 \theta} = \frac{-32}{V^2} \sec^2 \theta$$

where V is the release velocity and θ is the release angle. Let $\tan \theta = \alpha$. Then $\sec^2 \theta = 1 + \alpha^2$, and our differential equation becomes

$$y''(x) = -\frac{32(1 + \alpha^2)}{V^2}$$

Table 11.2   ENTER VALUES OF a and b 0YES
ENTER VALUES OF a and b 5, 8
ENTER VALUES OF T(a) and T(b) 90, 20
ENTER STEP SIZE H ? .1
ENTER INITIAL ESTIMATE OF T'(A) ? −35

USING T'(a) = −37.33333

r	T(r)	T'(r)
5.00	90.000	−37.333
5.10	86.340	−35.884
5.20	82.821	−34.517
5.30	79.434	−33.227
5.40	76.173	−32.007
5.50	73.030	−30.854
5.60	70.000	−29.762
5.70	67.076	−28.727
5.80	64.253	−27.745
5.90	61.525	−26.812
6.00	58.889	−25.926
6.10	56.339	−25.083
6.20	53.871	−24.280
6.30	51.481	−23.516
6.40	49.167	−22.786
6.50	46.923	−22.091
6.60	44.747	−21.426
6.70	42.637	−20.792
6.80	40.588	−20.185
6.90	38.599	−19.604
7.00	36.667	−19.048
7.10	34.789	−18.515
7.20	32.963	−18.004
7.30	31.187	−17.514
7.40	29.459	−17.044
7.50	27.778	−16.593
7.60	26.140	−16.159
7.70	24.545	−15.742
7.80	22.991	−15.341
7.90	21.477	−14.955
8.00	20.000	−14.583

Note that $\tan \theta = y'(0)$. Using a shooting method (which seems quite appropriate for this example), we select values of α and iterate until the end point result is satisfactory. In this example, the parameter α is a part of the differential equation itself, so the differential equation is modified after each iteration. There is no problem with this when we use an algorithm such as Algorithm 11.1. Using that algorithm with initial estimates 3 and 4 for $y'(0)$, we obtain the numerical solution shown in Table 11.3. The results indicate that we need $\tan \theta = 3.541121$, which

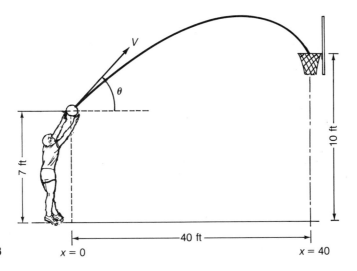

Figure 11.3 $x = 0$ $x = 40$

Table 11.3   ENTER RELEASE VELOCITY V (FT/SEC) ? 50
ENTER VALUES OF a and b 0, 40
ENTER VALUES OF X(a) and X(b) 7, 10
ENTER STEP SIZE H ? .5
ENTER INITIAL ESTIMATES OF y'(a) ? 3, 4

USING X'(A) = 3.541121

X	Y	YPRIME
0.00	7.00000	3.54112
2.00	13.73563	3.19451
4.00	19.77804	2.84790
6.00	25.12722	2.50129
8.00	29.78318	2.15467
10.00	33.74591	1.80806
12.00	37.01542	1.46145
14.00	39.59171	1.11484
16.00	41.47477	0.76823
18.00	42.66461	0.42161
20.00	43.16122	0.07500
22.00	42.96461	−0.27161
24.00	42.07478	−0.61822
26.00	40.49172	−0.96484
28.00	38.21543	−1.31145
30.00	35.24592	−1.65806
32.00	31.58319	−2.00467
34.00	27.22723	−2.35128
36.00	22.17805	−2.69790
38.00	16.43564	−3.04451
40.00	10.00001	−3.39112

corresponds to a release angle of 74.23°. The maximum height above the floor is 43.18 ft in this case.

    This output indicates that a release velocity of 50 ft/sec is higher than needed. If we repeat the computations using $V = 40$ ft/sec and initial estimates of 1 and 2 for $y'(0)$, we obtain the output shown in Table 11.4. The tangent of the release angle θ is

Table 11.4   USING X'(A) = 1.862372

X	Y	TIME
0.00	7.000	0.000
1.00	8.818	0.053
2.00	10.546	0.106
3.00	12.185	0.159
4.00	13.735	0.211
5.00	15.195	0.264
6.00	16.566	0.317
7.00	17.847	0.370
8.00	19.039	0.423
9.00	20.142	0.476
10.00	21.155	0.528
11.00	22.079	0.581
12.00	22.914	0.634
13.00	23.659	0.687
14.00	24.315	0.740
15.00	24.882	0.793
16.00	25.359	0.846
17.00	25.747	0.898
18.00	26.045	0.951
19.00	26.254	1.004
20.00	26.374	1.057
21.00	26.404	1.110
22.00	26.345	1.163
23.00	26.197	1.215
24.00	25.959	1.268
25.00	25.632	1.321
26.00	25.215	1.374
27.00	24.709	1.427
28.00	24.114	1.480
29.00	23.429	1.533
30.00	22.655	1.585
31.00	21.792	1.638
32.00	20.839	1.691
33.00	19.797	1.744
34.00	18.666	1.797
35.00	17.445	1.850
36.00	16.135	1.902
37.00	14.735	1.955
38.00	13.246	2.008
39.00	11.668	2.061
40.00	10.000	2.114

1.862372, so we need $\theta = 61.77°$. This will lead to a maximum height of 26.40 ft, as we see from the numerical output. It is interesting to note as a final observation that $x = V(\cos \theta)t$ at all times during the flight of the ball. It follows that the total flight time is the value of

$$\frac{x}{V \cos \theta} = \frac{x\sqrt{1 + y'(0)^2}}{V}$$

when $x = 40$ ft, $V = 40$ ft/sec, and $y'(0) = 1.8623772$. The resulting flight time is 2.114 seconds. We now have all the information our basketball player needs to know.

■ *Exercise Set 11.5*

1. Find an approximate solution to the hanging cable problem, as discussed in this section, by using Algorithm 11.2 with $h = 0.5$. Compare your results with the results shown in Table 11.1.

2. Find the sag in a cable such as the one in exercise 1 if the distance between the supports is 40 ft, the weight per unit length is 2 lb/ft, and the tension T_0 is 120 lb. Assume that $y_0 = y_1 = 20$, as in exercise 1. Estimate the tension at the support end of the cable.

3. Assume that $y(0) = 7$ in the basketball problem discussed in this section, but leave velocity V and θ unspecified.
   a. Show that the equation of the trajectory is

   $$y = 7 + \alpha x - \left(\frac{g}{2V^2}\right)(1 + \alpha^2)x^2$$

   where $\alpha = \tan \theta$.
   b. Set $g = 32$ ft/sec. Then show that $V = 40$ ft/sec and $y(40) = 10$ require that $\alpha = (5 \pm \sqrt{6})/4$.
   c. Given a value V, the value of α needed to achieve $y(40) = 10$ can be found by solving for α in the equation

   $$10 = 7 + \alpha x - \left(\frac{gx^2}{2V^2}\right)(1 + \alpha^2)$$

   with $x = 40$. Show that the minimum value of V that will produce a real solution for α is that V for which $V^2 = 3g + g\sqrt{1609}$. Show that $g = 32$ ft/sec implies that we need $V \geq 37.143$ ft/sec.
   d. Solve the basketball problem using $V = 37.2$ ft/sec with $y(0) = 7$ and $y(40) = 10$. Use Algorithm 11.1.
   e. Try to repeat part (d) using $V = 37$ ft/sec. Can you explain the numerical output?

4. Suppose that planet X has a radius R and no atmosphere. While an object is falling toward the surface, the altitude $x(t)$ above the surface satisfies the differential equation

   $$x''(t) = -\frac{g}{\left(1 + \dfrac{x}{R}\right)^2}$$

   where g is the acceleration of gravity at the surface of planet X. Suppose that $g = 18$ ft/sec^2 for this planet and that the radius is $R = 2000$ miles. Now suppose that a certain object falls to the surface from an unknown altitude A and that the descent lasts 5 minutes. Assuming that the initial value of $x'(t)$ was zero, find the initial altitude A and the impact velocity. Repeat for a descent time of 3 minutes. Try step sizes ranging from $h = 0.5$ to $h = 5$ seconds. What is a reasonable choice of h to produce useful numerical estimates?

5. If $x(t)$ represents the altitude at time t for an object falling through the earth's atmosphere at low speed, then

$$x''(t) = -g\left(1 - \left(\frac{x'(t)}{V}\right)^2\right)$$

where V is the terminal velocity. Assume that an object falls from an unknown initial altitude A and strikes the earth 2 minutes later. The terminal velocity is observed to be 30 mph. Find the altitude A assuming that the initial downward velocity $x'(0) = 0$. Also determine the impact velocity. Use $g = 32.16$ ft/sec^2 in the computations. Use Algorithm 11.6 with step sizes $h = 2$, 1, and 0.5. Can you see a physical reason why the numerical results obtained using $h = 2$ are suspect? Did reducing the step size from 1 to 0.5 improve the numerical estimates enough to justify the additional computation time required?

■ C H A P T E R ■

12 PARTIAL DIFFERENTIAL EQUATIONS

Equations containing only derivatives of a function of one variable are called ordinary differential equations. Equations containing partial derivatives of a function of two or more variables are called partial differential equations. Partial differential equations occur frequently in problems in the physical sciences: vibrations of a string or membrane, heat conduction in a solid, diffusion problems, potential problems, and many others. One of the fundamental equations in quantum mechanics is a partial differential equation known as the *Schrödinger equation*; and electromagnetic fields are described mathematically by a system of partial differential equations known as the *Maxwell equations*.

There are analytical methods for studying these equations, but no prior knowledge of such methods or the associated theory is assumed in this chapter. Since 1940, because of the widespread availability of modern computing machines, there has been a renewed interest in numerical methods for solving problems involving partial differential equations. The basic ideas behind these methods have been known for a long time, but the amount of computation required is so great that very little could be done in a practical sense prior to 1940. New methods have been developed and old ones have been explored in more detail. It is therefore appropriate to conclude a study of numerical analysis with some exploration of this relatively modern area of interest. An in-depth study, such as you will find in Ames (1977) or Smith (1978), will not be attempted here. After completing the material in this chapter, however, you should be prepared to proceed to such a study. In Section 12.1, we will survey some basic concepts needed to begin the study of numerical methods for solving partial differential equations. In the following sections, we will explore some of the better-known numerical methods for certain standard types of boundary value problems for partial differential equations.

12.1 Some Theory and Definitions

A partial differential equation that contains nth-order partial derivatives of the solution function, but no higher-order derivatives, is called a **partial differential equation (PDE) of order n**.

The equations

$$(x + y)u_x + 3xu_y = e^{xy} \cos 2y$$

and

$$(s + t^2)w_s + (s^2 - 2st)w_t = s^2 t^3$$

are first-order partial differential equations.

The equations

$$u_{xx} + u_{tt} = \sin xt \quad \text{and} \quad u_{xy} + xyu_y = x^2 + y$$

are second-order partial differential equations.

In this chapter, we will only consider numerical methods for second-order partial differential equations. The most useful partial differential equations in applied mathematics are of second order. Higher-order partial differential equations are rare. The methods used in this chapter, however, are easily modified to produce methods for estimating solutions of first-order partial differential equations as well as third-order or higher-order equations.

We will only consider *linear* equations. These are equations of the form

$$Au_{xx} + 2Bu_{xy} + Cu_{yy} + Eu_x + Fu_y + Gu = H$$

where A, B, C, E, F, G, and H are functions of x and y only. The equations

$$u_{xx} + xu_{yy} = x + y$$

and

$$(x + y)u_{xx} - xyu_y = \sin (xy)$$

are **linear second-order partial differential equations**. The equations

$$uu_{xx} + u_y = 3x$$

and

$$u_{xx}^2 + u_{yy}^2 - 2u_{xy} = x - 2y$$

are nonlinear equations.

Linear second-order equations are usually classified as follows:

1. If $B^2 - AC < 0$ in region R, the PDE is **elliptic** in R.
2. If $B^2 - AC = 0$ in region R, the PDE is **parabolic** in R.
3. If $B^2 - AC > 0$ in region R, the PDE is **hyperbolic** in R.

The three most important second-order partial differential equations are the Poisson, wave, and heat equations:

1. **Poisson equation** $\quad u_{xx} + u_{yy} = f(x, y)$
2. **Wave equation** $\quad c^2 u_{xx} = u_{tt}$
3. **Heat equation** $\quad k u_{xx} = u_t$

The Poisson equation is a standard representative of an elliptic equation. The wave equation is the standard example of a hyperbolic equation. The heat equation is a standard example of a parabolic equation. There exists a body of theory and methods for elliptic equations, another for hyperbolic equations, and still another for parabolic equations. For details, you can consult a standard text on partial differential equation, such as Epstein (1962) or Greenspan (1961).

The Poisson equation with $f = 0$ is called the **Laplace equation**. The Poisson equation and Laplace equation occur in the study of potential problems. The wave equation arises in the study of vibrations of strings or membranes and in wave propagation problems. The heat equation is also called the **diffusion equation** because it appears in the study of heat conduction in solids and in problems concerned with diffusion processes in gases and liquids.

All problems involving elliptic partial differential equations can be solved by similar numerical algorithms. Hyperbolic equations require different methods, and parabolic equations require yet another set of algorithms. Numerical methods for each of these types will be illustrated in this chapter by finding numerical solutions for the Poisson equation, the wave equation, and the heat equation.

The theory of uniqueness and existence of solutions for a partial differential equation on a region R in the plane is understandably much more complicated than existence and uniqueness theory for ordinary differential equations. Topological restrictions must be placed on the region R. The theorems require that we associate with each partial differential equation some particular combination of boundary conditions and initial conditions. Each combination of a partial differential equation with a set of boundary conditions and initial conditions defines a *boundary value problem*. So most uniqueness and existence theorems are actually concerned with the solution of a particular boundary value problem.

For example, suppose that we wish to find a solution for the Laplace equation

$$u_{xx} + u_{yy} = 0$$

on a region R with $u(x, y) = f(x, y)$ on the boundary of R. Suppose that the function f is continuous on the boundary of R. This is called a **Dirichlet problem**. It can be shown that a unique solution exists if R is a disk or a rectangular region. More complicated regions can be used without destroying the uniqueness and existence of a solution, but we will only consider the case of a rectangle in our discussion of the numerical solution of the Dirichlet problem.

A unique solution of the wave equation

$$c^2 u_{xx} = u_{tt}$$

exists that satisfies initial conditions

$$u(x, 0) = F(x) \quad \text{and} \quad u_t(x, 0) = G(x)$$

and boundary conditions

$$u(0, t) = 0, \quad u(a, t) = 0$$

if F and G are continuous on $0 < x < a$ with

$$F(0) = F(a) = 0$$

Such boundary conditions and initial conditions frequently arise in conjunction with applications involving the wave equation.

A unique solution of the heat equation

$$ku_{xx} = u_t$$

exists that satisfies initial condition

$$u(x, 0) = F(x)$$

with boundary conditions

$$u(0, t) = 0, \quad u(a, t) = 0$$

if F is continuous on $0 \leq x \leq a$ with

$$F(0) = F(a) = 0$$

These are typical of the boundary conditions and initial conditions that normally arise in conjunction with heat conduction problems. However, the end point conditions

$$u(0, t) = 0 \quad \text{and} \quad u(a, t) = 0$$

are sometimes replaced with conditions involving derivatives of u at the end points, which represent heat flow conditions at these points.

In this chapter, we will examine some of the more widely used numerical methods for solving standard boundary value problems for partial differential equations such as those mentioned here. All of the methods studied in this chapter are finite difference methods. These methods are easily extended to a variety of other problems involving slightly different partial differential equations or boundary conditions.

■ *Exercise Set 12.1*

1. Which of the following are second-order linear partial differential equations?
   a. $(x + y)u_{xx} + x^2 u_{xy} + e^x u_y = 0$

b. $u_{xy} + y^2 u_x + x^2 u_y = xy^2$

c. $xu_x + yu_y + (x^2 + y^2)u = 0$

d. $u_{yy} + (u_{xx})^2 + 2u_y = 0$

2. Show that if f and g are functions for which f'' and g'' exist, then

$$u = f(x + t) + g(x - t)$$

is a solution of the partial differential equation $u_{xx} - u_{tt} = 0$.

3. Find the general solution of the partial differential equation $u_{xy} = x + y$ by using partial integration.

4. Classify each equation as hyperbolic, parabolic, or elliptic over the plane.

a. $u_{xy} + u_x + u_y = 0$ **b.** $3u_{xx} + 2u_{yy} + u = 0$

c. $u_{xx} + 2u_{xy} + u_{yy} + xu_y = 0$ **d.** $2u_{xx} + 5u_{yy} + u_{xy} + x^2 u = 0$

5. Find values of a and b for which $u = e^{at} \sin bx$ is a solution of $u_{xx} + u_{tt} = 0$.

6. Show that

$$u = \frac{F(x + ct) + F(x - ct)}{2}$$

is the solution of the problem

$$c^2 u_{xx} = u_{tt} \quad u(x, 0) = F(x), \quad u_t(x, 0) = 0 \quad -\infty < x < \infty, \quad t > 0$$

7. Use the result in exercise 6 to find the solution of the problem

$$4u_{xx} = u_{tt} \quad u(x, 0) = \frac{1}{1 + x^2}, \quad u_t(x, 0) = 0 \quad -\infty < x < \infty, \quad t > 0$$

Find values of $u(1, 1)$, $u(3, 2)$, and $u(2, 5)$.

8. Find values of a and b for which

$$u = e^{-at} \sin b\pi x$$

is a solution of partial differential equation $ku_{xx} = u_t$.

9. Use the result in exercise 8 to find a solution of the boundary value problem

$$4u_{xx} = u_t \quad u(x, 0) = \sin \pi x, \quad u(0, t) = u(1, t) = 0$$

10. Verify that each of the following functions is a solution of the Laplace equation $u_{xx} + u_{yy} = 0$.

a. $u = xy$ **b.** $u = x^2 - y^2$ **c.** $u = \ln(x^2 + y^2)$

d. $u = e^{\pi x} \cos \pi y$ **e.** $u = \arctan(y/x)$

11. Suppose that $u_1(x, y)$ and $u_2(x, y)$ are solutions of the Laplace equation. Show that if c_1 and c_2 are any constants, then

$$u = c_1 u_1(x, y) + c_2 u_2(x, y)$$

is also a solution of the Laplace equation.

12. Show that the claim in exercise 11 is also true for solutions of the heat equation.

13. Show that the claim in exercise 11 is also true for solutions of the wave equation.

14. Show that the claim in exercise 11 is true for every linear second-order partial differential equation.

12.2   Hyperbolic Equations

In this section, we will develop an algorithm for approximating the solution of boundary value–initial-value problems of the following type:

$$c^2 u_{xx} = u_{tt} \qquad 0 \leq x \leq a, \quad t \geq 0$$

$$u(x, 0) = F(x) \quad \text{and} \quad u_t(x, 0) = G(x) \quad \text{on } [0, a]$$

$$u(0, t) = u(a, t) = 0$$

The partial differential equation in the problem is the wave equation. This is a hyperbolic equation (see Section 12.1). Boundary value–initial-value problems of this type are familiar to students of partial differential equations and occur frequently in the study of physical problems. A unique solution to such a problem exists if functions F and G have continuous second derivatives on interval $(0, a)$ and $F(a) = F(0) = 0$. Less stringent assumptions are possible, as shown in Chapter 4 of Greenspan (1961).

We wish to approximate the solution on the rectangle described by inequalities

$$0 \leq x \leq a \quad \text{and} \quad 0 \leq t \leq T$$

for some $T > 0$. The first step is to construct a grid on this rectangle. We choose two positive integers n and m. We then define $h = a/n$ and $k = T/m$. Let

$$x_i = ih \qquad i = 0, 1, 2, \ldots, n$$

$$t_j = jk \qquad j = 0, 1, 2, \ldots, m$$

and

$$u_{ij} = u(x_i, t_j)$$

We call (x_i, t_j) a *grid point* (some authors prefer *mesh point*). We will develop an algorithm for estimating values of $u(x, t)$ at the grid points.

We know from Section 9.1 that the second derivative of a function $y(x)$ can be estimated at a point a by use of the formula

$$y''(a) = \frac{y(a + h) - 2y(a) + y(a - h)}{h^2} + 0(h^2)$$

By applying this formula to $u(x, t)$ while holding t constant, we see that

$$u_{xx}(x_i, t_j) = \frac{u(x_i + h, t_j) - 2u(x_i, t_j) + u(x_i - h, t_j)}{h^2} + 0(h^2)$$

Similarly, holding x constant yields

$$u_{tt}(x_i, t_j) = \frac{u(x_i, t_j + k) - 2u(x_i, t_j) + u(x_i, t_j - k)}{k^2} + 0(k^2)$$

That is,

$$u_{xx}(x_i, t_j) = \frac{u_{i+1, j} - 2u_{ij} + u_{i-1, j}}{h^2} + 0(h^2)$$

and

$$u_{tt}(x_i, t_j) = \frac{u_{i, j+1} - 2u_{ij} + u_{i, j-1}}{k^2} + 0(k^2)$$

Substituting these results in the wave equation gives

$$\frac{u_{i+1, j} - 2u_{ij} + u_{i-1, j}}{h^2} = \frac{1}{c^2} \cdot \frac{u_{i, j+1} - 2u_{ij} + u_{i, j-1}}{k^2} + 0(h^2 + k^2)$$

Now let $\lambda = kc/h$. Then dropping the terms of order $0(h^2 + k^2)$, we find that

$$\lambda^2(u_{i+1, j} - 2u_{ij} + u_{i-1, j}) - (u_{i, j+1} - 2u_{ij} + u_{i, j-1}) \cong 0$$

So

$$u_{i, j+1} \cong \lambda^2 u_{i+1, j} + 2(1 - \lambda^2)u_{ij} + \lambda^2 u_{i-1, j} - u_{i, j-1}$$

If h and k are small, we can approximate u_{ij} by w_{ij}, where

$$w_{i, j+1} = \lambda^2 w_{i+1, j} + 2(1 - \lambda^2)w_{ij} + w_{i-1, j} - w_{i, j-1}$$
$$i = 1, 2, \ldots, n - 1, \quad j = 1, 2, \ldots, m - 1$$

We choose

$$w_{i0} = u_{i0} = u(x_i, 0) = F(x_i)$$

for each i. We need to estimate values

$$w_{i1} = u_{i1} = u(x_i, k)$$

We use boundary condition $u_t(x_i, 0) = G(x_i)$ to do this. Theoretically,

$$G(x_i) = u_t(x_i, 0) = \frac{u(x_i, k) - u(x_i, -k)}{2k} + 0(k^2)$$

We have to imagine $u(x, t)$ extended backward in time to make sense of the term $u(x_i, -k) = u_{i, -1}$. It then follows that

$$u_{i, -1} = u(x_i, -k) = u(x_i, k) - 2kG(x_i) + 0(k^2)$$

This result suggests that we use

$$w_{i,\,-1} = w_{i1} - 2kG(x_i)$$

in the iteration equation above. If we do this and set $j = 0$, we find that

$$w_{i1} = \lambda^2 w_{i+1,\,0} + 2(1 - \lambda^2)w_{i0} + \lambda^2 w_{i-1,\,0} - w_{i1} + 2kG(x_i)$$

Solving for w_{i1}, we find that

$$w_{i1} = \frac{\lambda^2}{2}(w_{i+1,\,0} + w_{i-1,\,0}) + (1 - \lambda^2)w_{i0} + kG(x_i) \qquad i = 1, 2, \ldots, n - 1$$

In summary, we have the following algorithm.

Algorithm 12.1

Step 1 Input a, T, c, n, and m.

Step 2 Set

$$h = \frac{a}{n}, \quad k = \frac{T}{m}, \quad \text{and} \quad L = \left(\frac{kc}{h}\right)^2$$

Step 3 Initialize variables:
For $j = 0, \ldots, m$, set $u(0, j) = u(n, j) = 0$.
For $i = 1, \ldots, n - 1$, set $u(i, 0) = F(ih)$.
For $i = 1, \ldots, n - 1$, set

$$u(i, 1) = \frac{L}{2}(u(i + 1, 0) + u(i - 1, 0)) + (1 - L)u(i, 0) + kG(ih)$$

Step 4 For $j = 1, \ldots, m$ and $i = 1, \ldots, n - 1$, set

$$u(i, j + 1) = 2(1 - L)u(i, j) + L(u(i + 1, j) + u(i - 1, j)) - u(i, j - 1)$$

Step 5 Stop.

We will now examine some machine output from Algorithm 12.1.

EXAMPLE 12.1 Find a numerical solution for the problem

$$4u_{xx} = u_{tt} \qquad 0 < x < 1, \quad 0 < t < 1$$

$$u(0, t) = u(1, t) = 0$$

$$u(x, 0) = \sin \pi x, \quad u_t(x, 0) = 0$$

Solution Using $n = 8$ and $m = 20$ with $a = 1$, $T = 1$, and $c = 2$ in Algorithm 12.1, the machine output is

T	X 0.125	0.250	0.375	0.500	0.625	0.750	0.875
0	0.383	0.707	0.924	1.000	0.924	0.707	0.383
0.050	0.364	0.673	0.879	0.951	0.879	0.673	0.364
0.100	0.310	0.573	0.748	0.810	0.748	0.573	0.310
0.150	0.226	0.417	0.545	0.590	0.545	0.417	0.226
0.200	0.119	0.220	0.288	0.312	0.288	0.220	0.119
0.250	0.001	0.003	0.003	0.004	0.003	0.003	0.001
0.300	−0.117	−0.216	−0.282	−0.305	−0.282	−0.216	−0.117
0.350	−0.223	−0.413	−0.539	−0.584	−0.539	−0.413	−0.223
0.400	−0.308	−0.570	−0.744	−0.806	−0.744	−0.570	−0.308
0.450	−0.363	−0.671	−0.877	−0.949	−0.877	−0.671	−0.363
0.500	−0.383	−0.707	−0.924	−1.000	−0.924	−0.707	−0.383
0.550	−0.365	−0.674	−0.881	−0.954	−0.881	−0.674	−0.365
0.600	−0.312	−0.576	−0.752	−0.814	−0.752	−0.576	−0.312
0.650	−0.228	−0.421	−0.550	−0.595	−0.550	−0.421	−0.228
0.700	−0.122	−0.225	−0.295	−0.319	−0.295	−0.225	−0.122
0.750	−0.004	−0.008	−0.010	−0.011	−0.010	−0.008	−0.004
0.800	0.114	0.211	0.275	0.298	0.275	0.211	0.114
0.850	0.221	0.408	0.534	0.578	0.534	0.408	0.221
0.900	0.307	0.567	0.740	0.801	0.740	0.567	0.307
0.950	0.362	0.669	0.875	0.947	0.875	0.669	0.362
1.000	0.383	0.707	0.924	1.000	0.924	0.707	0.383

∎

The theoretical solution to the problem in Example 12.1 is

$$u(x, t) = \sin \pi x \cos 2\pi t$$

This function satisfies the partial differential equation and the stated boundary conditions. This result can be used to evaluate the accuracy of the machine output. We find that some theoretical values of u are

$$u(0.250, 0.10) = \quad 0.5720614 \quad (0.573)$$

$$u(0.375, 0.45) = -0.8786617 \quad (-0.877)$$

$$u(0.625, 0.65) = -0.5430428 \quad (-0.550)$$

$$u(0.500, 0.95) = \quad 0.9510565 \quad (0.947)$$

The values in parentheses are the corresponding machine estimates in each case. As we can see, the machine results are comparable to the theoretical values and accurate enough for many purposes. However, we can obtain more accurate results. The accuracy of the output of this algorithm varies with the choice of ratio λ. Of course, λ is determined by our choice of integers n and m, which in turn determines values

Table 12.1

T	X 0.125	0.250	0.375	0.500	0.625	0.750	0.875
0	0.383	0.707	0.924	1.000	0.924	0.707	0.383
0.063	0.354	0.653	0.854	0.924	0.854	0.653	0.354
0.125	0.271	0.500	0.653	0.707	0.653	0.500	0.271
0.188	0.146	0.271	0.354	0.383	0.354	0.271	0.146
0.250	0.000	0.000	0.000	−0.000	−0.000	−0.000	−0.000
0.313	−0.146	−0.271	−0.354	−0.383	−0.354	−0.271	−0.146
0.375	−0.271	−0.500	−0.653	−0.707	−0.653	−0.500	−0.271
0.438	−0.354	−0.653	−0.854	−0.924	−0.854	−0.653	−0.354
0.500	−0.383	−0.707	−0.924	−1.000	−0.924	−0.707	−0.383
0.563	−0.354	−0.653	−0.854	−0.924	−0.854	−0.653	−0.354
0.625	−0.271	−0.500	−0.653	−0.707	−0.653	−0.500	−0.271
0.688	−0.146	−0.271	−0.354	−0.383	−0.354	−0.271	−0.146
0.750	0.000	0.000	0.000	0.000	−0.000	−0.000	−0.000
0.813	0.146	0.271	0.354	0.383	0.354	0.271	0.146
0.875	0.271	0.500	0.653	0.707	0.653	0.500	0.271
0.938	0.354	0.653	0.854	0.924	0.854	0.653	0.354
1.000	0.383	0.707	0.924	1.000	0.924	0.707	0.383

of the step sizes h and k. In the preceding numerical example, we were using $h = 0.125$ and $k = 0.05$. This yields ratio $\lambda = 0.8$, as you should verify. If we select $n = 8$ and $m = 16$, then $h = \frac{1}{8}$ and $k = \frac{1}{16}$, which yields $\lambda = 1$ in this case. The machine output seen in Table 12.1 results from these choices.

Now let us again compare machine output with theoretical values. We find the following:

$$u(0.250, 0.125) = 0.500000 \qquad (0.500)$$
$$u(0.375, 0.4375) = -0.8535534 \qquad (-0.854)$$
$$u(0.625, 0.625) = -0.6532814 \qquad (-0.653)$$
$$u(0.500, 0.9375) = 0.9238796 \qquad (0.924)$$

The agreement is much better now, although the step size k is larger than the value of k used in the preceding case. Actually, the results are even better than indicated. For page-formatting purposes, the machine output in Table 12.1 was restricted to rounded three-decimal values. If we remove that restriction, we find that the machine estimates are

$$u(0.250, 0.125) = 0.500000$$
$$u(0.375, 0.4375) = -0.853554$$
$$u(0.625, 0.625) = -0.653282$$
$$u(0.500, 0.9375) = 0.923880$$

The errors are very small for this choice of h and k. This is no accident. Best accuracy is always obtained from this algorithm when ratio $\lambda = kc/h$ is equal to 1 ($k = h/c$). Let us consider some error analysis before proceeding with this point.

Algorithm 12.1 is based on a finite difference approximation of the partial differential equation, derived by use of a finite difference approximation of the second derivatives of function u. It is a **finite difference method**. In fact, all the algorithms for finding numerical solutions of partial differential equations to be discussed in this chapter are finite difference methods. It is well to consider at this time the sources of error in such methods. The error in the numerical solution arises from two sources: (1) approximation of the partial differential equation by a difference equation and (2) round-off error in machine arithmetic. Following Ames (1977, p. 24), the error in values of the solution u arising from approximation of the partial differential equation by a difference equation will be called the **discretization error**. If $u_{ij} = u(x_i, t_j)$ is the exact value of the solution at the ijth grid point and $\hat{w}_{ij}$ is the numerical estimate arising from a finite difference method, then the total error at the ijth point is $u_{ij} - \hat{w}_{ij}$, which is the discretization error plus round-off error.

Definition 12.1

A finite difference method for a given partial differential equation is said to be **convergent** if the discretization error goes to zero at each grid point as h and k approach zero.

If a difference method is convergent, then the numerical estimates should improve as h and k are decreased. Eventually, of course, the benefit from decrease in step size is offset by the increase in round-off error that results from the increase in the amount of machine arithmetic required. Theoretical bounds for the discretization error can be established for some difference methods, and these can sometimes be used to prove convergence. Unfortunately, such bounds are not ordinarily useful in practice for actual estimation of the total error in the numerical solution.

Now let

$$PDE_{ij} = c^2 u_{xx}(x_i, t_j) - u_{tt}(x_i, t_j)$$

and let the difference equation approximation be

$$DIF_{ij} = c^2 \frac{u_{i+1,j} - 2u_{ij} + u_{i-1,j}}{h^2} - \frac{u_{i,j+1} - 2u_{ij} + u_{i,j-1}}{k^2}$$

The **local truncation error** for the finite difference employed in Algorithm 12.1 is

$$TE_{ij} = PDE_{ij} - DIF_{ij}$$

From the derivation of the equations, we know that in this case,

$$TE_{ij} = 0(h^2 + k^2)$$

In general, for finite difference methods, it is possible to derive theoretical expressions for truncation errors. The local truncation error should not be confused with the discretization error, however. If the local truncation error is small, we might reasonably expect the discretization error to be small also, but there is no guarantee of this. They are related but different errors.

Definition 12.2

If the local truncation error has a limit of zero as the step sizes h and k are decreased to zero, then we say that the difference equation is **consistent** with the partial differential equation.

Since $TE_{ij} = 0(h^2 + k^2)$ for the difference equation used in developing Algorithm 12.1, this truncation error has a limit of zero as h and k decrease to zero. Hence, this difference equation is consistent with the partial differential equation $C^2 u_{xx} = u_{tt}$. If $\lambda = 1$, however, the truncation error not only has a limit of zero but is in fact equal to zero. A proof of this fact requires some knowledge of Taylor series for functions of two variables. You may find such a proof in Johnson and Riess (1982, p. 486). So if $\lambda = 1$, then the only errors involved are round-off errors in the machine arithmetic. This explains the increase in accuracy achieved in Example 12.1 by using $\lambda = 1$.

Round-off error is difficult to analyze; in fact, it is not feasible to try to estimate the actual magnitude of such error. But we do need to be concerned with the rate of growth of round-off error as the computations proceed. This introduces the problem of *stability* of the finite difference algorithm.

EXAMPLE 12.2   Use Algorithm 12.1 to find a numerical solution for the problem in Example 12.1 using $n = 8$ and $m = 12$.

Solution   The machine results are as follows.

T	X 0.125	0.250	0.375	0.500	0.625	0.750	0.875
0	0.383	0.707	0.924	1.000	0.924	0.707	0.383
0.083	0.331	0.611	0.799	0.865	0.799	0.611	0.331
0.167	0.190	0.350	0.458	0.495	0.458	0.350	0.190
0.250	−0.003	−0.006	−0.007	−0.008	−0.007	−0.006	−0.003
0.333	−0.195	−0.360	−0.471	−0.509	−0.471	−0.360	−0.195
0.417	−0.334	−0.617	−0.806	−0.873	−0.806	−0.617	−0.334
0.500	−0.383	−0.707	−0.924	−1.000	−0.924	−0.707	−0.383
0.583	−0.328	−0.606	−0.791	−0.856	−0.791	−0.605	−0.328
0.667	−0.184	−0.340	−0.444	−0.482	−0.444	−0.341	−0.183
0.750	0.009	0.018	0.021	0.028	0.018	0.022	0.006
0.833	0.201	0.366	0.493	0.507	0.504	0.350	0.212
0.917	0.328	0.647	0.765	0.955	0.723	0.706	0.287
1.000	0.438	0.571	1.165	0.653	1.323	0.347	0.597

Clearly, something is wrong. Let us compare some machine estimates (in parentheses) with theoretical values:

$$u(0.25,\ 1.00) = 0.707107 \qquad (0.571)$$

$$u(0.25,\ 0.25) = 0 \qquad (-0.006)$$

$$u(0.75,\ 0.75) = 0 \qquad (0.022)$$

$$u(0.50,\ 1.00) = 1 \qquad (0.653)$$

These estimates are much less accurate than might be expected. The problem lies in the fact that $\lambda = \frac{4}{3}$ in this case. If $\lambda > 1$, the finite difference algorithm employed is *unstable*. This means that round-off error or any other error introduced into the estimates grows very rapidly as the computations proceed.   ∎

     The concepts of convergence, consistency, and stability of finite difference algorithms are important, but a rigorous, careful analysis is beyond the level of this text. A detailed analysis of these concepts and some basic theory can be found in Smith (1978, Chap. 3).

     If you are interested in pursuing methods and theory for the numerical solution of hyperbolic equations, consult Smith (1978) or Ames (1977, Chap. 4) for their discussions of numerical methods for partial differential equations.

■ *Exercise Set 12.2*

1. Use Algorithm 12.1 with the indicated data in each case to find a numerical solution for the boundary value–initial-value problem

$$c^2 u_{xx} = u_{tt} \qquad 0 \le x \le a,\ \ 0 \le t \le T$$

$$u(x,\ 0) = F(x),\quad u_t(x,\ 0) = G(x) = 0$$

$$u(0,\ t) = u(a,\ t) = 0$$

a. $c = 1,\ a = 1,\ T = 1,\ F(x) = x(1 - x),\ n = 4,\ m = 10$
b. $c = 1,\ a = 2,\ T = 1,\ F(x) = e^{-16(x-1)^2},\ n = 5,\ m = 10$
c. $c = \sqrt{2},\ a = 1,\ T = 1,\ F(x) = \begin{cases} 0, & 0 < x \le 0.5 \\ 0.5, & 0.5 < x < 1 \end{cases},\ n = 10,\ m = 25$

2. Consider the problem

$$u_{xx} = u_{tt} \qquad 0 \le x \le 1,\ \ 0 \le t \le 0.5$$

$$u(0,\ t) = u(1,\ t) = 0,\quad u(x,\ 0) = \sin \pi x,\quad u_t(x,\ 0) = 0$$

a. Verify that the solution is $u = \sin \pi x \cos \pi t$.
b. Use Algorithm 12.1 and hand computations with a small calculator to find a numerical solution. Use $n = 4$ and $m = 5$.
c. Compute the error at each interior grid point.

3. Use Algorithm 12.1 with machine computation and the following values of m and n to estimate the solution of the problem in exercise 2. Find the maximum error in the estimates of $u(x, t)$ in each case.

   a. $n = 5, m = 10$ **b.** $n = 5, m = 20$ **c.** $n = 10, m = 50$

4. Show that

$$u_{xx}(x_i, y_j) = \frac{u_{i+1, j} - 2u_{ij} + u_{i-1, j}}{h^2} - \frac{h^2}{12} \frac{\partial^4 u}{\partial x^4} (\xi_i, y_j)$$

and

$$u_{yy}(x_i, y_j) = \frac{u_{i, j+1} - 2u_{ij} + u_{i, j-1}}{k^2} - \frac{k^2}{12} \frac{\partial^4 u}{\partial x^4} (x_i, \eta_j)$$

5. Use the results in exercise 4 to find a theoretical expression for the local truncation error in the finite difference approximation to the wave equation used in this section.

6. Use the result in exercise 5 to show that the finite difference equation used in this section is consistent with partial differential equation $c^2 u_{xx} = u_{tt}$. (See Definition 12.2.)

7. If $L = 1$ and $G(x) = 0$, the formula used in step 3 of Algorithm 12.1 to compute values of $u(i, 1)$ becomes

$$u(i, 1) = \frac{u(i + 1, 0) + u(i - 1, 0)}{2} \qquad i = 1, \ldots, n - 1$$

Use this fact to find values of $u(i, 1)$ for the problem

$$u_{xx} = u_{tt} \qquad 0 \le x \le 1, \quad 0 \le t \le 1$$

$$u(0, t) = u(1, t) = 0, \quad F(x) = x(1 - x), \quad G(x) = 0$$

Use hand computation and rational arithmetic with step sizes $h = k = \frac{1}{3}$.

12.3   Parabolic Equations

In this section, we will study some algorithms for approximating the solution of the following type of boundary value problem.

$$cu_{xx} = u_t \qquad 0 \le x \le a, \quad t \ge 0$$

$$u(x, 0) = F(x) \qquad \text{on } [0, a]$$

$$u(0, t) = T_1, \quad u(a, t) = T_2 \qquad t > 0$$

The partial differential equation in this problem is usually called the heat equation (or diffusion equation), and it is a parabolic partial differential equation, as we know from Section 12.1. The boundary value problem is a standard one in the study of heat conduction or diffusion problems. It is known to have a unique solution if function F is continuous on the interval $[0, a]$. Function F can be viewed as representing the initial temperature distribution in a rod extending from $x = 0$ to $x = a$. In that case, constants T_1 and T_2 represent the end point temperatures.

We wish to approximate the solution on the rectangle described by inequalities

$$0 \le x \le a \quad \text{and} \quad 0 \le t \le T$$

for some $T > 0$. As in Section 12.2, the first step is to construct a grid on this rectangle. We choose two positive integers n and m. Then we define $h = a/n$ and $k = T/m$. Let

$$x_i = ih \qquad i = 0, 1, 2, \ldots, n$$
$$t_j = jk \qquad j = 0, 1, 2, \ldots, m$$

We will develop an algorithm for estimating values of the solution $u(x, t)$ at these mn grid points.

As pointed out in Section 12.2, the numerical estimation formula for the second derivative of a function, which was developed in Section 9.1, implies that

$$u_{xx}(x_i, t_j) = \frac{u(x_i + h, t) - 2u(x_i, t_j) + u(x_i - h, t_j)}{h^2} + 0(h^2)$$

Briefly, we see that

$$u_{xx}(x_i, t_j) = \frac{u_{i+1, j} - 2u_{ij} + u_{i-1, j}}{h^2} + 0(h^2)$$

It is also true that

$$u_t(x_i, t_j) = \frac{u(x_i, t_j + k) - u(x_i, t_j)}{k} + 0(k)$$

Briefly,

$$u_t(x_i, t_j) = \frac{u_{i, j+1} - u_{ij}}{k} + 0(k)$$

Substituting in the partial differential equation $cu_{xx} = u_t$ gives

$$\frac{u_{i+1, j} - 2u_{ij} + u_{i-1, j}}{h^2} = \frac{1}{c} \cdot \frac{u_{i, j+1} - u_{ij}}{k} + 0(h^2 + k)$$

Let $\lambda = kc/h$. Rearranging and dropping terms of order $0(h^2 + k)$, we obtain

$$u_{i, j+1} \cong \lambda u_{i+1, j} + (1 - 2\lambda)u_{ij} + \lambda u_{i-1, j}$$

It follows that if h and k are small, we can approximate u_{ij} by w_{ij}, where

$$w_{i, j+1} = \lambda w_{i+1, j} + (1 - 2\lambda)w_{ij} + \lambda w_{i-1, j}$$

$$i = 1, 2, \ldots, n - 1, \quad j = 0, 1, 2, \ldots, m$$

We choose

$$w_{i0} = u_{i0} = u(x_i, 0) = F(x_i) \qquad i = 1, 2, \ldots, n - 1$$

In summary, we have the following algorithm.

Algorithm 12.2

Step 1 Input a, T, c, n, and m.

Step 2 Set $h = a/n$, $k = T/m$, and $L = (kc/h^2)$.

Step 3 Initialize variables:

For $j = 0, \ldots, m$, set $u(0, j) = T_1$ and $u(n, j) = T_2$.

For $i = 1, \ldots, n - 1$, set $u(i, 0) = F(ih)$.

Step 4

For $j = 0, \ldots, m - 1$ and $i = 1, \ldots, n - 1$, set

$$u(i, j + 1) = Lu(i + 1, j) + (1 - 2L)u(i, j) + Lu(i - 1, j)$$

Step 5 Stop.

We will now examine some output from Algorithm 12.2.

EXAMPLE 12.3   Find a numerical solution for the problem

$$u_{xx} = u_t \qquad 0 \le x \le 1, \quad 0 \le t \le 0.5$$

$$u(0, t) = u(1, t) = 0, \quad u(x, 0) = \sin \pi x$$

Solution   Using $n = 5$ and $m = 50$ in Algorithm 12.2, we obtain the following machine results.

H = .2,	K = .01		L = .25	
T	0.200	0.400	0.600	0.800
0	0.588	0.951	0.951	0.588
0.010	0.532	0.860	0.860	0.532
0.020	0.481	0.778	0.778	0.481
0.030	0.435	0.704	0.704	0.435
0.040	0.393	0.637	0.637	0.393
0.050	0.356	0.576	0.576	0.356
0.060	0.322	0.521	0.521	0.322
0.070	0.291	0.471	0.471	0.291
0.080	0.263	0.426	0.426	0.263
0.090	0.238	0.385	0.385	0.238
0.100	0.215	0.349	0.349	0.215
0.110	0.195	0.315	0.315	0.195
0.120	0.176	0.285	0.285	0.176
0.130	0.159	0.258	0.258	0.159
0.140	0.144	0.233	0.233	0.144
0.150	0.130	0.211	0.211	0.130
0.160	0.118	0.191	0.191	0.118
0.170	0.107	0.173	0.173	0.107
0.180	0.097	0.156	0.156	0.097
0.190	0.087	0.141	0.141	0.087
0.200	0.079	0.128	0.128	0.079
0.210	0.071	0.116	0.116	0.071
0.220	0.065	0.105	0.105	0.065
0.230	0.058	0.095	0.095	0.058
0.240	0.053	0.086	0.086	0.053
0.250	0.048	0.077	0.077	0.048
0.260	0.043	0.070	0.070	0.043
0.270	0.039	0.063	0.063	0.039
0.280	0.035	0.057	0.057	0.035
0.290	0.032	0.052	0.052	0.032
0.300	0.029	0.047	0.047	0.029
0.310	0.026	0.042	0.042	0.026
0.320	0.024	0.038	0.038	0.024
0.330	0.021	0.035	0.035	0.021
0.340	0.019	0.031	0.031	0.019
0.350	0.018	0.028	0.028	0.018
0.360	0.016	0.026	0.026	0.016
0.370	0.014	0.023	0.023	0.014
0.380	0.013	0.021	0.021	0.013
0.390	0.012	0.019	0.019	0.012
0.400	0.011	0.017	0.017	0.011
0.410	0.010	0.016	0.016	0.010
0.420	0.009	0.014	0.014	0.009
0.430	0.008	0.013	0.013	0.008
0.440	0.007	0.011	0.011	0.007
0.450	0.006	0.010	0.010	0.006
0.460	0.006	0.009	0.009	0.006
0.470	0.005	0.009	0.009	0.005
0.480	0.005	0.008	0.008	0.005
0.490	0.004	0.007	0.007	0.004
0.500	0.004	0.006	0.006	0.004

■

The theoretical solution of the problem in Example 12.3 is

$$u(x, t) = e^{-\pi^2 t} \sin \pi x$$

You should verify this. We can use values computed from this exact solution to gain some idea of the accuracy of our numerical estimates. Some sample values are

$$u(0.4, 0.05) = 0.5806181 \qquad (0.576)$$
$$u(0.6, 0.06) = 0.5260504 \qquad (0.521)$$
$$u(0.2, 0.10) = 0.2190722 \qquad (0.215)$$
$$u(0.8, 0.14) = 0.1476164 \qquad (0.144)$$

The values in parentheses are the machine estimates obtained in each case. The estimates are comparable to the exact values, but not sufficiently accurate for many purposes.

There is also another difficulty with Algorithm 12.2. The numerical procedure is stable if λ is less than 0.5, but unstable otherwise. For a proof of this fact, see Burden and Faires (1985, p. 581). In Example 12.3, the values of h and k were selected so that $\lambda = 0.25$. Let us see what happens when a larger step size k is selected so that λ exceeds 0.5.

EXAMPLE 12.4   Find a numerical solution for the problem

$$u_{xx} = u_t \qquad 0 \le x \le 1, \quad 0 \le t \le 1$$
$$u(0, t) = u(1, t) = 0, \quad u(x, 0) = \sin \pi x$$

using Algorithm 12.2 with $n = 5$ and $m = 25$.

Solution   The machine results are as follows.

H = .2,   K = .04   L = .9999999

T	0.200	X 0.400	0.600	0.800
0	0.588	0.951	0.951	0.588
0.040	0.363	0.588	0.588	0.363
0.080	0.225	0.363	0.363	0.225
0.120	0.139	0.225	0.225	0.139
0.160	0.086	0.139	0.139	0.086
0.200	0.053	0.086	0.086	0.053
0.240	0.033	0.053	0.053	0.033
0.280	0.020	0.033	0.033	0.020
0.320	0.012	0.020	0.020	0.013
0.360	0.008	0.012	0.013	0.008
0.400	0.005	0.008	0.007	0.005
0.440	0.004	0.004	0.006	0.002

		X		
H = .2,		K = .04		L = .9999999

T	0.200	0.400	0.600	0.800
0.480	0.000	0.006	−0.000	0.004
0.520	0.006	−0.006	0.010	−0.004
0.560	−0.012	0.021	−0.019	0.013
0.600	0.033	−0.052	0.054	−0.033
0.640	−0.086	0.140	−0.139	0.087
0.680	0.225	−0.364	0.365	−0.226
0.720	−0.589	0.955	−0.955	0.591
0.760	1.544	−2.499	2.501	−1.546
0.800	−4.043	6.544	−6.545	4.046
0.840	10.587	−17.132	17.135	−10.592
0.880	−27.719	44.854	−44.859	27.727
0.920	72.573	−117.432	117.441	−72.587
0.960	−190.005	307.446	−307.460	190.028
1.000	497.451	−804.912	804.934	−497.488

∎

For t values from 0 to 0.4, the results in Example 12.4 are believable, although the accuracy is found to be poor upon closer examination. Clearly, the results for values of t greater than 0.5 are not believable. The solution values should never be negative in this example, and the large values near the end of the output clearly reveal the large errors produced by instability of the procedure for this choice of h and k. Ratio λ is 1 in this case and exceeds the limit of 0.5, of course, so the instability was predictable. If $\lambda = 0.5$, the numerical results are satisfactory, but some increase in accuracy results from using smaller values of k with consequent smaller values of λ. The necessity of using very small step sizes in the time direction is the principal fault of this method because of the increase in the amount of machine arithmetic required.

Algorithm 12.2 employs an explicit finite difference formula that permits us to compute each value on the $(j + 1)$th time line using only values on the jth time line. So we say, in this case, that Algorithm 12.2 is an explicit finite difference method. Algorithm 12.1 is also an explicit finite difference method. Such algorithms are simple to use but frequently suffer from instability problems.

There are also implicit finite difference methods for solving hyperbolic and parabolic partial differential equations. These methods use implicit difference formulas and necessitate solving a system of equations to determine the values of u on the $(j + 1)$th time line because the formulas involve more than one of these values in addition to values of u on preceding lines. Implicit methods do not suffer from instability problems, however. We will now examine an implicit method for solving the problem stated at the beginning of this section.

The following algorithm was introduced in Crank and Nicolson (1947). It has become the most widely used method for solving the heat equation.

We have

$$u_{xx}(x_i, t_j) = \frac{u_{i+1, j} - 2u_{ij} + u_{i-1, j}}{h^2} + 0(h^2)$$

Also,

$$u_{xx}(x_i, t_{j+1}) = \frac{u_{i+1, j+1} - 2u_{i, j+1} + u_{i-1, j+1}}{h^2} + 0(h^2)$$

We can estimate $u_{xx}(x, t)$ in the heat equation by averaging these results. The result is an estimate of $u_{xx}(x_i, t_j + k/2)$, of course. We find that

$$u_{xx}\left(x_i, t_j + \frac{k}{2}\right) = u_{xx}(x_i, t_{j+1/2})$$

$$= \frac{u_{i+1, j} + u_{i+1, j+1} - 2(u_{ij} + u_{i, j+1}) + (u_{i-1, j} + u_{i-1, j+1})}{2h^2} + 0(h^2)$$

Also,

$$u_t\left(x_i, t_j + \frac{k}{2}\right) = \frac{u(x_i, t_j + k) - u(x_i, t_j)}{k} + 0(k^2)$$

So the partial differential equation $cu_{xx} = u_t$ can be approximated at point $(x_i, t_j + k/2)$ by finite difference equation

$$\frac{u_{i, j+1} - u_{ij}}{ck} = \frac{u_{i+1, j} + u_{i+1, j+1} - 2(u_{ij} + u_{i, j+1}) + (u_{i-1, j} + u_{i-1, j+1})}{2h^2}$$

with local truncation error of order $0(h^2 + k^2)$. If we let $\lambda = kc/h^2$ and rearrange the terms, we find that

$$\frac{\lambda}{2} u_{i+1, j+1} - (\lambda + 1)u_{i, j+1} + \frac{\lambda}{2} u_{i-1, j+1} = -\frac{\lambda}{2} u_{i+1, j} + (\lambda - 1)u_{ij} - \frac{\lambda}{2} u_{i-1, j}$$

$$j = 0, 1, \ldots, m - 1, \quad i = 1, 2, \ldots, n - 1$$

This is the Crank-Nicolson difference equation, which forms the basis for the **Crank-Nicolson algorithm.** For each choice of j, we have $n - 1$ equations with $n - 1$ unknowns $u_{i, j+1}$, since $i = 1, 2, \ldots, n - 1$. Note that values of $u_{i, j+1}$ are known for $i = 0$ and $i = n$. For example, if $n = 5$, the system of equations determining values of u on the $(j + 1)$th line is

$$(\lambda + 1)u_{1, j+1} - \frac{\lambda}{2} u_{2, j+1} = b_1$$

$$-\frac{\lambda}{2} u_{1, j+1} + (\lambda + 1)u_{2, j+1} - \frac{\lambda}{2} u_{3, j+1} = b_2$$

$$-\frac{\lambda}{2} u_{2, j+1} + (\lambda + 1)u_{3, j+1} - \frac{\lambda}{2} u_{4, j+1} = b_3$$

$$-\frac{\lambda}{2} u_{3, j+1} + (\lambda + 1)u_{4, j+1} = b_4$$

where

$$b_1 = \frac{\lambda}{2} u_{2j} + (1 - \lambda)u_{1j} + \frac{\lambda}{2} (u_{0j} + u_{0, j+1})$$

$$b_2 = \frac{\lambda}{2} u_{3j} + (1 - \lambda)u_{2j} + \frac{\lambda}{2} u_{1j}$$

$$b_3 = \frac{\lambda}{2} u_{4j} + (1 - \lambda)u_{3j} + \frac{\lambda}{2} u_{2j}$$

$$b_4 = \frac{\lambda}{2} u_{3j} + (1 - \lambda)u_{4j} + \frac{\lambda}{2} (u_{5j} + u_{5, j+1})$$

For each j, the linear system to be solved is a **tridiagonal system**. Such systems can be solved readily by algorithms that are modified forms of Gauss elimination. Before we consider the details of such an algorithm, it is important to note that the linear systems encountered here can be put into another form that is better suited for computational procedures.

The Crank-Nicolson difference equation is equivalent to the difference equation

$$-u_{i-1, j+1} + \alpha u_{i, j+1} - u_{i+1, j+1} = u_{i+1, j} - \beta u_{ij} + u_{i-1, j}$$

where

$$\alpha = 2\left(1 + \frac{1}{\lambda}\right) \quad \text{and} \quad \beta = 2\left(1 - \frac{1}{\lambda}\right)$$

If we use this form of the difference equation, then to determine values of u on the $(j + 1)$th time line, we need to solve a tridiagonal linear system $\mathbf{Au} = \mathbf{b}$ in which matrix $\mathbf{A}$ has the simple form

$$\begin{bmatrix} \alpha & -1 & 0 & 0 & 0 & \cdots & 0 \\ -1 & \alpha & -1 & 0 & 0 & & 0 \\ 0 & -1 & \alpha & -1 & 0 & & 0 \\ 0 & 0 & -1 & \alpha & -1 & & 0 \\ \vdots & \vdots & \vdots & & \ddots & & \vdots \\ \vdots & \vdots & \vdots & & & \alpha & -1 \\ 0 & 0 & 0 & \cdots & \cdots & -1 & \alpha \end{bmatrix}$$

and the $\mathbf{b}$ matrix has elements given by

$$b_1 = u_{2j} - \beta u_{1j} + u_{0j} + u_{0, j+1}$$
$$b_2 = u_{3j} - \beta u_{2j} + u_{1j}$$
$$b_3 = u_{4j} - \beta u_{3j} + u_{2j}$$
$$\vdots \quad \vdots \quad \vdots \quad \vdots$$
$$b_{n-1} = u_{nj} - \beta u_{n-1, j} + u_{n-2, j} + u_{n, j+1}$$

In general, if **A** is such a matrix and we wish to solve the system $\mathbf{Ax} = \mathbf{b}$, we can use the following algorithm.

Algorithm 12.3

Step 1 Input $\alpha, b_1, b_2, b_3, \ldots, b_{n-1}$, and n.
Step 2 Set $m_1 = -1/\alpha$ and $a_1 = \alpha$.
Step 3 For $i = 2, 3, \ldots, n - 1$, compute

$$a_i = \alpha + m_{i-1} \quad \text{and} \quad b_i = b_i - m_{i-1}b_{i-1}$$

Then compute $m_i = -1/a_i$.

Step 4 Set $x_{n-1} = \dfrac{b_{n-1}}{a_{n-1}}$.

Step 5 For $i = n - 2, n - 3, \ldots, 1$, compute

$$x_i = \frac{b_i + x_{i+1}}{a_i}$$

EXAMPLE 12.5   Solve the linear system $\mathbf{Ax} = \mathbf{b}$ for which the matrix **A** is 4×4 and has the preceding form with $\alpha = 2$, $b_1 = 0$, $b_2 = 3$, $b_3 = -7$, and $b_4 = 7$.

Solution   Applying Algorithm 12.3, we find that

$$m_1 = -0.5$$
$$a_1 = 2$$
$$b_1 = 0$$
$$a_2 = \alpha + m_1 = 1.5$$
$$b_2 = b_2 - m_1 b_1 = 3 + 0.5(0) = 3$$
$$m_2 = -\tfrac{2}{3}$$
$$a_3 = \alpha + m_2 = 2 - \tfrac{2}{3} = \tfrac{4}{3}$$
$$b_3 = b_3 - m_2 b_2 = -7 + (\tfrac{2}{3})(3) = -5$$
$$m_3 = -\tfrac{3}{4}$$
$$a_4 = \alpha + m_3 = 2 - 0.75 = 1.25$$
$$b_4 = b_4 - m_3 b_3 = 7 + 0.75(-5) = 3.25$$
$$m_4 = -0.8$$
$$x_4 = \frac{b_4}{a_4} = \frac{3.25}{1.25} = 2.6$$

$$x_3 = \frac{b_3 + x_4}{a_3} = \frac{-5 + 2.6}{\frac{4}{3}} = -1.8$$

$$x_2 = \frac{b_2 + x_3}{a_2} = \frac{3 - 1.8}{1.5} = 0.8$$

$$x_1 = \frac{b_1 + x_2}{a_1} = \frac{0 + 0.8}{2} = 0.4$$

∎

We now have assembled all of the necessary parts for the following algorithm for solving the problem posed at the beginning of this section. Required input values are a, c, T = max value of t, T_1, T_2, and positive integers n and m.

Algorithm 12.4

Crank-Nicolson Solution of Parabolic Partial Differential Equation

Step 1 Compute $h = a/n$ and $k = T/m$.

Step 2 Compute $L = h^2/kc$.

Step 3 For $j = 0,\ldots,m$, set

$$u(0, j) = T_1 \quad \text{and} \quad u(n, j) = T_2$$

Step 4 For $i = 0,\ldots,n$, set

$$u(i, 0) = F(ih)$$

Step 5 Compute

$$A = 2(1 + L) \quad \text{and} \quad \beta = 2(1 - L)$$

Set $j = 0$.

Step 6 Compute

$$b_1 = u(2, j) - \beta u(1, j) + u(0, j) + u(0, j + 1)$$

Step 7 For $i = 2,\ldots,n - 2$, compute

$$b_i = u(i + 1, j) - \beta u(i, j) + u(i - 1, j)$$

Step 8 Compute

$$b_{n-1} = u(n, j) + u(n, j + 1) - \beta u(n - 1, j) + u(n - 2, j)$$

Step 9 Set $m_1 = -1/A$ and $A_1 = A$.

(continued)

Algorithm 12.4, continued

Step 10 For $i = 2, \ldots, n - 1$, compute

$$A_i = a + m_{i-1}, \quad b_i = b_i - m_{i-1}b_{i-1}, \quad \text{and} \quad m_i = \frac{-1}{A_i}$$

Step 11 Set

$$u(n - 1, j + 1) = \frac{b_{n-1}}{A_{n-1}}$$

Step 12 For $i = n - 2, n - 3, \ldots, 1$, compute

$$u(i, j + 1) = \frac{b + u(i + 1, j + 1)}{A_i}$$

Step 13 If $j = m$, then go to step 15.
Step 14 Set $j = j + 1$ and return to step 6.
Step 15 Stop.

We will now examine some machine output from this algorithm.

EXAMPLE 12.6   Find a numerical solution for the problem

$$0.25u_{xx} = u_t \qquad 0 \le x \le 2, \quad 0 \le t \le 0.3$$

$$u(0, t) = u(2, t) = 0$$

$$u(x, 0) = \sin \pi x$$

Solution   Using the Crank-Nicolson algorithm with $n = 8$ and $m = 30$, we obtain machine results as follows.

				X			
T	0.250	0.500	0.750	1.000	1.250	1.500	1.750
0	0.70711	1.00000	0.70711	0.00000	−0.70711	−1.00000	−0.70711
0.010	0.69073	0.97684	0.69073	0.00000	−0.69073	−0.97684	−0.69073
0.020	0.67473	0.95422	0.67473	0.00000	−0.67473	−0.95422	−0.67473
0.030	0.65911	0.93212	0.65911	0.00000	−0.65911	−0.93212	−0.65911
0.040	0.64384	0.91053	0.64384	0.00000	−0.64384	−0.91053	−0.64384
0.050	0.62893	0.88944	0.62893	0.00000	−0.62893	−0.88944	−0.62893
0.060	0.61436	0.86884	0.61436	0.00000	−0.61436	−0.86884	−0.61436
0.070	0.60013	0.84872	0.60013	0.00000	−0.60013	−0.84872	−0.60014
0.080	0.58624	0.82906	0.58624	0.00000	−0.58623	−0.82906	−0.58624
0.090	0.57266	0.80986	0.57266	0.00000	−0.57266	−0.80986	−0.57266
0.100	0.55940	0.79110	0.55940	0.00000	−0.55939	−0.79110	−0.55940
0.110	0.54644	0.77278	0.54644	0.00000	−0.54644	−0.77278	−0.54644
0.120	0.53378	0.75488	0.53378	0.00000	−0.53378	−0.75488	−0.53378

T	X 0.250	0.500	0.750	1.000	1.250	1.500	1.750
0.130	0.52142	0.73740	0.52142	0.00000	−0.52142	−0.73740	−0.52142
0.140	0.50935	0.72032	0.50935	0.00000	−0.50934	−0.72032	−0.50935
0.150	0.49755	0.70364	0.49755	0.00000	−0.49755	−0.70364	−0.49755
0.160	0.48603	0.68734	0.48603	0.00000	−0.48603	−0.68734	−0.48603
0.170	0.47477	0.67143	0.47477	0.00000	−0.47477	−0.67142	−0.47477
0.180	0.46377	0.65587	0.46377	0.00000	−0.46377	−0.65587	−0.46377
0.190	0.45303	0.64068	0.45303	0.00000	−0.45303	−0.64068	−0.45303
0.200	0.44254	0.62585	0.44254	0.00000	−0.44254	−0.62585	−0.44254
0.210	0.43229	0.61135	0.43229	0.00000	−0.43229	−0.61135	−0.43229
0.220	0.42228	0.59719	0.42228	0.00000	−0.42228	−0.59719	−0.42228
0.230	0.41250	0.58336	0.41250	0.00000	−0.41250	−0.58336	−0.41250
0.240	0.40295	0.56985	0.40295	0.00000	−0.40295	−0.56985	−0.40295
0.250	0.39361	0.55665	0.39361	0.00000	−0.39361	−0.55665	−0.39361
0.260	0.38450	0.54376	0.38450	0.00000	−0.38450	−0.54376	−0.38450
0.270	0.37559	0.53117	0.37559	0.00000	−0.37559	−0.53117	−0.37559
0.280	0.36689	0.51887	0.36689	0.00000	−0.36689	−0.51887	−0.36689
0.290	0.35840	0.50685	0.35840	0.00000	−0.35840	−0.50685	−0.35840
0.300	0.35010	0.49511	0.35010	0.00000	−0.35010	−0.49511	−0.35010

You should verify that

$$u(x, t) = e^{-\pi^2 t/4} \sin \pi x$$

satisfies all the conditions of the problem, so this is the unique solution. Using this theoretical solution to compute numerical values, we obtain

$$u(0.75, 0.05) = 0.625037 \qquad (0.62893) \qquad \text{Error} = 0.00389$$

$$u(0.50, 0.20) = 0.610498 \qquad (0.62585) \qquad \text{Error} = 0.015$$

$$u(0.25, 0.10) = 0.552494 \qquad (0.55940) \qquad \text{Error} = 0.0069$$

The numbers in parentheses are the Crank-Nicolson estimates. The errors are of the order of 10^{-2} or 10^{-3}. Smaller errors can be obtained by decreasing either h or k. If we apply the Crank-Nicolson algorithm again with $n = 16$ while keeping $m = 30$, we obtain the following results.

T	X 0.125	0.250	0.375	0.500	0.625	0.750	0.875
0	0.38268	0.70711	0.92388	1.00000	0.92388	0.70711	0.38268
0.010	0.37347	0.69009	0.90165	0.97593	0.90165	0.69009	0.37347
0.020	0.36449	0.67348	0.87995	0.95245	0.87995	0.67348	0.36449
0.030	0.35571	0.65727	0.85877	0.92953	0.85877	0.65728	0.35572
0.040	0.34715	0.64146	0.83810	0.90716	0.83810	0.64146	0.34715
0.050	0.33880	0.62602	0.81793	0.88533	0.81794	0.62602	0.33880
0.060	0.33065	0.61095	0.79825	0.86402	0.79825	0.61096	0.33065
0.070	0.32269	0.59625	0.77904	0.84323	0.77904	0.59625	0.32269

(continued)

T	X 0.125	0.250	0.375	0.500	0.625	0.750	0.875
0.080	0.31492	0.58190	0.76029	0.82294	0.76029	0.58190	0.31492
0.090	0.30734	0.56790	0.74200	0.80313	0.74200	0.56790	0.30735
0.100	0.29995	0.55423	0.72414	0.78380	0.72414	0.55423	0.29995
0.110	0.29273	0.54089	0.70671	0.76494	0.70671	0.54089	0.29273
0.120	0.28569	0.52788	0.68971	0.74653	0.68971	0.52788	0.28569
0.130	0.27881	0.51517	0.67311	0.72857	0.67311	0.51517	0.27881
0.140	0.27210	0.50278	0.65691	0.71103	0.65691	0.50278	0.27210
0.150	0.26555	0.49068	0.64110	0.69392	0.64110	0.49068	0.26555
0.160	0.25916	0.47887	0.62567	0.67722	0.62567	0.47887	0.25916
0.170	0.25292	0.46734	0.61061	0.66092	0.61061	0.46734	0.25293
0.180	0.24684	0.45610	0.59592	0.64502	0.59592	0.45610	0.24684
0.190	0.24090	0.44512	0.58158	0.62950	0.58158	0.44512	0.24090
0.200	0.23510	0.43441	0.56758	0.61435	0.56758	0.43441	0.23510
0.210	0.22944	0.42395	0.55392	0.59956	0.55392	0.42396	0.22944
0.220	0.22392	0.41375	0.54059	0.58513	0.54059	0.41375	0.22392
0.230	0.21853	0.40380	0.52758	0.57105	0.52758	0.40380	0.21853
0.240	0.21327	0.39408	0.51489	0.55731	0.51489	0.39408	0.21327
0.250	0.20814	0.38459	0.50250	0.54390	0.50250	0.38459	0.20814
0.260	0.20313	0.37534	0.49040	0.53081	0.49040	0.37534	0.20313
0.270	0.19824	0.36631	0.47860	0.51803	0.47860	0.36631	0.19824
0.280	0.19347	0.35749	0.46708	0.50557	0.46708	0.35749	0.19347
0.290	0.18882	0.34889	0.45584	0.49340	0.45584	0.34889	0.18882
0.300	0.18427	0.34049	0.44487	0.48153	0.44487	0.34049	0.18427

Owing to space limitations, only the numerical results for $x \leq 0.875$ are shown. We can use this sample of the output to evaluate the amount of decrease in the errors. We now find the following:

$$u(0.75, 0.05) = 0.625037 \qquad (0.62602) \qquad \text{Error} = 0.00098$$

$$u(0.50, 0.20) = 0.610498 \qquad (0.61435) \qquad \text{Error} = 0.00385$$

$$u(0.25, 0.10) = 0.552494 \qquad (0.55423) \qquad \text{Error} = 0.00174$$

Now the errors are of the order of 10^{-3} or 10^{-4}, and the improvement is clear. If more accurate results are desired, even smaller values of h and k can be used. The following table shows the results of decreasing the step sizes to $h = 0.05$ and $k = 0.005$.

H = .05   K = .005

T = 0.30000

X	U ESTIMATE	U(X, T)	ERROR
0.050	0.0747331	0.0746206	0.0001125
0.100	0.1476260	0.1474038	0.0002221
0.150	0.2168839	0.2165575	0.0003264
0.200	0.2808013	0.2803788	0.0004225
0.250	0.3378044	0.3372962	0.0005082
0.300	0.3864896	0.3859083	0.0005813

X	U ESTIMATE	U(X, T)	ERROR
0.350	0.4256582	0.4250180	0.0006402
0.400	0.4543457	0.4536624	0.0006833
0.450	0.4718457	0.4711362	0.0007096
0.500	0.4777273	0.4770089	0.0007184
0.550	0.4718457	0.4711362	0.0007095
0.600	0.4543456	0.4536625	0.0006832
0.650	0.4256582	0.4250181	0.0006400
0.700	0.3864896	0.3859085	0.0005811
0.750	0.3378043	0.3372964	0.0005079
0.800	0.2808012	0.2803790	0.0004222
0.850	0.2168838	0.2165577	0.0003261
0.900	0.1476260	0.1474041	0.0002220
0.950	0.0747332	0.0746209	0.0001123
1.000	0.0000003	0.0000003	−0.0000000
1.050	−0.0747327	−0.0746204	−0.0001123
1.100	−0.1476256	−0.1474036	−0.0002220
1.150	−0.2168833	−0.2165572	−0.0003262
1.200	−0.2808008	−0.2803786	−0.0004222
1.250	−0.3378039	−0.3372960	−0.0005079
1.300	−0.3864892	−0.3859082	−0.0005810
1.350	−0.4256578	−0.4250179	−0.0006399
1.400	−0.4543454	−0.4536623	−0.0006831
1.450	−0.4718454	−0.4711361	−0.0007093
1.500	−0.4777271	−0.4770089	−0.0007181
1.550	−0.4718455	−0.4711362	−0.0007093
1.600	−0.4543455	−0.4536626	−0.0006829
1.650	−0.4256580	−0.4250183	−0.0006397
1.700	−0.3864894	−0.3859086	−0.0005808
1.750	−0.3378041	−0.3372966	−0.0005075
1.800	−0.2808010	−0.2803792	−0.0004218
1.850	−0.2168836	−0.2165580	−0.0003256
1.900	−0.1476258	−0.1474044	−0.0002214
1.950	−0.0747330	−0.0746211	−0.0001119

Numerical estimates of the values of $u(x, t)$ for $t = 0.3$ are shown in the second column. Theoretical values of the solution are given in the third column. The resulting errors are shown in the last column. All errors are of the order of 10^{-4}. ∎

Exercise Set 12.3

1. Use Algorithm 12.2 with the indicated data in each case to find a numerical solution for the boundary value–initial-value problem

$$cu_{xx} = u_t \qquad 0 \leq x \leq a, \quad 0 \leq t \leq T$$

$$u(x, 0) = F(x)$$

$$u(0, t) = u(a, t) = 0$$

Compute the value of λ used in each case.

    a. $c = 1$, $a = 1$, $T = 0.5$, $F(x) = x(1 - x)$, $n = 4$, $m = 20$
    b. $c = 1$, $a = 2$, $T = 1$, $F(x) = (1 + x^2)^{-2}$, $n = 5$, $m = 20$
    c. $c = 0.25$, $a = 1$, $T = 0.4$, $F(x) = \begin{cases} 0, & 0 \le x < 0.5 \\ 0.5, & 0.5 \le x \le 1 \end{cases}$, $n = 10$, $m = 40$
    d. $c = 0.125$, $a = 5$, $T = 2$, $F(x) = 10$, $n = 20$, $m = 20$

2. Repeat exercise 1 using the Crank-Nicolson algorithm (Algorithm 12.4).

3. Consider the problem

$$0.5u_{xx} = u_t \qquad 0 \le x \le 1, \quad 0 \le t \le 0.5$$

$$u(0, t) = u(1, t) = 0, \quad u(x, 0) = \sin \pi x, \quad u_t(x, 0) = 0$$

    a. Verify that the solution is $u = e^{-\pi^2 t/2} \sin \pi x$.
    b. Use Algorithm 12.2 and hand computations with a small calculator to find a numerical solution. Use $n = 4$ and $m = 5$.
    c. Compute the error at each interior grid point.

4. Use Algorithm 12.2 with machine computation and the following values of m and n to estimate the solution of the problem in exercise 3. Find the maximum error in the estimates of $u(x, t)$ in each case.
    a. $n = 5$, $m = 10$
    b. $n = 5$, $m = 20$
    c. $n = 10$, $m = 50$

5. If Algorithm 12.2 is to be used to solve the problem in exercise 3, what is the smallest value of m that can be used with the given value of n to ensure that the algorithm will be stable?
    a. $n = 5$ **b.** $n = 8$ **c.** $n = 10$

6. Repeat exercise 4 using the Crank-Nicolson algorithm.

7. Use the results in exercise 4 of Exercise Set 12.2 to find a theoretical expression for the local truncation error in the finite difference equation approximation used in this section to produce Algorithm 12.2.

8. Use the result in exercise 7 to show that the finite difference equation used in Algorithm 12.2 is consistent with the partial differential equation $cu_{xx} = u_t$. (See Definition 12.2.)

9. Use Algorithm 12.2 to find values of $u(i, 1)$ for the problem

$$0.2u_{xx} = u_t \qquad 0 < x < 1, \quad 0 < t < 1$$

$$u(0, t) = u(1, t) = 0, \quad F(x) = x(1 - x)$$

Use hand computation and rational arithmetic with step sizes $h = 0.2$ and $k = 0.02$.

10. Repeat exercise 9 with $F(x) = 10$.

11. Use Algorithm 12.3 to solve the tridiagonal system with the values of n, α, $b_1, b_2, \ldots, b_{n-1}$ given.
    a. $n = 5$, $\alpha = 2$, $b_1 = 2$, $b_2 = -2.6$, $b_3 = 0.95$, $b_4 = -0.5$
    b. $n = 4$, $\alpha = 5$, $b_1 = -13$, $b_2 = 16.5$, $b_3 = -0.5$
    c. $n = 6$, $\alpha = 4$, $b_1 = b_2 = b_3 = b_4 = b_5 = 1$

12.4 Elliptic Equations

In this section, we will develop an algorithm for approximating the solution of the following boundary value problem:

$$u_{xx} + u_{yy} = f(x, y) \qquad 0 \le x \le a, \ \ 0 \le y \le b$$

$$u(x, 0) = G_1(x) \quad \text{and} \quad u(x, b) = G_2(x) \qquad 0 \le x \le a$$

$$u(0, y) = G_3(y) \quad \text{and} \quad u(a, y) = G_4(y) \qquad 0 \le y \le b$$

The partial differential equation in this problem is the Poisson equation, which is elliptic, as we know from Section 12.1. The boundary value problem is a Dirichlet problem for a rectangular region. Such problems occur frequently in the study of potential problems. Such a problem has a unique solution if functions G_1, G_2, G_3, and G_4 are continuous on their respective domains and function f is continuous on the rectangle. Function f sometimes represents a distribution of charge or mass in the rectangle described by the inequalities

$$0 \le x \le a \quad \text{and} \quad 0 \le y \le b$$

We want to approximate the solution $u(x, y)$ on that rectangle.

Again, as in Sections 12.2 and 12.3, we begin by constructing a grid on this rectangle. We choose two positive integers n and m. We define $h = a/n$ and $k = b/m$. Let

$$x_i = ih \qquad i = 0, 1, 2, \dots, n$$

$$y_j = jk \qquad j = 0, 1, 2, \dots, m$$

We need to develop some algorithm for estimating values of the solution $u(x, t)$ at these mn grid points.

As pointed out in the previous two sections, the second derivative u_{xx} can be approximated by using the fact that

$$u_{xx}(x_i, y_j) = \frac{u(x_i + h, y_j) - 2u(x_i, y_j) + u(x_i - h, y_j)}{h^2} + 0(h^2)$$

Briefly, we see that

$$u_{xx}(x_i, y_j) = \frac{u_{i+1, j} - 2u_{ij} + u_{i-1, j}}{h^2} + 0(h^2)$$

Similarly, it is true that

$$u_{yy}(x_i, y_j) = \frac{u_{i, j+1} - 2u_{ij} + u_{i, j-1}}{k^2} + 0(k^2)$$

Substituting these results in the Poisson equation gives the finite difference equation approximation to the partial differential equation,

$$\frac{u_{i+1,j} - 2u_{ij} + u_{i-1,j}}{h^2} + \frac{u_{i,j+1} - 2u_{ij} + u_{i,j-1}}{k^2} = f(x_i, y_j)$$

for which the local truncation error has order $0(h^2 + k^2)$.

Let $\lambda = k/h$. Then u_{ij} can be approximated by w_{ij}, where w_{ij} is the solution of the difference equation

$$\lambda^2 w_{i+1,j} - 2(1 + \lambda^2)w_{ij} + \lambda^2 w_{i-1,j} + w_{i,j+1} + w_{i,j-1} = k^2 f_{ij}$$

for $i = 1, 2, \ldots, n - 1$ and $j = 0, 1, 2, \ldots, m$. These equations form a linear system of $(n - 1)(m - 1)$ equations with $(n - 1)(m - 1)$ unknowns w_{ij}, $i = 1, 2, \ldots, n - 1$ and $j = 1, 2, \ldots, m - 1$. Note that we have

$$w_{i0} = u_{i0} = u(x_i, 0) = G(x_i)$$

and

$$w_{im} = u_{im} = u(x_i, b) = G_2(x_i)$$

for $i = 1, 2, \ldots, n - 1$. Similarly,

$$w_{0j} = u_{0j} = u(0, y_j) = G_3(y_j)$$

and

$$w_{nj} = u_{nj} = u(a, y_j) = G_4(y_j)$$

for $j = 0, 1, \ldots, m$. Values of $u_{ij} = w_{ij}$ are known on the boundary of the rectangle. We only need to determine values at interior points.

The solution of the linear system can be estimated by use of the Gauss-Seidel iteration algorithm or some modification of that method. Solving for w_{ij}, we find that

$$w_{ij} = \frac{\lambda^2(w_{i+1,j} + w_{i-1,j}) + w_{i,j+1} + w_{i,j-1} - k^2 f_{ij}}{2(1 + \lambda^2)}$$

Initial estimates of the grid point values $w_{ij} = u_{ij}$ are needed. We will begin by setting all values at interior grid points equal to zero and then iterate using the preceding equation for w_{ij} until the iterations cease to change significantly in value.

The preceding argument leads us to the following algorithm.

Algorithm 12.5

Step 1 Input a, b, n, m, and TOL.

Step 2 Set $h = a/n$, $k = b/m$, and $L = k^2/h^2$. Set $D = 2(1 + L)$.

Step 3 Initialize variables. For $j = 0, \ldots, m$, set

$$u(0, j) = G_3(jk) \quad \text{and} \quad u(n, j) = G_4(jk)$$

For $i = 1, \ldots, n - 1$, set

$$u(i, 0) = G_1(ih) \quad \text{and} \quad u(i, m) = G_2(ih)$$

Step 4 For $j = 1, \ldots, m - 1$ and $i = 1, \ldots, n - 1$, set

$$u(i, j) = w(i, j) = 0$$

Step 5 For $j = 1, \ldots, m - 1$ and $i = 1, \ldots, n - 1$, compute

$$u(i, j) = \frac{L[u(i - 1, j) + u(i + 1, j)] + u(i, j + 1) + u(i, j - 1) - k^2 f(ih, jk)}{D}$$

Step 6 For $i = 1, \ldots, n - 1$ and $j = 1, \ldots, m - 1$, compute

$$C(i, j) = \text{abs} \, ((u(i, j) - w(i, j))$$

Step 7 If max $C(i, j) <$ TOL, then go to step 10.

Step 8 For $i = 1, \ldots, n - 1$ and $j = 1, \ldots, m - 1$, set

$$w(i, j) = u(i, j)$$

Step 9 Go to step 5.

Step 10 Stop.

We will now examine some output from Algorithm 12.5.

EXAMPLE 12.7   Estimate the solution of the problem

$$u_{xx} + u_{yy} = xe^y \qquad 0 \le x \le 2, \quad 0 \le y \le 1$$

$$u(x, 0) = x, \quad u(x, 1) = ex$$

$$u(0, y) = 0, \quad u(2, y) = 2e^y$$

Solution   Using $n = 6$ and $m = 10$ with TOL $= 0.00001$ in Algorithm 12.5, we find machine results as follows.

YJ	XI	U(X, Y)
0.100	0.333	0.36841
0.100	0.667	0.73681
0.100	1.000	1.10522
0.100	1.333	1.47362
0.100	1.667	1.84200
0.200	0.333	0.40716
0.200	0.667	0.81433
0.200	1.000	1.22149
0.200	1.333	1.62864
0.200	1.667	2.03576
0.300	0.333	0.44999
0.300	0.667	0.89998
0.300	1.000	1.34997
0.300	1.333	1.79995
0.300	1.667	2.24989
0.400	0.333	0.49732
0.400	0.667	0.99464
0.400	1.000	1.49196
0.400	1.333	1.98926
0.400	1.667	2.48652
0.500	0.333	0.54962
0.500	0.667	1.09925
0.500	1.000	1.64887
0.500	1.333	2.19847
0.500	1.667	2.74802
0.600	0.333	0.60742
0.600	0.667	1.21485
0.600	1.000	1.82227
0.600	1.333	2.42967
0.600	1.667	3.03702
0.700	0.333	0.67130
0.700	0.667	1.34260
0.700	1.000	2.01389
0.700	1.333	2.68517
0.700	1.667	3.35640
0.800	0.333	0.74188
0.800	0.667	1.48377
0.800	1.000	2.22565
0.800	1.333	2.96752
0.800	1.667	3.70936
0.900	0.333	0.81989
0.900	0.667	1.63978
0.900	1.000	2.45967
0.900	1.333	3.27955
0.900	1.667	4.09941

NUMBER OF ITERATIONS: 112

The theoretical solution of this problem is $u(x, y) = xe^y$, as you should verify. Using this solution to obtain some accurate comparison values, we find

$$u(1.000, 0.100) = 1.10517 \qquad (1.10522)$$

$$u(0.667, 0.200) = 0.814269 \qquad (0.81433)$$

$$u(0.333, 0.500) = 0.549574 \qquad (0.54962)$$

$$u(1.667, 0.700) = 3.356255 \qquad (3.35640)$$

The machine estimates are in parentheses in each case. The errors are quite small considering the small values of n and m used to obtain these results. If we use $n = 12$ and $m = 10$, we find that the estimates at these same comparison points are now

$$u(1.000, 0.100) = 1.10519 \qquad \text{Error} = 2.00E - 05$$

$$u(0.667, 0.200) = 0.814270 \qquad \text{Error} = 1.00E - 06$$

$$u(0.333, 0.500) = 0.549570 \qquad \text{Error} = 4.00E - 06$$

$$u(1.667, 0.700) = 3.35638 \qquad \text{Error} = 1.25E - 04$$

Simply doubling n produces a noticeable increase in accuracy. These estimates are quite acceptable for most purposes, although the values of n and m are not large. A large number of iterations are required to produce these results, however. Using $m = 10$ with $n = 6$, the number of iterations was 112. Increasing n to 12 only increased the number of iterations to 113 surprisingly. ∎

A considerable decrease in the required number of iterations can be achieved by using better initial estimates. The preceding results were produced by setting all interior values of u equal to zero initially (step 4 of Algorithm 12.5). It would seem wiser to use the mean of all the values at grid points on the boundary as the initial value at each interior point. This procedure does, in fact, produce a dramatic decrease in the number of iterations required for the problem in Example 12.7. We only need to precede step 4 of Algorithm 12.5 by a computation of the average of the boundary points. Call this B. Then we replace step 4 by

---

Step 4 For $j = 1, \ldots, m - 1$ and $i = 1, \ldots, n - 1$, set

$$u(i, j) = w(i, j) = B$$

Using $n = 6$, $m = 10$, and $\text{TOL} = 0.00001$ again for the problem in Example 12.7, the machine output for Algorithm 12.5 is shown in Table 12.2. We find that the

Table 12.2

YJ	XI	U(X, Y)
0.100	0.333	0.36843
0.100	0.667	0.73684
0.100	1.000	1.10525
0.100	1.333	1.47364
0.100	1.667	1.84201
0.200	0.333	0.40720
0.200	0.667	0.81438
0.200	1.000	1.22154
0.200	1.333	1.62867
0.200	1.667	2.03578
0.300	0.333	0.45004
0.300	0.667	0.90006
0.300	1.000	1.35004
0.300	1.333	1.79999
0.300	1.667	2.24991
0.400	0.333	0.49738
0.400	0.667	0.99472
0.400	1.000	1.49203
0.400	1.333	1.98931
0.400	1.667	2.48654
0.500	0.333	0.54968
0.500	0.667	1.09933
0.500	1.000	1.64894
0.500	1.333	2.19852
0.500	1.667	2.74805
0.600	0.333	0.60747
0.600	0.667	1.21492
0.600	1.000	1.82233
0.600	1.333	2.42971
0.600	1.667	3.03704
0.700	0.333	0.67134
0.700	0.667	1.34265
0.700	1.000	2.01394
0.700	1.333	2.68520
0.700	1.667	3.35642
0.800	0.333	0.74191
0.800	0.667	1.48381
0.800	1.000	2.22569
0.800	1.333	2.96754
0.800	1.667	3.70937
0.900	0.333	0.81990
0.900	0.667	1.63980
0.900	1.000	2.45969
0.900	1.333	3.27956
0.900	1.667	4.09942

NUMBER OF ITERATIONS: 65
BAVE IS 2.004391

number of iterations required drops to only 65 if we use the mean of the boundary values as the initial value at each interior point. Otherwise, the machine results are virtually the same as before.

The procedure used in Algorithm 12.5 is an implicit finite difference method. To determine the value of the solution at any interior point, we have to estimate all the values at interior points simultaneously by solving a linear system. The requirement of solving a linear system poses no real difficulty, since the Gauss-Seidel algorithm can be used to find a solution readily in most problems. This procedure can be improved somewhat by use of the **successive overrelaxation method** (SOR method), a modification of the Gauss-Seidel iterations that sometimes produces more rapid convergence. For details, you may wish to consult Ames (1977, pp. 119–134) or Todd (1962, pp. 392–394).

The implicit algorithm described in this section does not suffer from stability problems. Accuracy of the numerical solution increases as the values of h and k are decreased, but no values of the ratio λ produce instability. It seems sensible, then, to choose a value of λ that facilitates the computations. Examination of the difference equation used in the algorithm reveals that $\lambda = 1$ is such a value. In that case, computation of L and D may be omitted in step 2 of Algorithm 12.5. Then step 5 is replaced by

---

Step 5 For $j = 1, \ldots, m - 1$ and $i = 1, \ldots, n - 1$, compute

$$u(i, j) = \frac{u(i - 1, j) + u(i + 1, j) + u(i, j + 1) + u(i, j - 1) - kf(ih, jk)}{4}$$

In the case of the Laplace equation $u_{xx} + u_{yy} = 0$ for which $f(x, y) = 0$, the iteration equation in step 5 becomes

$$u(i, j) = \frac{u(i - 1, j) + u(i + 1, j) + u(i, j + 1) + u(i, j - 1)}{4}$$

so $u(i, j)$ is estimated by the mean of the values at the four neighboring points.

For Dirichlet problems requiring solution of the Laplace equation with boundary values given on the boundary of the rectangle $0 < x < a$ and $0 < y < b$, we have the following algorithm derived from Algorithm 12.5 by using the preceding observations.

Algorithm 12.6   Laplace Equations $u_{xx} + u_{yy} = 0$

Step 1 Input a, b, n, and δ.
Step 2 Set $h = a/n$ and $m = b/h$. Set $S = 0$ and $p = 2(m + n + 1)$.
Step 3 For $j = 0, \ldots, m$, set

$$u(0, j) = G(jk), \quad u(n, j) = G(jk), \quad \text{and} \quad S = S + u(0, j) + u(n, j)$$

Step 4 For $i = 1, \ldots, n - 1$, set

$$u(i, 0) = G(ih), \quad u(i, m) = G(ih), \quad \text{and} \quad S = S + u(i, 0) + u(i, n)$$

Step 5 Set $B = S/p$.
Step 6 For $j = 1, \ldots, m - 1$ and $i = 1, \ldots, n - 1$, set

$$u(i, j) = w(i, j) = B$$

Step 7 For $j = 1, \ldots, m - 1$ and $i = 1, \ldots, n - 1$, compute

$$u(i, j) = \frac{u(i - 1, j) + u(i + 1, j) + u(i, j + 1) + u(i, j - 1)}{4}$$

Step 8 Set $M = 0$. For $i = 1, \ldots, n - 1$ and $j = 1, \ldots, m - 1$, compute

$$C(i, j) = \text{abs}((u(i, j) - w(i, j))$$

If $C(i, j) > M$, then set $M = C(i, j)$.
Step 9 If $M < \delta$, then go to step 12.
Step 10 For $i = 1, \ldots, n - 1$ and $j = 1, \ldots, m - 1$, set

$$w(i, j) = u(i, j)$$

Step 11 Go to step 7.
Step 12 Stop.

EXAMPLE 12.8   Find a numerical solution for the Dirichlet problem defined by

$$u_{xx} + u_{yy} = 0 \qquad 0 < x < 1, \quad 0 < y < 1$$

$$u(x, 0) = 0, \quad u(x, 1) = x$$

$$u(0, y) = 0, \quad u(1, y) = y$$

Machine output using Algorithm 12.6 with $n = 5$ is as follows.

YJ	XI	U(X, Y)
0.200	0.200	0.04000
0.200	0.400	0.08000
0.200	0.600	0.12000
0.200	0.800	0.16000
0.400	0.200	0.08000
0.400	0.400	0.16000
0.400	0.600	0.24000
0.400	0.800	0.32000
0.600	0.200	0.12000
0.600	0.400	0.24000
0.600	0.600	0.36000
0.600	0.800	0.48000
0.800	0.200	0.16000
0.800	0.400	0.32000
0.800	0.600	0.48000
0.800	0.800	0.64000

NUMBER OF ITERATIONS: 26
BAVE WAS .3

The exact solution of this problem is easily verified to be $u = xy$, so the machine output is seen to be errorless. The local truncation error for this problem is clearly zero because all partials of u higher than the second order are zero. Hence, all the error is round-off error. There is an insignificant amount of round-off error in this machine output.   ■

Solutions of the Laplace equation are called **harmonic functions**. So we can regard Algorithm 12.6 as a numerical procedure for finding the values of a harmonic function from the values on the boundary of a rectangle.

EXAMPLE 12.9   Estimate values in the interior of the rectangle described by $0 \le x \le 2$ and $0 \le y \le 3$ for the unique harmonic function $u(x, y)$ defined by

$$u(x, 0) = x^2 + 1, \quad u(x, 3) = x^2 - 8$$

$$u(0, y) = 1 - y^2, \quad u(2, y) = 5 - y^2$$

Here are some sample values resulting from using $a = 2$, $b = 3$, and $n = 10$ with TOL $= 0.00001$ in Algorithm 12.6.

x	y	u(x, y)
1.4	0.20	2.92000
0.8	0.40	1.47999
0.6	0.80	0.71999
0.6	0.60	1.00000
0.4	1.00	0.15999
0.2	1.60	−1.52000
0.8	2.40	−4.12000
0.6	2.80	−6.48000

In this case,

$$u(x, y) = x^2 - y^2 + 1$$

so it is easy to verify that the values shown are correct except for a very small amount of error, which is primarily round-off error in the machine arithmetic.   ■

■ *Exercise Set 12.4*

1. Use Algorithm 12.5 with the indicated data in each case to find a numerical solution for the boundary value problem

$$u_{xx} + u_{yy} = f(x, y) \qquad 0 < x < a, \quad 0 < y < b$$

$$u(x, 0) = G_1(x), \quad u(x, b) = G_2(x)$$

$$u(0, y) = G_3(y), \quad u(a, y) = G_4(y)$$

Use $n = 5$ and $m = 10$ in each case.
   a. $u(x, 0) = x^3, \quad u(x, 1) = 1 + x^3$
      $u(0, y) = y^2, \quad u(1, y) = 1 + y^2$
      $f(x, y) = 6x + 2$
   b. $u(x, 0) = \sin \pi x, \quad u(x, 2) = \sin \pi x$
      $u(0, y) = 0, \qquad u(2, y) = 0$
      $f(x, y) = -2\pi^2 \sin \pi x \cos \pi y$
   c. $u(x, 0) = 0, \quad u(x, 1) = x^3/6$
      $u(0, y) = 0, \quad u(1, y) = y^3/6$
      $f(x, y) = xy(x^2 + y^2)$

2. Use Algorithm 12.6 to find a numerical solution of the Laplace equation with the given boundary values. Use $n = m = 5$ in each case.
   a. $u(x, 0) = 2 \ln x, \quad u(x, 1) = \ln (1 + x^2)$
      $u(0, y) = 2 \ln y, \quad u(1, y) = \ln (1 + y^2)$
   b. $u(x, 0) = e^x, \qquad u(x, \pi) = -e^x$
      $u(0, y) = \cos y, \quad u(1, y) = e \cos y$
   c. $u(x, 0) = x^2, \qquad u(x, 1) = x^2 + 2x - 1$
      $u(0, y) = -y^2 \quad u(2, y) = 4 + 4y - y^2$

3. Consider the problem

$$u_{xx} + u_{yy} = 2y(y^2 + 3x^2) \qquad 0 < x < 1, \quad 0 < y < 1$$

$$u(0, y) = u(x, 0) = 0, \quad u(x, 1) = x^2, \quad u(1, y) = y^2$$

   a. Verify that the solution is $u = x^2 y^3$.
   b. Use Algorithm 12.5 with $n = 5$ and $m = 5$ to find a numerical solution.
   c. Compute the error at each interior grid point.

4. Use Algorithm 12.5 with machine computation and the following values of m and n to estimate the solution of the problem in exercise 3. Find the maximum error in the estimates of $u(x, t)$ in each case.
   a. $n = 5, m = 8$
   b. $n = 5, m = 20$
   c. $n = 10, m = 10$
   d. $n = 20, m = 5$

5. The basic idea of the SOR method is to accelerate convergence of Algorithm 12.5 by replacing step 5 with the pair of statements

$$v = \frac{L[u(i-1, j) + u(i+1, j)] + u(i, j+1) + u(i, j-1) - kf(ih, jk)}{D}$$

$$u(i, j) = wv + (1 - w)u(i, j)$$

If $n = m$, the recommended value of w is

$$\frac{2}{1 + \sin(\pi/n)}$$

Use this modified form of Algorithm 12.5 with $n = 10$ to find a numerical solution for the problem

$$u_{xx} + u_{yy} = xy(x^2 + y^2)$$

$$u(x, 0) = u(0, y) = 0$$

$$u(x, 1) = \frac{x^3}{6}, \quad u(1, y) = \frac{y^3}{6}$$

Compare these results with the solution obtained using Algorithm 12.5 without modification. Did the modified algorithm converge significantly faster?

6. Repeat exercise 5, first using $w = 1.1$ and then using $w = 1.75$. Which value gives more rapid convergence?

7. Use the results in exercise 4 of Exercise Set 12.2 to find a theoretical expression for the local truncation error in the finite difference equation approximation used in this section to produce Algorithm 12.5.

8. Use the result in exercise 7 to show that the finite difference equation used in Algorithm 12.5 is consistent with the partial differential equation

$$u_{xx} + u_{yy} = f(x, y)$$

(See Definition 12.2.)

12.5   Applications

In this section, we will consider the numerical solution of some problems that occur in the study of physical phenomena. Our selection is necessarily somewhat limited because we have only considered hyperbolic and parabolic equations that involve one space dimension. In these cases, we are limited to one-dimensional physical problems. Some of the most interesting physical problems that lead to parabolic or hyperbolic partial differential equations involve two or three dimensions. There are well-known numerical methods for such problems that are relatively straightforward extensions of the one-dimensional methods in this chapter. The amount of computation required can be considerable, however. In fact, even with modern high-speed computers, three-dimensional problems are notoriously difficult to handle. These problems require both a large amount of computer memory and a large amount of computer time. Such problems are beyond the scope of an introductory course. An examination of some one-dimensional problems is of value, however, toward gaining some understanding of the possible uses of numerical solutions of partial differential equations.

We shall begin with a modern look at a very old problem known as the vibrating string problem. We want to study vibrations of a metal string, such as a steel guitar string. We will assume that the string is attached to rigid supports at each end, so there is no movement there. We assume that the string is of length L with linear density (mass per unit length) ρ and that the string tension is T. Let the displacement at time t at the point x units from one end of the string be denoted by $u(x, t)$. Under these conditions, it can be shown that

$$c^2 u_{xx} = u_{tt}$$

where

$$c^2 = \frac{T}{\rho}$$

if the displacement $u(x, t)$ is small. For a derivation of this equation, you may wish to consult Kreyszig (1967, pp. 489–490). This partial differential equation was studied by Euler d'Alembert, and others prior to 1750. Both d'Alembert and Euler managed to show that for any sufficiently differentiable functions ϕ and ψ, the function

$$u(x, t) = \phi(x + ct) + \psi(x - ct)$$

would satisfy the partial differential equation.

For the vibrating string problem, we have the boundary (end point) conditions

$$u(0, t) = 0 \quad \text{and} \quad u(L, t) = 0$$

To uniquely determine $u(x, t)$, we need to know the initial shape of the string; that is, we need to know $u(x, 0) = F(x)$. We also need to know how the points on the string are moving; that is, we need to know $u_t(x, 0) = G(x)$.

For our numerical example, suppose that we have a guitar string suspended between supports 60 cm apart. Suppose that the mass per unit length is 0.0225 g/cm. The tension in the string is adjusted to be 1.4×10^7 dynes. Suppose that the string is pulled slightly aside (plucked) at the point 30 cm from the end where $x = 0$, so that the initial shape is described by $u(x, 0) = F(x)$ where

$$F(x) = \begin{cases} 0.01x & \text{if } 0 < x \le 30 \\ 0.30 - 0.01(x - 30) & \text{if } 30 < x < 60 \end{cases}$$

Assume that there is no initial movement, so $u_t(x, 0) = 0$. We would like to have a table of values of $u(x, t)$ for

$$x = 30 \quad \text{and} \quad 0 < t < \text{TMAX}$$

where TMAX is selected to cover two or three periods for the vibration.

Practical solution of this problem requires a combined use of numerical analysis and some physical information. Physically speaking, we expect the principal mode of vibration of the string to be the so-called fundamental mode with frequency $f = c/2L$. There may be some contribution from higher modes (harmonics), with frequencies that are integer multiples of this fundamental frequency. In this case, we find

$$c = \sqrt{\frac{T}{\rho}} = 2.494 \times 10^4 \text{ cm/sec}$$

So the fundamental frequency is $f = c/120 = 207.9$ cps. The period for one full vibration is $1/f = 0.00481$ second. We will select TMAX equal to four periods, so TMAX $= 0.01924$ second. In the numerical solution, a time step smaller than one period is needed to get a recognizable representation of the vibration.

We can use Algorithm 12.1 with $n = 6$ and $m = 48$. This gives a time step

$$k = \frac{\text{TMAX}}{m} = 4.0083 \times 10^{-4}$$

Best numerical results are obtained if ratio λ is close to 1 in value. It can be shown that the local truncation error is zero when $\lambda = 1$ if the analytic solution of this problem is infinitely differentiable. Although that is not the case here, we still can expect to obtain best results by using $\lambda = 1$. The value of λ must be less than or equal to 1 for stability. For $\lambda = 1$, we need

$$k = \frac{h}{c} = \frac{10}{2.494 \times 10^4} = 4.0096 \times 10^{-4}$$

So selection of $m = 48$ will produce a satisfactory value of k. The machine results using Algorithm 12.1 are shown in Table 12.3.

Note that the initial displacement of 0.3 cm at $x = 30$ is repeated at intervals of 4.81070 msec in exact agreement with the physical theory. Also, note that the displacement at the same point is zero at times 1.20268, 3.60803, 6.01338, and 8.41873 msec. These times correspond to $\frac{1}{4}$, $\frac{3}{4}$, $\frac{5}{4}$, and $\frac{7}{4}$ of one full period, which again agrees with physical expectations. These "physical checks" give us some idea of the accuracy of the numerical results. The output of the algorithm for this problem is clearly sufficiently accurate for practical use.

You might wish to make a similar analysis of the results shown in Table 12.4 obtained by increasing the string tension to 6×10^7 dynes. This increases the frequency to about 430.33 cps. The numerical output again corresponds very closely to the expected physical results.

For our next application, let us consider the problem of heat conduction in a rod of heat-conducting material, such as iron. An excellent introduction to heat conduction problems can be found in Lin and Segel (1974, Chap. 4). We will be concerned with the problem of determining the distribution of temperature along a metal rod of length L and constant cross section that is insulated on the sides with

Table 12.3   STRING LENGTH IS 60 cm.
MASS PER UNIT LENGTH IS 0.0225 g/cm.
STRING TENSION IS 1.4E + 07 dynes.
VIBRATION FREQUENCY IS 207.8699 WITH PERIOD 4.810702E − 03 sec.

T(MSEC)	X 10.00	20.00	30.00	40.00	50.00
0	0.10000	0.20000	0.30000	0.20000	0.10000
0.40089	0.10000	0.20000	0.20000	0.20000	0.10000
0.80178	0.10000	0.10000	0.10000	0.10000	0.10000
1.20268	0.00000	0.00000	0.00000	−0.00000	−0.00000
1.60357	−0.10000	−0.10000	−0.10000	−0.10000	−0.10000
2.00446	−0.10000	−0.20000	−0.20000	−0.20000	−0.10000
2.40535	−0.10000	−0.20000	−0.30000	−0.20000	−0.10000
2.80624	−0.10000	−0.20000	−0.20000	−0.20000	−0.10000
3.20713	−0.10000	−0.10000	−0.10000	−0.10000	−0.10000
3.60803	0.00000	0.00000	−0.00000	−0.00000	−0.00000
4.00892	0.10000	0.10000	0.10000	0.10000	0.10000
4.40981	0.10000	0.20000	0.20000	0.20000	0.10000
4.81070	0.10000	0.20000	0.30000	0.20000	0.10000
5.21159	0.10000	0.20000	0.20000	0.20000	0.10000
5.61249	0.10000	0.10000	0.10000	0.10000	0.10000
6.01338	0.00000	0.00000	0.00000	−0.00000	−0.00000
6.41427	−0.10000	−0.10000	−0.10000	−0.10000	−0.10000
6.81516	−0.10000	−0.20000	−0.20000	−0.20000	−0.10000
7.21605	−0.10000	−0.20000	−0.30000	−0.20000	−0.10000
7.61695	−0.10000	−0.20000	−0.20000	−0.20000	−0.10000
8.01784	−0.10000	−0.10000	−0.10000	−0.10000	−0.10000
8.41873	0.00000	0.00000	−0.00000	−0.00000	−0.00000
8.81962	0.10000	0.10000	0.10000	0.10000	0.10000
9.22051	0.10000	0.20000	0.20000	0.20000	0.10000
9.62140	0.10000	0.20000	0.30000	0.20000	0.10000
10.02230	0.10000	0.20000	0.20000	0.20000	0.10000
10.42319	0.10000	0.10000	0.10000	0.10000	0.10000
10.82408	0.00000	0.00000	0.00000	−0.00000	−0.00000
11.22497	−0.10000	−0.10000	−0.10000	−0.10000	−0.10000
11.62586	−0.10000	−0.20000	−0.20000	−0.20000	−0.10000
12.02676	−0.10000	−0.20000	−0.30000	−0.20000	−0.10000
12.42765	−0.10000	−0.20000	−0.20000	−0.20000	−0.10000
12.82854	−0.10000	−0.10000	−0.10000	−0.10000	−0.10000
13.22943	0.00000	0.00000	−0.00000	−0.00000	−0.00000
13.63032	0.10000	0.10000	0.10000	0.10000	0.10000
14.03121	0.10000	0.20000	0.20000	0.20000	0.10000
14.43211	0.10000	0.20000	0.30000	0.20000	0.10000
14.83300	0.10000	0.20000	0.20000	0.20000	0.10000
15.23389	0.10000	0.10000	0.10000	0.10000	0.10000
15.63478	0.00000	0.00000	0.00000	−0.00000	−0.00000
16.03567	−0.10000	−0.10000	−0.10000	−0.10000	−0.10000
16.43657	−0.10000	−0.20000	−0.20000	−0.20000	−0.10000
16.83746	−0.10000	−0.20000	−0.30000	−0.20000	−0.10000
17.23835	−0.10000	−0.20000	−0.20000	−0.20000	−0.10000
17.63924	−0.10000	−0.10000	−0.10000	−0.10000	−0.10000
18.04013	0.00000	0.00000	−0.00000	−0.00000	−0.00000
18.44102	0.10000	0.10000	0.10000	0.10000	0.10000
18.84192	0.10000	0.20000	0.20000	0.20000	0.10000
19.24281	0.10000	0.20000	0.30000	0.20000	0.10000

Table 12.4 STRING LENGTH IS 60 cm.
MASS PER UNIT LENGTH IS 0.0225 g/cm.
STRING TENSION IS 6E + 07 dynes.
VIBRATION FREQUENCY IS 430.3315 WITH PERIOD 2.32379E − 03 sec.

	X				
T(MSEC)	10.00	20.00	30.00	40.00	50.00
0	0.10000	0.20000	0.30000	0.20000	0.10000
0.19365	0.10000	0.20000	0.20000	0.20000	0.10000
0.38730	0.10000	0.10000	0.10000	0.10000	0.10000
0.58095	0.00000	0.00000	0.00000	−0.00000	−0.00000
0.77460	−0.10000	−0.10000	−0.10000	−0.10000	−0.10000
0.96825	−0.10000	−0.20000	−0.20000	−0.20000	−0.10000
1.16189	−0.10000	−0.20000	−0.30000	−0.20000	−0.10000
1.35554	−0.10000	−0.20000	−0.20000	−0.20000	−0.10000
1.54919	−0.10000	−0.10000	−0.10000	−0.10000	−0.10000
1.74284	0.00000	0.00000	−0.00000	−0.00000	−0.00000
1.93649	0.10000	0.10000	0.10000	0.10000	0.10000
2.13014	0.10000	0.20000	0.20000	0.20000	0.10000
2.32379	0.10000	0.20000	0.30000	0.20000	0.10000
2.51744	0.10000	0.20000	0.20000	0.20000	0.10000
2.71109	0.10000	0.10000	0.10000	0.10000	0.10000
2.90474	0.00000	0.00000	0.00000	−0.00000	−0.00000
3.09839	−0.10000	−0.10000	−0.10000	−0.10000	−0.10000
3.29204	−0.10000	−0.20000	−0.20000	−0.20000	−0.10000
3.48568	−0.10000	−0.20000	−0.30000	−0.20000	−0.10000
3.67933	−0.10000	−0.20000	−0.20000	−0.20000	−0.10000
3.87298	−0.10000	−0.10000	−0.10000	−0.10000	−0.10000
4.06663	0.00000	0.00000	−0.00000	−0.00000	−0.00000
4.26028	0.10000	0.10000	0.10000	0.10000	0.10000
4.45393	0.10000	0.20000	0.20000	0.20000	0.10000
4.64758	0.10000	0.20000	0.30000	0.20000	0.10000
4.84123	0.10000	0.20000	0.20000	0.20000	0.10000
5.03488	0.10000	0.10000	0.10000	0.10000	0.10000
5.22853	0.00000	0.00000	0.00000	−0.00000	−0.00000
5.42218	−0.10000	−0.10000	−0.10000	−0.10000	−0.10000
5.61583	−0.10000	−0.20000	−0.20000	−0.20000	−0.10000
5.80947	−0.10000	−0.20000	−0.30000	−0.20000	−0.10000
6.00312	−0.10000	−0.20000	−0.20000	−0.20000	−0.10000
6.19677	−0.10000	−0.10000	−0.10000	−0.10000	−0.10000
6.39042	0.00000	0.00000	−0.00000	−0.00000	−0.00000
6.58407	0.10000	0.10000	0.10000	0.10000	0.10000
6.77772	0.10000	0.20000	0.20000	0.20000	0.10000
6.97137	0.10000	0.20000	0.30000	0.20000	0.10000
7.16502	0.10000	0.20000	0.20000	0.20000	0.10000
7.35867	0.10000	0.10000	0.10000	0.10000	0.10000
7.55232	0.00000	0.00000	0.00000	−0.00000	−0.00000
7.74597	−0.10000	−0.10000	−0.10000	−0.10000	−0.10000
7.93962	−0.10000	−0.20000	−0.20000	−0.20000	−0.10000
8.13326	−0.10000	−0.20000	−0.30000	−0.20000	−0.10000
8.32691	−0.10000	−0.20000	−0.20000	−0.20000	−0.10000
8.52056	−0.10000	−0.10000	−0.10000	−0.10000	−0.10000
8.71421	0.00000	0.00000	−0.00000	−0.00000	−0.00000
8.90786	0.10000	0.10000	0.10000	0.10000	0.10000
9.10151	0.10000	0.20000	0.20000	0.20000	0.10000
9.29516	0.10000	0.20000	0.30000	0.20000	0.10000

the end points held at fixed temperatures T_1 and T_2. Assume that the metal has heat conductivity κ, specific heat σ and density ρ. Let $u(x, t)$ be the temperature of the rod x units from one end ($x = 0$ end) at time t. It can be shown that $u(x, t)$ satisfies the partial differential equation

$$\rho\sigma \frac{\partial u}{\partial t} = \frac{\partial}{\partial x}\left(\kappa \frac{\partial u}{\partial x}\right)$$

If ρ, σ, and κ are constants (homogeneous rod), then we have

$$u_t = ku_{xx}$$

where

$$k = \frac{\kappa}{\rho\sigma}$$

A derivation of this fact can be found in Lin and Segel (1974). A unique solution of the heat equation can be found if the end point temperatures are given, together with a function F describing the initial temperatures at points along the rod; that is, we need to know $F(x) = u(x, 0)$.

Suppose that we are given a steel rod with a length of 20 cm. Let us assume that

$$u(x, 0) = F(x) = 50°\,C$$

Beginning at time $t = 0$, the end points are maintained at a temperature of $0°\,C$. We want to determine the temperature distribution along the rod for subsequent times.

We find that for the type of steel used in the bar, we have heat conductivity $\kappa = 0.120$, density $\rho = 8.0$ g/cc and specific heat $\sigma = 0.11$. This gives

$$k = \frac{\kappa}{\rho\sigma} = \frac{0.12}{8(0.11)} = 0.136364$$

Using this value in Algorithm 12.2, we obtain the numerical estimates of the desired temperatures shown in Table 12.5.

It is interesting to compare the numerical results for the steel rod with the results for a silver rod of the same length. Silver has a density of 10.6 g/cc, conductivity $\kappa = 1.04$, and specific heat $\sigma = 0.056$. These values lead to

$$k = \frac{\kappa}{\rho\sigma} = \frac{1.04}{10.6(0.056)} = 1.75202$$

Using this value in Algorithm 12.2, we obtain the results shown in Table 12.6. Apparently, the silver rod of the same size cools much more rapidly. It is interesting to note that in a theoretical treatment, such as the discussion found in Lin and Segel (1974), it is usually shown that the rate of decrease of the temperature at a point on the bar is greater for larger values of the constant k. It is that effect we are observing

Table 12.5   STEEL ROD
ROD LENGTH (CM) IS 20
DENSITY (G/CC) IS 8
SPECIFIC HEAT IS 0.11
CONDUCTIVITY IS 0.12

TIME STEP IS 1 LAMBDA $= 8.522728E - 03$

		X		
T	4.000	8.000	12.000	16.000
0	50.000	50.000	50.000	50.000
5.00	47.940	49.965	49.965	47.940
10.00	46.048	49.847	49.847	46.048
15.00	44.305	49.658	49.658	44.305
20.00	42.696	49.406	49.406	42.696
25.00	41.209	49.100	49.100	41.209
30.00	39.831	48.745	48.745	39.831
35.00	38.551	48.350	48.350	38.551
40.00	37.360	47.919	47.919	37.360
45.00	36.249	47.458	47.458	36.249
50.00	35.209	46.970	46.970	35.209
55.00	34.235	46.461	46.461	34.235
60.00	33.320	45.933	45.933	33.320
65.00	32.458	45.390	45.390	32.458
70.00	31.645	44.834	44.834	31.645
75.00	30.875	44.269	44.269	30.875
80.00	30.145	43.695	43.695	30.145
85.00	29.452	43.116	43.116	29.452
90.00	28.792	42.532	42.532	28.792
95.00	28.162	41.946	41.946	28.162
100.00	27.560	41.358	41.358	27.560
105.00	26.984	40.770	40.770	26.984
110.00	26.430	40.183	40.183	26.430
115.00	25.898	39.598	39.598	25.898
120.00	25.386	39.015	39.015	25.386
125.00	24.892	38.436	38.436	24.892
130.00	24.415	37.860	37.860	24.415
135.00	23.954	37.289	37.289	23.954
140.00	23.507	36.723	36.723	23.507
145.00	23.074	36.162	36.162	23.074
150.00	22.653	35.606	35.606	22.653
155.00	22.244	35.057	35.057	22.244
160.00	21.847	34.513	34.513	21.847
165.00	21.460	33.976	33.976	21.460
170.00	21.082	33.445	33.445	21.082
175.00	20.715	32.921	32.921	20.715
180.00	20.355	32.404	32.404	20.355
185.00	20.005	31.893	31.893	20.005
190.00	19.662	31.389	31.389	19.662
195.00	19.327	30.892	30.892	19.327
200.00	18.999	30.402	30.402	18.999
205.00	18.678	29.919	29.919	18.678
210.00	18.364	29.443	29.443	18.364
215.00	18.056	28.973	28.973	18.056
220.00	17.754	28.511	28.511	17.754
225.00	17.458	28.055	28.055	17.458
230.00	17.168	27.606	27.606	17.168
235.00	16.883	27.164	27.164	16.883
240.00	16.604	26.729	26.729	16.604

Table 12.6   SILVER ROD
ROD LENGTH (CM) IS 20
DENSITY (G/CC) IS 10.6
SPECIFIC HEAT IS 0.056
CONDUCTIVITY IS 1.04

TIME STEP IS 1 LAMBDA = .1095013

		X		
T	4.000	8.000	12.000	16.000
0	50.000	50.000	50.000	50.000
5.00	31.773	45.703	45.703	31.773
10.00	24.072	37.895	37.895	24.072
15.00	19.148	30.787	30.787	19.148
20.00	15.410	24.899	24.899	15.410
25.00	12.436	20.116	20.116	12.436
30.00	10.042	16.247	16.247	10.042
35.00	8.110	13.122	13.122	8.110
40.00	6.550	10.598	10.598	6.550
45.00	5.290	8.560	8.560	5.290
50.00	4.273	6.913	6.913	4.273
55.00	3.451	5.584	5.584	3.451
60.00	2.787	4.510	4.510	2.787
65.00	2.251	3.642	3.642	2.251
70.00	1.818	2.942	2.942	1.818
75.00	1.468	2.376	2.376	1.468
80.00	1.186	1.919	1.919	1.186
85.00	0.958	1.550	1.550	0.958
90.00	0.774	1.252	1.252	0.774
95.00	0.625	1.011	1.011	0.625
100.00	0.505	0.816	0.816	0.505
105.00	0.408	0.659	0.659	0.408
110.00	0.329	0.533	0.533	0.329
115.00	0.266	0.430	0.430	0.266
120.00	0.215	0.347	0.347	0.215
125.00	0.173	0.281	0.281	0.173
130.00	0.140	0.227	0.227	0.140
135.00	0.113	0.183	0.183	0.113
140.00	0.091	0.148	0.148	0.091
145.00	0.074	0.119	0.119	0.074
150.00	0.060	0.096	0.096	0.060

in our numerical output, since the silver bar has $k = 1.75202$, while the steel bar has $k = 0.136364$.

If a more accurate solution of the heat equation is desired, we can use the Crank-Nicolson algorithm. For comparison, Table 12.7 shows the resulting Crank-Nicolson solution for the case of the silver bar. The results are similar to those presented in Table 12.6. The Crank-Nicolson estimates should be more accurate, however.

Table 12.7   SILVER ROD
LENGTH (CM) 20
DENSITY (G/CC) 10.6
SPECIFIC HEAT 0.056
CONDUCTIVITY 1.04

TIME STEP = 1 SEC.

T	4.000	X cm 8.000	12.000	16.000
0	50.000	50.000	50.000	50.000
5.000	32.615	45.475	45.475	32.615
10.000	24.583	38.053	38.053	24.583
15.000	19.500	31.145	31.145	19.500
20.000	15.715	25.331	25.331	15.715
25.000	12.724	20.566	20.566	12.724
30.000	10.317	16.688	16.688	10.317
35.000	8.368	13.539	13.539	8.368
40.000	6.789	10.984	10.984	6.789
45.000	5.507	8.911	8.911	5.507
50.000	4.468	7.229	7.229	4.468
55.000	3.625	5.865	5.865	3.625
60.000	2.941	4.758	4.758	2.941
65.000	2.386	3.860	3.860	2.386
70.000	1.935	3.131	3.131	1.935
75.000	1.570	2.540	2.540	1.570
80.000	1.274	2.061	2.061	1.274
85.000	1.033	1.672	1.672	1.033
90.000	0.838	1.356	1.356	0.838
95.000	0.680	1.100	1.100	0.680
100.000	0.552	0.893	0.893	0.552
105.000	0.448	0.724	0.724	0.448
110.000	0.363	0.588	0.588	0.363
115.000	0.295	0.477	0.477	0.295
120.000	0.239	0.387	0.387	0.239
125.000	0.194	0.314	0.314	0.194
130.000	0.157	0.255	0.255	0.157
135.000	0.128	0.206	0.206	0.128
140.000	0.104	0.168	0.168	0.104
145.000	0.084	0.136	0.136	0.084
150.000	0.068	0.110	0.110	0.068

The partial differential equation that arises in the study of heat conduction also arises in the study of one-dimensional diffusion problems. For that reason, this partial differential equation is often called the diffusion equation, and the constant k is called the coefficient of diffusivity for the medium. The physics of diffusion is more difficult to understand than heat conduction, but it is interesting to note that the mathematics is essentially the same. So the numerical method illustrated here is also applicable to diffusion problems as well as to many other related problems. An example of a numerical solution of a diffusion problem may be found in Wheatley and Gerald (1984, pp. 461–464).

Heat flow problems in two dimensions require solution of the partial differential equation

$$k(u_{xx} + u_{yy}) = u_t$$

A finite difference equation approximation to this partial differential equation can be constructed by approximating each of the derivatives by finite difference approximations. This leads to a numerical algorithm requiring the solution of a system of equations again, but the number of equations is quite large. Some practical computational problems arise in the management and organization of these calculations, but the calculations can be performed. Steady-state temperature distributions in two dimensions only require the solution of the Laplace equation and can be handled easily by use of Algorithm 12.5.

Numerical methods are extremely valuable in the study of problems that lead to nonlinear hyperbolic or parabolic partial differential equations. The scientific problems in these cases are too difficult to include in this elementary introduction to numerical methods, but the numerical procedures are extensions of the methods discussed in this chapter and can be readily understood by anyone familiar with these simpler procedures.

■ *Exercise Set 12.5*

1. Find a numerical solution of the vibrating string problem for a string of length 60 cm with the indicated values of ρ and T, given that

$$F(x) = \begin{cases} 0.01x & 0 \le x \le 30 \\ 0.30 - 0.01(x - 30) & 30 \le x \le 60 \end{cases}$$

   a. $\rho = 0.0225$, $T = 1.4 \times 10^7$
   b. $\rho = 0.05$, $T = 1.4 \times 10^7$
   c. $\rho = 0.01$, $T = 1.4 \times 10^7$
   d. $\rho = 0.04$, $T = 3.0 \times 10^7$
   Use $h = 10$ cm and $k = 5\sqrt{\rho/T}$ in each case.

2. Repeat exercise 1 using

$$u(x, 0) = F(x) = \begin{cases} 0.02x & 0 \le x \le 15 \\ 0.30 - \dfrac{x - 15}{150} & 15 \le x \le 60 \end{cases}$$

   with $h = 15$ and $k = 2.5\sqrt{\rho/T}$.

3. Use Algorithm 12.2 to find a numerical solution of the heat conduction problem in the case of a rod with the indicated values of L, κ, ρ, σ, and initial temperature distribution $F(x)$. Assume that end point temperatures are $0°$ C. Use $0 \le t \le 10$ seconds.

   a. $L = 20$ cm, $\kappa = 0.150$, $\rho = 8.0$, $\sigma = 0.11$, $F(x) = 30$
   b. $L = 50$ cm, $\kappa = 0.150$, $\rho = 8.0$, $\sigma = 0.11$, $F(x) = 30$

c. $L = 20$ cm, $\kappa = 1.10$, $\rho = 2.7$, $\sigma = 0.22$, $F(x) = 0.5x(20 - x)$

d. $L = 100$ cm, $\kappa = 1.04$, $\rho = 10.6$, $\sigma = 0.06$,

$$F(x) = \begin{cases} 0.8x & 0 \le x \le 50 \\ 40 - 0.8(x - 50) & 50 \le x \le 100 \end{cases}$$

4. Repeat exercise 3 using the Crank-Nicolson algorithm.

5. Repeat exercise 3 using end point temperatures $u(0, t) = 0°$ C and $u(L, t) = 20°$ C.

6. Repeat exercise 5 using the Crank-Nicolson algorithm.

7. The steady-state temperatures in a rod of length L with constant end point temperatures

$$u(0, t) = T_1 \quad \text{and} \quad u(L, t) = T_2$$

are described by $u(x, t) = f(x)$, where function f is the solution of the two-point boundary value problem

$$f''(x) = 0 \qquad f(0) = T_1, \quad f(L) = T_2$$

Solve this problem to find a general form for $f(x)$.

8. Given a rod with $L = 20$ cm, $\kappa = 1.05$, $\rho = 10.6$, and $\sigma = 0.056$, suppose that $u(0, t) = 20°$ C and $u(20, t) = 30°$ C while $u(x, 0) = 50°$ C.
   a. Use the result in exercise 7 to find the steady-state temperature distribution $f(x)$ for this rod.
   b. Find numerical estimates of the temperatures $u(x, t)$ for $0 < t < \text{TMAX}$ by using the Crank-Nicolson algorithm. Select TMAX large enough to allow the temperatures to approach the steady-state values. Compare the numerical estimates for $t = \text{TMAX}$ with the values of $f(x)$ found in part (a).

9. The steady-state temperature distribution in a thin rectangular plate with prescribed edge temperatures and insulated faces can be found by solving the Laplace equation over the rectangular region $0 < x < a$ and $0 < y < b$, where a and b are the lengths of the edges of the plate. The boundary values are the edge temperatures. Suppose that a certain plate of this type is 50 cm by 40 cm, and we view this plate as the rectangular region $0 < x < 50$ and $0 < y < 40$. The edge temperatures are given by

$$u(x, 0) = 10 + 0.4x, \qquad u(x, 40) = 10 + \frac{x^2}{125}$$

$$u(0, y) = 10, \qquad u(50, y) = 30$$

Use $n = 10$ in Algorithm 12.5 to estimate the steady-state temperature distribution over the plate.

BIBLIOGRAPHY

Ames, W. F. 1977. *Numerical methods for partial differential equations*. New York: Academic Press.

Arfken, G. 1966. *Mathematical methods for physicists*. New York: Academic Press.

Bairstow, L. 1914. *Rep. Memor. Adv. Comm. Aero.* 154: 51–63.

Borrelli, R. L. 1987. *Differential equations: A modeling approach*. Englewood Cliffs, NJ: Prentice-Hall.

Burden, R. L., and J. D. Faires. 1989. *Numerical analysis*. Boston: Prindle, Weber & Schmidt.

Clenshaw, C. W., and A. R. Curtis. 1960. A method for numerical integration on an automatic computer. *Numerische Mathematik* 2: 197–205.

Conte, S. D. 1966. The numerical solution of linear boundary value problems. *SIAM Rev.* 8, no. 3 (July): 309–321.

Conte, S. D., and C. de Boor. 1972. *Elementary numerical analysis*. 2nd ed. New York: McGraw-Hill.

Corliss, G. 1977. Which root does the bisection algorithm find? *SIAM Rev.* 19, no. 2 (April): 325–327.

Courant, R., K. Friedrichs, and H. Lewy. 1928. Uber die partiellen differenzengleichungen der mathematischen Physik. *Mathematische Annalen* 100: 32–74.

Crank, J., and P. Nicolson. 1947. A practical method for numerical evaluation of solutions of partial differential equations of the heat-conduction type. *Proc. Cambridge Philos. Soc.* 43: 50–67.

Dennis, J. E., and R. B. Schnabel. 1979. Least change secant updates for quasi-Newton methods. *SIAM Rev.* 21, no. 4 (October): 443–459.

Dennison, D. M. 1931. *Rev. Mod. Physics* 3: 280.

Dickson, L. E. 1955. *New first course in the theory of equations*. New York: Wiley.

Epstein, B. 1962. *Partial differential equations—An introduction*. New York: McGraw-Hill.

Forsythe, G. E. 1967. Today's computational methods of linear algebra. *SIAM Rev.* 9, no. 3 (July): 489–515.

Forsythe, G. E. 1970. Pitfalls in computation, or why a math book is not enough. *Am. Math. Monthly* 77 (November): 931.

Forsythe, G. E., and W. R. Wasow. 1960. *Finite-difference methods for partial differential equations*. New York: Wiley.

Fox, L. 1965. *An introduction to numerical linear algebra*. New York: Oxford University Press.

Francis, J. G. F. 1961. The QR transformation, parts I and II. *Computing J.* 4: 265–271, 332–345.

Fraser, W., and M. W. Wilson. 1966. Remarks on the Clenshaw-Curtis quadrature scheme. *SIAM Rev.* 8, no. 3 (July): 322–327.

Greenspan, D. 1957/1958. On popular methods and extant problems in the solution of polynomial equations. *Math. Mag.* 31: 239–253.

Greenspan, D. 1961. *Introduction to partial differential equations*. New York: McGraw-Hill.

Henrici, P. 1956. *Elements of numerical analysis.* New York: Wiley.

Henrici, P. 1958. *The quotient-difference algorithm.* National Bureau of Standards Applied Mathematics Series, vol. 49, 23–46.

Henrici, P. 1979. Fast Fourier methods in computational analysis. *SIAM Rev.* 21, no. 4 (October): 481–527.

Hildebrand, F. B. 1956. *Introduction to numerical analysis.* New York: McGraw-Hill.

Homer, S., and M. L. Leibowitz. 1972. *Inside the yield book.* Englewood Cliffs: Prentice-Hall.

Householder, A. S., and F. L. Bauer. 1959. On certain methods for expanding the characteristic polynomial. *Numerische Mathematik* 1: 29–37.

Householder, A. S. 1964. *The theory of matrices in numerical analysis.* New York: Blaisdell.

Householder, A. S. 1970. *The numerical treatment of a single nonlinear equation.* New York: McGraw-Hill.

Householder, A. S., and G. W. Stewart. 1971. The numerical factorization of a polynomial. *SIAM Rev.* 13, no. 1 (January): 38–46.

Jenkins, M. A., and J. F. Traub. 1970. A three-stage algorithm for real polynomials using quadratic iteration. *SIAM J. Numer. Anal.* (December): 545–566.

Jensen, J. A., and J. H. Rowland. *Methods of computation.* Glenview, IL: Scott, Foresman.

Johnson, L. W., and R. D. Riess. 1982. *Numerical analysis.* Reading, MA: Addison-Wesley.

Keller, H. B. 1968. *Numerical methods for two-point boundary-value problems.* New York: Blaisdell.

Kreyszig, E. 1967. *Advanced engineering mathematics.* New York: Wiley.

Kulisch, U. W., and W. L. Miranker. 1981. *Computer arithmetic in theory and practice.* New York: Academic Press.

Lanczos, C. 1956. *Applied analysis.* Englewood Cliffs, NJ: Prentice-Hall.

Lehmer, D. H. 1961. A machine method for solving polynomial equations. *J. Assoc. Comput. Mach.* 2: 151–162.

Leontief, W. 1966. *Input/output economics.* New York: Oxford University Press.

Lin, C. C., and L. A. Segel. 1974. *Mathematics applied to deterministic problems in the natural sciences.* New York: Macmillan.

Lin, S. 1943. A method of finding roots of algebraic equations. *J. Math and Phys.* 22: 60–77.

Marden, M. 1956. *Geometry of polynomials.* Providence, RI: American Mathematical Society.

Margenau, H. and Murphy, G. M. 1943. *The Mathematics of Physics and Chemistry.* New York: Van Nostrand.

Mitchell, A. R. 1969. *Computational methods in partial differential equations.* New York: Wiley.

Morris, J. L. 1983. *Computational methods in elementary numerical analysis.* New York: Wiley.

Muller, D. 1956. A method for solving algebraic equations using an automatic computer. *Math. Tables Aids to Computing* 10: 208–215.

Olver, F. W. J. 1952. The evaluation of zeroes of high-degree polynomials. *Philos. Trans. Roy. Soc., London,* Series A, vol. 244, 385–415.

Ortega, J. M. 1972. *Numerical analysis—A second course.* New York: Academic Press.

Peters, G., and J. H. Wilkinson. 1979. Inverse iteration, ill-conditioned equations and Newton's method. *SIAM Rev.* 21, no. 3 (July): 339–360.

Poorman, A. P. 1949. *Applied mechanics.* New York: McGraw-Hill.

Ralston, A. 1965. *A first course in numerical analysis.* New York: McGraw-Hill.

Roberts, J. D. 1961. *Notes on molecular orbital calculations.* Redwood City, CA: Benjamin/Cummings.

Roberts, S. M., and J. S. Shipman. 1972. *Two-point boundary-value problems: Shooting methods.* New York: American Elsevier.

Rogers, J. W. 1983. Locations of roots of polynomials. *SIAM Rev.* 25, no. 3 (July): 327–342.

Rosser, J. B. 1967. A Runge-Kutta for all seasons. *SIAM Rev.* 9, no. 3 (July): 417–452.

Rutishauser, H. 1954. Der quotienten-differenzen-algorithmus. *Z. Angew. Math. Phys.* 5: 496–507.

Saff, E. B., and R. S. Varga. 1977. *Padé and rational approximations.* New York: Academic Press.

Schiff, L. I. 1955. *Quantum mechanics.* New York: McGraw-Hill.

Schweitzer, P. J. 1977. A definite integral of N. Bohr. *SIAM Rev.* 19, no. 1 (January): 147.

Searl, M. F., ed. 1973. Energy modeling. Papers presented at a seminar on energy modeling, Washington, DC, January 25–26, 1973. Washington, DC: Resources for the Future.

Smith, G. D. 1978. *Numerical solution of partial differential equations.* New York: Clarendon Press-Oxford University Press.

Stroud, A. H., and D. Secrist. 1966. *Gaussian quadrature formulas.* Englewood Cliffs, NJ: Prentice-Hall.

Todd, J. 1962. *A survey of numerical analysis.* New York: McGraw-Hill.

Todd, J. 1974. John Von Neumann and the national accounting machine. *SIAM Rev.* 16, no. 4 (October): 526–530.

Varga, R. S. 1962. *Matrix iterative analysis.* Englewood Cliffs, NJ: Prentice-Hall.

Wheatley, P. O., and C. F. Gerald. 1984. *Applied numerical analysis.* Reading, MA: Addison-Wesley.

Wilkinson, J. H. 1959. The evaluation of the zeroes of ill-conditioned polynomials, I, II. *Numerische Mathematik* 1: 150–166, 167–180.

Zill, D. G. 1986. *A first course in differential equations.* Boston: Prindle, Weber & Schmidt.

1 COMPLEX NUMBERS

A **complex number** is a number of the form $a + bi$, where a and b are real numbers and $i = \sqrt{-1}$, so $i^2 = -1$.

The arithmetic operations of addition and multiplication are performed according to the rules

$$(a_1 + ib_1) + (a_2 + ib_2) = (a_1 + a_2) + (b_1 + b_2)i$$

$$(a_1 + ib_1)(a_2 + ib_2) = (a_1 a_2 - b_1 b_2) + (a_1 b_2 + a_2 b_1)i$$

Let $z = x + iy$. The negative of z is

$$-z = -x - iy$$

The **conjugate** of z is

$$\bar{z} = x - iy$$

The **absolute value** of z is

$$|z| = \sqrt{x^2 + y^2}$$

Note that $z\bar{z} = |z|^2$. Consequently,

$$\frac{1}{z} = \frac{\bar{z}}{|z|^2}$$

Then it follows that

$$\frac{z_1}{z_2} = z_1 \frac{1}{z_2} = \frac{z_1 \bar{z}_2}{|z_2|^2}$$

EXAMPLE A1.1   Given $z = 3 + 4i$, we have $|z| = 5$ and $\bar{z} = 3 - 4i$, and the reciprocal of z is

$$\frac{\bar{z}}{|z|} = 0.6 - 0.8i$$

■

EXAMPLE A1.2   Given $z_1 = 1 + i$ and $z_2 = 2 + 3i$, we have

$$z_1 + z_2 = 3 + 4i, \quad z_2 - z_1 = 1 + 2i, \quad z_1 z_2 = -1 + 5i$$

and

$$\frac{z_2}{z_1} = \frac{2 + 3i}{1 + i} = \frac{z_2 \bar{z}_1}{|z_1|^2} = \frac{(2 + 3i)(1 - i)}{2}$$

$$= \frac{5 + i}{2} = 2.5 + 0.5i \qquad \blacksquare$$

If $z = x + iy$ is any nonzero complex number, let $r = \sqrt{x^2 + y^2}$ and let θ be the angle between the x axis and the line through the origin passing through the point (x, y). Angle θ is defined by the pair of equations

$$\cos \theta = \frac{x}{r} \quad \text{and} \quad \sin \theta = \frac{y}{r}$$

Number r is the **modulus**, or absolute value, of z, and θ is the **argument** of z. Clearly,

$$z = r \cos \theta + ir \sin \theta = r(\cos \theta + i \sin \theta)$$

We call this the **polar form** of z. Polar forms are useful when powers or roots of complex numbers must be computed.

Theorem A1.1   (De Moivre's theorem) If k is any positive integer, then

$$z^k = r^k(\cos \theta + i \sin \theta)^k = r^k(\cos k\theta + i \sin k\theta)$$

For example, if we wish to compute $(1 + i)^3$, we use

$$(1 + i)^3 = (\sqrt{2})^3(\cos 45° + i \sin 45°)^3$$
$$= 2\sqrt{2}(\cos 135° + i \sin 135°)$$
$$= 2\sqrt{2}\left(\frac{-1}{\sqrt{2}} + \frac{1}{\sqrt{2}} i\right) = -2 + 2i$$

If we wish to solve $z^3 = 1$ to find the *cube roots of unity*, we proceed as follows. Let

$$z = r(\cos \theta + i \sin \theta)$$

Then for $n = 0, 1, 2, \ldots,$

$$r^3(\cos 3\theta + i \sin 3\theta) = 1 = \cos 2\pi n + i \sin 2\pi n$$

Hence, $r = 1$ and $\theta = 2\pi n/3$. Three distinct roots arise from choosing $n = 0, 1,$ and 2. These roots are

$$\cos 0 + i \sin 0 = 1$$

$$\cos \frac{2\pi}{3} + i \sin \frac{2\pi}{3} = \frac{-1}{2} + i \frac{\sqrt{3}}{2}$$

$$\cos \frac{4\pi}{3} + i \sin \frac{4\pi}{3} = \frac{-1}{2} - i \frac{\sqrt{3}}{2}$$

2 ROLLE'S THEOREM AND THE MEAN VALUE THEOREM

Two theorems from calculus are frequently used in numerical analysis: Rolle's theorem and the mean value theorem.

Rolle's Theorem

Suppose that f is continuous on interval $[a, b]$ and differentiable on the open interval (a, b). If $f(a) = f(b) = 0$, then $f'(c) = 0$ for some c in (a, b).

It is not difficult to see why Rolle's theorem is true. If f is a constant function, then c can be any point in interval (a, b). Suppose that f is not constant. If $f(x) > 0$ somewhere on (a, b), then since f is continuous, f must assume a maximum value at some point c in $[a, b]$. Since $f(a) = f(b) = 0$, we must have c in (a, b). From calculus we know that $f'(c) = 0$ or $f'(c)$ does not exist. In our case, $f'(c)$ does exist, so we conclude that $f'(c) = 0$. Similarly, if $f(x) < 0$ on (a, b), then f must assume a minimum at some point c in (a, b), and again it follows that $f'(c) = 0$.

Clearly, the conclusion of Rolle's theorem is still valid if $f(a) = f(b) = m \neq 0$. In that case, $g(x) = f(x) - m$ satisfies all hypotheses in Rolle's theorem, so $g'(c) = 0$ for some c. But $f'(c) = g'(c)$, so we also have $f'(c) = 0$.

Geometrically speaking, Rolle's theorem states that for functions that satisfy the hypotheses of the theorem, the graph of f must have at least one point where the tangent line is parallel to the x axis. The next theorem is a generalization of this fact for functions not satisfying the condition $f(a) = f(b) = 0$.

Mean Value Theorem

If f is continuous on $[a, b]$ and differentiable on (a, b), then

$$f(b) - f(a) = f'(c)(b - a)$$

for some c in (a, b).

The conclusion of the mean value theorem can also be stated as

$$\frac{f(b) - f(a)}{b - a} = f'(c)$$

for some c in (a, b).

A proof of the mean value theorem can be found in any calculus text. The mean value theorem is a consequence of Rolle's theorem.

Geometrically speaking, the theorem states that for functions satisfying the hypotheses, there must be some point $(c, f(c))$, on that part of the graph of f connecting points $(a, f(a))$ and $(b, f(b))$, where the tangent line is parallel to the straight line passing through points $(a, f(a))$ and $(b, f(b))$.

Rolle's theorem and the mean value theorem are used frequently in numerical analysis. Rolle's theorem is the key to deriving the error term for the Lagrange interpolation polynomial in Chapter 7, for example. The mean value theorem can be used to derive Newton's method in Chapter 2. Most of the uses of these two theorems are concerned with bounding errors or deriving error terms for estimation formulas.

3 TAYLOR'S THEOREM

The theorem from calculus discussed here is useful in approximating values of a function near a point where the values of the function and its derivatives are known.

Suppose that values of $f(a), f'(a), \ldots, f^n(a)$ are known for the function f at point a in the domain of f. The polynomial

$$P_n(x) = f(a) + f'(a)(x - a) + \frac{f''(a)}{2!} (x - a)^2 + \cdots + \frac{f^n(a)}{n!} (x - a)^n$$

is known as the *Taylor polynomial of degree n* associated with function f at point a.

Normally, polynomial $P_n(x)$ is regarded as an approximation to f in the neighborhood of point a. In many cases, values of $P_n(x)$ for x near point a provide very good approximations to $f(x)$. The following theorem is helpful in determining how accurate such a polynomial approximation is.

Taylor's Theorem Suppose function f and the first n derivatives of f are continuous on interval $[a, b]$. Assume that derivative $f^{n+1}(x)$ exists on (a, b). Let $P_n(x)$ be the Taylor polynomial associated with f at point a. Then

$$f(b) = P_n(b) + \frac{f^{n+1}(\xi)}{(n + 1)!} (b - a)^{n+1}$$

for some ξ in (a, b).

If $n = 0$, this becomes the mean value theorem, so in that sense, the Taylor theorem is an extension of the mean value theorem. A proof can be found in any advanced calculus book.

In calculus classes, this theorem is used to approximate values for trigonometric functions, exponential functions, or logarithmic functions near points in the domain. If you are not familiar with such applications, you can consult a standard calculus text.

ANSWERS TO SELECTED EXERCISES

Chapter 1

Exercise Set 1.1

1. $1.414213562\ldots$

2. **a.** $x_1 = 1.444444$, $x_2 = 1.442253$, $x_3 = 1.442250$, $x_4 = 1.442249$, $x_5 = 1.442250$
   c. $5E - 7$
   d. $x_5 = 1.442250$; not any better

3. **a.** Quadratic formula gives 500000 and 0 as roots
   b. Quadratic formula gives 49999.5 and 0
   Results unsatisfactory for both equations

5. Exact solution: $x = -189/1309 = -0.144385$; $y = 182/187 = 0.973262$

Exercise Set 1.2

1. **a.** 10000000   **b.** 11001   **c.** 0.1
   d. 1000.001   **e.** 0.101001   **f.** $0.00001100110011\ldots$

2. **a.** 1.25   **b.** 0.25
   c. 0.34375   **d.** 0.625

3. **a.** 0.0010101   **b.** 0.0011101   **c.** 1.000001
   d. 0.0101010   **e.** 0.1000001

4. **a.** 1.125   **b.** 4   **c.** 1.375

5. **a.** 2562   **b.** 255   **c.** 8192   **d.** 65535

7. **a.** 65535   **b.** 53003   **c.** 2048

9. **a.** 0.3125   **b.** 0.1875   **c.** 0.5

Exercise Set 1.3

1. IBM-PC results were
   a. $3.386958E - 9$ **b.** $3.384336E - 9$ **c.** $3.385746E - 9$

2. Results using a pocket calculator were
   a. 0.0000004   **b.** 0.0000003   **c.** 0.0000003

4. $104348/33215 = 3.141592653921421\ldots$
   $\pi = 3.14159265358979$ to 14 decimal places
   Absolute error is $3.562E - 10$; relative error is $1.1338E - 10$

7. $L(0.3) = 0.262349$ Absolute error is $1.519919E - 5$
   $L(0.7) = 0.530136$ Absolute error is $4.922748E - 4$
   $L(1)\ \ = 0.691358$ Absolute error is $1.789033E - 3$

Exercise Set 1.4

1. a. 15   **b.** -0.712 **c.** 30.0625
2. a. $x_1 = 1.416667$ Error $= 0.0024534$
   $x_2 = 1.414216$ Error $= 0.0000024$
   $x_3 = 1.414214$ Error $= 0.0000000$
   $x_4 = 1.414214$ Error $= 0.0000000$
   The convergence is apparently quadratic.

5. N	XN	ERROR	E(N)/E(N-1)
1	2.000000	0.718282	0.000000
2	2.250000	0.468282	1.533866
3	2.370371	0.347911	1.345981
4	2.441406	0.276876	1.256562
5	2.488320	0.229962	1.204007
6	2.521626	0.196656	1.169361
7	2.546499	0.171783	1.144792
8	2.565785	0.152497	1.126466
9	2.581176	0.137106	1.112257
10	2.593743	0.124539	1.100910
11	2.604199	0.114083	1.091653
12	2.613036	0.105246	1.083968
13	2.620603	0.097679	1.077468
14	2.627150	0.091132	1.071841
15	2.632877	0.085405	1.067055
16	2.637929	0.080353	1.062868
17	2.642416	0.075866	1.059151
18	2.646428	0.071854	1.055837
19	2.650036	0.068246	1.052857
20	2.653293	0.064989	1.050128

6. b. $2347/113 = 20.7699115044247787610619469026548672566371681415929 2$
   c. $1/131 = 0.00763358778625954198473282442748091603053435114503$

Chapter 2

Exercise Set 2.1

1. a. $f(1) = -3, f(2) = 14$; root in $(1, 2)$
   b. $f(0) = -1, f(1) = 1 - 1/e$; root in $(0, 1)$
   c. $f(-2) = -21, f(-1) = 2$; root in $(-2, -1)$
   d. $f(0) = 1, f(0.1) = 8.99$; root in $(0, 0.1)$

    e. $f(1.29) = 1.85E - 3, f(1.30) = -5.05E - 3$; root in $(1.29, 1.30)$
    f. $f(0) = -1, f(1) = 1$; root in $(0, 1)$
    g. $f(0.5) = \sqrt{e} - 3 < 0, f(1) = e - 2 > 0$; root in $(0.5, 1)$

2. a. One root near 0.87 **c.** One root near 0.75
    e. One root near 1.4
    g. One positive root near 0.54; infinitely many negative roots near odd multiples of $-\pi/2$; the first four are near -1.45, -4.7, -7.85, and -11 to be more precise.

3. a. $f(1) = f'(1) = 0, f''(1) = 4$; root of order 2
    c. $f(1) = 0, f'(1) = 3$; root of order 1 (simple root)
    e. $f(0) = 0, f'(0) = 1$; root of order 1

4. a. Local minimum near 1.32
    c. Local maximum near 1.76
    e. Minimum at $x = 0$, maximum at 2π
    g. Local maximum near 1.3

Exercise Set 2.2

1. a. 0.7035 **b.** 1.3247 **c.** 4.9651
    d. 0.8767 **e.** 4.1231

3. a. $(3, 4), 3.82843$ **b.** $(2, 3), 2.09455$
    c. $(0, 1), 0.8521$ **d.** $(0.5, 1), 0.51097$

5. $457°$

Exercise Set 2.3

1. a. 1.15417 **b.** 1.15417 **c.** 1.15417

3. The root is near 2; it is a double root

5. a. 2.219107 **b.** 1.895494 **c.** 0.606555
    d. 0.442854 **e.** 0.641714 **f.** 1.327864
    g. 0.8526055 **h.** 0.426303

Exercise Set 2.4

1. a. If the graph were linear or concave up on the interval $(1.292, 1.293)$, we could conclude that the root is closer to 1.293; if the graph were concave down, we should conclude that the root is closer to 1.292
    b. The absolute error is approximately $| f(1.293)/f'(1.293)|$; this is $3.06E - 4$; we conclude that the root restimate 1.293 is correct to 3 decimal places

3. a. $f(x) = x^2 - 3$ with root estimate $r = 1.7320508$; the error is approximately $f(r)/f'(r) = -7.569E - 9$; hence, $\sqrt{3} = 1.7320508$ should be correct to 7 decimal places

5. a. -0.703467422498 **c.** 2.219107148914
    e. $1.164247938460, -1.77286555783, -3.39138238063$
    g. $-0.55051025722, -5.44948974283, 1.58578643763, -4.41421356237$
    All errors less than $1E - 12$.

Exercise Set 2.5

1. a. $x = \sqrt[3]{3x - 1}$ or $x = (x^3 + 1)/3$
    b. $x = e^{-x} \sin x$ or $x = \arcsin(xe^x)$
    c. $x = e^{\sin x}$ or $x = \arcsin(\ln x)$
    d. $x = \arccos(e^{-x})$ or $x = \ln(\sec x)$

3. a. Use $g(x) = \frac{1}{2}(x + 1/x)$; then $|g'(x)| < 1$ on the interval $(1/\sqrt{3}, \infty)$; $x_0 = 2$ is in this interval, so the iterations converge
   b. $x = 1$ is a fixed point in $(1/\sqrt{3}, \infty)$, so the iterations converge to 1

5. a. $g(x) = \tan x$ implies $g'(x) = \sec^2 x$. So $g'(x) > 1$ for all x; the iterations $x_{n+1} = \tan x_n$ do not converge except for $x_0 = 0$
   b. $g(x) = \arctan x$ implies $g'(x) = 1/(1 + x^2) < 1$ for $x > 0$; but $-\pi/2 < g(x) < \pi/2$ for all x, so the root near 1.5 cannot appear as the limit of the sequence $x_{n+1} = \arctan (x_n)$; this sequence converges to zero for every x_0
   c. $x_{n+1} = \pi + \arctan x_n$ with $x_0 = 4.5$
   d. Converges to 4.49340980

6. a. $x = \sin x$ if and only if $x = 0$
   b. $x = e^{-x}$ if $x = 0.5671432904$
   c. $x = \ln x$ never happens; no fixed points
   d. $x = 1 - x^2 \leftrightarrow x^2 + x - 1 = 0 \leftrightarrow x = (-1 \pm \sqrt{5})/2$

9. a. 2.2360680   **b.** 2.6457513   **c.** 6.0827625

Exercise Set 2.6

2. a. I. $g(x) = (2/x)^{1/2} \to g'(x) = -\dfrac{\sqrt{2}}{2} x^{-3/2}$

     The convergence factor is $|g'(\sqrt[3]{2}) = 0.5$
     II. $g(x) = \frac{1}{2}(x + 2/x^2) \to g'(x) = \frac{1}{2}(1 - 4/x^3)$
     The convergence factor is $|g'(\sqrt[3]{2})| = 0.5$
   b. Both iteration schemes required 15 iterations

3. a. Fails. $x_0 = 1 \to x_1 = 0.5$; $f'(0.5) = 0$
   b. The modified method yields estimate 1.414213562375 on the fifth iteration.

5. The first 6 iterates are
   0.6000000000000000
   0.5058823529411765
   0.5000228885328451
   0.5000000003492460
   0.5000000000000000
   0.5000000000000000
   Yes, these results indicate quadratic convergence.

6. a. $g(x) = \ln (2 + \sin x)$, $g'(x) = \cos x/(2 + \sin x)$
     Root $r \cong 1.054127$, $g'(r)$ is not zero; linear convergence

   c. $g(x) = 1 - x^{1/2}$, $g'(x) = \dfrac{-1}{2} x^{-1/2}$, root $r \cong 0.381966$

     $g'(r)$ is not zero; linear convergence

Exercise Set 2.7

1. a. 1.291925   **b.** Aitken est.   R   **c.** 1.292695

1.291925		
1.292476	1.292696	0.2860
1.292633	1.292695	0.2855
1.292678	1.292695	0.2858
1.292691	1.292695	0.2838
1.292695	1.292695	0.2897

3. a.

n	Xn	b. AITKEN EST
1	2.000000	2.482144
2	2.250000	2.543688
3	2.370371	2.579562
4	2.441406	2.603143
5	2.488320	2.619859
6	2.521626	2.632355
7	2.546499	2.642001
8	2.565785	2.649673
9	2.581176	2.655985
10	2.593743	2.661278
11	2.604199	2.665678
12	2.613036	2.669175
13	2.620603	2.672865
14	2.627150	2.675723
15	2.632877	2.678115
16	2.637929	2.680288
17	2.642416	2.682178
18	2.646428	2.683658
19	2.650036	2.687442
20	2.653293	2.685609

c. Sequence x_n does not converge linearly.

Exercise Set 2.8

1. The roots are 0, 1.29269571937, 4.721292755888, and 7.85359327997

3. The roots are 36.8960184525, 22.3411277525, -13.6667736021, and 10.4296273071

5. The root is 0.73396362147

7. Roots are 0 and 4.49340945791

9. Real roots are 1.0325 and 2

Exercise Set 2.9

1. a. $0.9 = x - 0.15 \sin x \rightarrow x = 1.028477$
   b. $0.01 = x - 0.9 \sin x \rightarrow x = 0.09856435$

3. $x^3 - 3x^2 + 2.4 = 0 \rightarrow x = 1.1341$ ft

5. $5.916°$

Chapter 3

Exercise Set 3.1

1. a. Possible rational roots are ± 1, ± 2, ± 3, ± 6; the only rational root is 2
   b. Possible rational roots are ± 1, ± 2, ± 3, ± 4, ± 6, ± 8, ± 12, ± 24; rational roots are -3, 2, 4
   c. Possible rational roots are ± 1, ± 2, ± 4, ± 8; rational roots are -4, -2, 1
   d. Possible rational roots are ± 1, ± 2, ± 3, ± 6; the only rational root is 2

3. There cannot be any change of sign of the polynomial or any of its derivatives on $(0, \infty)$; hence, no positive root of any order is possible

5. a. $x^3 - x - 1 = (x - 2)(x^2 + 2x + 3) + 5 > 0$ if $x > 2$
  b. $x^3 - 2x^2 - 3x + 6 = (x - 3)(x^2 + x) + 6 > 0$ if $x > 3$

7. a. $z = x + 1; z^3 - 4z + 10 = 0$
  b. $z = x + \frac{1}{3}; z^3 + \frac{5}{3}z - \frac{259}{27} = 0$

9. The other roots are roots of a quadratic equation that is approximately $x^2 + 2.33x + 0.4289 = 0$; these roots should be real and distinct; the roots are 2.330059, -2.128419, and -0.201639

Exercise Set 3.2

1. a. 1.840625   **b.** 1.325307   **c.** 4.129794
  d. 2.094563   **e.** 7.079945

3. a. The product of the roots is -0.001
  b. If x is small (and a root), then $2x - 0.001 \cong 0$, since x^2, x^4, and x^5 are small; so $x \cong 0.0005$
  c. A single iteration with $x_0 = 0.0005$ gives $x_1 = 5.003757E - 4$, with error estimate $5.83E - 11$

4. a. $-0.2016398, -2.128419$ **b.** $-1.801938, -0.4450419$
  c. $-0.6527036, -2.879385$ **d.** $-1.772865, -3.391383$

6. a. $1.2469796037, -0.4450418679$, and -1.8019377358
  b. $-0.6457513111, 4.6457513111$, and $-2 \pm 3i$
  c. $0.3892873314 \pm 1.07067577749i, 0.6368829168$, and $-0.7077287898 \pm 0.8419548540i$
  d. ± 0.4494897428 and $2 \pm 3i$
  e. 7.0446667867 and $-3.5223333934 \pm 1.1047613326i$
  f. $4.6457513111, -0.6457513111$, and $-2 \pm 2i$

Exercise Set 3.3

1. a. $(-1 \pm i\sqrt{11})/2$ **b.** $1 \pm \sqrt{3}$
  c. $(1 \pm i)/\sqrt{2}, (-1 \pm i)/\sqrt{2}$
  d. $(-1 \pm i\sqrt{3})/2$

3. a. $-3.414214, -0.5857855, -0.2360689$, and 4.236068
  b. $-2.414214, -1.44949, 0.4142136$, and 3.44949
  c. $-5.645752, -0.3542473, 1.585785$, and 4.414214
  d. Real roots are -3.645751 and 1.645751; the other roots are the complex roots of the quadratic equation $x^2 - 2x + 10 = 0$; these are $1 \pm 3i$
  e. Real roots are -3.44949 and 1.449491; the other roots are the complex roots of the quadratic equation $x^2 - 2x + 4 = 0$; these are $1 \pm \sqrt{3}i$
  f. Real roots are -1.732051 and 1.732051; the complex roots are roots of $x^2 + 1 = 0$; these roots are $\pm i$

4. The real roots are 0.1002509 and -0.09975094; the complex roots are roots of the quadratic equation $x^2 - 0.0005x + 0.01 = 0$; these are $0.00025 \pm 0.099999687i$

Exercise Set 3.4

3. a. $-0.3435608 + 0.4553467i, 0.3435608 - 2.455346i$
  c. $0.618034i, -1.618034i$

5. a. $1, -1, \pm 0.3090169944 \pm 0.9510565163i$,
   and $\pm 0.8090169944 \pm 0.5877852523i$
 b. $-1, \pm i, \pm 0.7071067812, \pm 0.7071067812i$
 c. $-0.5 \pm 1.6583123952i, \pm 2.2360679775i$
 d. $0.5292492773, 0.1282480075, -1.3075794946$,
   $-0.6756593109, 1.3240677126 \pm 0.3936561450i$,
   $-0.3052975933 \pm 1.5909044644i, -1.2026956545 \pm 0.8859144439$,
   and $0.8467962956 \pm 1.3306134481i$
 e. $1.5549173944, 2.1638139350, -3.5963049332$,
   $0.4230819882 \pm 0.6224286589i$, and $-0.4842951863 \pm 0.4275006098i$

6. a. $-0.8043634947, -6.0232381873, 0.1160774748 \pm 0.4686021498i$,
   $0.7830991096 \pm 0.3165825069i$, and $-0.4853757434 \pm 1.0028433556i$
 b. $-0.8041991304, -6.0133984404, 0.1160166095 \pm 0.4685893854i$,
   $0.7834433442 \pm 0.3161446555i$, and $-0.4856611682 \pm 1.0036857896i$

Exercise Set 3.5

1. The equation $x^3 - 3x + 2 = 0$ has roots $r_1 = r_2 = 1$ and $r_3 = -2$; we have $\varepsilon = -0.001$ and $q(x) = x$; the analysis given is not applicable to r_1 or r_2 because these are not simple roots; in the case of r_3, we find $z_3(\varepsilon) \cong -2 + 0.001(-\frac{2}{9}) = -2.0002222$; this result agrees with the machine output through 7 decimal places

2. The roots of $x^3 - 20x + 34.4265$ are $2.5804377734, 2.5835397110$, and -5.1639774844; the roots of $x^3 - 19.99x + 34.4265$ are -5.1631167735 and $2.5815583868 \pm 0.0577157932i$; two roots have moved off the real line into the complex plane, while the other root has moved closer to the origin

5. The quadratic factors are $x^2 - 1.99x + 1$ and $x^2 - 2.01x + 1$; the quadratic formula gives roots $0.995 \pm 0.0998749i, 1.1051249$, and 0.9048751

Exercise Set 3.6

1. a. $P_2(x) = x^4 - 30x^3 + 273x^2 - 820x + 576$
 b. $P_2(x) = x^4 - 52x^3 + 718x^2 - 1668x + 441$
 c. $P_2(x) = x^3 + 2x^2 + x - 1$

3. a. $0.7987, 4.7695, 24.9525$, and 526
 b. $1, 5, 20, 500$; they are all usable

5. The first 5 root estimates are $1, 1, 5, 3.4$, and 2.88235; these are estimates of root $r = 3$

6. a. The real roots of $x^3 + 4x^2 - 7 = 0$ are $-3.391383, -1.772865$, and 1.164248
 b. The real roots of $x^3 + x^2 - 2x - 1 = 0$ are $-1.801938, -0.4450419$, and 1.24698
 c. The real roots of $x^3 + 3x^2 - 1 = 0$ are $-2.879385, -0.6527036$, and 0.5320889
 d. The only real root of $x^3 - 1 = 0$ is 1
 e. The real roots of $x^3 - 7x^2 + 7x + 15 = 0$ are $-1, 3$, and 5

Exercise Set 3.7

1. The five installments are best
2. 2.59 ft
3. $(-1.24698, 1.554959), (0.4450419, 0.198062)$, and $(1.801938, 3.246981)$
4. There are two possible cylinders: $r = 0.466139$ with $h = 1.435638$ ft or $r = 0.617536$ with $h = 0.817997$ ft

Chapter 4

Exercise Set 4.1

1. a. Nonsingular   **b.** Singular   **c.** Singular
d. Nonsingular   **e.** Singular

3. a. $\begin{bmatrix} \frac{3}{5} & -\frac{2}{5} \\ -\frac{2}{5} & \frac{3}{5} \end{bmatrix}$ **b.** $\begin{bmatrix} \frac{1}{13} & \frac{2}{13} \\ -\frac{5}{13} & \frac{3}{13} \end{bmatrix}$ **c.** $\begin{bmatrix} \frac{18}{19} & -\frac{15}{19} \\ \frac{30}{19} & \frac{45}{38} \end{bmatrix}$

5. a. $\begin{bmatrix} -\frac{2}{27} & \frac{19}{27} & \frac{5}{27} \\ -\frac{7}{27} & -\frac{1}{27} & \frac{4}{27} \\ \frac{5}{27} & -\frac{7}{27} & \frac{1}{27} \end{bmatrix}$ **b.** The inverse does not exist.

c. $\begin{bmatrix} -1 & -1 & 2 \\ 1 & 1 & -1 \\ -2 & -1 & 2 \end{bmatrix}$

8. a. Rows dependent   **b.** Rows independent   **c.** Rows dependent

9. a. Yes   **b.** No (too many vectors)   **c.** No (too few vectors)

Exercise Set 4.2

1. a. $\begin{bmatrix} 1 & 2 & 1 & 1 \\ 0 & -5 & -2 & 1 \\ 0 & -5 & -4 & -3 \end{bmatrix} \begin{bmatrix} 1 & 2 & 1 & 1 \\ 0 & -5 & -2 & 1 \\ 0 & 0 & -2 & -4 \end{bmatrix}; \mathbf{x} = \begin{bmatrix} 1 \\ -1 \\ 2 \end{bmatrix}$

b. $\begin{bmatrix} 1 & 2 & 0 & 7 \\ 0 & -3 & 2 & -4 \\ 0 & 2 & 1 & 5 \end{bmatrix} \begin{bmatrix} 1 & 2 & 0 & 7 \\ 0 & -3 & 2 & -4 \\ 0 & 0 & \frac{7}{3} & \frac{7}{3} \end{bmatrix}; \mathbf{x} = \begin{bmatrix} 3 \\ 2 \\ 1 \end{bmatrix}$

c. $\begin{bmatrix} 0.12 & 0.25 & -0.01 \\ 0 & -0.1925 & 0.1925 \end{bmatrix}; \mathbf{x} = \begin{bmatrix} 2 \\ -1 \end{bmatrix}$

d. Final stage of Gauss elimination is

$\begin{bmatrix} 1 & 1 & -1 & 1 & 3 \\ 0 & -1 & 2 & 0 & -1 \\ 0 & 0 & 3 & -1 & -2 \\ 0 & 0 & 0 & -3 & -3 \end{bmatrix}; \mathbf{x} = \begin{bmatrix} \frac{4}{3} \\ \frac{1}{3} \\ -\frac{1}{3} \\ 1 \end{bmatrix}$

e. Final stage of Gauss elimination is

$\begin{bmatrix} 2 & -3 & 2 & 5 & 3 \\ 0 & \frac{1}{2} & 0 & -\frac{1}{2} & -\frac{1}{2} \\ 0 & 0 & -1 & 0 & 2 \\ 0 & 0 & 0 & -1 & -7 \end{bmatrix}; \mathbf{x} = \begin{bmatrix} -5 \\ 6 \\ -2 \\ 7 \end{bmatrix}$

3. a. Rows are independent; the only solution is the zero (trivial) solution

b.
$$\mathbf{x} = \begin{bmatrix} -\frac{6}{5} \\ -\frac{2}{5} \\ 1 \end{bmatrix} t$$

c.
$$\mathbf{x} = \begin{bmatrix} -3 \\ -2 \\ 2 \\ 1 \end{bmatrix} t$$

5. Solve $A + B \quad\;\; = 1$
$\quad\quad 4A + B + C = 7; \; A = 1, \, B = 0, \text{ and } C = 3$
$\quad\quad 4A - 2B - C = 1$

Exercise Set 4.3

1. a.
$$\begin{bmatrix} -5 \\ 6 \\ -2 \\ 7 \end{bmatrix}$$

b.
$$\begin{bmatrix} 3.00002 \\ 1.000007 \\ -0.000002 \\ -5.483503E - 6 \end{bmatrix}$$

c.
$$\begin{bmatrix} -3.347246 \\ 1.669449 \\ 1.008347 \end{bmatrix}$$

d.
$$\begin{bmatrix} -2 \\ 1 \end{bmatrix}$$

e.
$$\begin{bmatrix} -30.139470 \\ 44.794190 \\ 14.61356 \end{bmatrix}$$

f.
$$\begin{bmatrix} 1 \\ 1.000003 \\ -1 \end{bmatrix}$$

4.
$$\begin{bmatrix} 0.995227 \\ 1.007160 \\ 1.002387 \end{bmatrix}$$

Exercise Set 4.4

1. a.
$$\begin{bmatrix} -5 \\ 6 \\ -2 \\ 7 \end{bmatrix}$$

b.
$$\begin{bmatrix} 3.000002 \\ 1 \\ -9.536781E - 6 \\ -4.768392E - 6 \end{bmatrix}$$

c.
$$\begin{bmatrix} -3.347246 \\ 1.669449 \\ 1.008347 \end{bmatrix}$$

d.
$$\begin{bmatrix} -2 \\ 1 \end{bmatrix}$$

e.
$$\begin{bmatrix} -30.13947 \\ 44.79419 \\ 14.61356 \end{bmatrix}$$

f.
$$\begin{bmatrix} 1 \\ 1.000003 \\ -1 \end{bmatrix}$$

4.
$$\begin{bmatrix} 1 \\ 1 \\ 1 \\ 1 \end{bmatrix}$$

Exercise Set 4.5

1. a. L MATRIX IS:

1.000000	0.000000	0.000000
0.000000	1.000000	0.000000
1.000000	1.000000	1.000000

U MATRIX IS:

1.000000	0.000000	1.000000
0.000000	1.000000	0.000000
0.000000	0.000000	0.000000

b. L MATRIX IS:

1.000000	0.000000	0.000000
3.000000	1.000000	0.000000
2.000000	0.846154	1.000000

U MATRIX IS:

1.000000	-4.000000	1.000000
0.000000	13.000000	-1.000000
0.000000	0.000000	3.846154

c. L MATRIX IS:

1.000000	0.000000	0.000000	0.000000
2.000000	1.000000	0.000000	0.000000
1.000000	0.000000	1.000000	0.000000
1.000000	0.333333	0.066667	1.000000

U MATRIX IS:

1.000000	1.000000	−1.000000	1.000000
0.000000	−3.000000	5.000000	1.000000
0.000000	0.000000	5.000000	−1.000000
0.000000	0.000000	0.000000	−3.266667

d. L MATRIX IS:

1.000000	0.000000	0.000000	0.000000
2.000000	1.000000	0.000000	0.000000
3.000000	2.000000	1.000000	0.000000
4.000000	3.000000	1.000000	1.000000

U MATRIX IS:

1.000000	−1.000000	1.000000	−1.000000
0.000000	2.000000	1.000000	1.000000
0.000000	0.000000	−6.000000	5.000000
0.000000	0.000000	0.000000	0.000000

3.
$$\mathbf{y} = \begin{bmatrix} -1.869566 \\ 2.347826 \\ 0.608696 \end{bmatrix} \qquad \mathbf{x} = \begin{bmatrix} 2.094518 \\ -1.559547 \\ 0.281664 \end{bmatrix}$$

Exercise Set 4.6

1. a.
$$\begin{bmatrix} 2.701299 & -0.467532 & 0.038961 \\ -2.064935 & 0.376623 & 0.051948 \\ -1.467532 & 0.311688 & -0.025974 \end{bmatrix}$$

b.
$$\begin{bmatrix} 1.000000 & 0.000000 & -1.000000 & 0.000000 & 0.000000 \\ -1.181818 & 3.272727 & 3.909091 & 1.545455 & -1.636364 \\ 0.818182 & -0.727273 & -1.090909 & -0.454545 & 0.363636 \\ -0.545455 & -1.181818 & -1.272727 & -1.363636 & 1.090909 \\ -0.636364 & -0.545455 & 0.181818 & -0.090909 & 0.272727 \end{bmatrix}$$

c.
$$\begin{bmatrix} 1.000000 & 0.500000 & -0.300000 & 0.350000 \\ 0.000000 & 0.500000 & -0.100000 & -0.550000 \\ 0.000000 & 0.000000 & 2.000000 & -0.400000 \\ 0.000000 & 0.000000 & 0.000000 & 0.500000 \end{bmatrix}$$

2. a.
$$\mathbf{x} = \begin{bmatrix} 2 \\ -1 \\ 1 \end{bmatrix}$$
b.
$$\mathbf{x} = \begin{bmatrix} 0.125 \\ 1.25 \\ 1.50 \end{bmatrix}$$

Exercise Set 4.7

3. a. 0.57879   **b.** 0.21213

4. $\mathbf{R} = \begin{bmatrix} 0.730016470 \\ 0.610013008 \\ 0.170003891 \end{bmatrix}$ $\mathbf{E} = \begin{bmatrix} 0.01 \\ -0.02 \\ 0.01 \end{bmatrix}$

5. a. Residuals are: $\begin{bmatrix} -3.37340929 \\ -0.51954983 \\ 3.10764065 \end{bmatrix}$ and $\begin{bmatrix} -3.369439376 \\ -0.516899941 \\ 3.107429523 \end{bmatrix}$

b. The residuals are too close to decide this question; an accurate solution is

$$x = \begin{bmatrix} 0.8951071 \\ 0.7637075 \\ 0.6097442 \end{bmatrix}$$

Errors calculated from this solution indicate that the solution

$$\begin{bmatrix} 0.61 \\ 0.764 \\ 0.895 \end{bmatrix}$$

is slightly more accurate

6. $\mathbf{R} = \begin{bmatrix} -1.786720524E - 6 \\ -1.241047565E - 6 \\ -9.996828805E - 7 \end{bmatrix}$ $\|\mathbf{R}\| = 1.7867E - 6$

$\mathbf{E} = \begin{bmatrix} -0.00054 \\ 0.0014 \\ -0.00089 \end{bmatrix}$ $\|\mathbf{E}\| = 0.0014$

Exercise Set 4.8

1. a. 12   **b.** 7   **c.** 49

3. a. 210   **b.** 6   **c.** 20

6. a. $x = \begin{bmatrix} 1 \\ -1 \\ 1 \end{bmatrix}$

b. $\|\mathbf{R}\| = 0.07$, $\|\mathbf{E}\| = 0.02$ **c.** $K = 22.5$

Exercise Set 4.9

1. Rational solution $\begin{bmatrix} 1 \\ 1 \end{bmatrix}$; machine solution $\begin{bmatrix} 1.000000 \\ 1.000001 \end{bmatrix}$

2. a. $\begin{bmatrix} 1.014257 \\ 0.965531 \\ 0.955542 \\ 1.045046 \end{bmatrix}$ **b.** $\begin{bmatrix} 0.870754 \\ 1.256560 \\ 1.030933 \\ 0.833461 \end{bmatrix}$

Even small changes can produce relatively large shifts in solutions

3. a. $K > 8.0715$ **b.** $K = 40.17$

5.

The inverse is $\begin{bmatrix} -\frac{29}{3} & -\frac{8}{3} & -32 \\ 8 & \frac{5}{2} & \frac{51}{2} \\ \frac{8}{3} & \frac{2}{3} & 9 \end{bmatrix}$

$\|A\| = 121,$ $\|A\| = 133/3,$ and $K = 5364$

6. a. $K = 748$ **b.** $1.3625E - 3$

Exercise Set 4.10

1. a. GAUSS ELIMINATION COMPLETED.

X(2) = 0.9999974
X(1) = 2.000002

RESIDUAL VECTOR
−3.397478565148049D − 08
−1.513958238774649D − 07

ERROR VECTOR	NEW X
E(2) = −2.563001E − 06	X(2) = 1
E(1) = 2.384186E − 06	X(1) = 2

b. GAUSS ELIMINATION COMPLETED.

X(3) =   1.499999
X(2) =   2.000000
X(1) = −0.998698

RESIDUAL VECTOR
−3.853440284729004D − 04
 1.16884708404541D − 03
 6.344914436340332D − 04

ERROR VECTOR	NEW X
E(3) = −7.152554E − 07	X(3) = 1.5
E(2) = −2.384185E − 07	X(2) = 2
E(1) = 1.302063E − 03	X(1) = −1

c. GAUSS ELIMINATION COMPLETED.

X(3) = 0.1
X(2) = 0.100000
X(1) = 0.100000

RESIDUAL VECTOR
1.345574851541187D − 08
1.338124278049690D − 08
1.3306737045581940D − 08

ERROR VECTOR	NEW X
E(3) = 1.490116E − 09	X(3) = 0.1
E(2) = −5.960464E − 09	X(2) = 0.1
E(1) = 1.788139E − 08	X(1) = 0.1

d. GAUSS ELIMINATION COMPLETED.

X(4) = 1
X(3) = 1.000000
X(2) = 1.000003
X(1) = 0.999995

RESIDUAL VECTOR
3.844495921612179D − 08
4.26173296652621D − 08
−5.602839348028965D − 08
−1.545548538928188D − 06

ERROR VECTOR   NEW X
E(4) = −2.171311E − 07   X(4) = 1
E(3) =   5.126385E − 07   X(3) = 0.9999998
E(2) =   3.196784E − 06   X(2) = 0.9999998
E(1) = −4.904515E − 06   X(1) = 1

2. a.
$$\mathbf{R} = \begin{bmatrix} 0.01 \\ -0.01 \\ 0.02 \end{bmatrix} \quad \mathbf{E} = \begin{bmatrix} -8.571386E - 3 \\ 2.739005E - 8 \\ 7.142856E - 3 \end{bmatrix} \quad \mathbf{x} = \begin{bmatrix} 1.008571 \\ 1.000000 \\ 0.992857 \end{bmatrix}$$

3. a.
$$\mathbf{E} = \begin{bmatrix} 5.390860E - 7 \\ -2.412927E - 6 \\ 2.011617E - 6 \end{bmatrix}$$

Exercise Set 4.11

1. a. ENTER ESTIMATES X1, X2, X3, ... XN
I = 1   ? 0
I = 2   ? 0
I = 3   ? 0

X1	X2	X3
1.200000	0.933333	−0.680000
1.038667	0.909333	−1.010667
1.008000	0.996978	−0.986507
1.001652	0.996235	−1.000356
1.000341	0.999851	−0.999445
1.000070	0.999844	−1.000011
1.000015	0.999993	−0.999977
1.000003	0.999994	−1.000000
1.000001	1.000000	−0.999999
1.000000	1.000000	−1.000000

b. ENTER ESTIMATES X1, X2, X3, ... XN
I = 1   ? 0
I = 2   ? 0
I = 3   ? 0

X1	X2	X3
0.166667	0.250000	0.777778
−0.196759	0.163889	0.583333
−0.096065	0.426736	0.752315
−0.261912	0.359664	0.572724
−0.174101	0.475070	0.665268
−0.253035	0.413839	0.571884
−0.196394	0.468371	0.632212
−0.239225	0.428906	0.583037
−0.206390	0.458571	0.619238
−0.230817	0.435754	0.591813
−0.212169	0.452704	0.612631
−0.226170	0.439758	0.596999
−0.215565	0.449478	0.608858
−0.223568	0.442117	0.599923
−0.217523	0.447672	0.606680
−0.222090	0.443476	0.601579
−0.218641	0.446647	0.605432
−0.221247	0.444253	0.602521
−0.219279	0.446061	0.604720
−0.220765	0.444695	0.603059
−0.219643	0.445727	0.604313
−0.220491	0.444948	0.603366
−0.219850	0.445536	0.604082
−0.220334	0.445092	0.603541
−0.219969	0.445428	0.603949
−0.220245	0.445174	0.603641
−0.220036	0.445366	0.603874
−0.220194	0.445221	0.603698
−0.220075	0.445330	0.603831
−0.220164	0.445247	0.603730
−0.220097	0.445310	0.603806
−0.220148	0.445263	0.603749
−0.220109	0.445298	0.603792
−0.220138	0.445272	0.603760
−0.220116	0.445292	0.603784
−0.220133	0.445276	0.603766
−0.220120	0.445288	0.603780
−0.220130	0.445279	0.603769
−0.220123	0.445286	0.603777
−0.220128	0.445281	0.603771
−0.220124	0.445285	0.603776
−0.220127	0.445282	0.603772
−0.220125	0.445284	0.603775
−0.220127	0.445282	0.603773
−0.220125	0.445284	0.603774
−0.220126	0.445283	0.603773
−0.220125	0.445283	0.603774

3.

X1	X2	X3
1.000000	0.200000	0.008000
0.980000	0.200600	0.007994
0.979940	0.200603	0.007994

Exact solution is: $x_1 = \dfrac{19491}{19890} \cong 0.9799397$

$$x_2 = \frac{399}{1989} \cong 0.2006033$$

$$x_3 = \frac{159}{19890} \cong 0.007994$$

5. a.

X1	X2	X3
0.550000	−0.562500	0.768333
0.606250	−0.750979	0.886486
0.775881	−0.870186	0.943340
0.883168	−0.933963	0.971553
0.940567	−0.966651	0.985692
0.969986	−0.983196	0.992799
0.984877	−0.991539	0.996376
0.992385	−0.995740	0.998176
0.996166	−0.997856	0.999082
0.998070	−0.998921	0.999538
0.999029	−0.999457	0.999767
0.999511	−0.999727	0.999883
0.999754	−0.999862	0.999941
0.999876	−0.999931	0.999970
0.999938	−0.999965	0.999985
0.999969	−0.999983	0.999993
0.999984	−0.999991	0.999996
0.999992	−0.999996	0.999998
0.999996	−0.999998	0.999999
0.999998	−0.999999	0.999999
0.999999	−0.999999	1.000000

b.

X1	X2	X3
0.550000	−0.550000	0.733333
0.595000	−0.714167	0.760000
0.742750	−0.742750	0.863444
0.768475	−0.846926	0.862800
0.862234	−0.852939	0.928821
0.867645	−0.919292	0.920328
0.927363	−0.915124	0.963842
0.923612	−0.958338	0.953101
0.962504	−0.950419	0.982410
0.955377	−0.979193	0.971945
0.981274	−0.970609	0.992078

(continued)

(continued)

X1	X2	X3
0.973548	−0.990169	0.982903
0.991152	−0.982274	0.996973
0.984046	−0.995821	0.989362
0.996239	−0.989097	0.999341
0.990187	−0.998631	0.993233
0.998768	−0.993148	1.000396
0.993833	−0.999949	0.995596
0.999954	−0.995596	1.000788
0.996036	−1.000501	0.997070
1.000451	−0.997105	1.000862
0.997394	−1.000673	0.998010
1.000606	−0.998055	1.000796
0.998249	−1.000669	0.998622
1.000602	−0.998667	1.000680
0.998800	−1.000592	0.999031
1.000533	−0.999070	1.000555
0.999163	−1.000494	0.999309
1.000445	−0.999342	1.000441
0.999408	−1.000398	0.999502
1.000358	−0.999528	1.000344

NUMBER OF ITERATIONS EXCEEDS 30.

Exercise Set 4.12

1. The center is at the point $(-602.3152, 404.3281, -8386.376)$; the distance is 8428.71

3. $\begin{bmatrix} 39434.47 \\ 30066.76 \\ 32316.56 \end{bmatrix}$

4. 2.4 Btu/hr/ft^2

Chapter 5

Exercise Set 5.1

1. a. A graph of the equations indicates two solutions with approximate locations $(2.9, -0.3)$ and $(-2, 2.2)$
   b. Four solutions near the points $(2.5, \pm 2.5), (-2.5, \pm 2.5)$
   c. One solution near $(2.95, 0.02)$
   d. Four solutions near the points $(1.5, \pm 1.25), (-1.5, \pm 1.25)$
   e. Four solutions near the points $(0.04, 24), (-0.04, -24), (24, 0.04)$, and $(-24, -0.04)$
   f. One solution near $(.4, .4)$

2. a. Two   **b.** Four   **c.** Two   **d.** Infinitely many

3. Any solution must be an intersection point for the sphere, cone, and cylinder; inspection of the graphs shows that the cone and cylinder have no points in common, so there is no solution for the system

5. b. The sum of squares cannot be zero otherwise; the intersection points are $(1.5811388, -1.5811388)$ and $(-1.5811388, 1.5811388)$

Exercise Set 5.2

1. **a.** $x_1 = 1.083205$, $y_1 = 0.944530$ **b.** $x_1 = 0.094868$, $y_1 = 2.031623$
   $x_2 = 1.169176$, $y_2 = 0.893452$ $x_2 = 0.194666$, $y_2 = 2.037988$
   c. $x_1 = 1.959386$, $y_1 = 1.908619$ **d.** $x_1 = -0.911247$, $y_1 = 1.953925$
   $x_2 = 1.916904$, $y_2 = 1.818091$ $x_2 = -0.822848$, $y_2 = 1.907174$

3. **a.** $x = 1.80883235$ **b.** $x = 0.34322360$
   $y = 1.27187449$ $y = 1.97032900$
   c. $x = 1.0000000$ **d.** $x = -0.51394890$
   $y = 1.0000000$ $y = \ \ \ 1.70034900$

6. **a.** $x = \pm 1.9318517$ and $x = \pm .5176381$
   b. Solutions are (use same sign for x and y) $(\pm 0.5176381, \pm 1.9318517)$ and $(\pm 1.9318517, \pm 0.5176381)$
   c. $x_1 = 2.029208$, $y_1 = 0.507439$; the solution point is P: $(1.9318517, 0.5176381)$; (x_1, y_1) is further than (x_0, y_0) from P

Exercise Set 5.3

1. **a.** There are two intersection points for the two parabolas $y = x^2 - 2$ and $y^2 = 2(x - 1)$
   b. $x = 1.808833$, $y = \ \ \ 1.271875$
   c. $x = 1.181889$, $y = \ -0.603139$

3. **a.** No   **b.** Yes, $x = -2.805118$, $y = 3.131313$
   c. No   **d.** Yes, $x = 3$, $y = 2$ **e.** Yes, $x = 3.584428$, $y = -1.848126$

5. **b.** $x^4 - 22x^{22} + x + 114 = 0$
   Roots are: 3, 3.584428, -2.805118, and -3.77931
   Solutions are: $(3, 2)$, $(3.584428, -1.848124)$, $(-2.805118, 3.131313)$, $(-3.77931, -3.283184)$

7. Using $x_{n+1} = \sqrt{2x_n - 1 + y_n/10}$, $y_{n+1} = 1/x_n$, with $x_0 = 2$, $y_0 = 1 \Rightarrow x = 1.252485$, $y = 0.637462$
   Using $x_{n+1} = \sqrt{1/y_n}$, $y_{n+1} = 10(x_n - 1)^2$, with $x_0 = 0.25$, $y_0 = 15 \Rightarrow x = -0.252481$, $y = 15.687088$
   A graph shows there are no other solutions

Exercise Set 5.4

1. **a.** Two solutions near $(1.8, 1.25)$ and $(1.2, -0.5)$; Newton iterations gave solutions $(1.80883235, 1.27187449)$ and $(1.18188857, -0.60313941)$
   b. Two solutions near $(0.35, 1.95)$ and $(1.5, -1.4)$; Newton iterations gave solutions $(0.34322356, 1.97032931)$ and $(1.45677644, -1.37032931)$
   c. The only solution is $(1, 1)$
   d. Two solutions near $(-0.5, 1.5)$ and $(1.5, -2.5)$; Newton iterations gave $(-0.51394894, 1.70034849)$ and $(1.44124416, -2.62764371)$

2. **a.** $(2.96124969, -0.48062485)$ and $(-2.16124969, 2.08062485)$
   b. $(2.77746030, \pm 1.51185789)$, $(-2.77746030, \pm 1.51185789)$
   c. $(2.97452091, 0.00849303)$
   d. $(1.86052102, 1.56892908)$
   e. $(0.03774261, 26.49525572)$ and $(26.49525572, 0.03774261)$
   f. $(0.24871557, 0.25128774)$

4. **a.** $\mathbf{DT} = \begin{bmatrix} e^x + xe^x & 2y \\ 2xy & x^2 + \cos y \end{bmatrix}$ **c.** $\mathbf{DT} = \begin{bmatrix} e^x & 1/y \\ y \sin x + xy \cos x & x \sin x \end{bmatrix}$

6.
$$\mathbf{DT} = \begin{bmatrix} 1 & 1 & 1 \\ 2 & -2 & 4 \\ -2 & 2 & -1 \end{bmatrix} \quad \mathbf{DT}^{-1} = \begin{bmatrix} \frac{1}{2} & -\frac{1}{4} & -\frac{1}{2} \\ \frac{1}{2} & -\frac{1}{12} & \frac{1}{6} \\ 0 & \frac{1}{3} & \frac{1}{3} \end{bmatrix}$$

8. Solve system $x^2 - x + y^3 = 0$
$$3xy^2 + \cos y = 0$$

The solution is $x = -0.368306$, $y = -0.795788$; at this point, $f_{xx} f_{yy} - (f_{xy})^2 < 0$ and $f_{xx} < 0$; this point is a local maximum point

Exercise Set 5.5

1. a. Quotient $x^2 - 6x + 10$ with remainder zero
c. Quotient $x^2 - x - 3$ with remainder $22x - 17$

3. a. REAL ROOTS: 1   −1

COMPLEX ROOT PAIR:
    −0.3090169943749475 + 0.9510565162951536 I
AND −0.3090169943749475 − 0.9510565162951536 I

COMPLEX ROOT PAIR:
    −0.8090169943749475 + 0.5877852522924731 I
AND −0.8090169943749475 − 0.5877852522924731 I

COMPLEX ROOT PAIR:
    0.3090169943749475 + 0.9510565162951538 I
AND 0.3090169943749475 − 0.9510565162951538 I

COMPLEX ROOT PAIR:
    0.8090169943749475 + 0.5877852522924731 I
AND 0.8090169943749475 − 0.5877852522924731 I

b. COMPLEX ROOT PAIR:
    −0.7071067811865475 + 0.7071067811865475 I
AND −0.7071067811865475 − 0.7071067811865475 I

COMPLEX ROOT PAIR:
    1.129033603927077D − 17 + 1 I
AND 1.129033603927077D − 17 − 1 I

COMPLEX ROOT PAIR:
    0.7071067811865476 + 0.7071067811865474 I
AND 0.7071067811865476 − 0.7071067811865474 I

LAST ROOT IS: −1

c. COMPLEX ROOT PAIR:
    −0.5 + 1.6583123951777 I
AND −0.5 − 1.6583123951777 I

COMPLEX ROOT PAIR:
    0 + 2.23606797749979 I
AND 0 − 2.23606797749979 I

d. REAL ROOTS: 0.1282480074768157   −0.6756593109408803
REAL ROOTS: 0.5292492772591756   −1.307579494637267

COMPLEX ROOT PAIR:
    −0.3052975933063871 + 1.590904464417402 I
AND −0.3052975933063871 − 1.590904464417402 I

COMPLEX ROOT PAIR:
    −1.202695654496385 + 0.8859144439310917 I
AND −1.202695654496385 − 0.8859144439310917 I

COMPLEX ROOT PAIR:
    1.324067712601854 + 0.3936561450252883 I
AND 1.324067712601854 − 0.3936561450252883 I

COMPLEX ROOT PAIR:
    0.846796295621996 + 1.330613448132447 I
AND 0.846796295621996 − 1.330613448132447 I

e. COMPLEX ROOT PAIR:
    −0.0879436251413253 + 0.6679183537972317 I
AND −0.0879436251413253 − 0.6679183537972317 I

REAL ROOTS: 2.218618087667011   −0.6402461880448812

COMPLEX ROOT PAIR:
    1.096318614717445 + 0.1790950043285949 I
AND 1.096318614717445 − 0.1790950043285949 I

LAST ROOT IS: −3.595121878774369

f. COMPLEX ROOT PAIR:
    −1.5 + 2.39791576165636 I
AND −1.5 − 2.39791576165636 I

LAST ROOT IS: 1

5. COMPLEX ROOT PAIR:
    −1.684404053910686 + 3.431331350197692 I
AND −1.684404053910686 − 3.431331350197692 I

LAST ROOT IS: 1.368808107821373

7. a. Factors are $x^2 + x + 1$ and $x^2 - 2$
   b. Factors are $x^2 + 0.6502815398728848x + 0.09778255792239954$,
   $x^2 - 6.650281539872885x + 10.22677276241436$
   c. Factors are $x - 1.324717957244746$,
   $x^2 + 1.324717957244746x + 0.7548776662466927$, and $x^2 - 2x + 2$

Exercise Set 5.6

1. a. ENTER ALPHA ? 4.26
   ENTER VISUAL MAGNITUDES ? 0.5, 13.5
   ENTER PERIOD IN YEARS ? 40.2

PARALLAX	MASS#1	MASS#2
0.2881259	1.693597	5.534722E − 02
0.3013008	1.650875	5.395106E − 02
0.3038778	1.64286	5.368912E − 02
0.3043711	1.641338	5.363941E − 02
0.3044652	1.641048	5.362991E − 02

DISTANCE IN PARSECS IS 3.284448
DISTANCE IN LIGHTYEARS 10.70073

b. ENTER ALPHA ? 2
ENTER VISIAL MAGNITUDES ? 2, 2.8
ENTER PERIOD IN YEARS ? 420

PARALLAX	MASS #1	MASS #2
2.830439E − 02	4.297603	3.481752
1.799766E − 02	5.566611	4.509852
1.651053E − 02	5.847816	4.737674
1.624153E − 02	5.902968	4.782356
1.619079E − 02	5.913531	4.790913

DISTANCE IN PARSECS IS 61.76353
DISTANCE IN LIGHTYEARS 201.2256

c. ENTER ALPHA ? 4.3
ENTER VISUAL MAGNITUDES ? 2.1, 3.4
ENTER PERIOD IN YEARS ? 620

PARALLAX	MASS #1	MASS #2
4.693866E − 02	3.135221	2.226879
3.378799E − 02	3.783118	2.687067
3.173719E − 02	3.920933	2.784954
3.136091E − 02	3.947747	2.804
3.128974E − 02	3.952876	2.807642

DISTANCE IN PARSECS IS 31.95936
DISTANCE IN LIGHTYEARS 104.1236

2. RINFINITY = 109736.91 ELECTRON MASS = 0.00054225
MASS OF HYDROGEN NUCLEUS = 1.00760775
HYDROGEN MASS/ELECTRON MASS = 1858.21

3. a. Solve the system $2xy - 4y = 8$
$$4y^2 - x^2 = 0$$
Solutions are $(-2, -1)$ and $(4, 2)$.
b. Solve $x^2 + y^2 - 1 = 0$
$$2xy + y^2 + 2 = 0$$
No steady-state solution exists

Chapter 6

Exercise Set 6.1

3. a. $P(x) = x^2 - 2x - 3$; the eigenvalues are -1 and 3; $P(0) = -3 = \lambda_1 \lambda_2$
b. det $\mathbf{A} = -3 = \lambda_1 \lambda_2$ **c.** $a_{n-1} = a_1 = -2$ and $\lambda_1 + \lambda_2 = 2$

6. $P(x) = (1 - x)(x^2 - \sqrt{2}x + 1)$; the eigenvalues are 1 and $(1 \pm i)/\sqrt{2}$

8. a. $2, 4, 1, 3$ **b.** $1, 1, 1, 1$ **c.** $2, 8$, and $(1 \pm i\sqrt{7})/2$

9. For $\lambda = -2, -1$, and -3, respectively, eigenvectors are:

$$t\begin{bmatrix} -4 \\ -\frac{3}{2} \\ 1 \end{bmatrix}, \quad t\begin{bmatrix} -\frac{15}{4} \\ 3 \\ 1 \end{bmatrix}, \quad t\begin{bmatrix} -\frac{7}{2} \\ -3 \\ 1 \end{bmatrix}$$

Exercise Set 6.2

1. **a.** Eigenvalues are 2, $2 \pm \sqrt{2}$

   b. Eigenvalues are 2, $2 \pm \sqrt{17}$

   c. Eigenvalues are $3 \pm \sqrt{13}$ and $(3 \pm \sqrt{145})/2$

   d. Eigenvalues are 1, 3, and $3 \pm \sqrt{10}$

3. The eigenvalue is 0.8431 to four decimal places

5. **a.** $\mathbf{A} - \mathbf{I}$ is singular because row 2 is twice row 1

   b. 1 is an eigenvalue

   c. The determinant of $\mathbf{A} + \mathbf{I}$ is 24; the determinant of $\mathbf{A}$ is -1

   d. $P(x) = -x^3 + 13x^2 - 11x - 1$; the roots of $P(x) = 0$ are 1, $6 \pm \sqrt{37}$

7. **a.** Eigenvalues are $2 + 5 = 7$, $2 + \sqrt{2} + 5 = 7 + \sqrt{2}$, and $2 - \sqrt{2} + 5 = 7 - \sqrt{2}$

   b. Eigenvalues are $2 - 3 = -1$, $2 + \sqrt{17} - 3 = -1 + \sqrt{17}$, and $2 - \sqrt{17} - 3 = -1 - \sqrt{17}$

   d. Eigenvalues are $1 + 4 = 5$, $3 + 4 = 7$, and $7 \pm \sqrt{10}$

Exercise Set 6.3

1. **a.** $\lambda = 6$ **b.** $\lambda = 4$ **c.** $\lambda = 1$ **d.** $\lambda = 30.288690$

$$\mathbf{x} = \begin{bmatrix} 1 \\ -1 \\ 1 \end{bmatrix} \quad \mathbf{x} = \begin{bmatrix} 1 \\ -1 \\ 1 \end{bmatrix} \quad \mathbf{x} = \begin{bmatrix} 1 \\ 1 \\ 1 \end{bmatrix} \quad \mathbf{x} = \begin{bmatrix} 0.957629 \\ 0.688937 \\ 1.000000 \\ 0.943782 \end{bmatrix}$$

3.

EIGENVECTOR COMPONENTS			EIGENVALUE ESTIMATES
1.000000	0.821429	0.214286	5.600000
1.000000	0.829114	0.221519	5.642857
1.000000	0.832215	0.224832	5.658228

4. **a.**

EIGENVECTOR COMPONENTS				EIGENVALUE ESTIMATES
1.000000	1.000000	1.000000	1.000000	1.000000
1.000000	1.000000	1.000000	1.000000	1.000000
1.000000	1.000000	1.000000	1.000000	1.000000
1.000000	1.000000	1.000000	1.000000	1.000000

   b.

EIGENVECTOR COMPONENTS			EIGENVALUE ESTIMATES
1.000000	−0.250000	−0.250000	−0.250000	4.000000
1.000000	−0.315789	−0.315789	−0.315789	4.750000
1.000000	−0.329787	−0.329787	−0.329787	4.947368
1.000000	−0.332623	−0.332623	−0.332623	4.989362
1.000000	−0.333191	−0.333191	−0.333191	4.997868
1.000000	−0.333305	−0.333305	−0.333305	4.999573
1.000000	−0.333328	−0.333328	−0.333328	4.999915
1.000000	−0.333332	−0.333332	−0.333332	4.999984
1.000000	−0.333333	−0.333333	−0.333333	4.999996
1.000000	−0.333333	−0.333333	−0.333333	4.999999
1.000000	−0.333333	−0.333333	−0.333333	5.000001
1.000000	−0.333333	−0.333333	−0.333333	5.000001
1.000000	−0.333333	−0.333333	−0.333333	5.000000
1.000000	−0.333333	−0.333333	−0.333333	5.000001
1.000000	−0.333333	−0.333333	−0.333333	5.000000

c. EIGENVECTOR COMPONENTS   EIGENVALUE ESTIMATES

−0.250000	1.000000	−0.250000	−0.250000	4.000000
−0.315789	1.000000	−0.315789	−0.315789	4.750000
−0.329787	1.000000	−0.329787	−0.329787	4.947368
−0.332623	1.000000	−0.332623	−0.332623	4.989362
−0.333191	1.000000	−0.333191	−0.333191	4.997868
−0.333305	1.000000	−0.333305	−0.333305	4.999573
−0.333328	1.000000	−0.333328	−0.333328	4.999915
−0.333332	1.000000	−0.333332	−0.333332	4.999984
−0.333333	1.000000	−0.333333	−0.333333	4.999996
−0.333333	1.000000	−0.333333	−0.333333	4.999999
−0.333333	1.000000	−0.333333	−0.333333	5.000001
−0.333333	1.000000	−0.333333	−0.333333	5.000001

d. EIGENVECTOR COMPONENTS   EIGENVALUE ESTIMATES

−0.250000	−0.250000	1.000000	−0.250000	4.000000
−0.315789	−0.315789	1.000000	−0.315789	4.750000
−0.329787	−0.329787	1.000000	−0.329787	4.947369
−0.332623	−0.332623	1.000000	−0.332623	4.989362
−0.333191	−0.333191	1.000000	−0.333191	4.997868
−0.333305	−0.333305	1.000000	−0.333305	4.999573
−0.333328	−0.333328	1.000000	−0.333328	4.999915
−0.333332	−0.333332	1.000000	−o.333332	4.999984
−0.333333	−0.333333	1.000000	−0.333333	4.999996
−0.333333	−0.333333	1.000000	−0.333333	4.999999
−0.333333	−0.333333	1.000000	−0.333333	5.000000

e. EIGENVECTOR COMPONENTS   EIGENVALUE ESTIMATES

−0.250000	−0.250000	−0.250000	1.000000	4.000000
−0.315789	−0.315789	−0.315789	1.000000	4.750000
−0.329787	−0.329787	−0.329787	1.000000	4.947369
−0.332623	−0.332623	−0.332623	1.000000	4.989362
−0.333191	−0.333191	−0.333191	1.000000	4.997868
−0.333305	−0.333305	−0.333305	1.000000	4.999573
−0.333328	−0.333328	−0.333328	1.000000	4.999915
−0.333332	−0.333332	−0.333332	1.000000	4.999983
−0.333333	−0.333333	−0.333333	1.000000	4.999997
−0.333333	−0.333333	−0.333333	1.000000	4.999999
−0.333333	−0.333333	−0.333333	1.000000	5.000000

7. a. $\det A = 3$ **b.** $A − I$ is singular
d. $P(x)$ has integer coefficients, so $a + \sqrt{b}$ is a root of $P(x) = 0$ if $a − \sqrt{b}$ is a root

e. EIGENVECTOR COMPONENTS   EIGENVALUE ESTIMATES

0.750000	1.000000	1.000000	4.000000
0.733333	1.000000	1.000000	3.750000
0.732143	1.000000	1.000000	3.733333
0.732057	1.000000	1.000000	3.732143
0.732051	1.000000	1.000000	3.732057
0.732051	1.000000	1.000000	3.732051

Exercise Set 6.4

1. a.
$$\mathbf{P} = \begin{bmatrix} 0 & -1 & 0 \\ -1 & 0 & 0 \\ 0 & 0 & 1 \end{bmatrix}$$

c.
$$\mathbf{P} = \begin{bmatrix} \frac{7}{25} & -\frac{24}{25} \\ -\frac{24}{25} & -\frac{7}{24} \end{bmatrix}$$

3. a. INPUT MATRIX A IS

1.000000	−1.000000	1.000000	1.000000
−1.000000	1.000000	0.000000	1.000000
1.000000	0.000000	2.000000	1.000000
1.000000	1.000000	1.000000	3.000000

P MATRICES ARE:

K = 1

1.000000	0.000000	0.000000	0.000000
0.000000	−0.577350	0.577350	0.577350
0.000000	0.577350	0.788675	−0.211325
0.000000	0.577350	−0.211325	0.788675

K = 2

1.000000	0.000000	0.000000	0.000000
0.000000	1.000000	0.000000	0.000000
0.000000	0.000000	−0.707107	−0.707107
0.000000	0.000000	−0.707107	0.707107

AFTER HOUSEHOLDER TRANSFORMATION:

1.000000	1.732050	−0.000000	−0.000000
1.732050	2.000000	−1.414214	−0.000000
−0.000000	−1.414214	2.500000	−0.866025
−0.000000	−0.000000	−0.866025	1.500000

5. Householder transformation yields matrix $\begin{bmatrix} 1 & -2 \\ -2 & 1 \end{bmatrix}$

This is not **A**

7. a. INPUT MATRIX A IS

3.000000	−0.707107	2.828427	−2.000000
−0.707107	1.000000	0.000000	−0.707107
2.828427	0.000000	3.000000	−2.828427
−2.000000	−0.707107	−2.828427	3.000000

P MATRICES ARE:

K = 1

1.000000	0.000000	0.000000	0.000000
0.000000	−0.200000	0.800000	−0.565685
0.000000	0.800000	0.466667	0.377124
0.000000	−0.565685	0.377124	0.733333

K = 2

1.000000	0.000000	0.000000	0.000000
0.000000	1.000000	0.000000	0.000000
0.000000	0.000000	−0.385365	0.922764
0.000000	0.000000	0.922764	0.385365

AFTER HOUSEHOLDER TRANSFORMATION:

3.000000	3.535534	−0.000000	0.000000
3.535534	5.319999	−1.522367	0.000000
−0.000000	−1.522367	1.739718	−0.126922
0.000000	0.000000	−0.126922	−0.059717

b. $BB^t = I$, so B is orthogonal   **d.** Eigenvalues are $0, 2, 4 \pm \sqrt{17}$

Exercise Set 6.5

1. a. $s = -1/\sqrt{5}$ and $c = 2/\sqrt{5}$

$$P = \begin{bmatrix} 2/\sqrt{5} & 1/\sqrt{5} & 0 \\ -1/\sqrt{5} & 2/\sqrt{5} & 0 \\ 0 & 0 & 1 \end{bmatrix} \qquad PA = \begin{bmatrix} 5/\sqrt{5} & 7/\sqrt{5} & 3/\sqrt{5} \\ 0 & 9/\sqrt{5} & -4/\sqrt{5} \\ 4 & 2 & 1 \end{bmatrix}$$

c. $s = -2/\sqrt{5}$ and $c = 1/\sqrt{5}$

$$P = \begin{bmatrix} 1/\sqrt{5} & 0 & 2/\sqrt{5} \\ 0 & 1 & 0 \\ -2/\sqrt{5} & 0 & 1/\sqrt{5} \end{bmatrix} \qquad PA = \begin{bmatrix} 10/\sqrt{5} & 5/\sqrt{5} & 4/\sqrt{5} \\ 1 & 5 & -1 \\ 0 & 0 & -3/\sqrt{5} \end{bmatrix}$$

2. a. R MATRIX IS

2.449490	2.041242	0.408248
0.000000	1.354006	1.600189
0.000000	−0.000000	1.809068

Q MATRIX IS

0.408248	0.123091	0.904534
0.816497	−0.492366	−0.301511
0.408248	0.861640	−0.301511

b. R MATRIX IS

4.242641	2.121320	0.942809
0.000000	3.082207	0.324443
0.000000	0.000000	−2.829461

Q MATRIX IS

0.942809	−0.324443	−0.076472
0.235702	0.811107	−0.535303
0.235702	0.486664	0.841191

3. a. R MATRIX IS

2.000000	1.000000	−1.500000	−4.000000
−0.000000	4.123106	4.001838	3.395499
0.000000	0.000000	3.568655	−1.565923
−0.000000	0.000000	−0.000000	0.135925

Q MATRIX IS

0.500000	0.363803	0.082417	−0.781572
−0.500000	0.363803	0.782961	−0.067963
0.500000	0.606339	0.090659	0.611665
−0.500000	0.606339	−0.609886	−0.101944

b. R MATRIX IS

9.327378	5.038929	7.933633	8.040844
−0.000000	4.960766	1.617288	−5.950138
0.000000	−0.000000	8.569821	−2.937052
0.000000	−0.000000	−0.000000	4.161059

Q MATRIX IS

0.321634	0.076462	−0.895629	0.297579
0.536056	−0.342921	0.385274	0.668291
−0.214423	0.822546	0.159963	0.501849
0.750479	0.447187	0.154349	−0.461499

6. a. INPUT MATRIX A IS

3.000000	5.000000	1.000000
5.000000	−2.000000	6.000000
1.000000	6.000000	−4.000000

A1 =

5.400000	5.765848	0.284078
5.765849	−7.277438	1.218058
0.284078	1.218058	−1.122563

A2 =

4.470508	6.733672	0.038254
6.733670	−6.416720	0.145338
0.038254	0.145337	−1.053788

A3 =

3.120945	7.631146	0.004986
7.631147	−5.067611	0.017597
0.004986	0.017597	−1.053334

A4 =

1.453243	8.313215	0.000637
8.313215	−3.399907	0.002192
0.000637	0.002192	−1.053335

A5 =

−0.435697	8.643423	0.000079
8.643424	−1.510968	0.000280
0.000080	0.000280	−1.053335

A9 =

−6.916157	6.299257	−0.000000
6.299257	4.969493	−0.000000
0.000000	0.000000	−1.053335

A10 =

−7.798814	5.330158	0.000000
5.330157	5.852152	0.000000
0.000000	0.000000	−1.053335

b. INPUT MATRIX A IS

2.000000	3.000000	−1.000000	2.000000
3.000000	4.000000	5.000000	1.000000
−1.000000	5.000000	2.000000	3.000000
2.000000	1.000000	3.000000	4.000000

A1 =

4.888890	5.229525	−0.740018	−1.660955
5.229525	3.569986	3.761932	0.378567
−0.740017	3.761932	0.701646	−1.204052
−1.660955	0.378567	−1.204052	2.839481

A2 =

8.867618	3.638751	−0.426430	0.709294
3.638751	−0.819265	3.062219	−1.047438
−0.426430	3.062219	1.115087	0.550077
0.709293	−1.047438	0.550077	2.836563

A3 =

9.937468	1.583549	−0.173972	−0.226989
1.583550	−2.613917	2.618689	0.955865
−0.173972	2.618689	1.767769	−0.231880
−0.226989	0.955865	−0.231880	2.908685

A9 =

10.138670	0.007204	−0.000269	−0.000144
0.007205	−4.083358	0.616078	0.146335
−0.000268	0.616076	3.048300	0.158454
−0.000144	0.146336	0.158454	2.896381

A10 =

10.138680	0.002941	−0.000088	0.000041
0.002941	−4.106184	0.475458	−0.102185
−0.000087	0.475460	3.087891	−0.151877
0.000041	−0.102184	−0.151878	2.879613

8. a. 7.696575, −0.861971, 3.165399
   b. −3.907726, 0.657025, 5.766109, 6.484596
   c. 40.396805, −7.466894, −18.929930
   d. 22.668447, −1.456723, 0.042164, 5.746122

Exercise Set 6.6

1. a. 1, 2, 0   **b.** 28, 29, 27   **c.** −12, −11, −13

3. a. $P(x) = -x^3 - x^2 - 16x - 16$
   b. Roots of $P(x) = 0$ are −1, $4i$, and −$4i$
   c. Eigenvalues in Example 6.4 minus 2

5. See exercise 8 in Exercise Set 6.5

6. a. Eigenvalues are 4.40205, 6.894492, and −0.2965403; spectral radius is 6.894492
   b. Eigenvalues are 3.796537, −0.2850407, 9.203481, and −3.714976; spectral radius is 9.203481

8. b.
$$\mathbf{A}^t\mathbf{A} = \begin{bmatrix} 78 & 28 & 17 \\ 28 & 11 & 9 \\ 17 & 9 & 45 \end{bmatrix}$$

Eigenvalues are 0.66315, 95.3757, and 37.9613; spectral radius is 95.3757 and $\|\mathbf{A}\| = 9.766$

Exercise Set 6.7

1. Eigenvalues are ± 1.801938, ± 1.246980, and ± 0.445042; the values of x are the same; the six energy levels are $\alpha \pm x\beta$, where x can be any one of the six values above

2. **b.** $P_6(x) = x^6 - 5x^4 + 6x^2 - 1$
   c. Machine results were ± 0.4450418679, ± 1.8019377358, ± 1.2469796037

5. 36.896011, 22.340593, 10.428867, -13.665491

Chapter 7

Exercise Set 7.1

1. **a.** $L_0(x) = (x^2 - 2x)/3$, $L_1(x) = -(x^2 - x - 2)/2$,
   $L_2(x) = (x^2 + x)/6$, $P(x) = (7x^2 + x)/6$
   b. $P(x) = 5L_1(x) = -5(x + 1)(x - 6)/12$
   c. $P(x) = -L_0(x) - L_1(x) - L_2(x) = -1$
   d. $L(x) = (x^2 - x)/2$, $L_1(x) = -x^2 + 1$, $L(x) = (x^2 + x)/2$,
   $P(x) = 2x + 3$

3. $L_0(0.5) = 0.5625$, $L_1(0.5) = -0.0625$, $L_2(0.5) = 0.5625$, $L_3(0.5) = -0.0625$, $P(0.5) = -7.375$

5. $L_0(1.3) = -0.136364$, $L_1(1.3) = 0.869566$, $L_2(1.3) = 0.266798$, $f(1.3) \cong 1.14032$

7. $h(2) = 1380$, $h''(t) = -30$ ft/sec; not earth

9. **a.** Sales volume is a discrete random variable; it is not the value of a continuous function of time
   c. There is no continuous function of any variable whose value is the number of satellites of a planet

Exercise Set 7.2

1. $P_1(x) = y_0 \dfrac{(x - x_1)}{(x_0 - x_1)} + y_1 \dfrac{(x - x_0)}{(x_1 - x_0)}$

   $P_2(x) = y_0 \dfrac{(x - x_1)(x - x_2)}{(x_0 - x_1)(x_0 - x_2)} + y_1 \dfrac{(x - x_0)(x - x_2)}{(x_1 - x_0)(x_1 - x_2)} + y_2 \dfrac{(x - x_0)(x - x_1)}{(x_2 - x_0)(x_2 - x_1)}$

3. The error term factor $f^{(n+1)}(\xi) = 0$ if $f(x)$ is a polynomial of degree n or less.

5. **a.** 1.265461; $|\text{error}| < 0.0015$, true error $= 0.00055$
   b. 1.094131, $|\text{error}| < 0.003$, true error $= 0.00131$
   c. 1.379142, $|\text{error}| < 0.00225$, true error $= 0.000737$
   d. 1.203858, $|\text{error}| < 0.00077$, true error $= 0.0003$

7. The error bound is $(M/6)h^3 s(s - 1)(s - 2)$

Exercise Set 7.3

1. **a.** 0.84062 **b.** 0.84062 **c.** 0.844393 **d.** 0.844393

3. $P_2(x) = \dfrac{2}{\pi} x - \dfrac{4}{\pi^2} x\left(x - \dfrac{\pi}{2}\right)$

   $P_2(\pi/6) = \frac{5}{9} = 0.555556$, $P_2(\pi/3) = \frac{8}{9} = 0.888889$

4. $P_4(0.01) = 0.00989923$, $P_4(0.3) = 0.294876$, $P_4(1) = 0.841546$

5. Absolute error in $P_4(0.01)$ estimate is 0.000101; relative error is 0.01006
   Absolute error in $P_4(1)$ estimate is 0.000075; relative error is 0.000089
   $P_4(1)$ is the best estimate

7. a. $P_2(1.6) = 1.265461$; error estimate is 0.000438
   b. $P_2(1.2) = 1.094131$; error estimate is 0.00087
   c. $P_2(1.9) = 1.379142$; error estimate is 0.000657
   d. $P_2(1.45) = 1.203858$; error estimate is 0.000226

9. 0.2116094 with error estimate 6×10^{-7}

11. All parts:
   $$P(x) = 1 + 2(x - 1) + (x - 1)(x - 2) - 3(x - 1)(x - 2)(x - 3)$$

Exercise Set 7.4

1. b. $P(15.25) = 0.9638$, $P(15.75) = 0.96055$

3. $P(12.2167) = 0.211607$, error $= 0.000003$

5. $J_0(1.3) \cong P(1.3) = 0.62026$, $J_0(1.1) \cong P(1.1) = 0.71941$

9. The third-order differences are all equal to 12, and $h = 1$; $P(x) = 2x^3 + x - 5$

Exercise Set 7.5

1. $P(104.35) = 4.707938$

3. The quadratic estimate is 99.74101 with error estimate 0.00007

4. The quadratic estimate is 0.845292 with error estimate 0.000144, so $\log 7 = 0.845$ should be correct to 3 decimal places

6. a. The Aitken algorithm gives estimates
   $$\sqrt{2} \cong 1.414212, \sqrt{2.1} \cong 1.449134, \text{ and } \sqrt{2.2} \cong 1.483238$$
   b. 1.324825   **c.** 1.292685

Exercise Set 7.6

1. $f(1.6) = 6.5536$. The total error is 0.0216, and the theoretical error is 0.0216, so the computational errors are negligible.

3. The total error is 1.97×10^{-5}; the absolute theoretical error is bounded by 3.92×10^{-7}; the theoretical error is insignificant compared to the total error; the amount of computation involved is small, so most of the error is data error; the estimate is correct to at most 4 decimal places in that case; the approximation 0.276 for $\sin 0.28$ should be safe; the approximation 0.2763 leaves some doubt about the fourth decimal place.

5. a. $|\text{error}|$ is approximately 10^{-7}
   b. The theoretical error bound is 3.894×10^{-7}
   c. The total error is 5×10^{-8}; the computational errors seem to be insignificant, since the total error is less than the theoretical bound; all of the error can be explained as theoretical error

7. a. Max of $|\phi(s)|$ is $2/3\sqrt{3}$.
   b. The error bound is $Mh^3/9\sqrt{3}$
   c. $M = 1 \Rightarrow h^3 < 15.95 \times 10^{-5} \Rightarrow h < 1.1595 \times 10^{-2}$; choose $h < 1.16 \times 10^{-2}$

Exercise Set 7.7

1. a. $S_0 + 4S_1 + S_2 = 24$ **b.** $S_0 = S_2 = 0$ implies $S_1 = 6$ **c.** $P_0(x) = 2x^3 - 2x + 1$
$P_1(x) = -2(x - 0.5)^3 + 3(x - 0.5)^2 - 0.5(x - 0.5) + 0.25$

3.
$$0.1S_0 + 0.6S_1 + 0.2S_2 \qquad\qquad = -0.086100$$
$$0.2S_1 + 0.8S_2 + 0.2S_3 \qquad = -0.098701$$
$$0.2S_2 + 0.8S_3 + 0.2S_4 = -0.084299$$

4. $P_0(x) = -0.1993902(x - 2)^3 + 0.213894(x - 2) + 0.30103$
$P_1(x) = 0.04002916(x - 2.1)^3 - 0.059817(x - 2.1)^2$
$\qquad\qquad + 0.2079123(x - 2.1) + 0.32222$

7. $P_2(x) = 0.04090612(x - 1)^3 - 0.15028(x - 1)^2 - 0.4422643(x - 1) + 0.7652$

9.

X	Y	YTRUE	ERROR
−1.000	0.0000	0.0000	−0.0000
−0.900	−0.3106	−0.3090	−0.0016
−0.800	−0.5878	−0.5878	0.0000
−0.700	−0.8022	−0.8090	0.0068
−0.600	−0.9405	−0.9511	0.0106
−0.500	−0.9932	−1.0000	0.0068
−0.400	−0.9511	−0.9511	0.0000
−0.300	−0.8100	−0.8090	−0.0010
−0.200	−0.5878	−0.5878	0.0000
−0.100	−0.3084	−0.3090	0.0006
0.000	0.0000	0.0000	−0.0000
0.100	0.3084	0.3090	−0.0006
0.200	0.5878	0.5878	−0.0000
0.300	0.8100	0.8090	0.0010
0.400	0.9511	0.9511	−0.0000
0.500	0.9932	1.0000	−0.0068
0.600	0.9405	0.9511	−0.0106
0.700	0.8022	0.8090	−0.0068
0.800	0.5878	0.5878	−0.0000
0.900	0.3106	0.3090	0.0016

11. Altitude estimate is 22193 ft; velocity estimate is 1551 ft/sec

Exercise Set 7.8

1. a. 0.997497   **b.** 0.996432   **c.** 1.722109   **d.** 2.143858

3. a. 1.544194, 1.514137, 1.424833
    b. 1.555008, 1.434227, 1.514016   **c.** 0.4211

6. a. 4.043036, 10.384370
    b. 3.338147, 6.504689
    d. 1.524682, 2.169761
    e. 3.314048
    f. 7.773928

Chapter 8

Exercise Set 8.1

1. $P_4(x) = 1 - \dfrac{x^2}{2} + \dfrac{x^4}{4}$, $P_4(0.1) = 0.995025$, $P_4(0.8) = 0.7824$

3. $|R_5(x)| \le \dfrac{x^6}{720}$, $|R_5(0.1)| < 1.39\mathrm{E} - 9$, $|R_5(0.9)| < 7.38\mathrm{E} - 4$

4. **a.** 7   **b.** $P_7(0.5) = 1.648721$,   error $= 2\mathrm{E} - 7$; yes

5. **a.** 16   **b.** $P_{16}(3) = 20.085536488$,   error $= 4\mathrm{E} - 7$; yes

7. $P_5(x) = x + \dfrac{x^3}{3} + \dfrac{2x^5}{15}$, $P_5(0.1) = 0.1003347$, $P_5(0.8) = 1.0143573$

9. $|R_n(x)| \le \dfrac{x^{n+1}}{(n+1)!}$ The series $\sum_{n=0}^{\infty} x^n/n!$ is a convergent power series on the whole real line; it follows that the nth term $x^n/n!$ has a limit of zero for every x; theoretically, this means that for a given x, by choosing n sufficiently large, sin x can be approximated with any desired accuracy

11. **a.** $P_5(x) = x - \dfrac{x^3}{6} + \dfrac{x^5}{40}$

   b.

X	P(X)
0.1	0.0398278
0.2	0.0792597
0.3	0.1179117
0.4	0.1554237
0.5	0.1914715
0.6	0.2257790
0.7	0.2581296
0.8	0.2883789
0.9	0.3164658
1.0	0.3424255

Exercise Set 8.2

1. $T_5(x) = 16x^5 - 20x^3 + 5x$
   $T_6(x) = 32x^6 - 48x^4 + 18x^2 + 1$

3. $T(x) = 8x^4 - 8x^2 + 1$ assumes extreme values of ± 1 at points ± 1, $\pm 1/\sqrt{2}$, and 0

5. **a.** $-1.883472x^2 + 0.6361014$
   b. $0.5429006x^3 - 1.1741090x^2 + 0.9946152x + 0.0218969$
   c. $0.5429006x^2 + 0.9946151$
   d. $-0.4705882x^3 + 0.9411766x$
   e. $0.1294701x^3 - 0.2111643x^2 + 0.4908354x + 1.0117100$

7. **a.** 0.507339   **c.** 0.00804
   e. $|f^{\mathrm{iv}}(x)|$ is unbounded on $[-1, 1]$, so no error bound can be found in this case

9. **a.** $\pi x - \dfrac{\pi^3 x^3}{6}$ **d.** $\dfrac{959}{960} - \dfrac{19x^2}{120}$

11. **a.** 0.0082774 (exact value is 0)
    b. 0.26356   (exact value is 0.2642411)

Exercise Set 8.3

1. a. $-0.2354625x^2 + 1.091301x - 0.007465$
c. $0.5367214x^2 - 1.103638x + 0.996294$

3. $c_0 = -7, c_1 = 2, c_2 = 4,$ and $c_3 = 1$

5. a. $0.903506x$
c. $-0.0937658x^2 + 0.5281224x + 0.6479184$
d. $1.759283x^2 - 2.923147x + 1.813430$

7. a. Taylor: $x - \dfrac{x^3}{6}$

Chebyshev: $0.9989828x - 0.1585047x^3$
Legendre: $-0.0629627x^3 + 0.9035008x$

b.

x	Taylor	Chebyshev	Legendre	$\sin x$
0.1	0.0998333	0.09973984	0.089721	0.099833
0.5	0.4791667	0.4796783	0.436010	0.479426
0.9	0.7785000	0.7835345	0.762151	0.783327

c. Near $x = 0.1$, the Taylor polynomial gives the best results; over the interval $[0, 1]$ as a whole, the Chebyshev polynomial approximations seem to be better; the Taylor polynomial is easiest to construct

Exercise Set 8.4

1. a. $\dfrac{6x}{6 + x^2}$ **b.** $\dfrac{6x}{6 - x^2}$

c. $\dfrac{3x}{3 + x^2}$ **d.** $\dfrac{6x + 3x^2}{6 + 6x + x^2}$

3.

x	$R(x)$	Arctan x	Error
0.2	0.1973684	0.1973956	0.0000271
0.5	0.4615385	0.4636476	0.0021091
0.7	0.6017191	0.6107260	0.0090068

5. a. $Q(0.1) = \arctan(0.1) = 0.0953101, |\text{error}| < 5E - 8$
$Q(0.5) = 0.4053334, \arctan(0.5) = 0.4054651, |\text{error}| = 0.0001318$
b. $|\text{error}|$ at 0.1 is 0.0000001 $|\text{error}|$ at 0.5 is 0.0000188
c. Comparing errors over the entire interval $[0, 1]$, we find that overall, the Padé approximant performs better

7. 3.16667

Chapter 9

Exercise Set 9.1

1. a.

h	$f'(x)$	Error Bound
0.1	-0.757576	0.0751
0.2	-0.699301	0.1503
0.4	-0.606061	0.3006

3. a. −0.563705   **c.** −0.71616

4. a. Bound = 0.001667, true error = 0.0009375
   c. Bound = 0.001667, true error = 0.001196

5. a. −0.641026

d. $\dfrac{f(a+h)-f(a)}{h} = -\dfrac{1}{a(a+h)} = -\dfrac{1}{c^2}$

implies $c^2 = a(a+h)$.

e. $\dfrac{f(a+h)-f(a)}{h} = 2a + h = 2\left(a + \dfrac{h}{2}\right) = f'\left(a + \dfrac{h}{2}\right)$,

since $f'(x) = 2x$.

9. $f''(0.6) = -0.8247$, error bound = 0.000833, true error is 0.000636
   $f''(0.8) = -0.6962$, error bound = 0.000833, true error is 0.000507

Exercise Set 9.2

1. a.

H	DERIVATIVE	ERROR
0.010000	0.8775666	−1.591444E − 05
0.005000	0.8775801	−2.503395E − 06
0.001000	0.877574	−8.523464E − 06
0.000500	0.8776486	6.598235E − 05
0.000100	0.8778274	2.448559E − 04

b.

H	DERIVATIVE	ERROR
0.010000	−0.6065399	−9.179115E − 06
0.005000	−0.6065309	−2.384186E − 07
0.001000	−0.6065369	−6.198883E − 06
0.000500	−0.6065965	−6.580353E − 05
0.000100	−0.6067753	−2.446175E − 04

c.

H	DERIVATIVE	ERROR
0.010000	1.298529	8.249283E − 05
0.005000	1.298469	2.288818E − 05
0.001000	1.298428	−1.895428E − 05
0.000500	1.298487	4.065037E − 05
0.000100	1.298785	3.387928E − 04

d.

H	DERIVATIVE	ERROR
0.010000	0.3032908	2.545118E − 05
0.005000	0.3032714	6.079674E − 06
0.001000	0.3032535	−1.180172E − 05
0.000500	0.3032684	3.099442E − 06
0.000100	0.3032386	−2.670288E − 05

e.

H	DERIVATIVE	ERROR
0.010000	0.7999957	−4.351139E − 06
0.005000	0.8000016	1.609325E − 06
0.001000	0.7999986	−1.430512E − 06
0.000500	0.8000135	1.347065E − 05
0.000100	0.8001924	1.923442E − 04

7. a. Bound on total error is $E = \dfrac{Mh}{2} + \dfrac{\delta}{h}$

  b. E is minimized if $h = \sqrt{\dfrac{2\delta}{M}}$; minimum value is $\sqrt{2M\delta}$

Exercise Set 9.3

3. a.

DHF EST.	EXTRAPOLATED VALUES		
1.013662			
1.003353	0.999917		
1.000834	0.999995	1.000000	
1.000208	1.000000	1.000000	1.000000

  d.

DHF EST.	EXTRAPOLATED VALUES		
1.634331			
1.575528	1.555927		
1.561874	1.557323	1.557416	
1.55852	1.557403	1.557408	1.557408

5. $D_{0.04}\,f = 3.440625$ Extrapolated values
$D_{0.02}\,f = 3.429250$ 3.425458
$D_{0.01}\,f = 3.426500$ 3.425583   3.425592
The true value is 3.425519; the most accurate estimate is $D_{0.02}^{(1)}\,f = 3.425458$ with error
$6.1\text{E} - 5$. Errors in the data limit the accuracy attainable.

9. a.

DHF EST.	EXTRAPOLATED VALUES		
0.982034			
0.994595	0.998783		
0.998504	0.999808	0.999876	
0.999605	0.999973	0.999984	0.999985

  b.

DHF EST.	EXTRAPOLATED VALUES		
0.493727			
0.498362	0.499908		
0.499585	0.499993	0.499999	
0.499895	0.500000	0.500000	0.500000

  c.

DHF EST.	EXTRAPOLATED VALUES		
0.496411			
0.498951	0.499798		
0.499714	0.499969	0.499980	
0.499925	0.499996	0.499997	0.499998

  d.

DHF EST.	EXTRAPOLATED VALUES		
1.103658			
1.500948	1.633378		
1.546402	1.561553	1.556765	
1.554943	1.557791	1.557540	1.557552

11. a. -4.000000, true $= -4$
  b. 0.606531, true $= 0.6065307$
  c. -0.061851, true $= -0.061851$
  d. 17.999997, true $= 18.0000$

Exercise Set 9.4

5. $\left(\dfrac{h}{3}\right)(y_0 + 4y_1 + y_2)$

9. The error is $-\frac{3}{80}h^5 f^{(4)}(\xi)$; if $f(x)$ is a polynomial of degree 3 or less, then $f^{(4)}(x) = 0$

10. **b.** 0.944444   **c.** 0.946146
     d. The Simpson estimate is best; the quadratic used in the Simpson approximation coincides with the graph of $f(x)$ at the ends of the interval and at the middle; this provides a better fit than the one given by the polynomial in part (a), which coincides with the graph only at $x = 0$

11. **d.** The difference is $0.946146 - 0.944444 = 0.001702$; $\dfrac{h^5}{18} = \dfrac{(0.5)^5}{18} = 0.001736$

Exercise Set 9.5

1. **a.** 0.6937714   **b.** 5.265042   **c.** 0.7849815   **d.** 0.9456909

2. **a.** 0.6931503   **b.** 5.304634   **c.** 0.7853983   **d.** 0.9460833

3. **a.** 0.7466708, 0.7467997, 0.746818
    c. -0.248754, -0.2497702, -0.2499369
    e. 1.480085, 1.480043, 1.480038

5. 1.480434, error bound is 0.00333

8. **a.** 0.1560962

9. 1.49365, error bound is 0.0000133

11. We need $N > 27$

12. Minimum N is 34; the estimate is 0.6931469; the error is 3E $-$ 7, which is satisfactory

15. **a.** For $N = 100$, the error is approximately 0.0004253; for $N = 200$, the error is approximately 0.0001
    b. Error estimate is 0.000284; the theoretical error estimate is based on discretization error only; for values of N as large as 3200, round-off error can be so large that the error estimate is useless

17. **a.** T_{100} error estimate is 6.4E $-$ 6
      T_{200} error estimate is 2.3E $-$ 6
    b. T_{100} error estimate is 6.137E $-$ 6
      T_{200} error estimate is 1.533E $-$ 6
    c. The total errors in the T_{100} estimates are 6.130E $-$ 6 for single precision and 6.129E $-$ 6 for double precision; the corresponding errors for T_{200} are 1.33E $-$ 6 and 1.526E $-$ 6

18. Error estimate $\frac{16}{15}(S_{400} - S_{200}) = 5.109$E $-$ 12; we would expect ten digits to be correct; there is some doubt about the eleventh digit

19. Error estimate is 1.247E $-$ 11; this implies 10 correct digits

21. The value is 0.8862190592 correct to 10 decimal places

Exercise Set 9.6

1. a. 5.284075   **b.** 5.281044
   c. Exact value is $\frac{16}{3} = 5.333333$; the errors in (a) and (b), respectively, are 0.04926 and 0.05223

3. The estimate of the integral is 0.3333488; the error bound is 2.56E − 5

4. [0, 1]

5. [0, 1]

7. a. TOL = 0.001 gives 0.2913076   TOL = 0.00001 gives 0.2913902
   c. TOL = 0.001 gives 0.500211   TOL = 0.00001 gives 0.5000004
   e. TOL = 0.001 gives 0.08415348   TOL = 0.00001 gives 0.08408162

Exercise Set 9.7

1. a. 0.75, 0.708333, 0.697023, 0.694121
   b. $T_2^{(1)} = 0.694445$, $T_4^{(2)} = 0.693175$, $T_8^{(3)} = 0.693147$

3. a. $T_2 = 0.775$, $T_4 = 0.782794$, $T_4^{(1)} = 0.785392$, $S_4 = 0.7853921$
   b. $T_2 = 2.54308$, $T_4 = 2.399166$, $T_4^{(1)} = 2.351195$, $S_4 = 2.351195$

4. a. $T_{40}^{(3)} = 0.666393$ **b.** $T_{40}^{(3)} = 0.785011$

5. a. $S_{40} = 0.6663458$, $S_{80} = 0.6665532$
   b. $S_{40} = 0.7849438$, $S_{80} = 0.7852416$

7. a. $\hat{S}_{80} = 0.666567$, $T_{80}^{(2)} = 0.666553$
   b. $\hat{S}_{80} = 0.7852615$, $T_{80}^{(2)} = 0.785238$; $\hat{S}_{80}$ estimates are slightly better

Exercise Set 9.8

5. a. 2.350336929   **b.** 2.4435358009
   c. 1.5833333328   **d.** 3.1376196719

7. a. 2.3504023865   **b.** 2.4523820035
   c. 1.57079441216   **d.** 3.1359039098

9. a. 0.669179634   **b.** −0.251845708
   c. 2.423951707   **d.** 1.4791075538

10. a. 0.461220702   **b.** 0.285582632   **c.** 0.746803334

Exercise Set 9.9

1. 1960 growth rate = 0.00844, doubling time = 82.14 years; 1970 growth rate = 0.00778, doubling time = 89.1 years

3. $t = 2$ accel. = 41.504 + 6.801 − 2(20.575) = 7.15 ft/sec^2; $t = 4$ accel. = 105.56 + 41.504 − 2(69.771) = 7.52 ft/sec^2

5. a. The function is $T = f(x) =$ temperature as a function of distance x from the surface of the windowpane
   b. 1575 cal/sec = 22500 Btu/hr

7. Less than 0.01%; approximately 99.99% of the radiation occurs at frequencies below 10 Hz

9. The maximum distance is about 38.74 AU; the distance traveled is about 15.3925 billion miles

10.

x	$J_0(x)$
0	1.00000
0.1	0.9975016
0.2	0.9900252
0.3	0.9776261
0.4	0.9603982
0.5	0.9384701

11. This integral results from setting $t = \sin \theta$ in the integral form in exercise 10, and the result given is often quoted in math/physics texts; however, this new integral is a nonconvergent improper integral; attempts to evaluate it using numerical integration methods lead to nonsense results

Chapter 10

Exercise Set 10.1

2. a. $\frac{1}{2}e^{x^2} - \frac{1}{2}$
   c. $y(0) = 1 \Rightarrow C = 1$, $y(x) = \frac{1}{2} + \frac{1}{2}e^{-x^2}$
   d. $f(x, y) = x - 2xy$ is continuous everywhere; $f_y(x, y) = -2x$ is continuous everywhere. The solution is unique by Theorem 10.1

3. a. $y'(0) = 1$, $y''(0) = 2$, $y'''(0) = 8$ **b.** $1 + x + x^2 + \frac{4}{3}x^3 + \cdots$

5. $y'(0) = -1$ and $x > 0$, $y > 0 \Rightarrow y' < 0$. So $y(x)$ decreases as we move away from initial point $(0, 1)$; $y''(x) = (e^{2x} - e^x)y > 0$ if $x > 0$ and $y > 0$; the graph is concave up in quadrant I; the curve descends from $(0, 1)$ to the x axis where $y' = 0$; all further values of y are zero

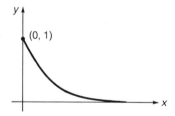

7. d. $y = \sin x$

Exercise Set 10.2

1. a.

X	Y
0.100	0.6000000
0.200	0.7200000
0.300	0.8640000
0.400	1.0368000
0.500	1.2441600

b.

X	Y
0.100	0.1000000
0.200	0.2010000
0.300	0.3050401
0.400	0.4143451
0.500	0.5315133

c.

X	Y
0.100	0.1800000
0.200	0.1720000
0.300	0.1748000
0.400	0.1873200
0.500	0.2085880

d.

X	Y
0.100	0.0000000
0.200	0.0100000
0.300	0.0310000
0.400	0.0640995
0.500	0.1105051

3.

H	X	Y
0.050000	1.200	0.8266159
0.010000	1.200	0.8320525
0.005000	1.200	0.8326964
0.001000	1.200	0.8332054
0.000500	1.200	0.8332682
0.000100	1.200	0.8333028
0.000050	1.200	0.8331597

4. c.

X	Y	ATAN
0.050	0.0500000	0.0499584
0.100	0.0998751	0.0996687
0.150	0.1493780	0.1488900
0.200	0.1982706	0.1973956
0.250	0.2463307	0.2449787
0.300	0.2933576	0.2914568
0.350	0.3391767	0.3366749
0.400	0.3836419	0.3805064
0.450	0.4266369	0.4228540
0.500	0.4680749	0.4636477
0.550	0.5078972	0.5028433
0.600	0.5460708	0.5404196
0.650	0.5825855	0.5763753
0.700	0.6174503	0.6107261
0.750	0.6506907	0.6435013
0.800	0.6823449	0.6747410
0.850	0.7124611	0.7044942
0.900	0.7410951	0.7328152
0.950	0.7683073	0.7597629
1.000	0.7941617	0.7853983

d.

X	Y	$\sqrt{1 + X}$
0.050	1.0250000	1.0246950
0.100	1.0493900	1.0488090
0.150	1.0732140	1.0723810
0.200	1.0965080	1.0954450
0.250	1.1193080	1.1180340
0.300	1.1416430	1.1401750

(continued)

(continued)

X	Y	$\sqrt{1 + X}$
0.350	1.1635410	1.1618950
0.400	1.1850280	1.1832160
0.450	1.2061240	1.2041600
0.500	1.2268520	1.2247450
0.550	1.2472290	1.2449900
0.600	1.2672730	1.2649110
0.650	1.2870010	1.2845230
0.700	1.3064260	1.3038410
0.750	1.3255620	1.3228760
0.800	1.3444220	1.3416410
0.850	1.3630170	1.3601470
0.900	1.3813590	1.3784050
0.950	1.3994570	1.3964240
1.000	1.4173210	1.4142140

5.

X	Y	$(X - 1)^{2/3}$
0.050	1.0333330	0.9663825
0.100	1.0661250	0.9321698
0.150	1.0984080	0.8973171
0.200	1.1302130	0.8617739
0.250	1.1615670	0.8254818
0.300	1.1924960	0.7883735
0.350	1.2230200	0.7503701
0.400	1.2531620	0.7113786
0.450	1.2829380	0.6712873
0.500	1.3123670	0.6299605
0.550	1.3414640	0.5872301
0.600	1.3702440	0.5428835
0.650	1.3987200	0.4966441
0.700	1.4269050	0.4481404
0.750	1.4548100	0.3968502
0.800	1.4824460	0.3419950
0.850	1.5098230	0.2823106
0.900	1.5369510	0.2154433
0.950	1.5638390	0.1357206
1.000	1.5904940	0.0000242
1.050	1.6169250	0.1357210
1.100	1.6431390	0.2154435
1.150	1.6691430	0.2823108
1.200	1.6949430	0.3419951
1.250	1.7205470	0.3968502
1.300	1.7459600	0.4481403
1.350	1.7711860	0.4966440
1.400	1.7962330	0.5428833
1.450	1.8211040	0.5872299
1.500	1.8458050	0.6299603
1.550	1.8703400	0.6712870
1.600	1.8947130	0.7113783
1.650	1.9189300	0.7503698
1.700	1.9429930	0.7883731

(continued)

X	Y	$(X - 1)^{2/3}$
1.750	1.9669060	0.8254814
1.800	1.9906740	0.8617734
1.850	2.0142990	0.8973166
1.900	2.0377860	0.9321692
1.950	2.0611360	0.9663819
2.000	2.0843540	0.9999994
2.050	2.1074430	1.0330610

Exercise Set 10.3

1. a. $1 + x + x^2$ **b.** $1 - x + x^2$

   c. $1 + 2x + 2x^2$ **d.** $1 - \dfrac{x^4}{24} + \dfrac{x^6}{180}$

4. a. 1.0525000, 1.1102530   **b.** 0.9525000, 0.9092976
   c. 1.1050000, 1.2221870   **d.** 1.000000, 0.9999974

6. a.

X	Y
0.050	1.0525000
0.100	1.1102530
0.150	1.1735290
0.200	1.2426090
0.250	1.3177930
0.300	1.3993930
0.350	1.4877360
0.400	1.5831700
0.450	1.6860580
0.500	1.7967810
0.550	1.9157410
0.600	2.0433600
0.650	2.1800820
0.700	2.3263740
0.750	2.4827260
0.800	2.6496530
0.850	2.8276980
0.900	3.0174300
0.950	3.2194480
1.000	3.4343820

b.

X	Y
0.050	1.0000000
0.100	0.9999974
0.150	0.9999828
0.200	0.9999409
0.250	0.9998504
0.300	0.9996846
0.350	0.9994116
0.400	0.9989947
0.450	0.9983934
0.500	0.9975635
0.550	0.9964578
0.600	0.9950271
0.650	0.9932206
0.700	0.9909866
0.750	0.9882736
0.800	0.9850304
0.850	0.9812072
0.900	0.9767562
0.950	0.9716324
1.000	0.9657934

d.

X	Y
0.050	0.0500000
0.100	0.0998750
0.150	0.1495001
0.200	0.1987513
0.250	0.2475054
0.300	0.2956404
0.350	0.3430358
0.400	0.3895731
0.450	0.4351359
0.500	0.4796102

(continued)

(continued)

X	Y
0.550	0.5228848
0.600	0.5648513
0.650	0.6054050
0.700	0.6444441
0.750	0.6818711
0.800	0.7175924
0.850	0.7515186
0.900	0.7835648
0.950	0.8136508
1.000	0.8417014

10. a. 1.1100000   **b.** 1.1102530   **c.** $3.373E - 4$
d. $y(0.1) = 1.1103418$, error $= 3.418E - 4$

Exercise Set 10.4

5. a. 1.0202000, 1.0408080   **b.** 0.01070710, 0.0219809
c. 0.5176758, 0.5357150   **d.** 2.0301000, 2.060403

7. a. 1.020201, 1.040811   **b.** 0.010652, 0.021920
c. 0.517677, 0.535718   **d.** 2.030101, 2.060404

9. a.

X	Y	**b.** X	Y
0.05	1.1050000	0.05	0.0579057
0.10	1.2210250	0.10	0.1228156
0.15	1.3492330	0.15	0.1926005
0.20	1.4909020	0.20	0.2665051
0.25	1.6474470	0.25	0.3440918
0.30	1.8204290	0.30	0.4250659
0.35	2.0115740	0.35	0.5092113
0.40	2.2227890	0.40	0.5963614
0.45	2.4561820	0.45	0.6863823
0.50	2.7140810	0.50	0.7791635
0.55	2.9990590	0.55	0.8746121
0.60	3.3139600	0.60	0.9726483
0.65	3.6619260	0.65	1.0732030
0.70	4.0464280	0.70	1.1762150
0.75	4.4713030	0.75	1.2816300
0.80	4.9407900	0.80	1.3893990
0.85	5.4595730	0.85	1.4994800
0.90	6.0328280	0.90	1.6118310
0.95	6.6662750	0.95	1.7264180
1.00	7.3662340	1.00	1.8432060

c.

X	Y
0.05	0.591971
0.10	0.694511
0.15	0.810003
0.20	0.941575
0.25	1.093424
0.30	1.271314
0.35	1.483379

(*continued*)

X	Y
0.40	1.741479
0.45	2.063589
0.50	2.478290
0.55	3.033822
0.60	3.818144
0.65	5.009110
0.70	7.022687
0.75	11.070870
0.80	22.263690
0.85	86.158110
0.90	3851.445000
0.95	7023343000.000000

d.

X	Y
0.05	2.152516
0.10	2.310408
0.15	2.474073
0.20	2.643917
0.25	2.820367
0.30	3.003863
0.35	3.194864
0.40	3.393846
0.45	3.601307
0.50	3.817765
0.55	4.043761
0.60	4.279861
0.65	4.526653
0.70	4.784754
0.75	5.054810
0.80	5.337495
0.85	5.633515
0.90	5.943609
0.95	6.268554
1.00	6.609161

Exercise Set 10.5

1. a. 0.199001   **b.** 0.834711

2. a.

T	X(t)	X'(t)
0.10	0.099833	0.995004
0.20	0.199001	0.979999
0.30	0.295833	0.955240
0.40	0.390049	0.920794
0.50	0.479992	0.877273
0.60	0.565503	0.824746
0.70	0.644941	0.764232
0.80	0.718350	0.695682
0.90	0.784078	0.620663
1.00	0.842482	0.538724

b. H = 0.1

T	X(t)	X'(t)
1.10	0.909091	−0.826446
1.20	0.834711	−0.696742
1.30	0.769743	−0.592504
1.40	0.716210	−0.512957
1.50	0.667151	−0.445091
1.60	0.627192	−0.393369
1.70	0.588477	−0.346306
1.80	0.557931	−0.311287
1.90	0.526220	−0.276908
2.00	0.502549	−0.252556

8. a.

T	X(t)	X'(t)
0.10	0.9950239	−0.099005
0.20	0.980394	−0.192158
0.30	0.956965	−0.27418
0.40	0.926071	−0.340858
0.50	0.889449	−0.38945
0.60	0.848931	−0.418718
0.70	0.806447	−0.429027
0.80	0.763805	−0.42209
0.90	0.722598	−0.400677
1.00	0.6841	−0.368201
1.10	0.649237	−0.328322
1.20	0.61857	−0.284569
1.30	0.59233	−0.240059
1.40	0.570464	−0.197301
1.50	0.552705	−0.158116
1.60	0.538635	−0.123634
1.70	0.527757	−0.094378
1.80	0.5195451	−0.070365
1.90	0.5134901	−0.051265
2.00	0.5091271	−0.036509

b.

T	X(t)	X'(t)
0.10	0.026726	0.479711
0.20	0.088468	0.744649
0.30	0.174198	0.965093
0.40	0.280716	1.162282
0.50	0.406631	1.344783
0.60	0.549944	1.516178
0.70	0.709803	1.679158
0.80	0.885618	1.8355
0.90	1.076775	1.986361
1.00	1.282769	2.132594

c.

T	X(t)	X'(t)
0.10	0.099669	0.990099
0.20	0.197395	0.961539
0.30	0.291457	0.917431
0.40	0.380506	0.862069
0.50	0.463636	0.800008
0.60	0.540426	0.735287
0.70	0.610759	0.671109
0.80	0.674804	0.609694
0.90	0.732901	0.5524
1.00	0.785499	0.499898

Exercise Set 10.6

5. Error estimate is 0.0013788

6. a. RUNGE-KUTTA STARTING VALUES

T	Y0
0.00	0.000000
0.05	0.050042
0.10	0.100335
0.15	0.151135

ADAMS-MOULTON ESTIMATES

T	Y0	Y	ERROR
0.20	0.202708	0.202710	0.0000001
0.25	0.255340	0.255342	0.0000002

b. RUNGE-KUTTA STARTING VALUES

T	Y0
0.05	0.050042
0.10	0.100335
0.15	0.151135

MILNE ESTIMATES

T	Y0	Y	ERROR
0.20	0.202708	0.202710	−0.0000001
0.25	0.255340	0.255342	−0.0000001

7. See answers to exercise 6(a) and (b)

9. RUNGE-KUTTA STARTING VALUES

T	Y0
0.00	0.000000
0.05	0.050042
0.10	0.100335
0.15	0.151135

HAMMING PREDICTOR-CORRECTOR

T	Y0	Y
0.20	0.202708	0.202710
0.25	0.255340	0.255342

10. a.

X	Y EST	Y TRUE
0.10	0.000000	0.000000
0.20	0.000000	0.000000
0.30	0.000000	0.000000
0.40	0.000000	0.000000
0.50	0.000000	0.000000
0.60	0.000000	0.000000
0.70	0.000000	0.000000
0.80	0.000000	0.000000
0.90	0.000000	0.000000
1.00	0.016667	0.000000
1.10	0.110243	0.095163
1.20	0.194915	0.181269
1.30	0.271529	0.259182
1.40	0.340852	0.329680
1.50	0.403578	0.393470
1.60	0.460335	0.451189
1.70	0.511691	0.503415
1.80	0.558160	0.550671
1.90	0.600206	0.593431
2.00	0.638252	0.632121

11. $y = \begin{cases} 0 & \text{if } x < 1 \\ 1 - e^{1-x} & \text{if } x \geq 1 \end{cases}$

See **YTRUE** column in answer to exercise 10(a) for values

Exercise Set 10.7

3. a.

T	X(t)	X'(t)	
0.10	3.000000	0.000000	Stable
0.20	3.000000	0.000000	
0.30	3.000000	0.000000	
0.40	3.000000	0.000000	
0.50	3.000000	0.000000	
0.60	3.000000	0.000000	
0.70	3.000000	0.000000	
0.80	3.000000	0.000000	
0.90	3.000000	0.000000	
1.00	3.000000	0.000000	
1.10	3.000000	0.000000	
1.20	3.000000	0.000000	
1.30	3.000000	0.000000	
1.40	3.000000	0.000000	
1.50	3.000000	0.000000	
1.60	3.000000	0.000000	
1.70	3.000000	0.000000	
1.80	3.000000	0.000000	
1.90	3.000000	0.000000	
2.00	3.000000	0.000000	

b.

T	X(t)	X'(t)	
0.10	1.190325	1.809675	Stable
0.20	1.361935	1.638065	
0.30	1.517938	1.482062	
0.40	1.658347	1.341653	
0.50	1.786269	1.213732	
0.60	1.901094	1.098906	
0.70	2.006050	0.993950	
0.80	2.099884	0.900116	
0.90	2.186073	0.813927	
1.00	2.262669	0.737331	
1.10	2.333539	0.666461	
1.20	2.395962	0.604039	
1.30	2.454347	0.545653	
1.40	2.505092	0.494908	
1.50	2.553328	0.446672	
1.60	2.594427	0.405574	
1.70	2.634443	0.365557	
1.80	2.667538	0.332462	
1.90	2.700936	0.299064	
2.00	2.727351	0.272650	

c.

T	XN	X'(t)	X(t)	
0.10	0.913613	−0.740838	0.913606	Unstable
0.20	0.851833	−0.555498	0.849604	
0.30	0.802513	−0.407539	0.802190	
0.40	0.770325	−0.310974	0.767065	
0.50	0.740318	−0.220954	0.741043	
0.60	0.726134	−0.178401	0.721766	
0.70	0.704638	−0.113914	0.707485	
0.80	0.703351	−0.110054	0.696906	
0.90	0.682627	−0.047881	0.689069	
1.00	0.693775	−0.081325	0.683262	
1.10	0.666362	0.000914	0.678961	
1.20	0.693958	−0.081873	0.675775	
1.30	0.649988	0.050037	0.673414	
1.40	0.703965	−0.111896	0.671665	
1.50	0.627608	0.117175	0.670370	
1.60	0.727400	−0.182200	0.669410	
1.70	0.591168	0.226495	0.668699	
1.80	0.772699	−0.318097	0.668172	
1.90	0.527549	0.417353	0.667782	
2.00	0.856170	−0.568510	0.667493	

d.

T	X(t)	X'(t)	
0.10	1.097541	0.951229	Stable
0.20	1.190246	0.904877	
0.30	1.278517	0.860742	
0.40	1.362394	0.818803	
0.50	1.442277	0.778861	
0.60	1.518167	0.740917	

(continued)

(*continued*)

T	X(t)	X'(t)	
0.70	1.590460	0.704770	Stable
0.80	1.659121	0.670440	
0.90	1.724548	0.637726	
1.00	1.786666	0.606667	
1.10	1.845882	0.577059	
1.20	1.902077	0.548961	
1.30	1.955674	0.522163	
1.40	2.006510	0.496745	
1.50	2.055023	0.472489	
1.60	2.101008	0.449496	
1.70	2.144922	0.427539	
1.80	2.186516	0.406742	
1.90	2.226271	0.386865	
2.00	2.263889	0.368056	

5. a. $A\left(\dfrac{1+\sqrt{5}}{2}\right)^n + B\left(\dfrac{1-\sqrt{5}}{2}\right)^n$ **b.** $A(\tfrac{1}{2})^n + B(\tfrac{1}{4})^n$ **c.** $A(3^n) + B(2^n)$

7. Improved Euler: $y_{n+1} = y_n + h\{\tfrac{1}{2}f(x_n, y_n) + \tfrac{1}{2}f(x_n + h, y_n + k_n)\}$
where $k_n = hf(x_n, y_n)$

$f(x, y) = -\lambda y \Rightarrow y_{n+1} = y_n + h\left\{-\dfrac{\lambda}{2}y_n + \tfrac{1}{2}[-\lambda(y_n - h\lambda y_n)]\right\}$

Simplifying $\Rightarrow y_{n+1} = y_n(1 - \lambda h + \tfrac{1}{2}\lambda^2 h^2)$

$\therefore \quad y_1 = y_0(1 - \lambda h + \tfrac{1}{2}\lambda^2 h^2)$

$\qquad y_2 = y_0(1 - \lambda h + \tfrac{1}{2}\lambda^2 h^2)^2$

$\qquad \vdots$

$\qquad y_n = y_0(1 - \lambda h + \tfrac{1}{2}\lambda^2 h^2)^n$

8. $1 - \lambda h + \tfrac{1}{2}h^2\lambda^2 > 1 \Leftrightarrow (h\lambda - 1)^2 > 1$
Since $\lambda > 0$, $h > 0$, then $\lambda h - 1 < -1$ is impossible,
so $\lambda h - 1 > 1$ and $\lambda h > 2$; thus,

$h > \dfrac{2}{\lambda} \Rightarrow 1 - \lambda h + \tfrac{1}{2}\lambda^2 h^2 > 1$;

for $\lambda = 10$, this $h > 0.2$

Exercise Set 10.8

3.

T	X	Y
0.10	9.512295	0.481610
0.20	9.048374	0.927840
0.30	8.607079	1.340710
0.40	8.187308	1.722133
0.50	7.788008	2.073922
0.60	7.408182	2.397795
0.70	7.046881	2.695378
0.80	6.703201	2.968214
0.90	6.376282	3.217761
1.00	6.065306	3.445402
1.10	5.769498	3.652446
1.20	5.488117	3.840131
1.30	5.220458	4.009631
1.40	4.965853	4.162056

(*continued*)

T	X	Y
1.50	4.723665	4.298455
1.60	4.493290	4.419822
1.70	4.274149	4.527097
1.80	4.065696	4.621170
1.90	3.867410	4.702881
2.00	3.678794	4.773024

5. a. $x_1' = x_2$, $x_2' = -x_1$; $x_1(0) = 0$, $x_2(0) = 1$

b.

T	X	Y
0.05	0.049979	0.998750
0.10	0.099833	0.995004
0.15	0.149438	0.988771
0.20	0.198669	0.980067
0.25	0.247404	0.968912
0.30	0.295520	0.955336
0.35	0.342898	0.939373
0.40	0.389418	0.921061
0.45	0.434966	0.900447
0.50	0.479426	0.877582
0.55	0.522687	0.852524
0.60	0.564642	0.825336
0.65	0.605186	0.796084
0.70	0.644218	0.764842
0.75	0.681639	0.731689
0.80	0.717356	0.696707
0.85	0.751280	0.659983
0.90	0.783327	0.621610
0.95	0.813415	0.581683
1.00	0.841471	0.540302

6. a.

T	X	X'
0.05	1.0036	0.1429
0.10	1.014	0.2732
0.15	1.0307	0.3931
0.20	1.0532	0.5044
0.25	1.0811	0.6087
0.30	1.114	0.7074
0.35	1.1517	0.8018
0.40	1.1941	0.8929
0.45	1.241	0.9818
0.50	1.2923	1.0691
0.55	1.3479	1.1559
0.60	1.4079	1.2426
0.65	1.4722	1.3299
0.70	1.5409	1.4184
0.75	1.6141	1.5086
0.80	1.6918	1.6011
0.85	1.7742	1.6961
0.90	1.8615	1.7942
0.95	1.9537	1.8958
1.00	2.0511	2.0013

b.

T	X	X'
0.05	1.0249	0.4999
0.10	1.0499	0.4998
0.15	1.0749	0.4994
0.20	1.0999	0.4986
0.25	1.1248	0.4972
0.30	1.1496	0.4951
0.35	1.1743	0.4922
0.40	1.1988	0.4883
0.45	1.2231	0.4831
0.50	1.2471	0.4767
0.55	1.2707	0.4687
0.60	1.2939	0.459
0.65	1.3166	0.4474
0.70	1.3387	0.4339
0.75	1.36	0.4181
0.80	1.3804	0.4
0.85	1.3999	0.3793
0.90	1.4183	0.356
0.95	1.4355	0.33
1.00	1.4513	0.301

c.

T	X	X'
0.05	1.9651	−0.3896
0.10	1.9615	0.2487
0.15	1.9906	0.9261
0.20	2.0549	1.6551
0.25	2.1572	2.4495
0.30	2.3012	3.3249
0.35	2.4914	4.299
0.40	2.7331	5.3918
0.45	3.0329	6.6262
0.50	3.3985	8.0285
0.55	3.8391	9.629
0.60	4.3653	11.4623
0.65	4.9898	13.5688
0.70	5.7275	15.9948
0.75	6.5955	18.794
0.80	7.6142	22.0288
0.85	8.8069	25.7713
0.90	10.2011	30.1051
0.95	11.8288	35.1272
1.00	13.7272	40.9505

7.

T	X1	X2	X3
5.000	996.90	2.55	0.55
10.000	995.50	3.25	1.26
15.000	993.71	4.14	2.15
20.000	991.45	5.26	3.29
25.000	988.57	6.68	4.74
30.000	984.94	8.48	6.59
35.000	980.35	10.73	8.92
40.000	974.57	13.56	11.87
45.000	967.33	17.07	15.60
50.000	958.30	21.41	20.28
55.000	947.11	26.74	26.15
60.000	933.36	33.19	33.45
65.000	916.60	40.91	42.50
70.000	896.42	49.98	53.60
75.000	872.45	60.43	67.12
80.000	844.45	72.18	83.38
85.000	812.34	84.98	102.68
90.000	776.31	98.43	125.26
95.000	736.82	111.95	151.23
100.000	694.65	124.82	180.52

Exercise Set 10.9

1. $x' = 0.9x - 0.075xy; x(0) = 50$
$y' = -0.7y + 0.03xy; y(0) = 5$

T	X	Y
1.00	65.935100	15.115990
2.00	25.062720	31.007520
3.00	8.324505	22.396930
4.00	5.461111	13.454160
5.00	6.185033	7.942194
6.00	9.522833	4.960753
7.00	17.080420	3.601176
8.00	32.215140	3.610413
9.00	56.264040	6.519985
10.00	57.601920	20.902810
11.00	17.789320	29.054100
12.00	7.129148	19.448010
13.00	5.617886	11.587480
14.00	7.063564	6.942562
15.00	11.445980	4.506919
16.00	20.901200	3.547786
17.00	38.897510	4.150514
18.00	61.883080	9.259921
19.00	42.309200	27.120020
20.00	12.568120	26.017570
21.00	6.345453	16.486300
22.00	5.982289	9.802892
23.00	8.298043	6.005561
24.00	14.088550	4.113652
25.00	26.041340	3.618636
26.00	47.011770	5.167060
27.00	62.791740	14.089930
28.00	27.186110	29.579970
29.00	9.243767	22.408880
30.00	5.976943	13.713020
31.00	6.632186	8.201905
32.00	10.027940	5.192280
33.00	17.693090	3.831784
34.00	32.763920	3.914528
35.00	55.543620	7.107946
36.00	53.715350	21.354290
37.00	17.233690	27.891340
38.00	7.357785	18.740290
39.00	6.017081	11.271730
40.00	7.673035	6.848850
41.00	12.451430	4.548002
42.00	22.568970	3.721456
43.00	41.144160	4.624611
44.00	61.395180	10.825080
45.00	36.146340	27.832400
46.00	11.482360	24.394350
47.00	6.429869	15.361010
48.00	6.458260	9.213036
49.00	9.228119	5.755491
50.00	15.814250	4.092466
51.00	29.034680	3.856604
52.00	50.589460	6.113535

3. Use $x'(t) = -0.08(x - 200)$; $x(0) = 200$

T	X	XPRIME
0.50	194.902600	−9.992209
1.00	190.005100	−9.600409
1.50	185.299600	−9.223971
2.00	180.778700	−8.862293
2.50	176.435000	−8.514798
3.00	172.261600	−8.180928
3.50	168.251900	−7.860149
4.00	164.399400	−7.551948
4.50	160.697900	−7.255832
5.00	157.141600	−6.971327
5.50	153.724700	−6.697977
6.00	150.441800	−6.435345
6.50	147.287600	−6.183011
7.00	144.257100	−5.940572
7.50	141.345500	−5.707638
8.00	138.548000	−5.483838
8.50	135.860200	−5.268813
9.00	133.277800	−5.062221
9.50	130.796600	−4.863728
10.00	128.412700	−4.673019

5. a. The time required is 202 seconds
b. The impact velocity is 1057 ft/sec

7.

X	CX	CXPRIME
0.05	0.050000	0.999997
0.10	0.099999	0.999950
0.15	0.149992	0.999747
0.20	0.199968	0.999200
0.25	0.249902	0.998048
0.30	0.299757	0.995953
0.35	0.349475	0.992506
0.40	0.398977	0.987227
0.45	0.448158	0.979567
0.50	0.496884	0.968913
0.55	0.544988	0.954595
0.60	0.592271	0.935897
0.65	0.638493	0.912067
0.70	0.683379	0.882333
0.75	0.726615	0.845924
0.80	0.767848	0.802096
0.85	0.806689	0.750155
0.90	0.842718	0.689498
0.95	0.875486	0.619650
1.00	0.904524	0.540302

Chapter 11

Exercise Set 11.1

1. a.

X	Y	YPRIME
1.00	3.00000	2.00000
1.05	3.11301	2.53050
1.10	3.25410	3.12400
1.15	3.42651	3.78350
1.20	3.63360	4.51200
1.25	3.87890	5.31250
1.30	4.16610	6.18800
1.35	4.49900	7.14150
1.40	4.88160	8.17600
1.45	5.31800	9.29450
1.50	5.81250	10.50000
1.55	6.36950	11.79550
1.60	6.99360	13.18400
1.65	7.68950	14.66850
1.70	8.46210	16.25200
1.75	9.31640	17.93750
1.80	10.25760	19.72801
1.85	11.29100	21.62651
1.90	12.42210	23.63601
1.95	13.65650	25.75951
2.00	15.00000	28.00002

c.

X	Y	YPRIME
0.00	1.00000	−1.21658
0.05	0.93801	−1.26098
0.10	0.87409	−1.29399
0.15	0.80881	−1.31540
0.20	0.74275	−1.32513
0.25	0.67649	−1.32320
0.30	0.61062	−1.30977
0.35	0.54570	−1.28508
0.40	0.48229	−1.24950
0.45	0.42092	−1.20350
0.50	0.36211	−1.14764
0.55	0.30631	−1.08255
0.60	0.25399	−1.00898
0.65	0.20555	−0.92769
0.70	0.16134	−0.83952
0.75	0.12169	−0.74535
0.80	0.08689	−0.64609
0.85	0.05715	−0.54266
0.90	0.03268	−0.43599
0.95	0.01359	−0.32700
1.00	0.00000	−0.21658

e.

X	Y	YPRIME
0.00	1.00000	2.00002
0.05	1.09430	1.78197
0.10	1.17898	1.61157
0.15	1.25598	1.47295
0.20	1.32665	1.35681
0.25	1.39194	1.25724
0.30	1.45258	1.17033
0.35	1.50914	1.09334
0.40	1.56205	1.02430
0.45	1.61168	0.96173
0.50	1.65831	0.90454
0.55	1.70220	0.85184
0.60	1.74356	0.80296
0.65	1.78255	0.75734
0.70	1.81934	0.71455
0.75	1.85405	0.67420
0.80	1.88680	0.63600
0.85	1.91768	0.59968
0.90	1.94679	0.56503
0.95	1.97421	0.53186
1.00	2.00000	0.50000

3. a.

X	Y EST	YPRIME
0.00	5.000000	4.000000
0.05	5.202500	4.100000
0.10	5.410000	4.200000
0.15	5.622500	4.300000
0.20	5.840000	4.400000
0.25	6.062500	4.500000
0.30	6.290000	4.600000
0.35	6.522500	4.700000
0.40	6.760000	4.799999
0.45	7.002500	4.899999
0.50	7.250000	4.999999
0.55	7.502500	5.099999
0.60	7.760000	5.199999
0.65	8.022500	5.299999
0.70	8.290000	5.399999
0.75	8.562500	5.499999
0.80	8.840000	5.599999
0.85	9.122501	5.699999
0.90	9.410001	5.799999
0.95	9.702501	5.899998
1.00	10.000000	5.999998

b. Same as part (a)
c. Same as part (a)

5. a. $y = 0$ implies $x = 1$, so $y(1) = 0$; $y = 1$ implies $x = 3$, so $y(3) = 1$
Differentiation gives

$$1 = 3y^2 y' + y'$$

so

$$y' = \frac{1}{1 + 3y^2}$$

Differentiation again gives

$$0 = 6y(y')^2 + 3y^2y'' + y''$$

So

$$0 = 6y(y')^2 + y''\left(\frac{1}{y'}\right) \quad \text{and} \quad 6y(y')^3 + y'' = 0$$

b. Algorithm 11.1 gives $y(1.5) = 0.42385$, $y(2) = 0.68233$, and $y(2.5) = 0.86122$; Substitution in $x = y^3 + y + 1$ gives, respectively, 1.4999942, 2.0000053, and 2.4999868, which agrees well with the true values

e. The algorithm produces overflow errors and no output

Exercise Set 11.2

1. a. USING X'(A) = 3.776482

TIME	X	XPRIME
1.05	3.212935	4.750221
1.10	3.475778	5.769276
1.15	3.790270	6.812747
1.20	4.157122	7.860362
1.25	4.576060	8.893131
1.30	5.045901	9.893791
1.35	5.564649	10.847080
1.40	6.129601	11.739900
1.45	6.737448	12.561280
1.50	7.384391	13.302430
1.55	8.066240	13.956540
1.60	8.778511	14.518680
1.65	9.516521	14.985630
1.70	10.275460	15.355670
1.75	11.050470	15.628430
1.80	11.836690	15.804700
1.85	12.629360	15.886270
1.90	13.423780	15.875710
1.95	14.215450	15.776330
2.00	15.000000	15.591940

c. USING X'(A) = −0.4571177

TIME	X	XPRIME
0.05	0.975872	−0.508411
0.10	0.949120	−0.562030
0.15	0.919637	−0.617563
0.20	0.887339	−0.674585
0.25	0.852161	−0.732658
0.30	0.814063	−0.791337
0.35	0.773025	−0.850176

(continued)

(*continued*)

TIME	X	XPRIME
0.40	0.729050	−0.908738
0.45	0.682163	−0.966591
0.50	0.632410	−1.023322
0.55	0.579856	−1.078538
0.60	0.524587	−1.131870
0.65	0.466706	−1.182977
0.70	0.406332	−1.231550
0.75	0.343598	−1.277316
0.80	0.278651	−1.320039
0.85	0.211648	−1.359519
0.90	0.142756	−1.395598
0.95	0.072147	−1.428158
1.00	0.000000	−1.457118

f. USING X′(A) = 0.9999992

TIME	X	XPRIME
1.05	2.054881	1.192970
1.10	2.119091	1.373553
1.15	2.192065	1.543856
1.20	2.273333	1.705555
1.25	2.362500	1.859999
1.30	2.459231	2.008284
1.35	2.563241	2.151303
1.40	2.674286	2.289796
1.45	2.792155	2.424376
1.50	2.916666	2.555555
1.55	3.047661	2.683767
1.60	3.185000	2.809375
1.65	3.328560	2.932691
1.70	3.478235	3.053980
1.75	3.633928	3.173470
1.80	3.795555	3.291358
1.85	3.963040	3.407816
1.90	4.136315	3.522992
1.95	4.315320	3.637016
2.00	4.500000	3.750001

3. a. USING X′(A) = 3.999998

TIME	X	XPRIME
0.10	5.410001	4.199999
0.20	5.840001	4.399999
0.30	6.290001	4.599999
0.40	6.760001	4.799999
0.50	7.250002	5.000000
0.60	7.760002	5.200000
0.70	8.290002	5.400000
0.80	8.840002	5.600001
0.90	9.410002	5.800001
1.00	10.000000	6.000001

b. USING X'(A) = 4.000001

TIME	X	XPRIME
0.05	5.202500	4.100001
0.10	5.410000	4.200001
0.15	5.622500	4.300000
0.20	5.840000	4.400000
0.25	6.062500	4.500000
0.30	6.290000	4.600000
0.35	6.522500	4.700000
0.40	6.760000	4.800000
0.45	7.002500	4.900000
0.50	7.250000	5.000000
0.55	7.502500	5.100000
0.60	7.760000	5.200000
0.65	8.022500	5.299999
0.70	8.290000	5.399999
0.75	8.562500	5.499999
0.80	8.840000	5.599999
0.85	9.122501	5.699999
0.90	9.410001	5.799999
0.95	9.702501	5.899999
1.00	10.000000	5.999999

d. $h = 0.05$ gives best results

e. No

5. $c_1 = 1, c_2 = -1; y(x) = x - e^{-2x}$

$h = 0.1$ $h = 0.05$

USING Y'(A) = 3   USING Y'(A) = 2.999999

X	Y	YPRIME	X	Y	YPRIME
0.10	−0.718733	2.637467	0.05	−0.854838	2.809674
0.20	−0.470324	2.340649	0.10	−0.718731	2.637461
0.30	−0.248817	2.097634	0.15	−0.590819	2.481636
0.40	−0.049335	1.898670	0.20	−0.470320	2.340640
0.50	0.132115	1.735771	0.25	−0.356531	2.213061
0.60	0.298800	1.602400	0.30	−0.248812	2.097623
0.70	0.453398	1.493205	0.35	−0.146586	1.993170
0.80	0.598099	1.403804	0.40	−0.049330	1.898658
0.90	0.734697	1.330607	0.45	0.043430	1.813139
1.00	0.864661	1.270679	0.50	0.132120	1.735759
1.10	0.989193	1.221614	0.55	0.217128	1.665742
1.20	1.109279	1.181443	0.60	0.298805	1.602388
1.30	1.225724	1.148553	0.65	0.377467	1.545063
1.40	1.339188	1.121626	0.70	0.453402	1.493193
1.50	1.450211	1.099579	0.75	0.526869	1.446260
1.60	1.559236	1.081529	0.80	0.598102	1.403792
1.70	1.666625	1.066750	0.85	0.667315	1.365366
1.80	1.772675	1.054651	0.90	0.734700	1.330597
1.90	1.877628	1.044744	0.95	0.800430	1.299137
2.00	1.981683	1.036634	1.00	0.864664	1.270670
			1.05	0.927542	1.244912
			1.10	0.989195	1.221606

(continued)

(*continued*)

X	Y	XPRIME
1.15	1.049740	1.200517
1.20	1.109281	1.181435
1.25	1.167913	1.164169
1.30	1.225725	1.148546
1.35	1.282793	1.134410
1.40	1.339188	1.121619
1.45	1.394975	1.110046
1.50	1.450211	1.099573
1.55	1.504949	1.090098
1.60	1.559236	1.081524
1.65	1.613115	1.073766
1.70	1.666625	1.066746
1.75	1.719801	1.060394
1.80	1.772674	1.054647
1.85	1.825274	1.049446
1.90	1.877627	1.044741
1.95	1.929756	1.040483
2.00	1.981682	1.036630

7.
X	Y	Y(X)	Errors
0.00	2.000000	2.000000	
0.05	2.052459	2.052459	
0.10	2.109675	2.109675	
0.15	2.171416	2.171416	
0.20	2.237461	2.237462	0.000001
0.25	2.307602	2.307602	
0.30	2.381637	2.381637	
0.35	2.459376	2.459376	
0.40	2.540640	2.540640	0.000000 $(< 10^{-6})$
0.45	2.625257	2.625257	
0.50	2.713062	2.713062	
0.55	2.803900	2.803900	
0.60	2.897623	2.897623	0.000000
0.65	2.994092	2.994092	
0.70	3.093171	3.093171	
0.75	3.194733	3.194734	
0.80	3.298658	3.298658	0.000000
0.85	3.404830	3.404830	
0.90	3.513139	3.513140	
0.95	3.623482	3.623482	
1.00	3.735759	3.735759	

9. **b.**
X	Y	Y(X)
0.00	−1.000000	−1.000000
0.05	−0.927675	−0.948771
0.10	−0.857042	−0.895171
0.15	−0.787690	−0.839333
0.20	−0.719274	−0.781397
0.25	−0.651508	−0.721509
0.30	−0.584159	−0.659816
0.35	−0.517042	−0.596475

(*continued*)

X	Y	Y(X)
0.40	−0.450018	−0.531643
0.45	−0.382986	−0.465482
0.50	−0.315882	−0.398157
0.55	−0.248674	−0.329837
0.60	−0.181359	−0.260693
0.65	−0.113958	−0.190897
0.70	−0.046517	−0.120624
0.75	0.020900	−0.050050
0.80	0.088212	0.020650
0.85	0.155322	0.091297
0.90	0.222120	0.161717
0.95	0.288487	0.231733
1.00	0.354293	0.301169
1.05	0.419404	0.369852
1.10	0.483677	0.437611
1.15	0.546967	0.504277
1.20	0.609125	0.569681
1.25	0.670001	0.633662
1.30	0.729445	0.696059
1.35	0.787307	0.756717
1.40	0.843438	0.815482
1.45	0.897692	0.872210
1.50	0.949925	0.926758
1.55	1.000000	0.978989

c.

T	X	X(t)
0.00	−1.000000	−1.000000
0.10	−0.893665	−0.895171
0.20	−0.778944	−0.781397
0.30	−0.656829	−0.659816
0.40	−0.528420	−0.531643
0.50	−0.394908	−0.398157
0.60	−0.257561	−0.260694
0.70	−0.117698	−0.120625
0.80	0.023317	0.020649
0.90	0.164102	0.161716
1.00	0.303266	0.301168
1.10	0.439430	0.437610
1.20	0.571238	0.569680
1.30	0.697377	0.696058
1.40	0.816586	0.815481
1.50	0.927672	0.926757

d.

T	X	X(t)
0.00	−1.000000	−1.000000
0.10	−0.895172	−0.895171
0.20	−0.781399	−0.781397
0.30	−0.659818	−0.659816
0.40	−0.531645	−0.531643

(*continued*)

(*continued*)

T	X	X(t)
0.50	−0.398159	−0.398157
0.60	−0.260695	−0.260693
0.70	−0.120626	−0.120624
0.80	0.020648	0.020650
0.90	0.161716	0.161717
1.00	0.301168	0.301169
1.10	0.437611	0.437611
1.20	0.569681	0.569681
1.30	0.696059	0.696059
1.40	0.815483	0.815482
1.50	0.926758	0.926758

e. 1.5 is a multiple of 0.05, while $\pi/2$ is not

Exercise Set 11.3

1. a.

X	Y
0.05	5.202499
0.10	5.409998
0.15	5.622497
0.20	5.839996
0.25	6.062495
0.30	6.289994
0.35	6.522493
0.40	6.759992
0.45	7.002492
0.50	7.249992
0.55	7.502491
0.60	7.759991
0.65	8.022491
0.70	8.289993
0.75	8.562492
0.80	8.839992
0.85	9.122495
0.90	9.409996
0.95	9.702498

c.

X	Y
0.05	0.194570
0.10	0.374316
0.15	0.544073
0.20	0.708027
0.25	0.869851
0.30	1.032833
0.35	1.199966
0.40	1.374038
0.45	1.557703
0.50	1.753548
0.55	1.964142
0.60	2.192091
0.65	2.440085
0.70	2.710939
0.75	3.007634
0.80	3.333361
0.85	3.691559
0.90	4.085959
0.95	4.520624

e.

X	Y
1.05	2.054916
1.10	2.119151
1.15	2.192144
1.20	2.273423
1.25	2.362598
1.30	2.459332
1.35	2.563343
1.40	2.674387
1.45	2.792252
1.50	2.916759
1.55	3.047748
1.60	3.185079
1.65	3.328632

(*continued*)

X	Y
1.70	3.478297
1.75	3.633981
1.80	3.795598
1.85	3.963073
1.90	4.136338
1.95	4.315332

3. b. See 1a

 c. $y(0.2)$ error is 0.000004; $y(0.5)$ and $y(0.8)$ errors are 0.000008

 d. Solutions in exercise 3 of Exercise Set 11.2 are more accurate

5. $h = 0.05$

X	Y	Y(X)	ERROR	
1.05	−0.055000	−0.055125	−0.000125	
1.10	−0.120752	−0.121000	−0.000248	
1.15	−0.198008	−0.198375	−0.000367	
1.20	−0.287519	−0.288000	−0.000481	
1.25	−0.390038	−0.390625	−0.000587	
1.30	−0.506316	−0.507000	−0.000684	
1.35	−0.637106	−0.637875	−0.000769	
1.40	−0.783158	−0.784000	−0.000842	max error
1.45	−0.945226	−0.946125	−0.000899	is 0.000962
1.50	−1.124061	−1.125000	−0.000939	
1.55	−1.320414	−1.321375	−0.000961	
1.60	−1.535038	−1.536000	−0.000962	
1.65	−1.768685	−1.769625	−0.000940	
1.70	−2.022106	−2.023000	−0.000894	
1.75	−2.296054	−2.296875	−0.000821	
1.80	−2.591279	−2.592000	−0.000720	
1.85	−2.908535	−2.909125	−0.000590	
1.90	−3.248572	−3.249001	−0.000428	
1.95	−3.612144	−3.612376	−0.000232	

$h = 0.1$

X	Y	Y(X)	ERROR	
1.10	−0.120000	−0.121000	−0.001000	
1.20	−0.286061	−0.288000	−0.001940	
1.30	−0.504242	−0.507000	−0.002758	
1.40	−0.780606	−0.784000	−0.003394	max error
1.50	−1.121212	−1.125000	−0.003788	is 0.003879
1.60	−1.532121	−1.536000	−0.003879	
1.70	−2.019394	−2.023000	−0.003607	
1.80	−2.589091	−2.592000	−0.002909	
1.90	−3.247273	−3.249001	−0.001728	

7. a. $y_{0.05}(1.3) = -0.506316$

 $y_{0.1}(1.3) = -0.504242$, $\hat{y}_{0.05}(1.3) = -0.507007$

 $\hat{y}_{0.05}(1.7) = -2.02301$

 Yes, extrapolation improved the results.

9.
$$\begin{bmatrix} -2.055556 & 1 & 0 & 0 \\ 1 & -2.040816 & 1 & 0 \\ 0 & 0 & -2.031250 & 1 \\ 0 & 0 & 1 & -2.024691 \end{bmatrix} \begin{bmatrix} y_1 \\ y_2 \\ y_3 \\ y_4 \end{bmatrix} = \begin{bmatrix} -2 \\ 0 \\ 0 \\ -4.5 \end{bmatrix}$$

Solution: $y_1 = 2.274788$, $y(1.2) = 2.273333$, $|\text{error}| = 0.001455$
$y_2 = 2.675954$, $y(1.4) = 2.674286$, $|\text{error}| = 0.001668$
$y_3 = 3.186341$, $y(1.6) = 3.185$, $|\text{error}| = 0.001341$
$y_4 = 3.796303$, $y(1.8) = 3.795555$, $|\text{error}| = 0.000747$

Exercise Set 11.4

1. a. USING X(A) = −4.133699

X	Y	YPRIME
1.00	−4.13370	1.00000
1.05	−4.07594	1.31211
1.10	−4.00229	1.63615
1.15	−3.91210	1.97381
1.20	−3.80465	2.32694
1.25	−3.67911	2.69755
1.30	−3.53457	3.08782
1.35	−3.36996	3.50019
1.40	−3.18414	3.93736
1.45	−2.97577	4.40232
1.50	−2.74338	4.89842
1.55	−2.48534	5.42941
1.60	−2.19979	5.99952
1.65	−1.88466	6.61350
1.70	−1.53762	7.27672
1.75	−1.15607	7.99528
1.80	−0.73706	8.77610
1.85	−0.27729	9.62704
1.90	0.22696	10.55708
1.95	0.77991	11.57647
2.00	1.38630	12.69696

c. USING X(A) = 6.999998

X	Y	YPRIME
1.00	7.00000	−3.00000
1.05	6.84237	−3.30750
1.10	6.66900	−3.63000
1.15	6.47912	−3.96750
1.20	6.27200	−4.32000
1.25	6.04687	−4.68750
1.30	5.80300	−5.07000
1.35	5.53962	−5.46750
1.40	5.25600	−5.88000
1.45	4.95137	−6.30750
1.50	4.62500	−6.75000
1.55	4.27612	−7.20750
1.60	3.90400	−7.68000
1.65	3.50787	−8.16750

(*continued*)

X	Y	YPRIME
1.70	3.08700	−8.67000
1.75	2.64062	−9.18750
1.80	2.16800	−9.71999
1.85	1.66837	−10.26749
1.90	1.14100	−10.82999
1.95	0.58512	−11.40749
2.00	−0.00000	−11.99999

e. USING X(A) = 4.763738

X	Y	YPRIME
0.00	4.76374	1.00000
0.05	4.81011	0.85726
0.10	4.84969	0.72810
0.15	4.88312	0.61123
0.20	4.91100	0.50548
0.25	4.93384	0.40980
0.30	4.95213	0.32322
0.35	4.96630	0.24488
0.40	4.97674	0.17399
0.45	4.98381	0.10985
0.50	4.98783	0.05182
0.55	4.98908	−0.00069
0.60	4.98784	−0.04821
0.65	4.98434	−0.09120
0.70	4.97879	−0.13010
0.75	4.97139	−0.16530
0.80	4.96231	−0.19715
0.85	4.95172	−0.22597
0.90	4.93976	−0.25205
0.95	4.92656	−0.27565
1.00	4.91224	−0.29700
1.05	4.89689	−0.31632
1.10	4.88063	−0.33379
1.15	4.86354	−0.34961
1.20	4.84570	−0.36392
1.25	4.82717	−0.37687
1.30	4.80803	−0.38859
1.35	4.78833	−0.39919
1.40	4.76813	−0.40878
1.45	4.74747	−0.41746
1.50	4.72640	−0.42532
1.55	4.70495	−0.43243
1.60	4.68317	−0.43886
1.65	4.66107	−0.44468
1.70	4.63871	−0.44994
1.75	4.61609	−0.45470
1.80	4.59324	−0.45901
1.85	4.57019	−0.46291
1.90	4.54696	−0.46644
1.95	4.52356	−0.46964
2.00	4.50000	−0.47253

h. USING X(A) = 1.999998

X	Y	YPRIME
1.00	2.00000	1.00000
1.05	2.05000	1.00000
1.10	2.10000	1.00000
1.15	2.15000	1.00000
1.20	2.20000	1.00000
1.25	2.25000	1.00000
1.30	2.30000	1.00000
1.35	2.35000	1.00000
1.40	2.40000	1.00000
1.45	2.45000	1.00000
1.50	2.50000	1.00000
1.55	2.55000	1.00000
1.60	2.60000	1.00000
1.65	2.65000	1.00000
1.70	2.70000	1.00000
1.75	2.75000	1.00000
1.80	2.80000	1.00000
1.85	2.85000	1.00000
1.90	2.90000	1.00000
1.95	2.95000	1.00000
2.00	3.00000	1.00000

3. USING X(A) = 8.672476E − 06

X	Y	YPRIME
2.00	0.00001	−12.00000
1.95	0.58513	−11.40750
1.90	1.14101	−10.83000
1.85	1.66838	−10.26750
1.80	2.16801	−9.72000
1.75	2.64063	−9.18750
1.70	3.08701	−8.67000
1.65	3.50788	−8.16750
1.60	3.90401	−7.68000
1.55	4.27613	−7.20750
1.50	4.62501	−6.75000
1.45	4.95138	−6.30750
1.40	5.25601	−5.88000
1.35	5.53963	−5.46749
1.30	5.80301	−5.06999
1.25	6.04688	−4.68749
1.20	6.27201	−4.31999
1.15	6.47913	−3.96749
1.10	6.66900	−3.62999
1.05	6.84238	−3.30749
1.00	7.00000	−2.99999

5. a.

X	Y
0.00	4.990855
0.10	5.400900
0.20	5.831057
0.30	6.281349
0.40	6.751803
0.50	7.242448
0.60	7.753321
0.70	8.284461
0.80	8.835918
0.90	9.407744
1.00	10.000000

b.

X	Y
0.00	4.960458
0.05	5.163008
0.10	5.370666
0.15	5.583444
0.20	5.801355
0.25	6.024411
0.30	6.252625
0.35	6.486013
0.40	6.724592
0.45	6.968377
0.50	7.217387
0.55	7.471642
0.60	7.731164
0.65	7.995975
0.70	8.266097
0.75	8.541559
0.80	8.822388
0.85	9.108614
0.90	9.400269
0.95	9.697385
1.00	10.000000

7.

X	Y	Y(X)	ERROR
1.00	0.0000153	0.0000000	−0.0000153
1.10	−0.1199852	−0.1210000	−0.0010149
1.20	−0.2860473	−0.2880001	−0.0019528
1.30	−0.5042320	−0.5069999	−0.0027679

(*continued*)

X	Y	Y(X)	ERROR	
1.40	−0.7805998	−0.7840000	−0.0034001	maximum
1.50	−1.1212120	−1.1250000	−0.0037882	error is
1.60	−1.5321210	−1.5360000	−0.0038791	0.0038791
1.70	−2.0193940	−2.0230000	−0.0036066	
1.80	−2.5890900	−2.5920000	−0.0029092	
1.90	−3.2472720	−3.2490010	−0.0017283	
2.00	−4.0000000	−4.0000000	0.0000000	

X	Y	Y(X)	ERROR	
1.00	0.0000509	0.0000000	−0.0000509	
1.05	−0.0549495	−0.0551249	−0.0001755	
1.10	−0.1207026	−0.1210000	−0.0002974	
1.15	−0.1979604	−0.1983750	−0.0004146	
1.20	−0.2874748	−0.2880001	−0.0005253	
1.25	−0.3899978	−0.3906250	−0.0006272	
1.30	−0.5062813	−0.5069999	−0.0007186	
1.35	−0.6370772	−0.6378750	−0.0007978	
1.40	−0.7831377	−0.7840000	−0.0008622	
1.45	−0.9452146	−0.9461252	−0.0009106	maximum
1.50	−1.1240600	−1.1250000	−0.0009403	error is
1.55	−1.3204130	−1.3213750	−0.0009615	0.0009626
1.60	−1.5350370	−1.5360000	−0.0009626	
1.65	−1.7686840	−1.7696250	−0.0009406	
1.70	−2.0221060	−2.0230000	−0.0008945	
1.75	−2.2960540	−2.2968750	−0.0008216	
1.80	−2.5912790	−2.5920000	−0.0007205	
1.85	−2.9085350	−2.9091250	−0.0005903	
1.90	−3.2485720	−3.2490010	−0.0004284	
1.95	−3.6121430	−3.6123760	−0.0002322	
2.00	−4.0000000	−4.0000000	0.0000000	

Exercise Set 11.5

2.

X	Y	YPRIME	
0.00	20.00000	−0.33954	
2.00	19.35600	−0.30452	
4.00	18.78170	−0.26984	
6.00	18.27645	−0.23546	
8.00	17.83970	−0.20134	
10.00	17.47095	−0.16744	
12.00	17.16981	−0.13373	
14.00	16.93594	−0.10017	
16.00	16.76907	−0.06672	Sag = $20 - 16.63569 \cong 3.36$ ft
18.00	16.66903	−0.03334	Tension = $120/\cos \alpha$
20.00	16.63569	0.00000	where $\tan \alpha = 0.33954$
22.00	16.66903	0.03334	Tension = 126.73 lb
24.00	16.76907	0.06672	
26.00	16.93594	0.10017	
28.00	17.16982	0.13373	

(*continued*)

(*continued*)

X	Y	YPRIME
30.00	17.47096	0.16744
32.00	17.83970	0.20134
34.00	18.27645	0.23546
36.00	18.78170	0.26984
38.00	19.35600	0.30452
40.00	20.00000	0.33954

4. Descent time 5 min $\Rightarrow$ initial altitude of 137.29 mi and impact velocity of 4941.6 ft/sec; $h = 1$ sec to 5 sec; all give good results; smaller values of h lead to time-consuming computations

5. $A = 5238.05$ ft; impact velocity $= 44$ ft/sec

Chapter 12

Exercise Set 12.1

1. **a.** Yes   **b.** Yes   **c.** No   **d.** No

3. $u = \dfrac{x^2 y + xy^2}{2} + F(y) + G(x)$

5. Any a and b for which $a = \pm b$

7. $u = \frac{1}{2}\left(\dfrac{1}{1 + (x + 2t)^2} + \dfrac{1}{1 + (x - 2t)^2}\right)$

   $u(1, 1) = 0.60000, u(3, 2) = 0.26000, u(2, 5) = 0.0111406$

8. $u = e^{-k\pi^2 b^2 t} \sin \pi b x$ is a solution for every real b

9. $u = e^{-4\pi^2 t} \sin \pi x$

Exercise Set 12.2

1. **a.** T	X 0.250	0.500	0.750
0	0.188	0.250	0.188
0.100	0.178	0.240	0.178
0.200	0.149	0.210	0.149
0.300	0.107	0.161	0.107
0.400	0.056	0.094	0.056
0.500	0.002	0.015	0.002
0.600	−0.050	−0.068	−0.050
0.700	−0.097	−0.146	−0.097
0.800	−0.136	−0.207	−0.136
0.900	−0.165	−0.246	−0.165
1.000	−0.180	−0.259	−0.180

b.

T	X 0.400	0.800	1.200	1.600
0	0.003	0.527	0.527	0.003
0.100	0.019	0.511	0.511	0.019
0.200	0.065	0.464	0.464	0.065
0.300	0.132	0.392	0.392	0.132
0.400	0.206	0.304	0.304	0.206
0.500	0.274	0.209	0.209	0.274
0.600	0.321	0.119	0.119	0.321
0.700	0.335	0.041	0.041	0.335
0.800	0.309	−0.018	−0.018	0.309
0.900	0.244	−0.057	−0.057	0.244
1.000	0.145	−0.077	−0.077	0.145

2. b, c.

T	X 0.250	0.500	0.750	
0	0.707	1.000	0.707	
0.100	0.674	0.953	0.674	
	−0.00147	−0.00208	−0.00147	
0.200	0.578	0.817	0.578	Errors are
	−0.00560	−0.00792	−0.00560	listed below
0.300	0.427	0.604	0.427	values of $u(x, t)$
	−0.01159	−0.01639	−0.01159	
0.400	0.237	0.335	0.237	
	−0.01822	−0.02577	−0.01822	
0.500	0.024	0.034	0.024	
	−0.02405	−0.03401	−0.02405	

3. a.

T	X 0.200	0.400	0.600	0.800	
0	0.588	0.951	0.951	0.588	
0.050	0.581	0.940	0.940	0.581	
	−0.00022	−0.00036	−0.00036	−0.00022	
0.100	0.560	0.906	0.906	0.560	
	−0.00087	−0.00141	−0.00141	−0.00087	
0.150	0.526	0.851	0.851	0.526	
	−0.00192	−0.00311	−0.00311	−0.00192	
0.200	0.479	0.775	0.775	0.479	
	−0.00332	−0.00537	−0.00537	−0.00332	
0.250	0.421	0.681	0.681	0.421	maximum error
	−0.00499	−0.00808	−0.00808	−0.00499	is 0.02299
0.300	0.352	0.570	0.570	0.352	(est. of $u(0.4, 0.5)$)
	−0.00686	−0.01110	−0.01110	−0.00686	
0.350	0.276	0.446	0.446	0.276	
	−0.00882	−0.01427	−0.01427	−0.00882	
0.400	0.192	0.311	0.311	0.192	
	−0.01077	−0.01743	−0.01743	−0.01077	
0.450	0.105	0.169	0.169	0.105	
	−0.01261	−0.02040	−0.02040	−0.01261	
0.500	0.014	0.023	0.023	0.014	
	−0.01421	−0.02299	−0.02298	−0.01420	

b. T	X			
	0.200	0.400	0.600	0.800
0	0.588	0.951	0.951	0.588
0.025	0.586	0.948	0.948	0.586
	−0.00006	−0.00009	−0.00009	−0.00006
0.050	0.581	0.940	0.940	0.581
	−0.00023	−0.00037	−0.00037	−0.00023
0.075	0.572	0.926	0.926	0.572
	−0.00052	−0.00084	−0.00084	−0.00052
0.100	0.560	0.906	0.906	0.560
	−0.00091	−0.00148	−0.00148	−0.00091
0.125	0.544	0.881	0.881	0.544
	−0.00141	−0.00229	−0.00229	−0.00141
0.150	0.526	0.851	0.851	0.526
	−0.00201	−0.00326	−0.00326	−0.00201
0.175	0.504	0.815	0.815	0.504
	−0.00270	−0.00437	−0.00437	−0.00270
0.200	0.479	0.775	0.775	0.479
	−0.00348	−0.00562	−0.00562	−0.00347
0.225	0.451	0.730	0.730	0.451
	−0.00432	−0.00699	−0.00699	−0.00432
0.250	0.421	0.681	0.681	0.421
	−0.00523	−0.00846	−0.00846	−0.00523
0.275	0.388	0.628	0.628	0.388
	−0.00619	−0.01001	−0.01001	−0.00619
0.300	0.353	0.571	0.571	0.353
	−0.00719	−0.01163	−0.01163	−0.00718
0.325	0.315	0.510	0.510	0.315
	−0.00821	−0.01328	−0.01328	−0.00821
0.350	0.276	0.447	0.447	0.276
	−0.00924	−0.01495	−0.01495	−0.00924
0.375	0.235	0.381	0.381	0.235
	−0.01027	−0.01662	−0.01662	−0.01027
0.400	0.193	0.312	0.312	0.193
	−0.01129	−0.01826	−0.01826	−0.01129
0.425	0.149	0.242	0.242	0.149
	−0.01227	−0.01985	−0.01985	−0.01227
0.450	0.105	0.170	0.170	0.105
	−0.01321	−0.02137	−0.02137	−0.01321
0.475	0.060	0.097	0.097	0.060
	−0.01408	−0.02279	−0.02279	−0.01408
0.500	0.015	0.024	0.024	0.015
	−0.01488	−0.02408	−0.02408	−0.01488

maximum error
is 0.02408
(est. of $u(0.4, 0.5)$)

Exercise Set 12.3

1. a.

	X		
T	0.250	0.500	0.750
0	0.188	0.250	0.188
0.025	0.138	0.200	0.138
0.050	0.108	0.150	0.108
0.075	0.082	0.116	0.082
0.100	0.063	0.088	0.063
0.125	0.048	0.068	0.048
0.150	0.037	0.052	0.037
0.175	0.028	0.040	0.028
0.200	0.022	0.030	0.022
0.225	0.016	0.023	0.016
0.250	0.013	0.018	0.013
0.275	0.010	0.014	0.010
0.300	0.007	0.010	0.007
0.325	0.006	0.008	0.006
0.350	0.004	0.006	0.004
0.375	0.003	0.005	0.003
0.400	0.003	0.004	0.003
0.425	0.002	0.003	0.002
0.450	0.001	0.002	0.001
0.475	0.001	0.002	0.001
0.500	0.001	0.001	0.001
0.525	0.001	0.001	0.001

b.

T	0.400	0.800	1.200	1.600
0	0.743	0.372	0.168	0.079
0.050	0.395	0.424	0.204	0.082
0.100	0.281	0.346	0.235	0.094
0.150	0.213	0.291	0.226	0.109
0.200	0.171	0.246	0.210	0.111
0.250	0.141	0.211	0.190	0.107
0.300	0.119	0.183	0.171	0.100
0.350	0.102	0.159	0.152	0.091
0.400	0.088	0.139	0.135	0.082
0.450	0.076	0.122	0.120	0.073
0.500	0.067	0.107	0.106	0.065
0.550	0.058	0.094	0.093	0.057
0.600	0.051	0.083	0.082	0.051
0.650	0.045	0.073	0.073	0.045
0.700	0.040	0.064	0.064	0.039
0.750	0.035	0.056	0.056	0.035
0.800	0.031	0.050	0.050	0.031
0.850	0.027	0.044	0.044	0.027
0.900	0.024	0.038	0.038	0.024
0.950	0.021	0.034	0.034	0.021
1.000	0.018	0.030	0.030	0.018
1.050	0.016	0.026	0.026	0.016

2. a. Lambda = 0.4

T	0.250	0.500	X 0.750
0	0.18750	0.25000	0.18750
0.025	0.14495	0.20213	0.14495
0.050	0.11394	0.16059	0.11394
0.075	0.08993	0.12707	0.08993
0.100	0.07104	0.10045	0.07104
0.125	0.05614	0.07939	0.05614
0.150	0.04436	0.06274	0.04436
0.175	0.03506	0.04958	0.03506
0.200	0.02770	0.03918	0.02770
0.225	0.02189	0.03096	0.02189
0.250	0.01730	0.02447	0.01730
0.275	0.01367	0.01934	0.01367
0.300	0.01081	0.01528	0.01081
0.325	0.00854	0.01208	0.00854
0.350	0.00675	0.00954	0.00675
0.375	0.00533	0.00754	0.00533
0.400	0.00421	0.00596	0.00421
0.425	0.00333	0.00471	0.00333
0.450	0.00263	0.00372	0.00263
0.475	0.00208	0.00294	0.00208
0.500	0.00164	0.00232	0.00164

b. Lambda = 0.3125

T	0.400	0.800	X 1.200	1.600
0	0.74316	0.37180	0.16797	0.07890
0.050	0.47923	0.38377	0.19742	0.08483
0.100	0.33802	0.34701	0.21160	0.09313
0.150	0.25439	0.30261	0.21106	0.09910
0.200	0.20044	0.26178	0.20157	0.10103
0.250	0.16315	0.22673	0.18758	0.09925
0.300	0.13592	0.19715	0.17182	0.09477
0.350	0.11516	0.17216	0.15580	0.08864
0.400	0.09878	0.15091	0.14035	0.08169
0.450	0.08550	0.13268	0.12587	0.07448
0.500	0.07450	0.11693	0.11254	0.06740
0.550	0.06524	0.10323	0.10040	0.06065
0.600	0.05733	0.09127	0.08944	0.05437
0.650	0.05051	0.08076	0.07959	0.04860
0.700	0.04459	0.07153	0.07077	0.04336
0.750	0.03941	0.06338	0.06289	0.03862
0.800	0.03488	0.05618	0.05586	0.03437
0.850	0.03089	0.04981	0.04961	0.03056
0.900	0.02737	0.04418	0.04405	0.02716
0.950	0.02426	0.03918	0.03910	0.02412
1.000	0.02151	0.03476	0.03471	0.02142

3. b.

T	X 0.250	0.500	0.750
0	0.707	1.000	0.707
0.100	0.376	0.531	0.376
	−0.055951	−0.079127	−0.055952
0.200	0.200	0.282	0.200
	−0.063889	−0.090353	−0.063889
0.300	0.106	0.150	0.106
	−0.054803	−0.077502	−0.054803
0.400	0.056	0.080	0.056
	−0.041851	−0.059188	−0.041851
0.500	0.030	0.042	0.030
	−0.030012	−0.042440	−0.030012
0.600	0.016	0.023	0.016
	−0.020690	−0.029265	−0.020690

c. Errors are indicated below $u(x, t)$ estimates in part (b)

4. a.

T	X 0.200	0.400	0.600	0.800
0	0.588	0.951	0.951	0.588
0.050	0.447	0.724	0.724	0.447
	−0.011798	−0.019090	−0.019090	−0.011799
0.100	0.341	0.551	0.551	0.341
	−0.018200	−0.029449	−0.029449	−0.018200
0.150	0.259	0.420	0.420	0.259
	−0.021058	−0.034073	−0.034073	−0.021059
0.200	0.197	0.319	0.319	0.197
	−0.021659	−0.035045	−0.035045	−0.021659
0.250	0.150	0.243	0.243	0.150
	−0.020886	−0.033794	−0.033794	−0.020886
0.300	0.114	0.185	0.185	0.114
	−0.019336	−0.031285	−0.031286	−0.019335
0.350	0.087	0.141	0.141	0.087
	−0.017404	−0.028161	−0.028160	−0.017404
0.400	0.066	0.107	0.107	0.066
	−0.015347	−0.024831	−0.024832	−0.015347
0.450	0.050	0.082	0.082	0.050
	−0.013322	−0.021556	−0.021555	−0.013322
0.500	0.038	0.062	0.062	0.038
	−0.011423	−0.018481	−0.018482	−0.011422
0.550	0.029	0.047	0.047	0.029
	−0.009696	−0.015689	−0.015688	−0.009696

b.

T	X 0.200	0.400	0.600	0.800
0	0.588	0.951	0.951	0.588
0.025	0.518	0.838	0.838	0.518
	−0.001940	−0.003139	−0.003139	−0.001940
0.050	0.456	0.738	0.738	0.456
	−0.003424	−0.005540	−0.005540	−0.003424
0.075	0.401	0.650	0.650	0.401
	−0.004531	−0.007331	−0.007331	−0.004531
0.100	0.354	0.572	0.572	0.354
	−0.005330	−0.008624	−0.008624	−0.005330
0.125	0.311	0.504	0.504	0.311
	−0.005878	−0.009512	−0.009512	−0.005879
0.150	0.274	0.444	0.444	0.274
	−0.006224	−0.010070	−0.010070	−0.006224
0.175	0.241	0.391	0.391	0.241
	−0.006406	−0.010366	−0.010366	−0.006407
0.200	0.213	0.344	0.344	0.213
	−0.006460	−0.010452	−0.010452	−0.006460
0.225	0.187	0.303	0.303	0.187
	−0.006412	−0.010375	−0.010375	−0.006412
0.250	0.165	0.267	0.267	0.165
	−0.006286	−0.010171	−0.010171	−0.006286
0.275	0.145	0.235	0.235	0.145
	−0.006100	−0.009871	−0.009871	−0.006101
0.300	0.128	0.207	0.207	0.128
	−0.005872	−0.009501	−0.009501	−0.005872
0.325	0.113	0.182	0.182	0.113
	−0.005612	−0.009081	−0.009081	−0.005612
0.350	0.099	0.160	0.160	0.099
	−0.005333	−0.008628	−0.008628	−0.005333
0.375	0.087	0.141	0.141	0.087
	−0.005041	−0.008157	−0.008157	−0.005041
0.400	0.077	0.124	0.124	0.077
	−0.004744	−0.007676	−0.007676	−0.004744
0.425	0.068	0.110	0.110	0.068
	−0.004447	−0.007199	−0.007196	−0.004448
0.450	0.060	0.097	0.097	0.060
	−0.004155	−0.006723	−0.006723	−0.004155
0.475	0.053	0.085	0.085	0.053
	−0.003870	−0.006261	−0.006261	−0.003870
0.500	0.046	0.075	0.075	0.046
	−0.003594	−0.005815	−0.005815	−0.003594

5. a. 3   **b.** 5   **c.** 6

9. 0.152, 0.232, 0.232, 0.152

11. a. 0.32, −1.36, −0.44, −0.47
   c. 0.3653846, 0.4615385, 0.4807692, 0.4615385, 0.36538746

Exercise Set 12.4

1. a. YJ	XI	U(X, Y)	c. YJ	XI	U(X, Y)
0.100	0.200	0.01798	0.100	0.200	−0.00001
0.100	0.400	0.07398	0.100	0.400	−0.00001
0.100	0.600	0.22598	0.100	0.600	0.00002
0.100	0.800	0.52199	0.100	0.800	0.00007
0.200	0.200	0.04797	0.200	0.200	−0.00001
0.200	0.400	0.10396	0.200	0.400	0.00005
0.200	0.600	0.25596	0.200	0.600	0.00025
0.200	0.800	0.55198	0.200	0.800	0.00066
0.300	0.200	0.09796	0.300	0.200	0.00000
0.300	0.400	0.15395	0.300	0.400	0.00024
0.300	0.600	0.30595	0.300	0.600	0.00093
0.300	0.800	0.60197	0.300	0.800	0.00228
0.400	0.200	0.16796	0.400	0.200	0.00005
0.400	0.400	0.22394	0.400	0.400	0.00063
0.400	0.600	0.37595	0.400	0.600	0.00226
0.400	0.800	0.67197	0.400	0.800	0.00543
0.500	0.200	0.25796	0.500	0.200	0.00013
0.500	0.400	0.31394	0.500	0.400	0.00128
0.500	0.600	0.46595	0.500	0.600	0.00445
0.500	0.800	0.76197	0.500	0.800	0.01064
0.600	0.200	0.36797	0.600	0.200	0.00026
0.600	0.400	0.42395	0.600	0.400	0.00226
0.600	0.600	0.57596	0.600	0.600	0.00773
0.600	0.800	0.87197	0.600	0.800	0.01841
0.700	0.200	0.49797	0.700	0.200	0.00043
0.700	0.400	0.55396	0.700	0.400	0.00362
0.700	0.600	0.70596	0.700	0.600	0.01232
0.700	0.800	1.00198	0.700	0.800	0.02925
0.800	0.200	0.64798	0.800	0.200	0.00067
0.800	0.400	0.70397	0.800	0.400	0.00544
0.800	0.600	0.85598	0.800	0.600	0.01841
0.800	0.800	1.15199	0.800	0.800	0.04368
0.900	0.200	0.81799	0.900	0.200	0.00096
0.900	0.400	0.87399	0.900	0.400	0.00776
0.900	0.600	1.02599	0.900	0.600	0.02623
0.900	0.800	1.32199	0.900	0.800	0.06220

NUMBER OF ITERATIONS: 61   NUMBER OF ITERATIONS: 36

2. a.

YJ	XI	U(X, Y)
0.200	0.200	−2.38402
0.200	0.400	−1.54917
0.200	0.600	−0.89326
0.200	0.800	−0.37778
0.400	0.200	−1.54917
0.400	0.400	−1.08681
0.400	0.600	−0.62445
0.400	0.800	−0.21080
0.600	0.200	−0.89326
0.600	0.400	−0.62445
0.600	0.600	−0.30693
0.600	0.800	0.01060
0.800	0.200	−0.37778
0.800	0.400	−0.21080
0.800	0.600	0.01060
0.800	0.800	0.25265

NUMBER OF ITERATIONS: 35

c.

YJ	XI	U(X, Y)
0.400	0.400	0.37636
0.400	0.800	1.19833
0.400	1.200	2.30561
0.400	1.600	3.71455
0.800	0.400	0.30712
0.800	0.800	1.47136
0.800	1.200	2.86954
0.800	1.600	4.55258
1.200	0.400	0.02076
1.200	0.800	1.51045
1.200	1.200	3.14864
1.200	1.600	5.06621
1.600	0.400	−0.29455
1.600	0.800	1.40106
1.600	1.200	3.14833
1.600	1.600	5.20364

NUMBER OF ITERATIONS: 31

3. b. / **c.**

YJ	XI	Uij	U(X, Y)	ERROR
0.200	0.200	0.00031	0.00032	−0.00001
0.200	0.400	0.00127	0.00128	−0.00001
0.200	0.600	0.00287	0.00288	−0.00001
0.200	0.800	0.00511	0.00512	−0.00001
0.400	0.200	0.00255	0.00256	−0.00001
0.400	0.400	0.01022	0.01024	−0.00002
0.400	0.600	0.02302	0.02304	−0.00002
0.400	0.800	0.04095	0.04096	−0.00001
0.600	0.200	0.00863	0.00864	−0.00001
0.600	0.400	0.03454	0.03456	−0.00002
0.600	0.600	0.07775	0.07776	−0.00001
0.600	0.800	0.13823	0.13824	−0.00001
0.800	0.200	0.02047	0.02048	−0.00001
0.800	0.400	0.08191	0.08192	−0.00001
0.800	0.600	0.18431	0.18432	−0.00001
0.800	0.800	0.32768	0.32768	−0.00000

5. $W = 1.527864$ SOR METHOD

YJ	XI	U(X, Y)	YJ	XI	U(X, Y)
0.100	0.100	0.00000	0.600	0.100	0.00004
0.100	0.200	0.00000	0.600	0.200	0.00029
0.100	0.300	0.00000	0.600	0.300	0.00097
0.100	0.400	0.00001	0.600	0.400	0.00230
0.100	0.500	0.00002	0.600	0.500	0.00450
0.100	0.600	0.00004	0.600	0.600	0.00778
0.100	0.700	0.00006	0.600	0.700	0.01235
0.100	0.800	0.00009	0.600	0.800	0.01843
0.100	0.900	0.00012	0.600	0.900	0.02624
0.200	0.100	0.00000	0.700	0.100	0.00006
0.200	0.200	0.00001	0.700	0.200	0.00046
0.200	0.300	0.00004	0.700	0.300	0.00154
0.200	0.400	0.00009	0.700	0.400	0.00366
0.200	0.500	0.00017	0.700	0.500	0.00715
0.200	0.600	0.00029	0.700	0.600	0.01235
0.200	0.700	0.00046	0.700	0.700	0.01961
0.200	0.800	0.00068	0.700	0.800	0.02927
0.200	0.900	0.00097	0.700	0.900	0.04167
0.300	0.100	0.00000	0.800	0.100	0.00009
0.300	0.200	0.00004	0.800	0.200	0.00068
0.300	0.300	0.00012	0.800	0.300	0.00230
0.300	0.400	0.00029	0.800	0.400	0.00546
0.300	0.500	0.00056	0.800	0.500	0.01067
0.300	0.600	0.00097	0.800	0.600	0.01843
0.300	0.700	0.00154	0.800	0.700	0.02927
0.300	0.800	0.00230	0.800	0.800	0.04369
0.300	0.900	0.00328	0.800	0.900	0.06221
0.400	0.100	0.00001	0.900	0.100	0.00012
0.400	0.200	0.00009	0.900	0.200	0.00097
0.400	0.300	0.00029	0.900	0.300	0.00328
0.400	0.400	0.00068	0.900	0.400	0.00778
0.400	0.500	0.00133	0.900	0.500	0.01519
0.400	0.600	0.00230	0.900	0.600	0.02624
0.400	0.700	0.00366	0.900	0.700	0.04167
0.400	0.800	0.00546	0.900	0.800	0.06221
0.400	0.900	0.00778	0.900	0.900	0.08857
0.500	0.100	0.00002			
0.500	0.200	0.00017			
0.500	0.300	0.00056			
0.500	0.400	0.00133			
0.500	0.500	0.00260			
0.500	0.600	0.00450			
0.500	0.700	0.00715			
0.500	0.800	0.01067			
0.500	0.900	0.01519			

NUMBER OF ITERATIONS: 28

Using Algorithm 12.5

YJ	XI	U(X, Y)	YJ	XI	U(X, Y)
0.100	0.100	−0.00001	0.600	0.100	0.00000
0.100	0.200	−0.00002	0.600	0.200	0.00023
0.100	0.300	−0.00003	0.600	0.300	0.00090
0.100	0.400	−0.00002	0.600	0.400	0.00222
0.100	0.500	−0.00001	0.600	0.500	0.00442
0.100	0.600	0.00000	0.600	0.600	0.00770
0.100	0.700	0.00003	0.600	0.700	0.01229
0.100	0.800	0.00007	0.600	0.800	0.01839
0.100	0.900	0.00011	0.600	0.900	0.02622
0.200	0.100	−0.00002	0.700	0.100	0.00003
0.200	0.200	−0.00003	0.700	0.200	0.00041
0.200	0.300	−0.00002	0.700	0.300	0.00148
0.200	0.400	0.00002	0.700	0.400	0.00359
0.200	0.500	0.00010	0.700	0.500	0.00708
0.200	0.600	0.00023	0.700	0.600	0.01229
0.200	0.700	0.00041	0.700	0.700	0.01956
0.200	0.800	0.00065	0.700	0.800	0.02924
0.200	0.900	0.00096	0.700	0.900	0.04166
0.300	0.100	−0.00003	0.800	0.100	0.00007
0.300	0.200	−0.00002	0.800	0.200	0.00065
0.300	0.300	0.00005	0.800	0.300	0.00226
0.300	0.400	0.00021	0.800	0.400	0.00542
0.300	0.500	0.00048	0.800	0.500	0.01062
0.300	0.600	0.00090	0.800	0.600	0.01839
0.300	0.700	0.00148	0.800	0.700	0.02924
0.300	0.800	0.00226	0.800	0.800	0.04367
0.300	0.900	0.00326	0.800	0.900	0.06220
0.400	0.100	−0.00002	0.900	0.100	0.00011
0.400	0.200	0.00002	0.900	0.200	0.00096
0.400	0.300	0.00021	0.900	0.300	0.00326
0.400	0.400	0.00059	0.900	0.400	0.00775
0.400	0.500	0.00124	0.900	0.500	0.01516
0.400	0.600	0.00222	0.900	0.600	0.02622
0.400	0.700	0.00359	0.900	0.700	0.04166
0.400	0.800	0.00542	0.900	0.800	0.06220
0.400	0.900	0.00775	0.900	0.900	0.08857
0.500	0.100	−0.00001			
0.500	0.200	0.00010	NUMBER OF ITERATIONS: 52		
0.500	0.300	0.00048			
0.500	0.400	0.00124			
0.500	0.500	0.00251			
0.500	0.600	0.00442			
0.500	0.700	0.00708			
0.500	0.800	0.01062			
0.500	0.900	0.01516			

Yes, the modified algorithm converged faster

Exercise Set 12.5

1. a. See Table 12.3

b.

	X				
T(MSEC)	10.00	20.00	30.00	40.00	50.00
0	0.10000	0.20000	0.30000	0.20000	0.10000
0.29881	0.10000	0.20000	0.20000	0.20000	0.10000
0.59761	0.10000	0.10000	0.10000	0.10000	0.10000
0.89642	0.00000	0.00000	0.00000	−0.00000	−0.00000
1.19523	−0.10000	−0.10000	−0.10000	−0.10000	−0.10000
1.49404	−0.10000	−0.20000	−0.20000	−0.20000	−0.10000
1.79284	−0.10000	−0.20000	−0.30000	−0.20000	−0.10000
2.09165	−0.10000	−0.20000	−0.20000	−0.20000	−0.10000
2.39046	−0.10000	−0.10000	−0.10000	−0.10000	−0.10000
2.68926	0.00000	0.00000	−0.00000	−0.00000	−0.00000
2.98807	0.10000	0.10000	0.10000	0.10000	0.10000
3.28688	0.10000	0.20000	0.20000	0.20000	0.10000
3.58569	0.10000	0.20000	0.30000	0.20000	0.10000
3.88449	0.10000	0.20000	0.20000	0.20000	0.10000
4.18330	0.10000	0.10000	0.10000	0.10000	0.10000
4.48211	0.00000	0.00000	0.00000	−0.00000	−0.00000
4.78091	−0.10000	−0.10000	−0.10000	−0.10000	−0.10000
5.07972	−0.10000	−0.20000	−0.20000	−0.20000	−0.10000
5.37853	−0.10000	−0.20000	−0.30000	−0.20000	−0.10000
5.67734	−0.10000	−0.20000	−0.20000	−0.20000	−0.10000
5.97614	−0.10000	−0.10000	−0.10000	−0.10000	−0.10000
6.27495	0.00000	0.00000	−0.00000	−0.00000	−0.00000
6.57376	0.10000	0.10000	0.10000	0.10000	0.10000
6.87256	0.10000	0.20000	0.20000	0.20000	0.10000
7.17137	0.10000	0.20000	0.30000	0.20000	0.10000
7.47018	0.10000	0.20000	0.20000	0.20000	0.10000
7.76899	0.10000	0.10000	0.10000	0.10000	0.10000
8.06779	0.00000	0.00000	0.00000	−0.00000	−0.00000
8.36660	−0.10000	−0.10000	−0.10000	−0.10000	−0.10000
8.66541	−0.10000	−0.20000	−0.20000	−0.20000	−0.10000
8.96421	−0.10000	−0.20000	−0.30000	−0.20000	−0.10000
9.26302	−0.10000	−0.20000	−0.20000	−0.20000	−0.10000
9.56183	−0.10000	−0.10000	−0.10000	−0.10000	−0.10000
9.86064	0.00000	0.00000	−0.00000	−0.00000	−0.00000
10.15944	0.10000	0.10000	0.10000	0.10000	0.10000
10.45825	0.10000	0.20000	0.20000	0.20000	0.10000
10.75706	0.10000	0.20000	0.30000	0.20000	0.10000
11.05586	0.10000	0.20000	0.20000	0.20000	0.10000
11.35467	0.10000	0.10000	0.10000	0.10000	0.10000
11.65348	0.00000	0.00000	0.00000	−0.00000	−0.00000
11.95229	−0.10000	−0.10000	−0.10000	−0.10000	−0.10000
12.25109	−0.10000	−0.20000	−0.20000	−0.20000	−0.10000
12.54990	−0.10000	−0.20000	−0.30000	−0.20000	−0.10000
12.84871	−0.10000	−0.20000	−0.20000	−0.20000	−0.10000
13.14751	−0.10000	−0.10000	−0.10000	−0.10000	−0.10000
13.44632	0.00000	0.00000	−0.00000	−0.00000	−0.00000
13.74513	0.10000	0.10000	0.10000	0.10000	0.10000
14.04394	0.10000	0.20000	0.20000	0.20000	0.10000
14.34274	0.10000	0.20000	0.30000	0.20000	0.10000

d.

	X				
T(MSEC)	10.00	20.00	30.00	40.00	50.00
0	0.10000	0.20000	0.30000	0.20000	0.10000
0.18257	0.10000	0.20000	0.20000	0.20000	0.10000
0.36515	0.10000	0.10000	0.10000	0.10000	0.10000
0.54772	0.00000	0.00000	0.00000	−0.00000	−0.00000
0.73000	−0.10000	−0.10000	−0.10000	−0.10000	−0.10000
0.91287	−0.10000	−0.20000	−0.20000	−0.20000	−0.10000
1.09545	−0.10000	−0.20000	−0.30000	−0.20000	−0.10000
1.27802	−0.10000	−0.20000	−0.20000	−0.20000	−0.10000
1.46059	−0.10000	−0.10000	−0.10000	−0.10000	−0.10000
1.64317	0.00000	0.00000	−0.00000	−0.00000	−0.00000
1.82574	0.10000	0.10000	0.10000	0.10000	0.10000
2.00832	0.10000	0.20000	0.20000	0.20000	0.10000
2.19089	0.10000	0.20000	0.30000	0.20000	0.10000
2.37346	0.10000	0.20000	0.20000	0.20000	0.10000
2.55604	0.10000	0.10000	0.10000	0.10000	0.10000
2.73861	0.00000	0.00000	0.00000	−0.00000	0.00000
2.92119	−0.10000	−0.10000	−0.10000	−0.10000	−0.10000
3.10376	−0.10000	−0.20000	−0.20000	−0.20000	−0.10000
3.28634	−0.10000	−0.20000	−0.30000	−0.20000	−0.10000
3.46891	−0.10000	−0.20000	−0.20000	−0.20000	−0.10000
3.65148	−0.10000	−0.10000	−0.10000	−0.10000	−0.10000
3.83406	0.00000	0.00000	−0.00000	−0.00000	−0.00000
4.01663	0.10000	0.10000	0.10000	0.10000	0.10000
4.19921	0.10000	0.20000	0.20000	0.20000	0.10000
4.38178	0.10000	0.20000	0.30000	0.20000	0.10000
4.56435	0.10000	0.20000	0.20000	0.20000	0.10000
4.74693	0.10000	0.10000	0.10000	0.10000	0.10000
4.92950	0.00000	0.00000	0.00000	−0.00000	−0.00000
5.11208	−0.10000	−0.10000	−0.10000	−0.10000	−0.10000
5.29465	−0.10000	−0.20000	−0.20000	−0.20000	−0.10000
5.47723	−0.10000	−0.20000	−0.30000	−0.20000	−0.10000
5.65980	−0.10000	−0.20000	−0.20000	−0.20000	−0.10000
5.84237	−0.10000	−0.10000	−0.10000	−0.10000	−0.10000
6.02495	0.00000	0.00000	−0.00000	−0.00000	−0.00000
6.20752	0.10000	0.10000	0.10000	0.10000	0.10000
6.39010	0.10000	0.20000	0.20000	0.20000	0.10000
6.57267	0.10000	0.20000	0.30000	0.20000	0.10000
6.75524	0.10000	0.20000	0.20000	0.20000	0.10000
6.93782	0.10000	0.10000	0.10000	0.10000	0.10000
7.12039	0.00000	0.00000	0.00000	−0.00000	−0.00000
7.30297	−0.10000	−0.10000	−0.10000	−0.10000	−0.10000
7.48554	−0.10000	−0.20000	−0.20000	−0.20000	−0.10000
7.66812	−0.10000	−0.20000	−0.30000	−0.20000	−0.10000
7.85069	−0.10000	−0.20000	−0.20000	−0.20000	−0.10000
8.03326	−0.10000	−0.10000	−0.10000	−0.10000	−0.10000
8.21584	0.00000	0.00000	−0.00000	−0.00000	−0.00000
8.39841	0.10000	0.10000	0.10000	0.10000	0.10000
8.58099	0.10000	0.20000	0.20000	0.20000	0.10000

3. ROD LENGTH (CM) IS 20
DENSITY (G/CC) IS 2.7
SPECIFIC HEAT IS 0.22
CONDUCTIVITY IS 1.1

TIME STEP IS 0.5 LAMBDA $= 5.787037E - 02$

	X			
T	4.000	8.000	12.000	16.000
0	32.000	48.000	48.000	32.000
0.50	31.074	47.074	47.074	31.074
1.00	30.202	46.148	46.148	30.202
1.50	29.377	45.225	45.225	29.377
2.00	28.594	44.308	44.308	28.594
2.50	27.849	43.399	43.399	27.849
3.00	27.137	42.499	42.499	27.137
3.50	26.455	41.610	41.610	26.455
4.00	25.801	40.733	40.733	25.801
4.50	25.172	39.869	39.869	25.172
5.00	24.566	39.018	39.018	24.566
5.50	23.981	38.182	38.182	23.981
6.00	23.415	37.360	37.360	23.415
6.50	22.867	36.553	36.553	22.867
7.00	22.336	35.761	35.761	22.336
7.50	21.820	34.984	34.984	21.820
8.00	21.319	34.222	34.222	21.319
8.50	20.832	33.476	33.476	20.832
9.00	20.358	32.744	32.744	20.358
9.50	19.897	32.027	32.027	19.897
10.00	19.447	31.325	31.325	19.447

5.

	X			
T	4.000	8.000	12.000	16.000
0	32.000	48.000	48.000	32.000
0.50	31.074	47.074	47.074	32.231
1.00	30.202	46.148	46.215	32.383
1.50	29.377	45.229	45.411	32.466
2.00	28.594	44.322	44.651	32.494
2.50	27.850	43.431	43.929	32.475
3.00	27.140	42.558	43.237	32.416
3.50	26.461	41.705	42.571	32.323
4.00	25.812	40.873	41.928	32.203
4.50	25.190	40.063	41.304	32.060
5.00	24.593	39.274	40.698	31.897
5.50	24.019	38.507	40.106	31.718
6.00	23.468	37.761	39.528	31.525
6.50	22.937	37.036	38.963	31.321
7.00	22.425	36.331	38.409	31.108
7.50	21.932	35.647	37.866	30.888
8.00	21.457	34.982	37.334	30.662
8.50	20.998	34.335	36.812	30.431
9.00	20.554	33.707	36.299	30.196
9.50	20.126	33.096	35.796	29.959
10.00	19.712	32.501	35.302	29.721

7. $f(x) = T_1 + x\left(\dfrac{T_2 - T_1}{L}\right)$

8. **a.** $f(x) = 20 + \dfrac{x}{2}$

b. LENGTH (CM.) 20
DENSITY (G/CC) 10.6
SPECIFIC HEAT 0.056
CONDUCTIVITY 1.05

TIME STEP = 1 SEC.

		X cm		
T	4.000	8.000	12.000	16.000
0	50.000	50.000	50.000	50.000
10.000	34.806	43.295	44.616	39.660
20.000	29.916	36.640	38.489	35.671
30.000	27.123	32.260	34.227	33.070
40.000	25.346	29.408	31.401	31.334
50.000	24.191	27.543	29.542	30.188
60.000	23.435	26.322	28.322	29.435
70.000	22.941	25.522	27.522	28.941
80.000	22.617	24.998	26.998	28.617
90.000	22.404	24.654	26.654	28.404
100.000	22.265	24.429	26.429	28.265
110.000	22.174	24.281	26.281	28.174
120.000	22.114	24.184	26.184	28.114
130.000	22.075	24.121	26.121	28.075
140.000	22.049	24.079	26.079	28.049
150.000	22.032	24.052	26.052	28.032
160.000	22.021	24.034	26.034	28.021
170.000	22.014	24.022	26.022	28.014
180.000	22.009	24.015	26.015	28.009
190.000	22.006	24.010	26.010	28.006
200.000	22.004	24.006	26.006	28.004
210.000	22.003	24.004	26.004	28.003
220.000	22.002	24.003	26.003	28.002
230.000	22.001	24.002	26.002	28.001
240.000	22.001	24.001	26.001	28.001
250.000	22.000	24.001	26.001	28.000
260.000	22.000	24.000	26.000	28.000
270.000	22.000	24.000	26.000	28.000
280.000	22.000	24.000	26.000	28.000
290.000	22.000	24.000	26.000	28.000
300.000	22.000	24.000	26.000	28.000

9.

YJ	XI	U(X, Y)	YJ	XI	U(X, Y)
5.000	5.000	11.95511	30.000	5.000	11.54677
5.000	10.000	13.91464	30.000	10.000	13.14168
5.000	15.000	15.88257	30.000	15.000	14.82444
5.000	20.000	17.86201	30.000	20.000	16.62317
5.000	25.000	19.85493	30.000	25.000	18.55429
5.000	30.000	21.86200	30.000	30.000	20.62316
5.000	35.000	23.88257	30.000	35.000	22.82442
5.000	40.000	25.91464	30.000	40.000	25.14166
5.000	45.000	27.95510	30.000	45.000	27.54676
10.000	5.000	11.90578	35.000	5.000	11.36881
10.000	10.000	13.82088	35.000	10.000	12.80914
10.000	15.000	15.75363	35.000	15.000	14.37511
10.000	20.000	17.71052	35.000	20.000	16.10188
10.000	25.000	19.69568	35.000	25.000	18.00876
10.000	30.000	21.71052	35.000	30.000	20.10188
10.000	35.000	23.75362	35.000	35.000	22.37510
10.000	40.000	25.82087	35.000	40.000	24.80913
10.000	45.000	27.90578	35.000	45.000	27.36880
15.000	5.000	11.84713	40.000	5.000	11.11930
15.000	10.000	13.70947	40.000	10.000	12.35097
15.000	15.000	15.60052	40.000	15.000	13.76498
15.000	20.000	17.53076	40.000	20.000	15.40047
15.000	25.000	19.50675	40.000	25.000	17.27699
15.000	30.000	21.53075	40.000	30.000	19.40047
15.000	35.000	23.60051	40.000	35.000	21.76497
15.000	40.000	25.70946	40.000	40.000	24.35096
15.000	45.000	27.84712	40.000	45.000	27.11930
20.000	5.000	11.77327	45.000	5.000	10.75744
20.000	10.000	13.56933	45.000	10.000	11.71044
20.000	15.000	15.40821	45.000	15.000	12.93336
20.000	20.000	17.30521	45.000	20.000	14.45802
20.000	25.000	19.26981	45.000	25.000	16.29826
20.000	30.000	21.30520	45.000	30.000	18.45802
20.000	35.000	23.40820	45.000	35.000	20.93336
20.000	40.000	25.56931	45.000	40.000	23.71044
20.000	45.000	27.77326	45.000	45.000	26.75743
25.000	5.000	11.67660			
25.000	10.000	13.38634			
25.000	15.000	15.15777			
25.000	20.000	17.01206			
25.000	25.000	18.96206			
25.000	30.000	21.01205			
25.000	35.000	23.15775			
25.000	40.000	25.38633			
25.000	45.000	27.67659			

NUMBER OF ITERATIONS: 100

BAVE WAS 21.175

INDEX